The Stanford–Cambridge Program is an innovative publishing venture resulting from the collaboration between Cambridge University Press and Stanford University and its Press.

The Program provides a new international imprint for the teaching and communication of pure and applied sciences. Drawing on Stanford's eminent faculty and associated institutions, books within the Program reflect the high quality of teaching and research at Stanford University.

The Program includes textbooks at undergraduate level, and research monographs, across a broad range of the sciences.

Cambridge University Press publishes and distributes books in the Stanford–Cambridge Program throughout the world.

Randomized Algorithms

Randomized Algorithms

Rajeev Motwani
Stanford University

Prabhakar Raghavan
IBM Thomas J. Watson Research Center

PUBLISHED BY THE PRESS SYNDICATE OF THE UNIVERSITY OF CAMBRIDGE
The Pitt Building, Trumpington Street, Cambridge, United Kingdom

CAMBRIDGE UNIVERSITY PRESS
The Edinburgh Building, Cambridge CB2 2RU, UK http://www.cup.cam.ac.uk
40 West 20th Street, New York, NY 10011-4211, USA http://www.cup.org
10 Stamford Road, Oakleigh, Melbourne 3166, Australia
Ruiz de Alarcón 13, 28014 Madrid, Spain

First published 1995
Reprinted 1997, 2000

Printed in the United States of America

Typeface Times 10.5/13pt *System* LaTeX [UPH]

A catalogue record for this book is available from the British Library

Library of Congress Cataloguing in Publication data

Motwani, Rajeev.
Randomized algorithms / Rajeev Motwani, Prabhakar Raghavan.
p. cm.
Includes bibliographical references and index.
ISBN 0 521 47465 5
1. Stochastic processes–Data processing. 2. Algorithms.
I. Raghavan, Prabhakar. II. Title.
QA274.M68 1995
004′.01′5192–dc20 94-44271

ISBN 0 521 47465 5 hardback

Contents

Preface ix

I Tools and Techniques 1

1 Introduction 3

1.1 A Min-Cut Algorithm 7
1.2 Las Vegas and Monte Carlo 9
1.3 Binary Planar Partitions 10
1.4 A Probabilistic Recurrence 15
1.5 Computation Model and Complexity Classes 16
Notes 23
Problems 25

2 Game-Theoretic Techniques 28

2.1 Game Tree Evaluation 28
2.2 The Minimax Principle 31
2.3 Randomness and Non-uniformity 38
Notes 40
Problems 41

3 Moments and Deviations 43

3.1 Occupancy Problems 43
3.2 The Markov and Chebyshev Inequalities 45
3.3 Randomized Selection 47
3.4 Two-Point Sampling 51
3.5 The Stable Marriage Problem 53
3.6 The Coupon Collector's Problem 57
Notes 63
Problems 64

4 Tail Inequalities 67

4.1 The Chernoff Bound 67

4.2 Routing in a Parallel Computer 74
4.3 A Wiring Problem 79
4.4 Martingales 83
Notes 96
Problems 97

5 The Probabilistic Method 101

5.1 Overview of the Method 101
5.2 Maximum Satisfiability 104
5.3 Expanding Graphs 108
5.4 Oblivious Routing Revisited 112
5.5 The Lovász Local Lemma 115
5.6 The Method of Conditional Probabilities 120
Notes 122
Problems 124

6 Markov Chains and Random Walks 127

6.1 A 2-SAT Example 128
6.2 Markov Chains 129
6.3 Random Walks on Graphs 132
6.4 Electrical Networks 135
6.5 Cover Times 137
6.6 Graph Connectivity 139
6.7 Expanders and Rapidly Mixing Random Walks 143
6.8 Probability Amplification by Random Walks on Expanders 151
Notes 155
Problems 156

7 Algebraic Techniques 161

7.1 Fingerprinting and Freivalds' Technique 162
7.2 Verifying Polynomial Identities 163
7.3 Perfect Matchings in Graphs 167
7.4 Verifying Equality of Strings 168
7.5 A Comparison of Fingerprinting Techniques 169
7.6 Pattern Matching 170
7.7 Interactive Proof Systems 172
7.8 PCP and Efficient Proof Verification 180
Notes 186
Problems 188

II Applications 195

8 Data Structures 197

8.1 The Fundamental Data-structuring Problem 197

8.2 Random Treaps 201
8.3 Skip Lists 209
8.4 Hash Tables 213
8.5 Hashing with O(1) Search Time 221
Notes 228
Problems 229

9 Geometric Algorithms and Linear Programming 234

9.1 Randomized Incremental Construction 234
9.2 Convex Hulls in the Plane 236
9.3 Duality 239
9.4 Half-space Intersections 241
9.5 Delaunay Triangulations 245
9.6 Trapezoidal Decompositions 248
9.7 Binary Space Partitions 252
9.8 The Diameter of a Point Set 256
9.9 Random Sampling 258
9.10 Linear Programming 262
Notes 273
Problems 275

10 Graph Algorithms 278

10.1 All-pairs Shortest Paths 278
10.2 The Min-Cut Problem 289
10.3 Minimum Spanning Trees 296
Notes 302
Problems 304

11 Approximate Counting 306

11.1 Randomized Approximation Schemes 308
11.2 The DNF Counting Problem 310
11.3 Approximating the Permanent 315
11.4 Volume Estimation 329
Notes 331
Problems 333

12 Parallel and Distributed Algorithms 335

12.1 The PRAM Model 335
12.2 Sorting on a PRAM 337
12.3 Maximal Independent Sets 341
12.4 Perfect Matchings 347
12.5 The Choice Coordination Problem 355
12.6 Byzantine Agreement 358
Notes 361
Problems 363

13 Online Algorithms 368

13.1 The Online Paging Problem 369
13.2 Adversary Models 372
13.3 Paging against an Oblivious Adversary 374
13.4 Relating the Adversaries 377
13.5 The Adaptive Online Adversary 381
13.6 The k-Server Problem 384
Notes 387
Problems 389

14 Number Theory and Algebra 392

14.1 Preliminaries 392
14.2 Groups and Fields 395
14.3 Quadratic Residues 402
14.4 The RSA Cryptosystem 410
14.5 Polynomial Roots and Factors 412
14.6 Primality Testing 417
Notes 426
Problems 427

Appendix A **Notational Index** 429
Appendix B **Mathematical Background** 433
Appendix C **Basic Probability Theory** 438

References 447
Index 467

Preface

THE last decade has witnessed a tremendous growth in the area of randomized algorithms. During this period, randomized algorithms went from being a tool in computational number theory to finding widespread application in many types of algorithms. Two benefits of randomization have spearheaded this growth: simplicity and speed. For many applications, a randomized algorithm is the simplest algorithm available, or the fastest, or both.

This book presents the basic concepts in the design and analysis of randomized algorithms at a level accessible to advanced undergraduates and to graduate students. We expect it will also prove to be a reference to professionals wishing to implement such algorithms and to researchers seeking to establish new results in the area.

Organization and Course Information

We assume that the reader has had undergraduate courses in Algorithms and Complexity, and in Probability Theory. The book is organized into two parts. The first part, consisting of seven chapters, presents basic tools from probability theory and probabilistic analysis that are recurrent in algorithmic applications. Applications are given along with each tool to illustrate the tool in concrete settings. The second part of the book also contains seven chapters, each focusing on one area of application of randomized algorithms. The seven areas of application we have selected are: data structures, graph algorithms, geometric algorithms, number theoretic algorithms, counting algorithms, parallel and distributed algorithms, and online algorithms. Naturally, some of the algorithms used for illustration in Part I do fall into one of these seven categories. The book is not meant to be a compendium of every randomized algorithm that has been devised, but rather a comprehensive and representative selection. The Appendices review basic material on probability theory.

We have taught several regular as well as short-term courses based on the material in this book, as have some of our colleagues. It is virtually impossible to cover all the material in the book in a single academic term or in a week's intensive course. We regard Chapters 1–4 as the core around which a course may be built. Following the treatment of this material, the instructor may continue with that portion of the remainder of Part I that supports the material of Part II (s)he wishes to cover. Chapters 5–13 depend only on material in Chapters 1–4, with the following exceptions:

1. Chapter 5 on Probabilistic Methods is a prerequisite for Chapters 6 (Random Walks) and 11 (Approximate Counting).
2. Chapter 6 on Random Walks is a prerequisite for Chapter 11 (Approximate Counting).
3. Chapter 7 on Algebraic Techniques is a prerequisite for Chapters 14 (Number Theory and Algebra) and 12 (Parallel and Distributed Algorithms).

We have included three types of problems in the book. **Exercises** occur throughout the text, and are designed to deepen the reader's understanding of the material being covered in the text. Usually, an exercise will be a variant, extension, or detail of an algorithm or proof being studied. **Problems** appear at the end of each chapter and are meant to be more difficult and involved than the **Exercises** in the text. In addition, **Research Problems** are listed in the Discussion section at the end of each chapter. These are problems that were open at the time we wrote the book; we offer them as suggestions for students (and of course professional researchers) to work on.

Based on our experience with teaching this material, we recommend that the instructor use one of the following course organizations:

- A comprehensive basic course: In addition to Chapters 1–4, this course would cover the material in Chapters 5, 6, and 7 (thus spanning all of Part 1).
- A course oriented toward algebra and number theory: Following Chapters 1–4, this course would cover Chapters 7, 14, and 12.
- A course oriented toward graphs, data structures, and geometry: Following Chapters 1–4, this course would cover Chapters 8, 9, and 10.
- A course oriented toward random walks and counting algorithms: Following Chapters 1–4, this course would cover Chapters 5, 6, and 11.

Each of these courses may be pruned and given in abridged form as an intensive course spanning 3–5 days.

Paradigms for Randomized Algorithms

A handful of general principles lies at the heart of almost all randomized algorithms, despite the multitude of areas in which they find application. We briefly survey these here, with pointers to chapters in which examples of these principles may be found. The following summary draws heavily from ideas in the survey paper by Karp [243].

Foiling an adversary. The classical adversary argument for a deterministic algorithm establishes a lower bound on the running time of the algorithm by constructing an input on which the algorithm fares poorly. The input thus constructed may be different for each deterministic algorithm. A randomized algorithm can be viewed as a probability distribution on a set of deterministic algorithms. While the adversary may be able to construct an input that foils one (or a small fraction) of the deterministic algorithms in the set, it is difficult to devise a single input that is likely to defeat a randomly chosen algorithm. While this paradigm underlies the success of *any* randomized algorithm, the most direct examples appear in Chapter 2 (in game tree evaluation), Chapter 7 (in efficient proof verification), and Chapter 13 (in online algorithms).

Random sampling. The idea that a random sample from a population is representative of the population as a whole is a pervasive theme in randomized algorithms. Examples of this paradigm arise in almost all the chapters, most notably in Chapters 3 (selection algorithms), 8 (data structures), 9 (geometric algorithms), 10 (graph algorithms), and 11 (approximate counting).

Abundance of witnesses. Often, an algorithm is required to determine whether an input (say, a number x) has a certain property (for example, "is x prime?"). It does so by finding a *witness* that x has the property. For many problems, the difficulty with doing this deterministically is that the witness lies in a search space that is too large to be searched exhaustively. However, by establishing that the space contains a large number of witnesses, it often suffices to choose an element at random from the space. The randomly chosen item is likely to be a witness; further, independent repetitions of the process reduce the probability that a witness is not found on any of the repetitions. The most striking examples of this phenomenon occur in number theory (Chapter 14).

Fingerprinting and hashing. A long string may be represented by a short *fingerprint* using a random mapping. In some pattern-matching applications, it can be shown that two strings are likely to be identical if their fingerprints are identical; comparing the short fingerprints is considerably faster than comparing the strings themselves (Chapter 7). This is also the idea behind *hashing*, whereby a small set S of elements drawn from a large universe is mapped into a smaller universe with a guarantee that distinct elements in S are likely to have distinct images. This leads to efficient schemes for deciding membership in S (Chapters 7 and 8) and has a variety of further applications in generating pseudo-random numbers (for example, two-point sampling in Chapter 3 and pairwise independence in Chapter 12) and complexity theory (for instance, algebraic identities and efficient proof verification in Chapter 7).

Random re-ordering. A striking use of randomization in a number of problems in data structuring and computational geometry involves randomly re-ordering the input data, followed by the application of a relatively naive algorithm. After the re-ordering step, the input is unlikely to be in one of the orderings that is pathological for the naive algorithm. (Chapters 8 and 9).

Load balancing. For problems involving choice between a number of resources, such as communication links in a network of processors, randomization can be used to "spread" the load evenly among the resources, as demonstrated in Chapter 4. This is particularly useful in a parallel or distributed environment where resource utilization decisions have to be made locally at a large number of sites without reference to the global impact of these decisions.

Rapidly mixing Markov chains. For a variety of problems involving counting the number of combinatorial objects with a given property, we have approximation algorithms based on randomly sampling an appropriately defined population. Such sampling is often difficult because it may require computing the size of the sample space, which is precisely the problem we would like to solve via sampling. In some cases, the sampling can be achieved by defining a Markov chain on the elements of the population and showing that a short random walk using this Markov chain is likely to sample the population uniformly (Chapter 11).

Isolation and symmetry breaking. In parallel computation, when solving a problem with many feasible solutions it is important to ensure that the different processors are working toward finding the same solution. This requires isolating a specific solution out of the space of all feasible solutions without actually knowing any single element of the solution space. A clever randomized strategy achieves *isolation*, by implicitly choosing a random ordering on the feasible solutions and then requiring the processors to focus on finding the solution of lowest rank. In distributed computation, it is often necessary for a collection of processors to break a deadlock and arrive at a consensus. Randomization is a powerful tool in such deadlock-avoidance, as shown in Chapter 12.

Probabilistic methods and existence proofs. It is possible to establish that an object with certain properties exists by arguing that a randomly chosen object has the properties with positive probability. Such an argument gives no clue as to how to find such an object. Sometimes, the method is used to guarantee the existence of an algorithm for solving a problem; we thus know that the algorithm exists, but have no idea what it looks like or how to construct it. This raises the issue of *non-uniformity* in algorithms (Chapters 2 and 5).

Conventions

Most of the conventions we use are described where they first arise. One worth mentioning here is the issue of *integer breakage:* as long as it does not materially affect the algorithm or analysis being considered (and the intent is unambiguous from the context), we omit ceilings and floors from numbers that strictly should be integers. Thus, we might say "choose $\sqrt{n}$ elements from the set of size n" even when n is not a perfect square. Our intent is to present the crux of the algorithm/analysis without undue notational clutter from ceilings and floors. The expression $\log x$ denotes $\log_2 x$, and the expression $\ln x$ denotes the natural logarithm of x.

Acknowledgements

This book would not have been possible without the guidance and tutelage of Dick Karp. It was he who taught us this field and gave us invaluable guidance at every stage of the book – from the initial planning to the feedback he gave us from using a preliminary version of the manuscript in a graduate course at Berkeley.

We thank the following colleagues, who carefully read portions of the manuscript and pointed out many errors in early versions: Pankaj Agarwal, Donald Aingworth, Susanne Albers, David Aldous, Noga Alon, Sanjeev Arora, Julien Basch, Allan Borodin, Joan Boyar, Andrei Broder, Bernard Chazelle, Ken Clarkson, Don Coppersmith, Cynthia Dwork, Michael Goldwasser, David Gries, Kazuyoshi Hayase, Mary Inaba, Sandy Irani, David Karger, Anna Karlin, Don Knuth, Tom Leighton, Mike Luby, Keju Ma, Karthik Mahadevan, Colin McDiarmid, Ketan Mulmuley, Seffi Naor, Daniel Panario, Bill Pulleyblank, Vijaya Ramachandran, Raimund Seidel, Tom Shiple, Alistair Sinclair, Joel Spencer, Madhu Sudan, Hisao Tamaki, Martin Tompa, Gert Vegter, Jeff Vitter, Peter Winkler, and David Zuckerman. We apologize in advance to any colleagues whose names we have inadvertently omitted.

Special thanks go to Allan Borodin and the students of his CSC 2421 class at the University of Toronto (Fall 1994), as well as to Gudmund Skovbjerg Frandsen, Prabhakar Ragde, and Eli Upfal for giving us detailed feedback from courses they taught using early versions of the manuscript. Their suggestions and advice have been invaluable in making this book more suitable for the classroom.

We thank Rao Kosaraju, Ron Rivest, Joel Spencer, Jeff Ullman, and Paul Vitanyi for providing us with much help and advice on the process of writing and improving the manuscript.

The first author is grateful to Stanford University for the environment and resources which made this effort possible. Several colleagues in the Computer Science Department provided invaluable advice and encouragement. Don Knuth played the role of mentor and his faith in this project was a tremendous source of encouragement. John Mitchell and Jeff Ullman were especially helpful with the mechanics of the publication process. This book owes a great deal to the students, teaching assistants, and other participants in the various offerings of the course CS 365 (Randomized Algorithms) at Stanford. The feedback from these people was invaluable in refining the lecture notes that formed a partial basis for this book. Steven Phillips made a significant contribution as a teaching assistant in CS 365 on two different occasions. Special thanks are due to Yossi Azar, Amotz Bar-Noy, Bob Floyd, Seffi Naor, and Boris Pittel for their guest lectures and help in preparing class notes. The following students transcribed some lecture notes, and their class participation was vital to the development of this material: Julien Basch, Trevor Bourget, Tom Chavez, Edith Cohen, Anil Gangolli, Michael Goldwasser, Bert Hackney, Alan Hu, Jim Hwang, Vasilis Kallistros, Anil Kamath, David Karger, Robert Kennedy, Sanjeev Khanna,

Daphne Koller, Andrew Kosoresow, Sherry Listgarten, Alan Morgan, Steve Newman, Jeffrey Oldham, Steven Phillips, Tomasz Radzik, Ram Ramkumar, Will Sawyer, Sunny Siu, Eric Torng, Theodora Varvarigou, Eric Veach, Alex Wang, and Paul Zhang.

The research and book-writing efforts of the first author have been supported by the following grants and awards: the Bergmann Award from the US-Israel Binational Science Foundation; an IBM Faculty Development Award; gifts from the Mitsubishi Corporation; NSF Grant CCR-9010517; the NSF Young Investigator Award CCR-9357849, with matching funds from IBM Corporation, Schlumberger Foundation, Shell Foundation, and Xerox Corporation; and various grants from the Office of Technology Licensing at Stanford University.

The second author is indebted to his colleagues at the Mathematical Sciences Department of the IBM Thomas J. Watson Research Center, and to the IBM Corporation for providing the facilities and environment that made it possible to write this book. He also thanks Sandeep Bhatt for his encouragement and support of a course on Randomized Algorithms taught by the author at Yale University; the class notes from that course formed a partial basis for this book.

We are indebted to Lauren Cowles of Cambridge University Press for her editorial help and advice in the preparation of the manuscript; this book has emerged much improved as a result of her untiring efforts.

Rajeev Motwani thanks his wife Asha for her love, encouragement, and cheerfulness; without her distractions this book would have been completed several months earlier. This task would not have been possible without the constant support and faith of his family over the years. Finally, the two mutts Tipu and Noori deserve special mention for giving company during the many late night editing sessions.

Prabhakar Raghavan thanks his wife Srilatha for her love and support, his parents for their inspiration, and his children Megha and Manish for ensuring that there was never a dull moment when writing this book.

World-Wide Web

Current information on this book may be found at the following address on the World-Wide Web:

http://www.cup.org/Reviews&blurbs/RanAlg/RanAlg.html

This address may be used for ordering information, reporting errors and viewing an edited list of errors found by other readers.

PART ONE
Tools and Techniques

CHAPTER 1
Introduction

CONSIDER sorting a set S of n numbers into ascending order. If we could find a member y of S such that half the members of S are smaller than y, then we could use the following scheme. We partition $S \setminus \{y\}$ into two sets S_1 and S_2, where S_1 consists of those elements of S that are smaller than y, and S_2 has the remaining elements. We recursively sort S_1 and S_2, then output the elements of S_1 in ascending order, followed by y, and then the elements of S_2 in ascending order. In particular, if we could find y in cn steps for some constant c, we could partition $S \setminus \{y\}$ into S_1 and S_2 in $n-1$ additional steps by comparing each element of S with y; thus, the total number of steps in our sorting procedure would be given by the recurrence

$$T(n) \leq 2T(n/2) + (c+1)n, \tag{1.1}$$

where $T(k)$ represents the time taken by this method to sort k numbers on the worst-case input. This recurrence has the solution $T(n) \leq c'n \log n$ for a constant c', as can be verified by direct substitution.

The difficulty with the above scheme in practice is in finding the element y that splits $S \setminus \{y\}$ into two sets S_1 and S_2 of the same size. Examining (1.1), we notice that the running time of $O(n \log n)$ can be obtained even if S_1 and S_2 are *approximately* the same size – say, if y were to split $S \setminus \{y\}$ such that neither S_1 nor S_2 contained more than $3n/4$ elements. This gives us hope, because we know that every input S contains at least $n/2$ candidate splitters y with this property. How do we quickly find one?

One simple answer is to choose an element of S at random. This does not always ensure a splitter giving a roughly even split. However, it is reasonable to hope that in the recursive algorithm we will be lucky fairly often. The result is an algorithm we call **RandQS**, for Randomized Quicksort.

Algorithm **RandQS** is an example of a *randomized algorithm* – an algorithm that makes random choices during execution (in this case, in Step 1). Let us assume for the moment that this random choice can be made in unit time; we will say more about this in the Notes section. What can we prove about the running time of **RandQS**?

Algorithm RandQS:

Input: A set of numbers S.

Output: The elements of S sorted in increasing order.

1. Choose an element y uniformly at random from S: every element in S has equal probability of being chosen.
2. By comparing each element of S with y, determine the set S_1 of elements smaller than y and the set S_2 of elements larger than y.
3. Recursively sort S_1 and S_2. Output the sorted version of S_1, followed by y, and then the sorted version of S_2.

As is usual for sorting algorithms, we measure the running time of **RandQS** in terms of the number of comparisons it performs since this is the dominant cost in any reasonable implementation. In particular, our goal is to analyze the *expected* number of comparisons in an execution of **RandQS**. Note that all the comparisons are performed in Step 2, in which we compare a randomly chosen partitioning element to the remaining elements. For $1 \leq i \leq n$, let $S_{(i)}$ denote the element of *rank* i (the ith smallest element) in the set S. Thus, $S_{(1)}$ denotes the smallest element of S, and $S_{(n)}$ the largest. Define the random variable X_{ij} to assume the value 1 if $S_{(i)}$ and $S_{(j)}$ are compared in an execution, and the value 0 otherwise. Thus, X_{ij} is a count of comparisons between $S_{(i)}$ and $S_{(j)}$, and so the total number of comparisons is $\sum_{i=1}^{n}\sum_{j>i} X_{ij}$. We are interested in the expected number of comparisons, which is clearly

$$\mathbf{E}[\sum_{i=1}^{n}\sum_{j>i} X_{ij}] = \sum_{i=1}^{n}\sum_{j>i} \mathbf{E}[X_{ij}]. \tag{1.2}$$

This equation uses an important property of expectations called *linearity of expectation*; we will return to this in Section 1.3.

Let p_{ij} denote the probability that $S_{(i)}$ and $S_{(j)}$ are compared in an execution. Since X_{ij} only assumes the values 0 and 1,

$$\mathbf{E}[X_{ij}] = p_{ij} \times 1 + (1 - p_{ij}) \times 0 = p_{ij}. \tag{1.3}$$

To facilitate the determination of p_{ij}, we view the execution of **RandQS** as a binary tree T, each node of which is labeled with a distinct element of S. The root of the tree is labeled with the element y chosen in Step 1, the left sub-tree of y contains the elements in S_1 and the right sub-tree of y contains the elements in S_2. The structures of the two sub-trees are determined recursively by the executions of **RandQS** on S_1 and S_2. The root y is compared to the elements in the two sub-trees, but no comparison is performed between an element of the left sub-tree and an element of the right sub-tree. Thus, there is a comparison between $S_{(i)}$ and $S_{(j)}$ if and only if one of these elements is an ancestor of the other.

The in-order traversal of T will visit the elements of S in a sorted order, and this is precisely what the algorithm outputs; in fact, T is a (random) binary search tree (we will encounter this again in Section 8.2). However, for the analysis we are interested in the level-order traversal of the nodes. This is the permutation π obtained by visiting the nodes of T in increasing order of the level numbers, and in a left-to-right order within each level; recall that the ith level of the tree is the set of all nodes at distance exactly i from the root.

To compute p_{ij}, we make two observations. Both observations are deceptively simple, and yet powerful enough to facilitate the analysis of a number of more complicated algorithms in later chapters (for example, in Chapters 8 and 9).

1. There is a comparison between $S_{(i)}$ and $S_{(j)}$ if and only if $S_{(i)}$ or $S_{(j)}$ occurs earlier in the permutation π than any element $S_{(\ell)}$ such that $i < \ell < j$. To see this, let $S_{(k)}$ be the earliest in π from among all elements of rank between i and j. If $k \notin \{i, j\}$, then $S_{(i)}$ will belong to the left sub-tree of $S_{(k)}$ while $S_{(j)}$ will belong to the right sub-tree of $S_{(k)}$, implying that there is no comparison between $S_{(i)}$ and $S_{(j)}$. Conversely, when $k \in \{i, j\}$, there is an ancestor–descendant relationship between $S_{(i)}$ and $S_{(j)}$, implying that the two elements are compared by **RandQS**.

2. Any of the elements $S_{(i)}, S_{(i+1)}, \ldots, S_{(j)}$ is equally likely to be the first of these elements to be chosen as a partitioning element, and hence to appear first in π. Thus, the probability that this first element is either $S_{(i)}$ or $S_{(j)}$ is exactly $2/(j-i+1)$.

We have thus established that $p_{ij} = 2/(j-i+1)$. By (1.2) and (1.3), the expected number of comparisons is given by

$$\begin{aligned}\sum_{i=1}^{n}\sum_{j>i} p_{ij} &= \sum_{i=1}^{n}\sum_{j>i}\frac{2}{j-i+1}\\ &\leq \sum_{i=1}^{n}\sum_{k=1}^{n-i+1}\frac{2}{k}\\ &\leq 2\sum_{i=1}^{n}\sum_{k=1}^{n}\frac{1}{k}.\end{aligned}$$

It follows that the expected number of comparisons is bounded above by $2nH_n$, where H_n is the *nth Harmonic number*, defined by $H_n = \sum_{k=1}^{n} 1/k$.

Theorem 1.1: *The expected number of comparisons in an execution of* **RandQS** *is at most* $2nH_n$.

From Proposition B.4 (Appendix B), we have that $H_n \sim \ln n + \Theta(1)$, so that the expected running time of **RandQS** is $O(n \log n)$.

Exercise 1.1: Consider the (random) permutation π of S induced by the level-order traversal of the tree T corresponding to an execution of **RandQS**. Is π *uniformly* distributed over the space of all permutations of the elements $S_{(1)}, \ldots, S_{(n)}$?

It is worth examining carefully what we have just established about **RandQS**. The expected running time *holds for every input.* It is an expectation that depends only on the random choices made by the algorithm, and *not* on any assumptions about the distribution of the input. The behavior of a randomized algorithm can vary even on a single input, from one execution to another. The running time becomes a random variable, and the running-time analysis involves understanding the distribution of this random variable.

We will prove bounds on the performances of randomized algorithms that rely solely on their random choices and not on any assumptions about the inputs. It is important to distinguish this from the *probabilistic analysis of an algorithm,* in which one assumes a distribution on the inputs and analyzes an algorithm that may itself be deterministic. In this book we will generally not deal with such probabilistic analysis, except occasionally when illustrating a technique for analyzing randomized algorithms.

Note also that we have proved a bound on the *expected* running time of the algorithm. In many cases (including **RandQS**, see Problem 4.14), we can prove an even stronger statement: that *with very high probability* the running time of the algorithm is not much more than its expectation. Thus, on almost every execution, independent of the input, the algorithm is shown to be fast.

The randomization involved in our **RandQS** algorithm occurs only in Step 1, where a random element is chosen from a set. We define a randomized algorithm as an algorithm that is allowed access to a source of independent, unbiased, random bits; it is then permitted to use these random bits to influence its computation. It is easy to sample a random element from a set S by choosing $O(\log |S|)$ random bits and then using these bits to index an element in the set. However, some distributions cannot be sampled using only random bits. For example, consider an algorithm that picks a random real number from some interval. This requires infinitely many random bits. While we will usually not worry about the conversion of random bits to the desired distribution, the reader should keep in mind that random bits are a resource whose use involves a non-trivial cost. Moreover, there is sometimes a non-trivial computational overhead associated with sampling from a seemingly well-behaved distribution. For example, consider the problem of using a source of unbiased random bits to sample uniformly from a set S whose cardinality is *not* a power of 2 (see Problem 1.2).

There are two principal advantages to randomized algorithms. The first is performance – for many problems, randomized algorithms run faster than the best known deterministic algorithms. Second, many randomized algorithms are simpler to describe and implement than deterministic algorithms of comparable performance. The randomized sorting algorithm described above is an example. This book presents many other randomized algorithms that enjoy these advantages.

In the next few sections, we will illustrate some basic ideas from probability theory using simple applications to randomized algorithms. The reader wishing to review some of the background material on the analysis of algorithms or on elementary probability theory is referred to the Appendices.

1.1. A Min-Cut Algorithm

Two events $\mathcal{E}_1$ and $\mathcal{E}_2$ are said to be *independent* if the probability that they both occur is given by

$$\mathbf{Pr}[\mathcal{E}_1 \cap \mathcal{E}_2] = \mathbf{Pr}[\mathcal{E}_1] \times \mathbf{Pr}[\mathcal{E}_2] \tag{1.4}$$

(see Appendix C). In the more general case where $\mathcal{E}_1$ and $\mathcal{E}_2$ are not necessarily independent,

$$\mathbf{Pr}[\mathcal{E}_1 \cap \mathcal{E}_2] = \mathbf{Pr}[\mathcal{E}_1 \mid \mathcal{E}_2] \times \mathbf{Pr}[\mathcal{E}_2] = \mathbf{Pr}[\mathcal{E}_2 \mid \mathcal{E}_1] \times \mathbf{Pr}[\mathcal{E}_1], \tag{1.5}$$

where $\mathbf{Pr}[\mathcal{E}_1 \mid \mathcal{E}_2]$ denotes the *conditional probability* of $\mathcal{E}_1$ given $\mathcal{E}_2$. Sometimes, when a collection of events is not independent, a convenient method for computing the probability of their intersection is to use the following generalization of (1.5).

$$\mathbf{Pr}[\cap_{i=1}^{k} \mathcal{E}_i] = \mathbf{Pr}[\mathcal{E}_1] \times \mathbf{Pr}[\mathcal{E}_2 \mid \mathcal{E}_1] \times \mathbf{Pr}[\mathcal{E}_3 \mid \mathcal{E}_1 \cap \mathcal{E}_2] \cdots \mathbf{Pr}[\mathcal{E}_k \mid \cap_{i=1}^{k-1} \mathcal{E}_i]. \tag{1.6}$$

Consider a graph-theoretic example. Let G be a connected, undirected multigraph with n vertices. A *multigraph* may contain multiple edges between any pair of vertices. A *cut* in G is a set of edges whose removal results in G being broken into two or more components. A *min-cut* is a cut of minimum cardinality. We now study a simple algorithm for finding a min-cut of a graph.

We repeat the following step: pick an edge uniformly at random and merge the two vertices at its end-points (Figure 1.1). If as a result there are several edges between some pairs of (newly formed) vertices, retain them all. Edges between vertices that are merged are removed, so that there are never any self-loops. We refer to this process of merging the two end-points of an edge into a single vertex as the *contraction* of that edge. With each contraction, the number of vertices of G decreases by one. The crucial observation is that an edge contraction does not reduce the min-cut size in G. This is because every cut in the graph at any intermediate stage is a cut in the original graph. The algorithm continues the contraction process until only two vertices remain; at this point, the set of edges between these two vertices is a cut in G and is output as a candidate min-cut.

Does this algorithm always find a min-cut? Let us analyze its behavior after first reviewing some elementary definitions from graph theory.

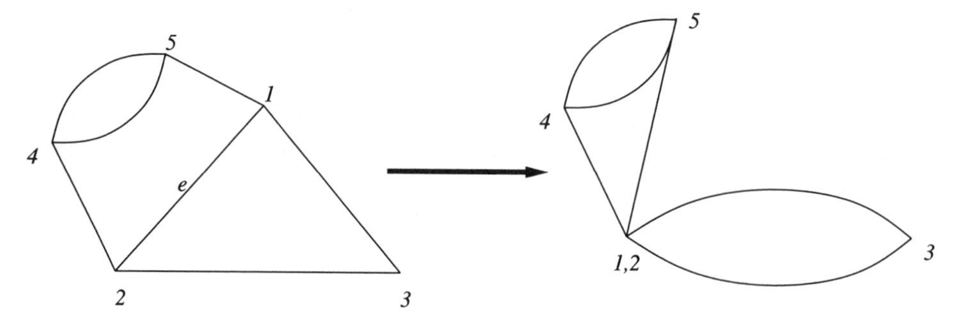

Figure 1.1: A step in the min-cut algorithm; the effect of contracting edge $e = (1, 2)$ is shown.

► **Definition 1.1:** For any vertex v in a multigraph G, the *neighborhood* of v, denoted $\Gamma(v)$, is the set of vertices of G that are adjacent to v. The *degree* of v, denoted $d(v)$, is the number of edges incident on v. For a set S of vertices of G, the neighborhood of S, denoted $\Gamma(S)$, is the union of the neighborhoods of the constituent vertices.

Note that $d(v)$ is the same as the cardinality of $\Gamma(v)$ when there are no self-loops or multiple edges between v and any of its neighbors.

Let k be the min-cut size. We fix our attention on a particular min-cut C with k edges. Clearly G has at least $kn/2$ edges; otherwise there would be a vertex of degree less than k, and its incident edges would be a min-cut of size less than k. We will bound from below the probability that no edge of C is ever contracted during an execution of the algorithm, so that the edges surviving till the end are exactly the edges in C.

Let $\mathcal{E}_i$ denote the event of *not* picking an edge of C at the ith step, for $1 \leq i \leq n-2$. The probability that the edge randomly chosen in the first step is in C is at most $k/(nk/2) = 2/n$, so that $\mathbf{Pr}[\mathcal{E}_1] \geq 1-2/n$. Assuming that $\mathcal{E}_1$ occurs, during the second step there are at least $k(n-1)/2$ edges, so the probability of picking an edge in C is at most $2/(n-1)$, so that $\mathbf{Pr}[\mathcal{E}_2 \mid \mathcal{E}_1] \geq 1 - 2/(n-1)$. At the ith step, the number of remaining vertices is $n-i+1$. The size of the min-cut is still at least k, so the graph has at least $k(n-i+1)/2$ edges remaining at this step. Thus, $\mathbf{Pr}[\mathcal{E}_i \mid \cap_{j=1}^{i-1} \mathcal{E}_j] \geq 1 - 2/(n-i+1)$. What is the probability that no edge of C is ever picked in the process? We invoke (1.6) to obtain

$$\mathbf{Pr}[\cap_{i=1}^{n-2} \mathcal{E}_i] \geq \prod_{i=1}^{n-2} \left(1 - \frac{2}{n-i+1}\right) = \frac{2}{n(n-1)}.$$

The probability of discovering a particular min-cut (which may in fact be the unique min-cut in G) is larger than $2/n^2$. Thus our algorithm may err in declaring the cut it outputs to be a min-cut. Suppose we were to repeat the above algorithm $n^2/2$ times, making independent random choices each time. By (1.4), the probability that a min-cut is not found in any of the $n^2/2$

attempts is at most

$$\left(1 - \frac{2}{n^2}\right)^{n^2/2} < 1/e.$$

By this process of repetition, we have managed to reduce the probability of failure from $1-2/n^2$ to a more respectable $1/e$. Further executions of the algorithm will make the failure probability arbitrarily small – the only consideration being that repetitions increase the running time.

Note the extreme simplicity of the randomized algorithm we have just studied. In contrast, most deterministic algorithms for this problem are based on network flows and are considerably more complicated. In Section 10.2 we will return to the min-cut problem and fill in some implementation details that have been glossed over in the above presentation; in fact, it will be shown that a variant of this algorithm has an expected running time that is significantly smaller than that of the best known algorithms based on network flow.

Exercise 1.2: Suppose that at each step of our min-cut algorithm, instead of choosing a random edge for contraction we choose two vertices at random and coalesce them into a single vertex. Show that there are inputs on which the probability that this modified algorithm finds a min-cut is exponentially small.

1.2. Las Vegas and Monte Carlo

The randomized sorting algorithm and the min-cut algorithm exemplify two different types of randomized algorithms. The sorting algorithm *always* gives the correct solution. The only variation from one run to another is its running time, whose distribution we study. We call such an algorithm a *Las Vegas algorithm.*

In contrast, the min-cut algorithm may sometimes produce a solution that is incorrect. However, we are able to bound the probability of such an incorrect solution. We call such an algorithm a *Monte Carlo algorithm.* In Section 1.1 we observed a useful property of a Monte Carlo algorithm: if the algorithm is run repeatedly with independent random choices each time, the failure probability can be made arbitrarily small, at the expense of running time. Later, we will see examples of algorithms in which both the running time and the quality of the solution are random variables; sometimes these are also referred to as Monte Carlo algorithms. For decision problems (problems for which the answer to an instance is YES or NO), there are two kinds of Monte Carlo algorithms: those with *one-sided error*, and those with *two-sided error.* A Monte Carlo algorithm is said to have two-sided error if there is a non-zero probability that it errs when it outputs either YES or NO. It is said to have one-sided error if the probability that it errs is zero for at least one of the possible outputs (YES/NO) that it produces.

We will see examples of all three types of algorithms – Las Vegas, Monte Carlo with one-sided error, and Monte Carlo with two-sided error – in this book.

Which is better, Monte Carlo or Las Vegas? The answer depends on the application – in some applications an incorrect solution may be catastrophic. A Las Vegas algorithm is by definition a Monte Carlo algorithm with error probability 0. The following exercise gives us a way of deriving a Las Vegas algorithm from a Monte Carlo algorithm. Note that the efficiency of the derivation procedure depends on the time taken to verify the correctness of a solution to the problem.

Exercise 1.3: Consider a Monte Carlo algorithm A for a problem Π whose expected running time is at most $T(n)$ on any instance of size n and that produces a correct solution with probability $\gamma(n)$. Suppose further that given a solution to Π, we can verify its correctness in time $t(n)$. Show how to obtain a Las Vegas algorithm that always gives a correct answer to Π and runs in expected time at most $(T(n) + t(n))/\gamma(n)$.

In attempting Exercise 1.3 the reader will have to use a simple property of the *geometric random variable* (Appendix C). Consider a biased coin that, on a toss, has probability p of coming up HEADS and $1 - p$ of coming up TAILS. What is the expected number of (independent) tosses up to and including the first head? The number of such tosses is a random variable that is said to be *geometrically distributed.* The expectation of this random variable is $1/p$. This fact will prove useful in numerous applications.

Exercise 1.4: Let $0 < \epsilon_2 < \epsilon_1 < 1$. Consider a Monte Carlo algorithm that gives the correct solution to a problem with probability at least $1 - \epsilon_1$, regardless of the input. How many independent executions of this algorithm suffice to raise the probability of obtaining a correct solution to at least $1 - \epsilon_2$, regardless of the input?

We say that a Las Vegas algorithm is an *efficient Las Vegas* algorithm if on any input its expected running time is bounded by a polynomial function of the input size. Similarly, we say that a Monte Carlo algorithm is an *efficient Monte Carlo* algorithm if on any input its worst-case running time is bounded by a polynomial function of the input size.

1.3. Binary Planar Partitions

We now illustrate another very useful and basic tool from probability theory: *linearity of expectation.* For random variables $X_1, X_2, \ldots$,

$$\mathbf{E}[\sum_i X_i] = \sum_i \mathbf{E}[X_i]. \tag{1.7}$$

(See Proposition C.5.) We have implicitly used this tool in our analysis of **RandQS**. A point that cannot be overemphasized is that (1.7) holds *regardless* of any dependencies between the X_i.

► **Example 1.1:** A ship arrives at a port, and the 40 sailors on board go ashore for revelry. Later at night, the 40 sailors return to the ship and, in their state of inebriation, each chooses a random cabin to sleep in. What is the expected number of sailors sleeping in their own cabins?

The inefficient approach to this problem would be to consider all 40^{40} arrangements of sailors in cabins. The solution to this example will involve the use of a simple and often useful device called an *indicator variable*, together with linearity of expectation. Let X_i be 1 if the ith sailor chooses her own cabin, and 0 otherwise. Thus X_i indicates whether or not a certain event occurs, and is hence called an indicator variable. We wish to determine the expected number of sailors who get their own cabins, which is $\mathbf{E}[\sum_{i=1}^{40} X_i]$. By linearity of expectation, this is $\sum_{i=1}^{40} \mathbf{E}[X_i]$. Since the cabins are chosen at random, the probability that the ith sailor gets her own cabin is $1/40$, so $\mathbf{E}[X_i] = 1/40$. Thus the expected number of sailors who get their own cabins is $\sum_{i=1}^{40} 1/40 = 1$.

Our next illustration is the construction of a *binary planar partition* of a set of n disjoint line segments in the plane, a problem with applications to computer graphics. A binary planar partition consists of a binary tree together with some additional information, as described below. Every internal node of the tree has two children. Associated with each node v of the tree is a region $r(v)$ of the plane. Associated with each internal node v of the tree is a line $\ell(v)$ that intersects $r(v)$. The region corresponding to the root is the entire plane. The region $r(v)$ is partitioned by $\ell(v)$ into two regions $r_1(v)$ and $r_2(v)$, which are the regions associated with the two children of v. Thus, any region r of the partition is bounded by the partition lines on the path from the root to the node corresponding to r in the tree.

Given a set $S = \{s_1, s_2, \ldots, s_n\}$ of non-intersecting line segments in the plane, we wish to find a binary planar partition such that every region in the partition contains at most one line segment (or a portion of one line segment). Notice that the definition allows us to divide an input line segment s_i into several segments $s_{i1}, s_{i2}, \ldots$, each of which lies in a different region. The example of Figure 1.2 gives such a partition for a set of three line segments (dark lines).

Exercise 1.5: Show that there exists a set of line segments for which no binary planar partition can avoid breaking up some of the segments into pieces, if each segment is to lie in a different region of the partition.

Binary planar partitions have two applications in computer graphics. Here, we describe one of them, the problem of *hidden line elimination* in computer

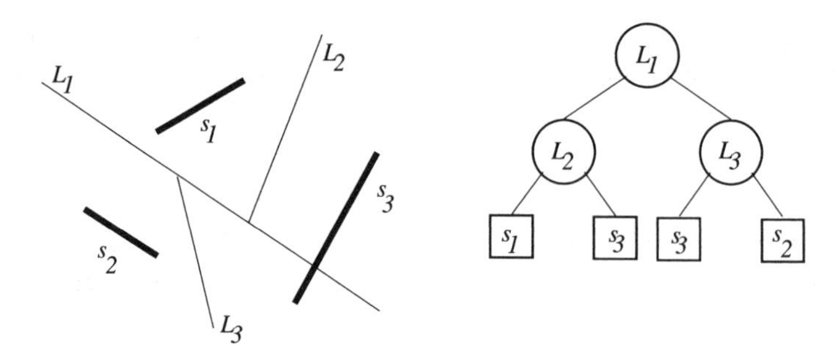

Figure 1.2: An example of a binary planar partition for a set of segments (dark lines). Each leaf is labeled by the line segment it contains. The labels $r(v)$ are omitted for clarity.

graphics. The second application has to do with the *constructive solid geometry* (or CSG) representation of a polyhedral object.

In rendering a scene on a graphics terminal, we are often faced with a situation in which the scene remains fixed, but it is to be viewed from several directions (for instance, in a flight simulator, where the simulated motion of the plane causes the viewpoint to change). The hidden line elimination problem is the following: having adopted a viewpoint and a direction of viewing, we want to draw only the portion of the scene that is visible, eliminating those objects that are obscured by other objects "in front" of them relative to the viewpoint. In such a situation, we might be prepared to spend some computational effort preprocessing the scene so that given a direction of viewing, the scene can be rendered quickly with hidden lines eliminated.

One approach to this problem uses a binary partition tree. In this chapter we consider the simple case where the scene lies entirely in the plane, and we view it from a point in the same plane. Thus, the output is a one-dimensional projected "picture." We can assume that the input scene consists of non-intersecting line segments, since any line that is intersected by another can be broken up into segments, each of which touches other lines only at its endpoints (if at all). Once the scene has been thus decomposed into line segments, we construct a binary planar partition tree for it. Now, given the direction of viewing, we use an idea known as the *painter's algorithm* to render the scene: first draw the objects that are furthest "behind," and then progressively draw the objects that are in front. Given the binary planar partition tree, the painter's algorithm can be implemented by recursively traversing the tree as follows. At the root of the tree, determine which side of the partitioning line L_1 is "behind" from the viewpoint and render all the objects in that sub-tree (recursively). Having completely rendered the portion of the tree corresponding to that sub-tree, do the same for the portion in "front" of L_1, "painting over" objects already drawn.

The time it takes to render the scene depends on the size of the binary planar partition tree. We therefore wish to construct a binary planar partition that is as small as possible. Notice that since the tree must be traversed completely to render the scene, the depth of the tree is immaterial in this application. Because the construction of the partition can break some of the input segments s_i into smaller pieces, the size of the partition need not be n; in fact, it is not clear that a partition of size $O(n)$ always exists.

In this chapter we consider only the planar case just described; in Chapter 9 we generalize the idea of a binary planar partition to handle the rendition of a three-dimensional scene on a two-dimensional screen (a far more interesting case for computer graphics).

For a line segment s, let $l(s)$ denote the line obtained by extending (if necessary) s on both sides to infinity. For the set $S = \{s_1, s_2, \ldots s_n\}$ of line segments, a simple and natural class of partitions is the set of *autopartitions*, which are formed by only using lines from the set $\{l(s_1), l(s_2), \ldots l(s_n)\}$ in constructing the partition. We only consider autopartitions from here on.

Algorithm RandAuto:

Input: A set $S = \{s_1, s_2, \ldots, s_n\}$ of non-intersecting line segments.

Output: A binary autopartition P_π of S.

1. Pick a permutation π of $\{1, 2, \ldots, n\}$ uniformly at random from the $n!$ possible permutations.
2. **while** a region contains more than one segment, cut it with $l(s_i)$ where i is first in the ordering π such that s_i cuts that region.

In the partition resulting from an execution of **RandAuto**, a segment may lie on the boundary between two regions of the partition. We declare such a segment to lie in one region or the other in any convenient way.

Theorem 1.2: *The expected size of the autopartition produced by* **RandAuto** *is* $O(n \log n)$.

PROOF: For line segments u and v, define $index(u, v)$ to be i if $l(u)$ intersects $i - 1$ other segments before hitting v, and $index(u, v) = \infty$ if $l(u)$ does not hit v. Since a segment u can be extended in two directions, it is possible that $index(u, v) = index(u, w)$ for two different lines v and w (in Figure 1.3, $index(u, v_1) = index(u, v_2) = 2$).

Let us denote by $u \dashv v$ the event that $l(u)$ cuts v in the constructed partition. Let $index(u, v) = i$, and let $u_1, u_2, \ldots u_{i-1}$ be the segments that $l(u)$ intersects before hitting v. The event $u \dashv v$ happens only if u occurs before any of $\{u_1, u_2, \ldots u_{i-1}, v\}$ in the randomly chosen permutation π. The probability that this happens is $1/(i + 1)$.

Let $C_{u,v}$ be an indicator variable that is 1 if $u \dashv v$ and 0 otherwise; clearly, $\mathbf{E}[C_{u,v}] = \mathbf{Pr}[u \dashv v] \leq 1/(index(u, v) + 1)$. The size of P_π equals n plus the number of intersections due to cuts. Thus, its expectation is $n + \mathbf{E}[\sum_u \sum_v C_{u,v}]$ and by

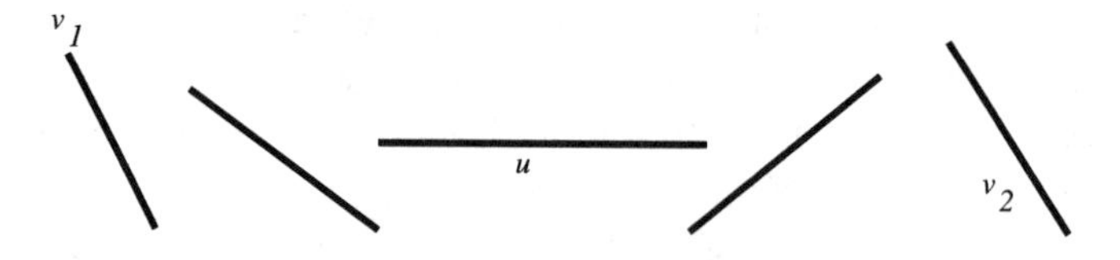

Figure 1.3: An illustration of *index*(u, v).

linearity of expectation this equals

$$n + \sum_u \sum_{v \neq u} \mathbf{Pr}[u \dashv v] \leq n + \sum_u \sum_{v \neq u} \frac{1}{index(u, v) + 1}. \tag{1.8}$$

For any line segment u and any finite positive integer i, there are at most two vertices v and w such that $index(u, v)$ and $index(u, w)$ equals i. This is because the extension of the segment u along either of the two possible directions will meet any other line segment at most once. Thus, in each of the two directions, there is a total ordering on the points of intersection with other segments and the index values increase monotonically. This implies that

$$\sum_{v \neq u} \frac{1}{index(u, v) + 1} \leq \sum_{i=1}^{n-1} \frac{2}{i + 1}.$$

Combining this with (1.8) implies that the expected size of P_π is bounded above by

$$n + 2 \sum_u \sum_{i=1}^{n-1} \frac{1}{i + 1} \leq n + 2nH_n,$$

which is $O(n \log n)$. □

Note that in computing the expected number of intersections, we only made use of linearity of expectation. We do not require any independence between the events $u \dashv v$ and $u \dashv w$, for segments u, v, and w. Indeed, these events need not be independent in general.

One way of interpreting Theorem 1.2 is as follows: since the expected size of the binary planar partition constructed by the algorithm is $O(n \log n)$ on *any* input, there *must exist* a binary autopartition of size $O(n \log n)$ for every input. This follows from the simple fact that any random variable assumes at least one value that is no greater than its expectation (and, indeed, one that is no less than its expectation). Thus we have used a probabilistic argument to assert that a combinatorial object – in this case a binary autopartition of size $O(n \log n)$ – *exists with absolute certainty* rather than with some probability. This is an example of the *probabilistic method in combinatorics*. We will study the probabilistic method in greater detail in Chapter 5.

1.4. A Probabilistic Recurrence

Frequently, we express a random variable of interest as a recurrence in terms of other random variables. In this section, we study one such situation using the **Find** algorithm analyzed in detail in Problem 1.9. The material in this section, although useful, is not an essential prerequisite for subsequent topics and may be omitted in the first reading.

The **Find** algorithm for selecting the kth smallest of a set S of n elements works as follows. We pick a random element y and partition $S \setminus \{y\}$ into two sets S_1 and S_2 (elements smaller and larger than y respectively) as in **RandQS**. Suppose $|S_1| = k-1$; then y is the desired element and we are done. Otherwise, if $|S_1| \geq k$, we recursively find the kth smallest element of S_1; else we recursively find the $(k - |S_1| - 1)$th smallest element in S_2.

The expected number of comparisons made by the **Find** algorithm is the subject of Problem 1.9. Suppose instead that we were to ask the following question: what is the expected number of times we make the recursive call in the algorithm? Equivalently, what is the expected number of times we pick a random element in the algorithm? While this question may not be especially important for the **Find** algorithm, it is the kind of question that arises in the analysis of a number of parallel and geometric algorithms. Intuitively, we expect that the size of the residual problem in the **Find** algorithm is divided by a constant factor at each recursive level, so that we expect that the number of recursive invocations is $O(\log n)$. Below, we show that this intuition can be formalized in a general setting.

Let $g(x)$ be a monotone non-decreasing function from the positive reals to the positive reals. Consider a particle whose position changes at discrete time steps and is always at a positive integer. If the particle is currently at position $m > 1$, it proceeds at the next step to the position $m - X$, where X is a random variable ranging over the integers $1, \ldots, m-1$. All we know about X is that $\mathbf{E}[X] \geq g(m)$, and that X is chosen independently of the past. It is clear that the particle will always reach position 1 and the process terminates in that state. The interesting question is, assuming that the particle starts at position n, what is the expected number of steps before it reaches position 1? The reader may associate the position of the particle with the size of the problem in a recursive call of the **Find** algorithm. Although we have more information about the distribution of X in the case of **Find**'s analysis, it turns out that the bound on the expected size of the residual problem suffices for proving the following result.

Theorem 1.3: *Let T be the random variable denoting the number of steps in which the particle reaches the position 1. Then,* $\mathbf{E}[T] \leq \int_1^n dx/g(x)$.

PROOF: The proof is by induction on n; let us suppose the theorem holds for values of m smaller than n. Let $f(m) = \int_1^m dx/g(x)$ for $m \geq 1$. We wish to show that $\mathbf{E}[T] \leq f(n)$.

Consider the first step, during which the particle proceeds from position n to position $n - X$, where X is chosen from a distribution for which $\mathbf{E}[X] \geq g(n)$. We have

$$\begin{aligned}
\mathbf{E}[T] &\leq 1 + \mathbf{E}[f(n - X)] && (1.9)\\
&= 1 + \mathbf{E}[\int_1^n \frac{dy}{g(y)} - \int_{n-X}^n \frac{dy}{g(y)}] && (1.10)\\
&= 1 + f(n) - \mathbf{E}[\int_{n-X}^n \frac{dy}{g(y)}] && (1.11)\\
&\leq 1 + f(n) - \mathbf{E}[\int_{n-X}^n \frac{dy}{g(n)}] && (1.12)\\
&= 1 + f(n) - \frac{\mathbf{E}[X]}{g(n)} && (1.13)\\
&\leq f(n). && (1.14)
\end{aligned}$$

The inequality (1.12) follows from the assumption that $g(y)$ is non-decreasing, while (1.14) follows from the lower bound on $\mathbf{E}[X]$. □

Exercise 1.6: If X were to range over all integers having value at most $m-1$ (possibly including negative integers), how would the statement and proof of Theorem 1.3 change?

For the **Find** algorithm, we can show (following the analysis of Problem 1.9) that $g(m) \geq m/4$. We may then apply the above theorem to bound the expected number of recursive calls to **Find** by $4 \ln n$.

Exercise 1.7: What prevents us from using Theorem 1.3 to bound the expected number of levels of recursion in the **RandQS** algorithm?

1.5. Computation Model and Complexity Classes

In this section we discuss models of computation used in this book, and follow this with a review of complexity classes.

1.5.1. RAMs and Turing Machines

Following common practice, throughout this book we use the *Turing machine* model to discuss complexity-theory issues. As is common, however, we switch to the *RAM* (random access machine) as the model of computation when describing and analyzing algorithms (except in the study of parallel and distributed algorithms in Chapter 12, where we define a version of the RAM model for machines working in parallel). We begin by defining the Turing machine, which is an abstract model of an algorithm.

► **Definition 1.2:** A *deterministic Turing machine* is a quadruple $M = (S, \Sigma, \delta, s)$. Here S is a finite set of states, of which $s \in S$ is the machine's *initial state*. The machine uses a finite set of symbols, denoted Σ; this set includes special symbols BLANK and FIRST. The function δ is the *transition function* of the Turing machine, mapping $S \times \Sigma$ to $(S \cup \{\text{HALT,YES,NO}\}) \times \Sigma \times \{\leftarrow, \rightarrow, \text{STAY}\}$. The machine has three halting states HALT (the *halting state*), YES (the *accepting state*), and NO (the *rejecting state*) (these are states, but formally not in S).

The input to the Turing machine is generally thought of as being written on a *tape*; unless otherwise specified, the machine may read from and write on this tape. We assume that HALT, YES, and NO, as well as the symbols $\leftarrow, \rightarrow$, and STAY, are not in $S \cup \Sigma$. The machine begins in the initial state s with its *cursor* at the first symbol of the input x (i.e., the left end of the tape); this symbol is always FIRST. The rest of the input is a string of finite length from $(\Sigma \backslash \{\text{BLANK, FIRST}\})^*$; the left-most BLANK on the tape identifies the end of the input string.

The transition function dictates the actions of the machine, and may be thought of as its *program.* In each step, the machine reads the symbol α of the input currently pointed to by the cursor; based on this symbol and the current state of the machine, it chooses a next state, a symbol β to be overwritten on α and a cursor motion direction from $\{\leftarrow, \rightarrow, \text{STAY}\}$ (here $\leftarrow$ and $\rightarrow$ specify a motion by one step to the left and right, respectively, while STAY specifies that the cursor remain in its present position). The transition function is designed to ensure that the cursor never falls off the left end of the input, identified by FIRST. The machine may of course overwrite the BLANK symbol.

If the machine halts in the YES state, we say that it has *accepted* the input x. If the machine halts in the NO state, we say that it has *rejected* the input x. The third halting state, HALT, is for the computation of functions whose range is not Boolean; in such cases, the output of the function computation is written onto the tape. An algorithm corresponds to a Turing machine that always halts.

A *probabilistic Turing machine* is a Turing machine augmented with the ability to generate an unbiased coin flip in one step. It corresponds to a randomized algorithm. On any input x, a probabilistic Turing machine accepts x with some probability, and we study this probability.

In the light of these definitions, we may speak of an algorithm accepting or rejecting an input (we visualize the Turing machine underlying the algorithm as accepting or rejecting), and similarly speak of a randomized algorithm accepting or rejecting an input with some probability.

In the RAM model, we have a machine that can perform the following types of operations involving registers and main memory: input–output operations, memory–register transfers, indirect addressing, branching, and arithmetic operations. Each register or memory location may hold an integer that can be accessed as a unit, but an algorithm has no access to the representation of the number.

The arithmetic instructions permitted are $+, -, \times, /$. In addition, an algorithm can compare two numbers, and evaluate the square root of a positive number.

Two types of RAM models are defined based on the cost used for measuring the running time of a program. In the *unit-cost RAM* (sometimes also called the *uniform RAM*), each instruction can be performed in one time step. This model is believed to be much too powerful since there is no known polynomial-time simulation of this model by Turing machines. This situation arises because the unit-cost RAM, unlike the more restricted Turing machine, is able to use multiplication to quickly compute extremely large integers. However, if we disallow all arithmetic operations besides addition and subtraction, then it is possible to show that the resulting model is equivalent to Turing machines under polynomial-time simulations.

A more realistic version of the RAM is the so-called *log-cost RAM* where each instruction requires time proportional to the logarithm of the size of its operands. It turns out that the log-cost RAM with the complete arithmetic instruction set is equivalent to Turing machines under polynomial-time simulations.

For simplicity, we will work with the general unit-cost RAM model. At the same time, we will avoid misuse of its power by ensuring that in all algorithms under consideration the size of the operands is polynomially bounded in the input size. Thus, our algorithm can be transformed to the log-cost RAM model with only a small (logarithmic in the input size) multiplicative slow-down in the running time. We also assume that the RAM can in a single step choose an element uniformly at random from a set of cardinality polynomial in the size of the problem input. Standard texts on automata and complexity (see the Notes section) give proofs of the following basic fact.

Proposition 1.4: *Any Turing machine computation of length polynomial in the size of the input can be simulated by a RAM computation of length polynomial in the size of the input. Any RAM computation of length polynomial in the size of the input can be simulated by a Turing machine computation of length polynomial in the size of the input.*

1.5.2. Complexity Classes

We now define some basic complexity classes focusing on those involving randomized algorithms. For these definitions, the underlying model of computation is assumed to be the Turing machine, but by the preceding discussion it could be substituted by a log-cost RAM or the restricted form of the unit-cost RAM.

In complexity theory, it is common to concentrate on the decision problem derived from some hard optimization problem. This enables the development of an elegant theoretical framework, and the decision problem is usually not significantly different in structure from its optimization counterpart. For instance, consider the *satisfiability* problem, in which an instance consists of a set of clauses in conjunctive normal form (CNF). Because the satisfiability problem appears at various points in this book, we define some terminology relating to it. The Boolean inputs are called *variables*, which may appear in either uncomplemented or complemented form in a clause. The uncomplemented or complemented variables in a clause are known as *literals* (respectively, *unnegated* and *negated* literals). A clause is said to be satisfied if at least one of the literals in it is TRUE. A solution consists either of an assignment of Boolean values to the variables that ensures that every clause is satisfied (such an assignment is known as a *truth assignment*), or a negative answer that it is not possible to assign inputs so as to satisfy all the clauses simultaneously. The decision version of this problem, commonly abbreviated SAT, seeks only a YES or NO answer depending on whether or not all the clauses can simultaneously be satisfied, without demanding an assignment of values to the inputs (in case the answer is YES).

► **Example 1.2:** Consider the following instance of satisfiability:

$$(x_1 \vee \overline{x}_2 \vee x_4) \wedge (\overline{x}_3 \vee \overline{x}_4 \vee x_5) \wedge (\overline{x}_1 \vee x_2 \vee x_4 \vee \overline{x}_5).$$

In this example, there are three clauses. The first stipulates that either x_1 should be TRUE, or x_2 should be FALSE, or x_4 should be TRUE. The literal $\overline{x}_2$ denotes that one way of satisfying the first clause is to set x_2 FALSE. The first two clauses have three literals each, while the third has four. The assignments x_1 = TRUE, x_3 = FALSE, and x_5 = FALSE suffice to satisfy all the clauses (regardless of the values assigned to x_2 and x_4). Thus the solution to this instance for the decision question (SAT) is YES.

Any decision problem can be treated as a language recognition problem. Fix a finite alphabet Σ, usually $\Sigma = \{0, 1\}$, and let Σ^* be the set of all possible strings over this alphabet. Denote by $|s|$ the length of a string s. A *language* $L \subseteq \Sigma^*$ is any collection of strings over Σ. The corresponding *language recognition* problem is to decide whether a given string x in Σ^* belongs to L. An algorithm solves a language recognition problem for a specific language L by *accepting* (output YES) any input string contained in L, and *rejecting* (output NO) any input string not contained in L. The SAT problem can easily be cast in the form of a language recognition problem by devising a suitable encoding of formulas as bit-strings.

A complexity class is a collection of languages all of whose recognition problems can be solved under prescribed bounds on the computational resources. We are primarily interested in various forms of efficient algorithms, where efficient is defined as being *polynomial time*. Recall that an algorithm has polynomial running time if it halts within $n^{O(1)}$ time on any input of length n. The following definitions list some interesting complexity classes.

► **Definition 1.3:** The class ***P*** consists of all languages L that have a polynomial-time algorithm A such that for any input $x \in \Sigma^*$,

- $x \in L \Rightarrow A(x)$ accepts.
- $x \notin L \Rightarrow A(x)$ rejects.

► **Definition 1.4:** The class ***NP*** consists of all languages L that have a polynomial-time algorithm A such that for any input $x \in \Sigma^*$,

- $x \in L \Rightarrow \exists y \in \Sigma^*$, $A(x, y)$ accepts, where $|y|$ is bounded by a polynomial in $|x|$.
- $x \notin L \Rightarrow \forall y \in \Sigma^*$, $A(x, y)$ rejects.

A useful view of ***P*** and ***NP*** is the following. The class ***P*** consists of all languages L such that for any x in L a proof of the membership x in L (represented by the string y) can be *found* and *verified* efficiently. On the other hand, ***NP*** consists of all languages L such that for any x in L, a proof of the membership of x in L can be *verified* efficiently. Obviously, $\boldsymbol{P} \subseteq \boldsymbol{NP}$, but it is not known whether $\boldsymbol{P} = \boldsymbol{NP}$. If $\boldsymbol{P} = \boldsymbol{NP}$, the existence of an efficiently verifiable proof implies that it is possible to actually find such a proof efficiently.

For any complexity class $\mathcal{C}$, we define the complementary class co-$\mathcal{C}$ as the set of languages whose complement is in the class $\mathcal{C}$. That is,

$$\text{co-}\mathcal{C} = \{L \mid \overline{L} \in \mathcal{C}\}.$$

It is obvious that $\boldsymbol{P} = \text{co-}\boldsymbol{P}$ and $\boldsymbol{P} \subseteq \boldsymbol{NP} \cap \text{co-}\boldsymbol{NP}$. We do not know whether $\boldsymbol{P} = \boldsymbol{NP} \cap \text{co-}\boldsymbol{NP}$ or whether $\boldsymbol{NP} = \text{co-}\boldsymbol{NP}$, although both statements are widely believed to be false.

Likewise, we can define deterministic and non-deterministic complexity classes for different bounds on the running time. Let *exponential time* denote a running time which is $2^{p(n)}$ for some polynomial $p(n)$ in the input size. Allowing exponential time instead of polynomial time in Definitions 1.3 and 1.4 gives us the complexity classes ***EXP*** and ***NEXP***. Clearly, $\boldsymbol{EXP} \subseteq \boldsymbol{NEXP}$, but once again we do not know whether this inclusion is strict. On the other hand, we do know that if $\boldsymbol{P} = \boldsymbol{NP}$, then $\boldsymbol{EXP} = \boldsymbol{NEXP}$.

We can also define *space* complexity classes by leaving the running time unconstrained and instead placing a bound on the space used by an algorithm. In the case of Turing machines, the space used is determined by the number of distinct positions on the tape that are scanned during an execution; for RAMs, the space requirement is simply the number of words of memory required by an algorithm. In Definitions 1.3 and 1.4, requiring polynomial space instead of polynomial time yields the definition of the class ***PSPACE*** and ***NPSPACE***. A ***PSPACE*** algorithm may run for super-polynomial time. These classes behave differently from the time complexity classes; for example, we know that $\boldsymbol{PSPACE} = \boldsymbol{NPSPACE}$ and $\boldsymbol{PSPACE} = \text{co-}\boldsymbol{PSPACE}$.

We next review the notions of polynomial reductions and completeness for a complexity class.

► **Definition 1.5:** A *polynomial reduction* from a language $L_1 \subseteq \Sigma^*$ to a language $L_2 \subseteq \Sigma^*$ is a function $f : \Sigma^* \to \Sigma^*$ such that:

1. There is a polynomial-time algorithm that computes f.
2. For all $x \in \Sigma^*$, $x \in L_1$ if and only if $f(x) \in L_2$.

Exercise 1.8: Show that if there is a polynomial reduction from L_1 to L_2, then $L_2 \in \boldsymbol{P}$ implies that $L_1 \in \boldsymbol{P}$.

► **Definition 1.6:** A language L is *NP-hard* if, for all $L' \in \boldsymbol{NP}$, there is a polynomial reduction from L' to L.

Thus, if any ***NP***-hard decision problem can be solved in polynomial time, then so can all problems in ***NP***.

► **Definition 1.7:** A language L is *NP-complete* if it is in ***NP*** and is ***NP***-hard.

Intuitively the decision problems corresponding to ***NP***-complete languages are the "hardest" problems in ***NP***. Note that the notion of ***NP***-completeness applies only to decision problems; the optimization problem corresponding to an ***NP***-complete decision problem is ***NP***-hard, but is not ***NP***-complete because it is not in ***NP*** by definition. As with ***NP***, the notions of hardness and completeness can be generalized to any class $\mathcal{C}$, for an appropriate notion of reduction. Unless otherwise specified, the default notion of a reduction is a polynomial reduction, and this is typically used for defining hardness and completeness in complexity classes that are a superset of ***P***, such as ***PSPACE***.

We generalize these classes to allow for randomized algorithms. The basic idea is to replace the existential and universal quantifiers in the definition of ***NP*** by probabilistic requirements.

► **Definition 1.8:** The class ***RP*** (for Randomized Polynomial time) consists of all languages L that have a randomized algorithm A running in worst-case polynomial time such that for any input x in Σ^*,

- $x \in L \Rightarrow \mathbf{Pr}[A(x) \text{ accepts}] \geq \frac{1}{2}$.
- $x \notin L \Rightarrow \mathbf{Pr}[A(x) \text{ accepts}] = 0$.

The choice of the bound on the error probability $1/2$ is arbitrary. In fact, as was observed in the case of the min-cut algorithm, independent repetitions of the algorithm can be used to go from the case where the probability of success is *polynomially small* to the case where the probability of error is *exponentially small* while changing only the degree of the polynomial that bounds the running time. Thus, the success probability can be changed to an inverse polynomial function of the input size without significantly affecting the definition of ***RP***.

Observe that an ***RP*** algorithm is a Monte Carlo algorithm that can err only when $x \in L$. This is referred to as *one-sided error*. The class co-***RP*** consists of languages that have polynomial-time randomized algorithms erring only in the case when $x \notin L$. A problem belonging to both ***RP*** and co-***RP*** can be solved by a randomized algorithm with *zero-sided error*, i.e., a Las Vegas algorithm.

► **Definition 1.9:** The class ***ZPP*** (for Zero-error Probabilistic Polynomial time) is the class of languages that have Las Vegas algorithms running in expected polynomial time.

Exercise 1.9: Show that $ZPP = RP \cap \text{co-}RP$.

Consider now the class of problems that have randomized Monte Carlo algorithms making *two-sided errors.*

► **Definition 1.10:** The class ***PP*** (for Probabilistic Polynomial time) consists of all languages L that have a randomized algorithm A running in worst-case polynomial time such that for any input x in Σ^*,

- $x \in L \Rightarrow \mathbf{Pr}[A(x) \text{ accepts}] > \frac{1}{2}$.
- $x \notin L \Rightarrow \mathbf{Pr}[A(x) \text{ accepts}] < \frac{1}{2}$.

To reduce the error probability of a two-sided error algorithm, we can perform several independent iterations on the same input and produce the output (accept or reject) that occurs in the majority of these iterations. Unfortunately, the definition of the class ***PP*** is rather weak: because we have no bound on how far from 1/2 the probabilities are, it may not be possible to use a small number of repetitions of an algorithm A with such two-sided error probability to obtain an algorithm with significantly smaller error probability.

Exercise 1.10: Consider a randomized algorithm with two-sided error probabilities as in the definition of ***PP***. Show that a polynomial number of independent repetitions of this algorithm need not suffice to reduce the error probability to 1/4. (Consider the case where the error probability is $1/2 - 1/2^n$.)

A more useful class of two-sided error randomized algorithms corresponds to the following complexity class.

► **Definition 1.11:** The class ***BPP*** (for Bounded-error Probabilistic Polynomial time) consists of all languages L that have a randomized algorithm A running in worst-case polynomial time such that for any input x in Σ^*,

- $x \in L \Rightarrow \mathbf{Pr}[A(x) \text{ accepts}] \geq \frac{3}{4}$.
- $x \notin L \Rightarrow \mathbf{Pr}[A(x) \text{ accepts}] \leq \frac{1}{4}$.

In a later chapter (see Problem 4.8) we will show that for this class of algorithms the error probability can be reduced to $1/2^n$ with only a polynomial number of iterations. In fact, the probability bounds $3/4$ and $1/4$ can be changed to $1/2 + 1/p(n)$ and $1/2 - 1/p(n)$, respectively, for any polynomially bounded function $p(n)$ without affecting this error reduction property or the definition of the class ***BPP*** to a significant extent.

The reader is referred to Problems 1.11–1.14 for several basic relationships between these complexity classes. There are several interesting open questions regarding the relationships between these randomized complexity classes, for example:

1. Is $\boldsymbol{RP} = \text{co-}\boldsymbol{RP}$?
2. Is $\boldsymbol{RP} \subseteq \boldsymbol{NP} \cap \text{co-}\boldsymbol{NP}$? (Note that since $\text{co-}\boldsymbol{RP} \subseteq \text{co-}\boldsymbol{NP}$, showing that $\boldsymbol{RP} = \text{co-}\boldsymbol{RP}$ would imply $\boldsymbol{RP} \subseteq \boldsymbol{NP} \cap \text{co-}\boldsymbol{NP}$.)
3. Is $\boldsymbol{BPP} \subseteq \boldsymbol{NP}$?

Although these classes are defined in terms of decision problems, they can be used to classify the complexity of a broader class of problems such as search or optimization problems. We will overload our notation a bit by using the complexity class labels for referring to algorithms. For example, **RandQS** will be called a ***ZPP*** algorithm.

Consider the following decision version of the min-cut problem: given a graph G and integer K, verify that the min-cut size in G equals K. Assume that we have modified (by incorporating sufficiently many repetitions) the Monte Carlo min-cut algorithm to reduce its probability of error below $1/4$. This algorithm can solve the decision problem by computing a cut value k and comparing it with K. This gives a ***BPP*** algorithm. In the case where K is indeed the min-cut value, the algorithm may not come up with the right value and, hence, may reject the input. Conversely, if the min-cut value is smaller than K, the algorithm may only find cuts of size K and, hence, may accept the input.

We may modify this decision problem: given G and K, verify that the min-cut size in G is *at most* K. Now, the algorithm described above translates into an ***RP*** algorithm for this problem. In the case where the actual min-cut size C is larger than K, the algorithm will never accept the input. This is because it can only find cuts of size k no smaller than C and hence greater than K.

Notes

The ideas underlying randomized algorithms can be traced back to *Monte Carlo methods* used in numerical analysis, statistical physics, and simulation. In the context of computability theory, the notion of a probabilistic Turing machine was proposed by de Leeuw, Moore, Shannon, and Shapiro [122] and further explored in the pioneering work of Rabin [340] and Gill [166]. Berlekamp [57], Rabin [341], and Solovay and Strassen [382] gave early examples of concrete randomized algorithms. Rabin [341] proposed randomized algorithms for problems in computational geometry and in number theory. Around the same time, Solovay and Strassen [382] gave a randomized Monte

Carlo algorithm for testing for primality; this problem is explored further in Chapter 14, as is the randomized algorithm for factoring polynomials due to Berlekamp [57].

In the last twenty years, the array of techniques for devising and analyzing randomized algorithms has grown. We develop these techniques in the chapters to follow. Karp [243], Maffioli, Speranza, and Vercellis [289], and Welsh [415] give excellent surveys of randomized algorithms. Johnson [220] surveys the probabilistic (or "average-case") analysis of algorithms (sometimes also referred to as "distributional complexity"), contrasting it with randomized algorithms surveyed in his following bulletin [221].

Our **RandQS** algorithm is based on Hoare's algorithm [201]. The min-cut algorithm of Section 1.1, together with many variations and extensions, is due to Karger [231].

Monte Carlo methods have been popular in the sciences for over a hundred years now. The classic experiment on approximating the value of π by dropping needles on a sheet of paper with parallel lines is described in an eighteenth-century paper by Buffon [86] (see also Hall [190]). The origin of the modern theory of Monte Carlo methods in the physical sciences is widely attributed to Ulam, von Neumann, and Fermi [116]. The term *Las Vegas algorithm* was introduced by Babai [37], although he uses the term in a slightly different sense. Our usage conforms to the currently accepted notion of a Las Vegas algorithm.

An important issue, alluded to in the discussion following the analysis of **RandQS** but otherwise not covered in detail in this book, is the generation of random samples from various types of distributions. First, there is the question of generating randomness within the inherently deterministic computers that will implement our randomized algorithms. This leads into the area of pseudo-random number generation, which is surveyed in the article by Boppana and Hirschfeld [73] and in Knuth's book [259]. Even if we assume that a source of truly random bits is available, there is the issue of converting this into the various types of distributions that may be required in randomized algorithms (for example, see Problems 1.2 and 1.3). This problem is studied in the context of Monte Carlo simulations, for example in the work of von Neumann [409, 410], and Knuth [259] covers this in great detail. A comprehensive study of this important family of problems in terms of its computational complexity was undertaken by Knuth and Yao [264]. The complexity of random sampling of combinatorial structures, such as graphs with specified properties, has been studied by Pruhs and Manber [338]; as discussed in Chapter 11, the problem of counting the number of combinatorial structures with specified properties, often a difficult computational problem, can sometimes be reduced to random sampling.

The idea of using independent iterations to reduce the error probability of Monte Carlo algorithms has an analog for Las Vegas algorithms. Alt, Guibas, Mehlhorn, Karp, and Wigderson [25] study the possibility of reducing the probability that the running time of a Las Vegas algorithm substantially exceeds its expected value by employing the following strategy: choose a sequence (T_i) and use independent iterations of the Las Vegas algorithm, aborting the ith iteration in T_i steps, until one of the iterations terminates successfully within the allotted time. These results were strengthened by Luby, Sinclair, and Zuckerman [286], who also considered the minimization of the expected total running time of such strategies.

The material of Section 1.3 is drawn from Paterson and Yao [329]. The **Find** algorithm described in Section 1.4 is due to Hoare [200]. Theorem 1.3 is given in a paper by Karp, Upfal and Wigderson [250]. Karp [244] gives a number of additional results on probabilistic recurrence relations.

The reader is referred to introductory texts on algorithms and complexity such as those by Aho, Hopcroft, and Ullman [5, 6] and Papadimitriou [326] for more details on the Turing machine model and the RAM model. It is known, for instance, that sorting n numbers requires $\Omega(n \log n)$ operations in the RAM model of computation. The books by Bovet and Crescenzi [81] and by Papadimitriou [326] contain a more detailed treatment of the complexity classes described in this chapter.

Problems

1.1 (Due to J. von Neumann [409].)

(a) Suppose you are given a coin for which the probability of HEADS, say p, is *unknown*. How can you use this coin to generate unbiased (i.e., **Pr**[HEADS] $=$ **Pr**[TAILS] $= 1/2$) coin-flips? Give a scheme for which the expected number of flips of the biased coin for extracting one unbiased coin-flip is no more than $1/[p(1-p)]$. (**Hint** : Consider two consecutive flips of the biased coin.)

(b) Devise an extension of the scheme that extracts the largest possible number of independent, unbiased coin-flips from a given number of flips of the biased coin.

1.2 (Due to D.E. Knuth and A. C-C. Yao [264].)

(a) Suppose you are provided with a source of unbiased random bits. Explain how you will use this to generate uniform samples from the set $S = \{0, \ldots, n-1\}$. Determine the expected number of random bits required by your sampling algorithm.

(b) What is the *worst-case* number of random bits required by your sampling algorithm? Consider the case when n is a power of 2, as well as the case when it is not.

(c) Solve (a) and (b) when, instead of unbiased random bits, you are required to use as the source of randomness uniform random samples from the set $\{0, \ldots, p-1\}$; consider the case when n is a power of p, as well as the case when it is not.

1.3 (Due to D.E. Knuth and A. C-C. Yao [264].) Suppose you are provided with a source of unbiased random bits. Provide efficient (in terms of expected running time and expected number of random bits used) schemes for generating samples from the distribution over the set $\{2, 3, \ldots, 12\}$ induced by rolling two unbiased dice and taking the sum of their outcomes.

1.4 (a) Suppose you are required to generate a random permutation of size n. Assuming that you have access to a source of independent and unbiased random bits, suggest a method for generating random permutations of size n. Efficiency is measured in terms of both time and number of random bits. What lower bounds can you prove for this task?

(b) Consider the following method for generating a random permutation of size n. Pick n random values X_1, ..., X_n independently from the uniform distribution over the interval $[0, 1]$. Now, the permutation that orders the random variables in ascending order is claimed to be a random permutation, and it can be determined by sorting the random values. Is the claim correct? How efficient is this scheme?

(c) Consider the following "lazy" implementation of the scheme suggested in (b). The binary representation of the fraction X_i is a sequence of unbiased and independent random bits. At any given stage of the sorting algorithm, we would have chosen only as many bits of each X_i as necessary to resolve all the comparisons performed up to that point. When comparing X_i to X_j, if the current prefixes of their binary expansions do not determine the outcome of the comparisons, then we extend their prefixes by choosing further random bits until this happens. Compute tight bounds on the *expected* number of random bits used by this implementation.

1.5 Consider the problem of using a source of unbiased random bits to generate samples from the set $S = \{0, \ldots, n-1\}$ such that the element i is chosen with probability p_i. Show how to perform this sampling using $O(\log n)$ random bits per sample, regardless of the values of p_i. Use the result from part (c) of Problem 1.4.

1.6 Consider a sequence of n flips of an unbiased coin. Let H_i denote the absolute value of the excess of the number of HEADS over the number of TAILS seen in the first i flips. Define $H = \max_i H_i$. Show that $\mathbf{E}[H_i] = \Theta(\sqrt{i})$, and that $\mathbf{E}[H] = \Theta(\sqrt{n})$.

1.7 Suppose we choose a permutation π of the ordered set $N = \{1, 2, \ldots n\}$ uniformly at random from the space of all permutations of N. Let $L(\pi)$ denote the length of the longest increasing subsequence in permutation π.

(a) For large n and some positive constant c, prove that $\mathbf{E}[L(\pi)] \geq c\sqrt{n}$.

(b) Is the bound in (a) tight?

1.8 Consider adapting the min-cut algorithm of Section 1.1 to the problem of finding an *s-t min-cut* in an undirected graph. In this problem, we are given an undirected graph G together with two distinguished vertices s and t. An s-t cut is a set of edges whose removal from G disconnects s from t; we seek an s-t cut of minimum cardinality. As the algorithm proceeds, the vertex s may get amalgamated into a new vertex as a result of an edge being contracted; we call this vertex the s-vertex (initially the s-vertex is s itself). Similarly, we have a t-vertex. As we run the contraction algorithm, we ensure that we never contract an edge between the s-vertex and the t-vertex.

(a) Show that there are graphs in which the probability that this algorithm finds an s-t min-cut is exponentially small.

(b) How large can the number of s-t min-cuts in an instance be?

1.9 Consider the **Find** algorithm described in Section 1.4 for selecting the kth smallest of a set S of n elements. Show that the algorithm finds the kth smallest element in S in expected time $O(n)$.

1.10 Consider the setting of Example 1.1. Show that the probability that no sailor returns to her own cabin approaches $1/e$ as the number of sailors grows large.

1.11 Verify the following inclusions:

$$\boldsymbol{P} \subseteq \boldsymbol{RP} \subseteq \boldsymbol{NP} \subseteq \boldsymbol{PSPACE} \subseteq \boldsymbol{EXP} \subseteq \boldsymbol{NEXP}.$$

It is not known whether these inclusions are strict.

1.12 Verify the following inclusions:

$$\boldsymbol{RP} \subseteq \boldsymbol{BPP} \subseteq \boldsymbol{PP}.$$

It is not known whether these inclusions are strict.

1.13 Show that $\boldsymbol{PP} = \text{co-}\boldsymbol{PP}$ and $\boldsymbol{BPP} = \text{co-}\boldsymbol{BPP}$.

1.14 Show that $\boldsymbol{NP} \subseteq \boldsymbol{PP} \subseteq \boldsymbol{PSPACE}$.

1.15 (Due to K-I. Ko [265].) Show that $\boldsymbol{NP} \subseteq \boldsymbol{BPP}$ implies $\boldsymbol{NP} = \boldsymbol{RP}$.

CHAPTER 2

Game-Theoretic Techniques

IN this chapter we study several ideas that are basic to the design and analysis of randomized algorithms. All the topics in this chapter share a game-theoretic viewpoint, which enables us to think of a randomized algorithm as a probability distribution on deterministic algorithms. This leads to the *Yao's Minimax Principle*, which can be used to establish a lower bound on the performance of a randomized algorithm.

2.1. Game Tree Evaluation

We begin with another simple illustration of linearity of expectation, in the setting of *game tree evaluation*. This example will demonstrate a randomized algorithm whose expected running time is smaller than that of any deterministic algorithm. It will also serve as a vehicle for demonstrating a standard technique for deriving a *lower bound* on the running time of *any* randomized algorithm for a problem.

A *game tree* is a rooted tree in which internal nodes at even distance from the root are labeled MIN and internal nodes at odd distance are labeled MAX. Associated with each leaf is a real number, which we call its *value*. The *evaluation* of the game tree is the following process. Each leaf *returns* the value associated with it. Each MAX node returns the largest value returned by its children, and each MIN node returns the smallest value returned by its children. Given a tree with values at the leaves, the evaluation problem is to determine the value returned by the root.

The evaluation of game trees plays a central role in artificial intelligence, particularly in game-playing programs. The reader may readily associate the children of a node with the options available to one of the two players in a game. The leaves represent the value of the game for either player. One player seeks to maximize this value, while the other tries to minimize it. At each step, an evaluation algorithm chooses a leaf and reads its value.

We study the number of such steps taken by an algorithm for evaluating a game tree. We do not charge the algorithm for any other computation.

We will limit our discussion to the special case in which the values at the leaves are bits, 0 or 1. Thus, each MIN node can be thought of as a Boolean AND operation and each MAX node as a Boolean OR operation. This special case is of interest in its own right, having applications in mechanical theorem proving. Let $T_{d,k}$ denote a uniform tree in which the root and every internal node has d children and every leaf is at distance $2k$ from the root. Thus, any root-to-leaf path passes through k AND nodes (including the root itself) and k OR nodes, and there are d^{2k} leaves. An *instance* of the evaluation problem consists of the tree $T_{d,k}$ together with a Boolean value for each of the d^{2k} leaves. Given an algorithm, we study the maximum number of steps it takes to evaluate any instance of $T_{d,k}$.

An algorithm begins by specifying a leaf whose value is to be read at the first step. Thereafter, it specifies such a leaf at each step, based on the values it has read on previous steps. In a deterministic algorithm, the choice of the next leaf to be read is a deterministic function of the values at the leaves read so far. For a randomized algorithm, this choice may be randomized.

In Problem 2.1, the reader is asked to show that for any deterministic evaluation algorithm, there is an instance of $T_{d,k}$ that forces the algorithm to read the values on all d^{2k} leaves.

We now give a simple randomized algorithm and study the expected number of leaves it reads on any instance of $T_{d,k}$. To simplify our presentation, we restrict ourselves to the case $d = 2$. Any deterministic algorithm for this case can be made to read all $2^{2k} = 4^k$ leaves on some instance of $T_{2,k}$. Our randomized algorithm is based on the following simple observation. Consider a single AND node with two leaves. If the node were to return 0, at least one of the leaves must contain 0. A deterministic algorithm inspects the leaves in a fixed order, and an adversary can therefore always "hide" the 0 at the second of the two leaves inspected by the algorithm. Reading the leaves in a random order foils this strategy. With probability $1/2$, the algorithm chooses the hidden 0 on the first step, so its expected number of steps is $3/2$, which is better than the worst case for any deterministic algorithm. Similarly, in the case of an OR node, if it were to return a 1, then a randomized order of examining the leaves will reduce the expected number of steps to $3/2$.

The reader may wonder how the randomized algorithm can benefit if the AND node were to return 1, or if the OR node were to return a 0. If the two children of these nodes are leaves, then clearly both leaves must be examined. The point is that at an internal AND node in a tree returning a 1, examining the two OR children (and evaluating their sub-trees) in a random order is still beneficial. The two OR children of an AND node must also return 1, and this is the easy case for the OR nodes. Similarly, at an internal OR node returning 0, the two AND children must return 0, and this is the easy case for the AND nodes. To explain this better, we specify the complete algorithm.

To evaluate an AND node v, the algorithm chooses one of its children (a sub-tree rooted at an OR node) at random and evaluates it by recursively invoking the algorithm. If 1 is returned by the sub-tree, the algorithm proceeds to evaluate the other child (again by recursive application). If 0 is returned, the algorithm returns 0 for v. To evaluate an OR node, the procedure is the same with the roles of 0 and 1 interchanged. We now argue by induction on k that the expected cost of evaluating any instance of $T_{2,k}$ is at most 3^k.

The basis ($k = 1$) is an easy extension of our illustration above. Assume now that the expected cost of evaluating any instance of $T_{2,k-1}$ is at most 3^{k-1}. We establish the inductive step. Consider first a tree T whose root is an OR node, each of whose children is the root of a copy of $T_{2,k-1}$. If the root of T were to evaluate to 1, at least one of its children returns 1. With probability $1/2$ this child is chosen first, incurring (by the inductive hypothesis) an expected cost of at most 3^{k-1} in evaluating T. With probability at most $1/2$ both sub-trees are evaluated, incurring a net cost of at most $2 \times 3^{k-1}$. Putting these observations together, the expected cost of determining the value of T is at most

$$\frac{1}{2} \times 3^{k-1} + \frac{1}{2} \times 2 \times 3^{k-1} = \frac{3}{2} \times 3^{k-1}. \tag{2.1}$$

If on the other hand the OR were to evaluate to 0, both children must be evaluated, incurring a cost of at most $2 \times 3^{k-1}$.

Consider next the root of the tree $T_{2,k}$, an AND node. If it evaluates to 1, then both its sub-trees rooted at OR nodes return 1. By the discussion in the previous paragraph and by linearity of expectation, the expected cost of evaluating $T_{2,k}$ to 1 is at most $2 \times (3/2) \times 3^{k-1} = 3^k$. On the other hand, if the instance of $T_{2,k}$ evaluates to 0, at least one of its sub-trees rooted at OR nodes returns 0. With probability $1/2$ it is chosen first, and so the expected cost of evaluating $T_{2,k}$ is at most

$$2 \times 3^{k-1} + \frac{1}{2} \times \frac{3}{2} \times 3^{k-1} \leq 3^k.$$

Here the first term bounds the cost of evaluating both sub-trees of the OR node that returns 0; the second term accounts for the fact that with probability $1/2$, an additional cost of $(3/2)3^{k-1}$ may be incurred in evaluating its sibling that returns 1.

Theorem 2.1: *Given any instance of $T_{2,k}$, the expected number of steps for the above randomized algorithm is at most 3^k.*

Since $n = 4^k$ the expected running time of our randomized algorithm is at most $n^{\log_4 3}$, which we bound by $n^{0.793}$. Thus, the expected number of steps is smaller than the worst case for any deterministic algorithm. We will see other instances in later chapters. Note that the algorithm above is a Las Vegas algorithm and always produces the correct answer.

	Scissors	Paper	Stone
Scissors	0	1	−1
Paper	−1	0	1
Stone	1	−1	0

Figure 2.1: Matrix for scissors-paper-stone.

2.2. The Minimax Principle

The randomized algorithm of the preceding section has an expected running time of at most $n^{0.793}$ on any uniform binary AND-OR tree with n leaves. Can we establish that *no randomized algorithm* can have a lower expected running time? We are thus seeking a lower bound on the running time of any randomized algorithm for this problem. As a first step toward this end, we introduce a standard technique for proving such lower bounds: the *minimax principle*. Indeed, it is the only known general technique for proving lower bounds on the running times of randomized algorithms. This technique only applies to algorithms that terminate in finite time on all inputs and sequences of random choices. In Section 2.2.3, we will apply this technique to the game tree evaluation problem. We begin with a review of some elementary concepts in game theory. Note that the notion of *game* theory is not directly related to the *game* tree evaluation problem studied above. Rather, the game theory studied below yields the minimax principle, a general tool, which we will then apply to randomized algorithms for the game tree evaluation problem.

2.2.1. Game Theory

Consider the following game. Roberta and Charles put their hands behind their backs and make a sign for one of the following: stone (closed fist), paper (open palm), and scissors (two fingers). They then simultaneously display their chosen sign. The winner is determined by the following rules: paper beats stone by wrapping it, scissors beats paper by cutting it, and stone beats scissors by dulling it. The loser pays \$1 to the winner, and the outcome is a draw when the two players choose the same sign. We can represent this game by the matrix in Figure 2.1. The rows of the matrix represent Roberta's choices; the columns, Charles' choices. The entries in the matrix are the amounts to be paid by Charles to Roberta.

This is an instance of a *two-person zero-sum game*, and the matrix is called the *payoff matrix*. It is called a zero-sum game because the net amount won by Roberta and Charles is always exactly zero. In general, any two-person zero-sum game can be represented by an $n \times m$ payoff matrix $\boldsymbol{M}$ with real entries. (Throughout this book, we use boldface to denote vectors and matrices;

	Scissors	Paper	Stone
Scissors	0	1	2
Paper	−1	0	1
Stone	−2	−1	0

Figure 2.2: Matrix for modified scissors-paper-stone.

generally, vectors will be lower-case symbols, and matrices upper-case symbols. For a vector $\boldsymbol{x}$, we denote by x_i its ith component. All vectors are column vectors unless otherwise specified.) The set of possible strategies of the row player R is in correspondence with the rows of $\boldsymbol{M}$, and likewise for the strategies of the column player C. The entry M_{ij} is the amount paid by C to R when R chooses strategy i and C chooses strategy j.

Naturally, the goal of the row (column) player is to maximize (minimize) the payoff. Assume that this is a zero-information game, in that neither player has any information about the opponent's strategy. If R chooses strategy i, then she is guaranteed a payoff of $\min_j M_{ij}$, regardless of C's strategy. An *optimal strategy* for R is an i that maximizes $\min_j M_{ij}$. Let $V_R = \max_i \min_j M_{ij}$ denote the lower bound on the value of the payoff to R when she uses an optimal strategy. An optimal strategy for C is a j that gives the best possible upper bound on the payoff from C to R. A similar argument establishes that C's optimal strategy ensures that his payoff to R is at most $V_C = \min_j \max_i M_{ij}$.

Exercise 2.1: Show that the following inequality is valid for all payoff matrices.

$$\max_i \min_j M_{ij} \leq \min_j \max_i M_{ij}.$$

In general, the inequality in Exercise 2.1 is strict; for example, in scissors-paper-stone, $V_R = -1$ and $V_C = 1$. When these two quantities are equal, the game is said to have a solution and the *value* of the game is $V = V_R = V_C$. The solution (or the *saddle-point*) is the specific choice of (optimal) strategies that lead to this payoff. For games with a solution, let ρ and γ denote optimal strategies for R and C, respectively; clearly, $V = M_{\rho\gamma}$. In general, a player could have more than one optimal strategy.

Figure 2.2 shows a modified version of the scissors-paper-stone game, where the amount to be paid in certain cases is changed. It is easy to verify that this game has value $V = 0$ and the solution is $\rho = 1$ and $\gamma = 1$. (Do you see why the other diagonal entries do not correspond to saddle-points?)

What happens when a game has no solution? Then there is no clear-cut optimal strategy for any player. In fact, any knowledge of the opponent's strategy can be used to improve the payoff, unlike the case of games with saddle-points. An interesting way to get around this is to introduce randomization in the choice of strategies. So far we have been talking about deterministic or *pure* strategies, but now we focus on randomized or *mixed* strategies. A mixed strategy is a probability distribution on the set of possible strategies. The row player picks a vector $\boldsymbol{p} = (p_1, \ldots, p_n)$, which is a probability distribution on the rows of $\boldsymbol{M}$, i.e., p_i is the probability that R will choose strategy i; similarly, the column player has a vector $\boldsymbol{q} = (q_1, \ldots, q_m)$, which is a probability distribution on the columns of $\boldsymbol{M}$. The payoff is now a random variable, and its expectation is given by

$$\mathbf{E}[\text{payoff}] = \boldsymbol{p}^T \boldsymbol{M} \boldsymbol{q} = \sum_{i=1}^{n} \sum_{j=1}^{m} p_i M_{ij} q_j.$$

As before, using V_R to denote the best possible lower bound on the expected payoff to R that can be ensured by choosing a strategy $\boldsymbol{p}$, and using V_C to denote the best possible upper bound on the expected payoff by C by choosing a strategy $\boldsymbol{q}$, we obtain

$$\begin{aligned} V_R &= \max_{\boldsymbol{p}} \min_{\boldsymbol{q}} \boldsymbol{p}^T \boldsymbol{M} \boldsymbol{q} \\ V_C &= \min_{\boldsymbol{q}} \max_{\boldsymbol{p}} \boldsymbol{p}^T \boldsymbol{M} \boldsymbol{q}. \end{aligned}$$

Here, the min and max range over all possible distributions. The well-known Minimax Theorem of von Neumann implies that this game always has a solution and that $V_R = V_C$.

Theorem 2.2 (von Neumann's Minimax Theorem): *For any two-person zero-sum game specified by a matrix* $\boldsymbol{M}$,

$$\max_{\boldsymbol{p}} \min_{\boldsymbol{q}} \boldsymbol{p}^T \boldsymbol{M} \boldsymbol{q} = \min_{\boldsymbol{q}} \max_{\boldsymbol{p}} \boldsymbol{p}^T \boldsymbol{M} \boldsymbol{q}.$$

In other words, the largest expected payoff that R can guarantee by choosing a mixed strategy is equal to the smallest expected payoff that C can guarantee using a mixed strategy. This common expected payoff value, called the value of the game, is denoted by V. A pair of mixed strategies $(\widehat{\boldsymbol{p}}, \widehat{\boldsymbol{q}})$ which respectively maximize the left-hand side and minimize the right–hand side of the equation in Theorem 2.2 is called a saddle-point, and the two distributions are called optimal mixed strategies.

Observe that once $\boldsymbol{p}$ is fixed, $\boldsymbol{p}^T \boldsymbol{M} \boldsymbol{q}$ is a linear function of $\boldsymbol{q}$ and is minimized by setting to 1 the q_j with the smallest coefficient in this linear function. The implications of this observation are rather interesting. If C knows the distribution $\boldsymbol{p}$ being used by R, then his optimal strategy is a pure strategy. A similar comment applies in the other direction. Also, this observation leads to a simplified version of the minimax theorem. Let $\boldsymbol{e}_k$ denote a unit vector with a 1 in the kth position and 0s elsewhere.

Theorem 2.3 (Loomis' Theorem): *For any two-person zero-sum game specified by a matrix* $\boldsymbol{M}$,

$$\max_{\boldsymbol{p}} \min_{j} \boldsymbol{p}^T \boldsymbol{M} \boldsymbol{e}_j = \min_{\boldsymbol{q}} \max_{i} \boldsymbol{e}_i^T \boldsymbol{M} \boldsymbol{q}.$$

2.2.2. Yao's Technique

We now describe the application of the above game-theoretic results to proving lower bounds on the performance of randomized algorithms. The idea is to view the algorithm designer as the column player C and the adversary choosing the input as the row player R. The columns correspond to the set of all possible algorithms; the rows correspond to the set of all possible inputs (of a fixed size). It is important to keep in mind that each column corresponds to a deterministic algorithm that always produces a correct solution. The payoff from C to R is some real-valued measure of the performance of an algorithm, such as the running time, the quality of the solution obtained, communication cost, or space. (In all the examples we will encounter in this book, the entries in the payoff matrix will be positive integers.) For the sake of concreteness, we assume in this chapter that the payoff refers to the running time, but it should be obvious that the following observations apply to any other measure. The algorithm designer would like to choose an algorithm that minimizes the payoff, while the adversary would like to maximize the payoff.

Consider a problem where the number of distinct inputs of a fixed size is finite, as is the number of distinct (deterministic, terminating, and always correct) algorithms for solving that problem. A pure strategy for C corresponds to the choice of a deterministic algorithm, while a pure strategy for R corresponds to a specific input. Notice that an optimal pure strategy for C corresponds to an optimal deterministic algorithm, and V_C is the worst-case running time of any deterministic algorithm for the problem, which we call the deterministic complexity of the problem. (The meaning of V_R is related to the non-deterministic complexity of the problem. If the game has a solution, then the non-deterministic and deterministic complexities coincide.)

Our interest is in the interpretation of the mixed strategies for the algorithm designer and the adversary. A mixed strategy for C is a probability distribution over the space of (always correct) deterministic algorithms, so it is a Las Vegas randomized algorithm. An optimal mixed strategy for C is an optimal Las Vegas algorithm. A mixed strategy for R is a distribution over the space of all inputs.

Let us define the *distributional complexity* of the problem at hand as the expected running time of the best deterministic algorithm for the worst distribution on the inputs. This complexity is smaller than the deterministic complexity, since the algorithm knows the input distribution.

Theorem 2.3 implies that the distributional complexity equals the least possible expected running time achievable by any randomized algorithm. (We reiterate that these observations apply only to scenarios where the number of algorithms is finite.) We restate von Neumann's and Loomis's theorems in the language of algorithms as follows.

Corollary 2.4: *Let Π be a problem with a finite set $\mathcal{I}$ of input instances (of a fixed size), and a finite set of deterministic algorithms $\mathcal{A}$. For input $I \in \mathcal{I}$ and algorithm $A \in \mathcal{A}$, let $C(I,A)$ denote the running time of algorithm A on input I.*

For probability distributions $\boldsymbol{p}$ *over* $\mathcal{I}$ *and* $\boldsymbol{q}$ *over* $\mathcal{A}$, *let* $I_{\boldsymbol{p}}$ *denote a random input chosen according to* $\boldsymbol{p}$ *and* $A_{\boldsymbol{q}}$ *denote a random algorithm chosen according to* $\boldsymbol{q}$. *Then,*

$$\max_{\boldsymbol{p}} \min_{\boldsymbol{q}} \mathbf{E}[C(I_{\boldsymbol{p}}, A_{\boldsymbol{q}})] = \min_{\boldsymbol{q}} \max_{\boldsymbol{p}} \mathbf{E}[C(I_{\boldsymbol{p}}, A_{\boldsymbol{q}})]$$

and

$$\max_{\boldsymbol{p}} \min_{A \in \mathcal{A}} \mathbf{E}[C(I_{\boldsymbol{p}}, A)] = \min_{\boldsymbol{q}} \max_{I \in \mathcal{I}} \mathbf{E}[C(I, A_{\boldsymbol{q}})].$$

From this corollary, we obtain the following proposition, which provides the desired lower bound technique.

Proposition 2.5 (Yao's Minimax Principle): *For all distributions* $\boldsymbol{p}$ *over* $\mathcal{I}$ *and* $\boldsymbol{q}$ *over* $\mathcal{A}$,

$$\min_{A \in \mathcal{A}} \mathbf{E}[C(I_{\boldsymbol{p}}, A)] \leq \max_{I \in \mathcal{I}} \mathbf{E}[C(I, A_{\boldsymbol{q}})].$$

In other words, the expected running time of the optimal deterministic algorithm for an arbitrarily chosen input distribution $\boldsymbol{p}$ is a lower bound on the expected running time of the optimal (Las Vegas) randomized algorithm for Π. Thus, to prove a lower bound on the randomized complexity, it suffices to choose any distribution $\boldsymbol{p}$ on the input and prove a lower bound on the expected running time of deterministic algorithms for that distribution. The power of this technique lies in the flexibility in the choice of $\boldsymbol{p}$ and, more importantly, the reduction to a lower bound on deterministic algorithms. It is important to remember that the deterministic algorithm "knows" the chosen distribution $\boldsymbol{p}$.

The above discussion dealt only with lower bounds on the performance of Las Vegas algorithms. We conclude this section with a brief discussion of Monte Carlo algorithms with error probability $\epsilon \in [0, 1/2]$. Let us define the distributional complexity with error ϵ, denoted $\min_{A \in \mathcal{A}} \mathbf{E}[C_\epsilon(I_{\boldsymbol{p}}, A)]$, to be the minimum expected running time of any deterministic algorithm that errs with probability at most ϵ under the input distribution $\boldsymbol{p}$. Similarly, we denote by $\max_{I \in \mathcal{I}} \mathbf{E}[C_\epsilon(I, A_{\boldsymbol{q}})]$ the expected running time (under the worst input) of any randomized algorithm that errs with probability at most ϵ (again, the randomized algorithm is viewed as a probability distribution $\boldsymbol{q}$ on deterministic algorithms). Analogous to Proposition 2.5, we then have:

Proposition 2.6: *For all distributions* $\boldsymbol{p}$ *over* $\mathcal{I}$ *and* $\boldsymbol{q}$ *over* $\mathcal{A}$ *and any* $\epsilon \in [0, 1/2]$,

$$\frac{1}{2}(\min_{A \in \mathcal{A}} \mathbf{E}[C_{2\epsilon}(I_{\boldsymbol{p}}, A)]) \leq \max_{I \in \mathcal{I}} \mathbf{E}[C_\epsilon(I, A_{\boldsymbol{q}})].$$

A pointer to the source of Proposition 2.6 is given in the Notes section.

2.2.3. Lower Bound for Game Tree Evaluation

We now apply Yao's Minimax Principle to the problem of game tree evaluation. The lower bound that results only applies to algorithms that terminate in a finite number of steps on any input and sequence of random choices. Note that a randomized algorithm for game tree evaluation can in fact be viewed as a probability distribution over deterministic algorithms, because the length of the computation as well as the number of choices at each step are both finite. We may imagine that all of these coins are tossed before the beginning of the execution.

Once again, we limit our attention to instances of the AND-OR tree $T_{2,k}$. While we could continue our discussion in the language of alternating levels of AND and OR nodes, the following exercise will lead to a slightly more compact representation.

Exercise 2.2: Show that the tree $T_{2,k}$ is equivalent to a balanced binary tree all of whose leaves are at distance $2k$ from the root, and all of whose internal nodes compute the NOR function: a node returns the value 1 if both inputs are 0, and 0 otherwise.

We proceed with the analysis of this tree of NORs of depth $2k$. In order to prove a lower bound on the expected number of leaves evaluated by any randomized algorithm, we have to specify a distribution on instances (values for the leaves), and then prove a lower bound on the expected running time of any deterministic algorithm on such inputs. It is important to distinguish between the expected running time of the randomized algorithm (which is over the random choices made by the algorithm), and the expected running time of the deterministic algorithm when proving the lower bound (this being over the random instances). We also remind the reader that our lower bound will only apply to Las Vegas randomized algorithms that always evaluate the tree correctly.

Let $p = (3 - \sqrt{5})/2$. Each leaf of the tree is independently set to 1 with probability p. Note that if each input to a NOR node is independently 1 with probability p, then the probability that its output is 1 is the probability that both its inputs are 0, which is

$$\left(\frac{\sqrt{5}-1}{2}\right)^2 = \frac{3-\sqrt{5}}{2} = p.$$

Thus the value of every node of the NOR tree is 1 with probability p, and the value of a node is independent of the values of all the other nodes on the same level. Consider a deterministic algorithm that is evaluating a tree furnished with such random inputs; let v be a node of the tree whose value the algorithm is trying to determine. Intuitively, the algorithm should determine the value of one child of v before inspecting any leaf of the other sub-tree. By doing so, it can try to maximize the benefit of information obtained by inspecting leaves. An alternative view of this process is that the deterministic algorithm inspects leaves visited in a depth-first search of the tree, except of course that it ceases to visit sub-trees of a node v once the value of v has been determined. Let us call such algorithms *depth-first pruning* algorithms, referring to the order of traversal and the fact that sub-trees that supply no additional information are "pruned" away without being inspected.

Proposition 2.7: *Let T be a* NOR *tree each of whose leaves is independently set to 1 with probability q for a fixed value $q \in [0,1]$. Let $W(T)$ denote the minimum, over all deterministic algorithms, of the expected number of steps to evaluate T. Then, there is a depth-first pruning algorithm whose expected number of steps to evaluate T is $W(T)$.*

A formal proof of Proposition 2.7 by induction is omitted here and can be found in the reference given at the end of this chapter.

Proposition 2.7 tells us that for the purposes of our lower bound, we may restrict our attention to depth-first pruning algorithms. We return to a NOR tree with n leaves, each of which is set to 1 independently with probability $p = (3-\sqrt{5})/2$. For a depth-first pruning algorithm evaluating this tree, let $W(h)$ be the expected number of leaves it inspects in determining the value of a node at distance h from the leaves. Clearly

$$W(h) = W(h-1) + (1-p) \times W(h-1),$$

where the first term represents the work done in evaluating one of the sub-trees of the node, and the second term represents the work done in evaluating the other sub-tree (which will be necessary if the first sub-tree returns the value 0, an event occurring with probability $1-p$). Letting h be $\log_2 n$ and solving, we get $W(h) \geq n^{0.694}$.

Theorem 2.8: *The expected running time of any randomized algorithm that always evaluates an instance of $T_{2,k}$ correctly is at least $n^{0.694}$, where $n = 2^{2k}$ is the number of leaves.*

We note that our lower bound of $n^{0.694}$ is less than the upper bound of $n^{0.793}$ that follows from Theorem 2.1. Could it be that our lower bound technique is weak? Corollary 2.4 precludes this possibility, since the identity it gives is an equality; thus for any lower bound on the expected running time there must be a distribution on the inputs such that the running time of the best deterministic algorithm matches this lower bound. One possibility is that we have not chosen the best possible probability distribution for the values of the leaves. Indeed, in the NOR tree if both inputs to a node are 1, no reasonable algorithm will read leaves of both sub-trees of that node. Thus, to prove the best lower bound, we have to choose a distribution on the inputs that precludes the possibility that both inputs to a node will be 1; in other words, the values of the inputs are chosen at random but not independently. This stronger (and considerably harder) analysis shows that our algorithm of Section 2.1 is optimal.

2.3. Randomness and Non-uniformity

A basic issue in the study of randomized algorithms is the extent to which randomization is necessary for solving a problem. When is it possible to remove the randomization in a randomized algorithm? The answer depends on a number of aspects of the problem being solved. The goal of this section is to show that this question is more subtle than appears at first, and touches on the issue of *uniformity* in algorithms. We now study the notion of a randomized circuit, and a general technique by which randomization can be removed in polynomial-sized randomized circuits.

A *Boolean circuit with n inputs* is a directed acyclic graph with the following properties:

1. There are n vertices of in-degree 0; these are called the *inputs* to the circuit and are labeled $x_1, x_2, \ldots, x_n$. There is one vertex with out-degree 0; this is called the *output* of the circuit.

2. Every vertex v that is not an input or the output is labeled with one Boolean function $b(v)$ from the set $\{\text{AND}, \text{OR}, \text{NOT}\}$. A vertex labeled NOT has in-degree 1.

3. Every input to the circuit is assigned a Boolean value. Under such an assignment of input values, each vertex v computes the Boolean function $b(v)$ of the values on the incoming edges, and assigns this value to its outgoing edges. The value of the output is thus a Boolean function of $x_1, x_2, \ldots, x_n$; the circuit is said to compute this function.

4. The *size* of a circuit is the number of vertices in it.

A *randomized circuit* is very similar, except that there may be more than n vertices of in-degree 0, and these are partitioned into two classes: (1) *random inputs*, each of which is assigned an independent random value from $\{0, 1\}$, and (2) the *n circuit inputs*, which are labeled $x_1, x_2, \ldots, x_n$. A randomized circuit is said to compute a function f of the inputs $x_1, x_2, \ldots, x_n$ if the following properties hold:

1. For inputs $x_1, x_2, \ldots, x_n$ for which $f(x_1, \ldots, x_n) = 0$, the output of the circuit is 0 regardless of the values of random inputs.

2. If, on the other hand, $f(x_1, \ldots, x_n) = 1$, the output of the circuit is 1 with probability at least $1/2$.

Consider a Boolean function $f : \{0, 1\}^* \to \{0, 1\}$. We denote by f_n the function f restricted to inputs from $\{0, 1\}^n$. A sequence $C = C_1, C_2, \ldots$ of circuits is a *circuit family for f* if C_n has n inputs and computes $f_n(x_1, x_2, \ldots, x_n)$ at its output for all n-bit inputs $(x_1, \ldots, x_n)$. The family C is said to be *polynomial-sized* if the size of C_n is bounded above by $p(n)$ for every n, where $p(.)$ is a polynomial. A *randomized circuit family for f* is a circuit family for f that, in addition to the n inputs $x_1, \ldots, x_n$, takes m random bits $r_1, \ldots, r_m$, each of which is equiprobably 0 or 1. In addition, for every n, circuit C_n must satisfy two properties:

1. If $f_n(x_1,\ldots,x_n) = 0$, then the output of the circuit is 0 regardless of the values of random inputs $r_1,\ldots,r_m$.
2. If $f_n(x_1,\ldots,x_n) = 1$, then the output of the circuit is 1 with probability at least 1/2. In other words, at least one half of the 2^m choices of the bits $r_1,\ldots,r_m$ will result in the circuit evaluating to 1. We will refer to such m-tuples $r_1,\ldots,r_m$ as *witnesses* for $(x_1,\ldots,x_n)$, in that they testify to the correct value of $f_n(x_1,\ldots,x_n)$ when it is 1.

Theorem 2.9 below asserts that randomization can be eliminated in polynomial-sized circuits.

Theorem 2.9 (Adleman's Theorem): *If a Boolean function has a randomized, polynomial-sized circuit family, then it has a polynomial-sized circuit family.*

PROOF: The proof is by a simple counting argument.

We show how to turn a given randomized polynomial-sized circuit C_n for $f_n(x_1, \ldots, x_n)$ using random inputs $r_1,\ldots,r_m$, into a deterministic polynomial-sized circuit D_n that computes $f_n(x_1,\ldots,x_n)$.

Form a matrix $\boldsymbol{M}$ with 2^n rows, one for each possible input from $\{0,1\}^n$. The matrix has 2^m columns, one for each of the possible m-tuples from $\{0,1\}^m$ that the r_i can assume. The entry M_{jk} is 1 if the setting of the $r_1,\ldots,r_m$ corresponding to column k is a witness for the input $x_1,\ldots,x_n$ corresponding to row j; otherwise, the entry is 0. Eliminate all rows of $\boldsymbol{M}$ corresponding to inputs for which f_n evaluates to 0.

By definition, at least half the entries of every surviving row of $\boldsymbol{M}$ equal 1. Therefore, there must be a column with at least half its entries 1; in other words, there is an assignment of 0s and 1s to the r_i that serves as a witness to at least half of the possible inputs. Let this witness be $r_1(1),\ldots,r_m(1)$. Build a circuit T_1, which is a copy of C_n with the random inputs "hard-wired" to $r_1(1),\ldots,r_m(1)$. Delete the column in $\boldsymbol{M}$ corresponding to $r_1(1),\ldots,r_m(1)$, and all rows that had 1s in this column. Thus T_1 computes the correct value of $f_n(x_1,\ldots,x_n)$ whenever the input corresponds to one of the rows we have just eliminated.

The matrix that remains still has the property that every row has at least half its entries equal to 1, since the string $r_1(1),\ldots,r_m(1)$ was not a witness for any of these rows whereas half the entries in these rows are guaranteed to be 1s. Repeat the construction above, picking a second string $r_1(2),\ldots,r_m(2)$ that is a witness for at least half the remaining inputs and building a circuit T_2. Continuing in this manner, we will have deleted all the rows of $\boldsymbol{M}$ while building at most n circuits $T_1,\ldots,T_n$.

Now we take the OR of the outputs of the circuits $T_1,\ldots,T_n$, and this is a (deterministic) circuit whose size is $n+1$ times that of the randomized circuit we started with. □

The technique in Theorem 2.9 is the first example we have seen of *derandomization* – where we take a randomized algorithm or computation, and diminish or entirely remove the randomness in it. This is often a useful technique for the design of deterministic algorithms. Does Theorem 2.9 mean that randomization is dispensable in all polynomial-time computations? The answer is no, and has to do with the issue of non-uniformity in computation. The deterministic circuit generated by the above process is one that works for a particular value of n. Indeed, the circuit it produces for n inputs may have very little resemblance to the circuit it produces for $n+1$ inputs, even if the original randomized circuits were similar. Any "practical" algorithm or circuit will in fact exhibit this property of similarity, which is formalized in the literature under the name *uniformity*.

Complexity theory formalizes this intuition by classifying algorithms as being *uniform* or *non-uniform* as follows. Let $a(n)$ be a function from the positive integers to strings in Σ^*. An algorithm A is said to use *advice* a if on an input of length n it is given the string $a(n)$ on a read-only tape. We say that A decides a language L with advice a if on an input x it uses the read-only string $a(|x|)$ to decide the membership of x in L. In other words, a single advice string $a(n)$ enables the algorithm A to decide the membership of x in L for all inputs x having length n. Uniform algorithms are those that use no advice strings at all, whereas non-uniform algorithms are those that use such advice. For the complexity class $\boldsymbol{P}$, we define the class $\boldsymbol{P}/\text{poly}$ to consist of all languages L that have a non-uniform polynomial-time algorithm A such that the length of the advice string $a(n)$ is bounded by a polynomial in n. Likewise, we may define the class $\boldsymbol{RP}/\text{poly}$.

Exercise 2.3: Consider any language $L \subseteq \{0,1\}^*$. We define a Boolean function f corresponding to the language L as follows. For any positive integer n, let f_n be the Boolean function such that for any $x \in \{0,1\}^n$, $f_n(x)$ assumes the value 1 if $x \in L$ and 0 otherwise. If there is a circuit family for f, we refer to it as a circuit family for L. Show that $L \in \boldsymbol{P}/\text{poly}$ if and only if it has a polynomial-sized circuit family.

In an analogous fashion, we may speak of a language L as having a randomized circuit family. Clearly, $L \in \boldsymbol{RP}/\text{poly}$ if and only if it has a randomized polynomial-sized circuit family. In the light of this discussion, we may interpret Theorem 2.9 as proving that $\boldsymbol{RP}/\text{poly} \subseteq \boldsymbol{P}/\text{poly}$. We thus have:

Corollary 2.10: $\boldsymbol{RP} \subseteq \boldsymbol{P}/poly$.

In summary, the removal of randomness in Theorem 2.9 only shows that this can be done in principle; it is not known how to do this in any uniform or practical way.

Notes

The material of Section 2.1 is based on a paper of Snir [381].

Most of the material in Section 2.2 is covered in textbooks on game theory. Some good sources are the books by Wang [213], Luce and Raiffa [287], and von Neumann and Morgenstern [411]. Theorem 2.2 is due to von Neumann [408], and Theorem 2.3

is due to Loomis [279]. The application of the minimax theorems to proving lower bounds on randomized algorithms was pointed out by Yao [419]. Proposition 2.6 is also from [419]. In fact, for proving lower bounds, we do not require the equality established in Corollary 2.4; all we require is the inequality of Proposition 2.5. It is possible to give a direct proof of the inequality (not the equality) without resorting to game theory; the reader can find this in the paper of Fich, Meyer auf der Heide, Ragde, and Wigderson [147].

In our lower bound for game tree evaluation, the principle that any deterministic algorithm may as well determine the value of one sub-tree before inspecting any leaves of its sibling (used in Section 2.2.3) is due to Tarsi [393]. Saks and Wigderson [362] refined the lower bound of Section 2.2.3 to show that Snir's algorithm is optimal among all randomized algorithms.

Theorem 2.9 is due to Adleman [1]; a version of this theorem applicable to circuit families with two-sided error is due to Gill [166]. The notion of non-uniformity is studied in depth in the paper by Karp and Lipton [245]. The reader interested in the material of Sections 2.2.2 and 2.3 may wish to explore recent related work of Althöfer [26] and of Lipton and Young [278].

Problems

2.1 Show that for any deterministic evaluation algorithm, there is an instance of $T_{d,k}$ that forces the algorithm to read the values on all d^{2k} leaves.

2.2 Generalize the randomized algorithm and analysis of Section 2.1 to trees $T_{d,k}$ for $d > 2$.

2.3 (Due to R. Boppana [362].) Consider a uniform rooted tree of height h – every leaf is at distance h from the root. The root, as well as any internal node, has three children. Each leaf has a Boolean value associated with it. Each internal node returns the value returned by the majority of its children. The evaluation problem consists of determining the value of the root; at each step, an algorithm can choose one leaf whose value it wishes to read.

(a) Show that for any deterministic algorithm, there is an instance (a set of Boolean values for the leaves) that forces it to read all $n = 3^h$ leaves.

(b) Consider the recursive randomized algorithm that evaluates two sub-trees of the root chosen at random. If the values returned disagree, it proceeds to evaluate the third sub-tree. Show that the expected number of leaves read by this algorithm (on any instance) is at most $n^{0.9}$.

2.4 Determine the value V_R of the following 2×2 matrix game and give optimal mixed strategies for the two players.

$$\begin{pmatrix} 5 & 6 \\ 7 & 4 \end{pmatrix}$$

2.5 (Due to R.M. Karp.) Let (a_{ij}) be a $m \times n$ matrix, let the vector $(p_1, p_2, \ldots, p_m)$ consist of reals in $[0, 1]$ such that $\sum_{i=1}^{m} p_i = 1$, and let $(q_1, q_2, \ldots, q_n)$ consist of reals in $[0, 1]$ such that $\sum_{i=1}^{n} q_i = 1$. Prove algebraically that $\max_q \min_i \sum_{j=1}^{n} a_{ij} q_j \leq \min_p \max_j \sum_{i=1}^{m} p_i a_{ij}$.

2.6 Use Yao's Minimax Principle to prove a lower bound on the expected running time of any Las Vegas algorithm for sorting n numbers.

2.7 (Due to R.M. Karp.) You are given an array A containing n numbers in sorted order. In one step, an algorithm may specify an integer $i \in [1, n]$, and is given the value of $A[i]$ in return. Determine lower and upper bounds on the expected number of steps taken by a Las Vegas randomized algorithm to determine whether or not a given key k is present in the array.

2.8 (Due to R.M. Karp.) In a graph with n vertices, where n is even, a perfect matching is a set of $n/2$ edges, no two of which meet at a common vertex. Consider a randomized algorithm that takes an n-vertex graph as input and correctly determines whether the graph has a perfect matching. At each step the algorithm asks a question of the form "Is there an edge between vertex i and vertex j?" The complexity of the algorithm is defined as the maximum, over all n-vertex graphs G, of the expected number of questions $C(n)$ asked when the input graph is G. Prove: $C(n) = \Omega(n^2)$.

2.9 (Due to R.M. Karp.) Give lower bounds on the expected number of steps for Las Vegas algorithms for the following problems:

(a) Given a string of n bits, the algorithm must determine whether the string contains three consecutive 1s. In one step, it is allowed to inspect one bit of the string. All other computation is free.

(b) Given a graph on n vertices, the algorithm must determine whether the graph contains a vertex of degree 0. In one step, it specifies two vertices and is told whether there is an edge between the specified vertices (just as in Problem 2.8). All other computation is free.

2.10 (Due to R.M. Karp.) Given a list of n values $v_1, v_2, \ldots, v_n$, the *majority element problem* is to determine the index i, if one exists, such that the value v_i occurs more than $n/2$ times in the list. Determine lower and upper bounds on the expected running time of any Las Vegas algorithm that solves the majority element problem under the assumption that the algorithm can at each step specify two indices, and is told whether or not the corresponding list entries are equal.

2.11 What happens to the proof of Theorem 2.9 if in the second condition in the definition of a randomized circuit we were to replace "at least half" by "at least $1/k$ for $k > 2$"?

2.12 Show that $\boldsymbol{BPP} \subseteq \boldsymbol{P}/\text{poly}$.

CHAPTER 3

Moments and Deviations

IN Chapters 1 and 2, we bounded the expected running times of several randomized algorithms. While the expectation of a random variable (such as a running time) may be small, it may frequently assume values that are far higher. In analyzing the performance of a randomized algorithm, we often like to show that the behavior of the algorithm is good almost all the time. For example, it is more desirable to show that the running time is small with high probability, not just that it has a small expectation. In this chapter we will begin the study of general methods for proving statements of this type. We will begin by examining a family of stochastic processes that is fundamental to the analysis of many randomized algorithms: these are called *occupancy problems.* This motivates the study (in this chapter and the next) of general bounds on the probability that a random variable deviates far from its expectation, enabling us to avoid such custom-made analyses. The probability that a random variable deviates by a given amount from its expectation is referred to as a *tail probability* for that deviation. Readers wishing to review basic material on probability and distributions may consult Appendix C.

3.1. Occupancy Problems

We begin with an example of an *occupancy problem.* In such problems we envision each of m indistinguishable objects ("balls") being randomly assigned to one of n distinct classes ("bins"). In other words, each ball is placed in a bin chosen independently and uniformly at random. We are interested in questions such as: what is the maximum number of balls in any bin? what is the expected number of bins with k balls in them? Such problems are at the core of the analyses of many randomized algorithms ranging from data structures to routing in parallel computers. Later, in Section 3.6, we will encounter a variant of the occupancy problem, known as the *coupon collector's problem*; in

Chapter 4, we will apply sophisticated techniques to various random variables arising in occupancy problems.

Our discussion of the occupancy problem will illustrate a recurrent tool in the analysis of randomized algorithms: that *the probability of the union of events is no more than the sum of their probabilities.* This is a special case of the Boole-Bonferroni Inequalities (Proposition C.2) and can be formally stated as follows: for arbitrary events $\mathcal{E}_1, \mathcal{E}_2, \ldots, \mathcal{E}_n$, not necessarily independent,

$$\mathbf{Pr}[\cup_{i=1}^{n} \mathcal{E}_i] \leq \sum_{i=1}^{n} \mathbf{Pr}[\mathcal{E}_i].$$

This principle is extremely useful because it assumes nothing about the dependencies between the events. Thus, it enables us to analyze phenomena involving events with very complicated interactions, without having to unravel the interactions.

Consider first the case $m = n$. For $1 \leq i \leq n$, let X_i be the number of balls in the ith bin. Following Example 1.1, we have $\mathbf{E}[X_i] = 1$ for all i. Yet we do not expect that during a typical experiment every bin receives exactly one ball. Rather, we expect some bins to have no balls at all, and others to have many more than one.

Let us try now to make a statement of the form "with very high probability, no bin receives more than k balls," for a suitably chosen k. Let $\mathcal{E}_j(k)$ denote the event that bin j has k or more balls in it. We concentrate on analyzing $\mathcal{E}_1(k)$. The probability that bin 1 receives exactly i balls is

$$\binom{n}{i}\left(\frac{1}{n}\right)^i\left(1-\frac{1}{n}\right)^{n-i} \leq \binom{n}{i}\left(\frac{1}{n}\right)^i \leq \left(\frac{ne}{i}\right)^i\left(\frac{1}{n}\right)^i = \left(\frac{e}{i}\right)^i.$$

The second inequality results from an upper bound for binomial coefficients (Proposition B.2). Thus,

$$\mathbf{Pr}[\mathcal{E}_1(k)] \leq \sum_{i=k}^{n}\left(\frac{e}{i}\right)^i \leq \left(\frac{e}{k}\right)^k\left(1+\frac{e}{k}+\left(\frac{e}{k}\right)^2+\cdots\right). \tag{3.1}$$

Let $k^* = \lceil (3 \ln n)/\ln \ln n \rceil$. Then,

$$\mathbf{Pr}[\mathcal{E}_1(k^*)] \leq \left(\frac{e}{k^*}\right)^{k^*}\frac{1}{1-e/k^*} \leq n^{-2}.$$

The same computation tells us that this upper bound applies to $\mathbf{Pr}[\mathcal{E}_i(k^*)]$ for all i, but can we say that *no bin* is likely to have more than k^* balls in it? For this we invoke the principle mentioned at the beginning of this section: the probability of the union of the events $\mathcal{E}_i(k^*)$ is no more than their sum. We obtain that

$$\mathbf{Pr}[\cup_{i=1}^{n} \mathcal{E}_i(k^*)] \leq \sum_{i=1}^{n} \mathbf{Pr}[\mathcal{E}_i(k^*)] \leq \frac{1}{n}.$$

Thus we have established:

Theorem 3.1: *With probability at least* $1 - 1/n$, *no bin has more than* $k^* = (e \ln n)/\ln \ln n$ *balls in it.*

Interestingly, when m is of the order of $n \log n$, the bin with the most balls has about the same number of balls as the expected number of balls in any bin. This phenomenon is exploited in a number of randomized algorithms (see, for instance, Section 4.2).

Exercise 3.1: For $m = n \log n$, show that with probability $1 - o(1)$ every bin contains $O(\log n)$ balls.

We turn to a classic combinatorial problem. Suppose that m balls are randomly assigned to n bins. We study the probability of the event that they all land in distinct bins. The special case $n = 365$ is popular in mathematical lore as the *birthday problem.* The interpretation is that the 365 days of the year correspond to 365 bins, and the birthday of each of m people is chosen independently and uniformly from all 365 days (ignoring leap years). How large must m be before two people in the group are likely to share their birthdays?

Consider the assignment of the balls to the bins as a sequential process: we throw the first ball into a random bin, then the second ball, and so on. For $2 \leq i \leq m$, let $\mathcal{E}_i$ denote the event that the ith ball lands in a bin not containing any of the first $i-1$ balls. We will bound $\mathbf{Pr}[\cap_{i=2}^m \mathcal{E}_i]$ from above. From (1.6), we can write

$$\mathbf{Pr}[\cap_{i=2}^m \mathcal{E}_i] = \mathbf{Pr}[\mathcal{E}_2]\mathbf{Pr}[\mathcal{E}_3 \mid \mathcal{E}_2]\mathbf{Pr}[\mathcal{E}_4 \mid \mathcal{E}_2 \cap \mathcal{E}_3] \cdots \mathbf{Pr}[\mathcal{E}_m \mid \cap_{i=2}^{m-1} \mathcal{E}_i].$$

Now, it is easy to compute $\mathbf{Pr}[\mathcal{E}_i \mid \cap_{j=2}^{i-1} \mathcal{E}_j]$: this is simply the probability that the ith ball lands in an empty bin given that the first $i-1$ all fell into distinct bins, and is thus $1 - (i-1)/n$. Making use of the fact that $1 - x \leq e^{-x}$, we have

$$\mathbf{Pr}[\cap_{i=2}^m \mathcal{E}_i] \leq \prod_{i=2}^m \left(1 - \frac{i-1}{n}\right) \leq \prod_{i=2}^m e^{-(i-1)/n} = e^{-m(m-1)/2n}.$$

Thus, we see that for m equal to $\lceil \sqrt{2n} + 1 \rceil$, the probability that all m balls land in distinct bins is at most $1/e$; as m increases beyond this value, the probability drops rapidly.

3.2. The Markov and Chebyshev Inequalities

We have seen above that making statements about the probability that a random variable deviates far from its expectation may involve a detailed, problem-specific analysis. Often, one can avoid such detailed analyses by resorting to general inequalities on such tail probabilities.

We begin with the Markov inequality, a fundamental tool we will invoke repeatedly when we develop more sophisticated bounding techniques. Let X be a discrete random variable and $f(x)$ be any real-valued function. Then the expectation of $f(X)$ is given by (see Appendix C)

$$\mathbf{E}[f(X)] = \sum_x f(x)\mathbf{Pr}[X = x].$$

Theorem 3.2 (Markov Inequality): *Let Y be a random variable assuming only non-negative values. Then for all $t \in \mathbb{R}^+$,*

$$\mathbf{Pr}[Y \geq t] \leq \frac{\mathbf{E}[Y]}{t}.$$

Equivalently,

$$\mathbf{Pr}[Y \geq k\mathbf{E}[Y]] \leq \frac{1}{k}.$$

PROOF: Define a function $f(y)$ by $f(y) = 1$ if $y \geq t$, and 0 otherwise. Then $\mathbf{Pr}[Y \geq t] = \mathbf{E}[f(Y)]$. Since $f(y) \leq y/t$ for all y,

$$\mathbf{E}[f(Y)] \leq \mathbf{E}\left[\frac{Y}{t}\right] = \frac{\mathbf{E}[Y]}{t},$$

and the theorem follows. □

This is the tightest possible bound when we know only that Y is non-negative and has a given expectation. Unfortunately, the Markov inequality by itself is often too weak to yield useful results. The following exercise may help the reader appreciate this; it shows that the Markov inequality is tight only for rather uninteresting distributions.

Exercise 3.2: Given a positive integer k, describe a random variable X assuming only non-negative values, such that

$$\mathbf{Pr}[X \geq k\mathbf{E}[X]] = \frac{1}{k}.$$

The following generalization of Markov's inequality underlies its usefulness in deriving stronger bounds.

Exercise 3.3: Let Y be any random variable and h any non-negative real function. Show that for all $t \in \mathbb{R}^+$,

$$\mathbf{Pr}[h(Y) \geq t] \leq \frac{\mathbf{E}[h(Y)]}{t}.$$

We now show that the Markov inequality can be used to derive better bounds on the tail probability by using more information about the distribution of the random variable. The first of these is the Chebyshev bound, which is based on the knowledge of the variance of the distribution; we will apply this to the analysis of a simple randomized selection algorithm.

For a random variable X with expectation μ_X, its *variance* σ_X^2 is defined to be $\mathbf{E}[(X-\mu_X)^2]$. The *standard deviation* of X, denoted σ_X, is the positive square root of σ_X^2. (See Appendix C.)

Theorem 3.3 (Chebyshev's Inequality): *Let X be a random variable with expectation μ_X and standard deviation σ_X. Then for any $t \in \mathbb{R}^+$,*

$$\mathbf{Pr}[|X-\mu_X| \geq t\sigma_X] \leq \frac{1}{t^2}.$$

PROOF: First, note that

$$\mathbf{Pr}[|X-\mu_X| \geq t\sigma_X] = \mathbf{Pr}[(X-\mu_X)^2 \geq t^2\sigma_X^2].$$

The random variable $Y = (X-\mu_X)^2$ has expectation σ_X^2, and applying the Markov inequality to Y bounds this probability from above by $1/t^2$. □

3.3. Randomized Selection

We now consider the use of random sampling for the problem of selecting the kth smallest element in a set S of n elements drawn from a totally ordered universe. We assume that the elements of S are all distinct, although it is not very hard to modify the following analysis to allow for multisets. Let $r_S(t)$ denote the rank of an element t (the kth smallest element has rank k) and let $S_{(i)}$ denote the ith smallest element of S. We extend the use of this notation to subsets of S as well. Thus we seek to identify $S_{(k)}$.

In Step 1 (see following page), we sample with replacement: for instance, if an element s of S is chosen to be in R on the first of our $n^{3/4}$ drawings, the remaining $n^{3/4}-1$ drawings are all as likely to pick s again as any other element in S. This style of sampling appears to be wasteful, but we employ it here because it keeps our analysis clean. Sampling without replacement would result in a marginally sharper analysis, but in practice this may be slightly harder to implement: throughout the sampling process, we would have to keep track of the elements chosen so far.

Figure 3.1 illustrates Step 3, where small elements are at the left end of the picture and large ones at the right. Determining (in Step 4) whether $S_{(k)} \in P$ is easy since we know the ranks $r_S(a)$ and $r_S(b)$ and we compare either or both of these to k, depending on which of the three **if** statements in Step 4 we execute. The sorting in Step 5 can be performed in $O(n^{3/4}\log n)$ steps.

Algorithm LazySelect:

Input: A set S of n elements from a totally ordered universe, and an integer k in $[1, n]$.

Output: The kth smallest element of S, $S_{(k)}$.

1. Pick $n^{3/4}$ elements from S, chosen independently and uniformly at random with replacement; call this multiset of elements R.
2. Sort R in $O(n^{3/4} \log n)$ steps using any optimal sorting algorithm.
3. Let $x = kn^{-1/4}$. For $\ell = \max\{\lfloor x - \sqrt{n} \rfloor, 1\}$ and $h = \min\{\lceil x + \sqrt{n} \rceil, n^{3/4}\}$, let $a = R_{(\ell)}$ and $b = R_{(h)}$. By comparing a and b to every element of S, determine $r_S(a)$ and $r_S(b)$.
4. **if** $k < n^{1/4}$, **then** $P = \{y \in S \mid y \leq b\}$;
   else if $k > n - n^{1/4}$, let $P = \{y \in S \mid y \geq a\}$;
   else if $k \in [n^{1/4}, n - n^{1/4}]$, let $P = \{y \in S \mid a \leq y \leq b\}$;
   Check whether $S_{(k)} \in P$ *and* $|P| \leq 4n^{3/4} + 2$. If not, repeat Steps 1–3 until such a set P is found.
5. By sorting P in $O(|P| \log |P|)$ steps, identify $P_{(k - r_S(a) + 1)}$, which is $S_{(k)}$.

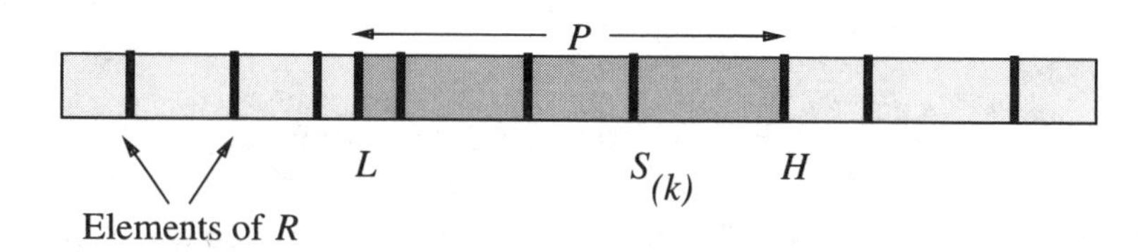

Figure 3.1: The **LazySelect** algorithm.

Thus the idea of the algorithm is to identify two elements a and b in S such that both of the following statements hold with high probability:

1. The element $S_{(k)}$ that we seek is in P.
2. The set P of elements between a and b is not very large, so that we can sort P inexpensively in Step 5.

We examine how either of these requirements could fail. We focus on the most interesting case when $k \in [n^{1/4}, n - n^{1/4}]$, so that $P = \{y \in S \mid a \leq y \leq b\}$; the analysis for the other two cases of Step 4 is similar and in fact somewhat simpler.

If the element a is greater than $S_{(k)}$ (or if b is smaller than $S_{(k)}$), we fail because P does not contain $S_{(k)}$. For this to happen, fewer than ℓ of the samples in R should be smaller than $S_{(k)}$ (respectively, at least h of the random samples should be smaller than $S_{(k)}$). We will bound the probability that this happens using the Chebyshev bound.

The second type of failure occurs when P is too big. To study this, we define $k_\ell = \max\{1, k - 2n^{3/4}\}$ and $k_h = \min\{k + 2n^{3/4}, n\}$. To obtain an upper bound on the probability of this kind of failure, we will be pessimistic and say that failure occurs if either $a < S_{(k_\ell)}$ or $b > S_{(k_h)}$. We prove that this is also unlikely, again using the Chebyshev bound. Before we perform this analysis, we establish an important property of independent random variables. Recall the definition of a joint density function $p(x, y)$ for random variables X and Y (Definition C.9).

► **Definition 3.1:** Let X and Y be random variables and $f(x, y)$ be a function of two real variables. Then,

$$\mathbf{E}[f(X,Y)] = \sum_{x,y} f(x,y)p(x,y).$$

For independent random variables X and Y we have from Proposition C.6

$$\mathbf{E}[XY] = \mathbf{E}[X]\mathbf{E}[Y]. \tag{3.2}$$

Lemma 3.4: *Let $X_1, X_2, \ldots, X_m$ be independent random variables. Let $X = \sum_{i=1}^m X_i$. Then $\sigma_X^2 = \sum_{i=1}^m \sigma_{X_i}^2$.*

PROOF: Let μ_i denote $\mathbf{E}[X_i]$, and $\mu = \sum_{i=1}^m \mu_i$. The variance of X is given by

$$\mathbf{E}[(X-\mu)^2] = \mathbf{E}[(\sum_{i=1}^m (X_i - \mu_i))^2].$$

Expanding the latter and using linearity of expectations, we obtain

$$\mathbf{E}[(X-\mu)^2] = \sum_{i=1}^m \mathbf{E}[(X_i-\mu_i)^2] + 2\sum_{i<j} \mathbf{E}[(X_i-\mu_i)(X_j-\mu_j)].$$

Since all pairs X_i, X_j are independent, so are the pairs $(X_i - \mu_i), (X_j - \mu_j)$. By (3.2), each term in the latter summation can be replaced by $\mathbf{E}[(X_i - \mu_i)]$ $\mathbf{E}[(X_j - \mu_j)]$. Since $\mathbf{E}[(X_i - \mu_i)] = \mathbf{E}[X_i] - \mu_i = 0$, the latter summation vanishes. It follows that

$$\mathbf{E}[(X-\mu)^2] = \sum_{i=1}^m \mathbf{E}[(X_i-\mu_i)^2] = \sum_{i=1}^m \sigma_{X_i}^2.$$

□

As in the analysis of **RandQS** in Chapter 1, we measure the running time of **LazySelect** in terms of the number of comparisons performed by it.

Theorem 3.5: *With probability $1 - O(n^{-1/4})$,* **LazySelect** *finds $S_{(k)}$ on the first pass through Steps 1–5, and thus performs only $2n + o(n)$ comparisons.*

PROOF: The time bound is easily established by examining the algorithm; Step 3 requires $2n$ comparisons, and all other steps perform $o(n)$ comparisons, provided the algorithm finds $S_{(k)}$ on the first pass through Steps 1–5. We now consider the first mode of failure listed above: $a > S_{(k)}$ because fewer than ℓ of the samples in R are less than or equal to $S_{(k)}$ (so that $S_{(k)} \notin P$). Let $X_i = 1$ if the ith random sample is at most $S_{(k)}$, and 0 otherwise; thus $\mathbf{Pr}[X_i = 1] = k/n$, and $\mathbf{Pr}[X_i = 0] = 1 - k/n$. Let $X = \sum_{i=1}^{n^{3/4}} X_i$ be the number of samples of R that are at most $S_{(k)}$. Note that we really do mean the number of samples, and not the number of distinct elements. The random variables X_i are *Bernoulli trials* (Appendix C): each may be thought of as the outcome of a coin toss. Then, using Lemma 3.4 and the variance of a Bernoulli trial with success probability p

$$\mu_X = \frac{kn^{3/4}}{n} = kn^{-1/4},$$

and

$$\sigma_X^2 = n^{3/4}\left(\frac{k}{n}\right)\left(1 - \frac{k}{n}\right) \leq \frac{n^{3/4}}{4}.$$

This implies that $\sigma_X \leq n^{3/8}/2$. Applying the Chebyshev bound to X,

$$\mathbf{Pr}[|X - \mu_X| \geq \sqrt{n}] \leq \mathbf{Pr}[|X - \mu_X| \geq 2n^{1/8}\sigma_X] = \mathrm{O}\left(n^{-1/4}\right).$$

An essentially identical argument shows that

$$\mathbf{Pr}[b < S_{(k)}] = \mathrm{O}\left(n^{-1/4}\right).$$

Since the probability of the union of events is at most the sum of their probabilities, the probability that either of these events occurs (causing $S_{(k)}$ to lie outside P) is $\mathrm{O}(n^{-1/4})$.

Now for the second mode of failure – that P contains more than $4n^{3/4} + 2$ elements. For this, the analysis is very similar to that above in studying the first mode of failure, with k_ℓ and k_h playing the role of k. The analysis shows that $\mathbf{Pr}[a < S_{(k_\ell)}]$ and $\mathbf{Pr}[b > S_{(k_h)}]$ are both $\mathrm{O}(n^{-1/4})$ (the reader should verify these details). Adding up the probabilities of all of these failure modes, we find that the probability that Steps 1–3 fail to find a suitable set P is $\mathrm{O}(n^{-1/4})$. □

Exercise 3.4: The failure probability can be driven down further at the expense of increased running time. For a suitable definition of the $\mathrm{o}(n)$ term, give an upper bound on the probability that the algorithm does not find $S_{(k)}$ in $cn + \mathrm{o}(n)$ steps for $c > 2$.

Exercise 3.5: Theorem 3.5 tells us that the probability that **LazySelect** terminates in $2n + \mathrm{o}(n)$ steps goes to 1 as $n \to \infty$. Suggest a modification in the algorithm that brings the constant in the linear term down to 1.5 from 2. We will refine this further in Problem 4.15.

This adds to the significance of **LazySelect**: the best known deterministic selection algorithms use $3n$ comparisons in the worst case and are quite complicated to implement. Further, it is known that any deterministic algorithm for finding the median requires at least $2n$ comparisons, so we have a randomized algorithm that is both fast and has an expected number of comparisons that is provably smaller than that of any deterministic algorithm. The high probability bound of the previous exercise can be easily converted into a bound on the expected running time:

Exercise 3.6: Show that as a direct corollary of Theorem 3.5, the expected running time of the **LazySelect** algorithm is $2n + o(n)$.

Consider what happens when we modify **LazySelect** to be recursive as follows: in Step 5, instead of sorting P we recursively use **LazySelect** to find $P_{(k-r_S(a)+1)}$. In this recursive version, the size of the candidate set P in which we are seeking $S_{(k)}$ is shrinking as the recursion proceeds. Using our analysis we can prove that at a typical stage of recursion the probability of failure at that stage is $O(|P|^{-1/4})$. But $|P|$ is diminishing, so that this probability of failure is rising as the algorithm proceeds! Thus, when the candidate set is down to a constant size, for instance, the failure probability is up to a constant and there is very little we can do about it. This is a fundamental barrier, not a weakness of our analysis. This is a typical problem with recursive randomized algorithms, and rears its head again in parallel randomized algorithms (where we always try to break a problem into smaller sub-problems) as well. A standard solution is to stop the recursion when the problem size is down to a certain size, and switch to a different, more expensive but deterministic technique – as we did by sorting in Step 5 of **LazySelect**.

3.4. Two-Point Sampling

We have so far been making use of the fact that the variance of the sum of *independent* random variables equals the sum of their variances. In fact, we can make a stronger statement. Let X and Y be discrete random variables defined on the same probability space. The *joint density function* of X and Y is the function

$$p(x, y) = \mathbf{Pr}[\{X = x\} \cap \{Y = y\}].$$

Thus $\mathbf{Pr}[Y = y] = \sum_x p(x, y)$, and

$$\mathbf{Pr}[X = x \mid Y = y] = \frac{p(x, y)}{\mathbf{Pr}[Y = y]}.$$

These definitions extend to a set $X_1, X_2, \ldots$ of more than two random variables. Such a set of random variables is said to be *pairwise independent* if for all $i \neq j$, and $x, y \in \mathbb{R}$,

$$\mathbf{Pr}[X_i = x \mid X_j = y] = \mathbf{Pr}[X_i = x].$$

We will use the result from the following exercise.

Exercise 3.7: Let n be a prime number and $\mathbb{Z}_n$ denote the ring of integers modulo n. For a and b chosen independently and uniformly at random from $\mathbb{Z}_n$, let $Y_i = ai + b \bmod n$. Show that for $i \not\equiv j \pmod{n}$, Y_i and Y_j are uniformly distributed on $\mathbb{Z}_n$ and pairwise independent. (Make use of the fact that in the field $\mathbb{Z}_n$, given fixed values for y_i and y_j, we can solve $y_i \equiv ai+b \pmod{n}$ and $y_j \equiv aj+b \pmod{n}$ uniquely for a and b.)

The following exercise is similar to Lemma 3.4.

Exercise 3.8: Let $X_1, X_2, \ldots, X_m$ be *pairwise* independent random variables, and $X = \sum_{i=1}^{m} X_i$. Show that $\sigma_X^2 = \sum_{i=1}^{m} \sigma_{X_i}^2$.

We now consider an application of these concepts to the reduction of the number of random bits used by ***RP*** algorithms (see Definition 1.8). Consider an ***RP*** algorithm A for deciding whether input strings x belong to a language L. Given x, A picks a random number r from the range $\mathbb{Z}_n = \{0, \ldots, n-1\}$, for a suitable choice of a prime n, and computes a binary value $A(x,r)$ with the following properties:

- If $x \in L$, then $A(x,r) = 1$ for at least half the possible values of r.
- If $x \notin L$, then $A(x,r) = 0$ for all possible choices of r.

For a randomly chosen r, $A(x,r) = 1$ is conclusive proof that $x \in L$, while $A(x,r) = 0$ is evidence that $x \notin L$.

For any $x \in L$, we refer to the values of r for which $A(x,r) = 1$ as *witnesses* for x; clearly, at least $n/2$ of the n possible values of r are witnesses. Of course, for $x \notin L$, there are no witnesses at all. The definition allows different $x \in L$ to have different sets of witnesses. Generally, n will be too large for us to test efficiently all the n potential witnesses for a given input x. However, for any $x \in L$, a random choice of r is a witness with probability at least $1/2$.

The fear is that $x \in L$ but the randomly chosen value of r yields $A(x,r) = 0$. However, we can drive down this probability of incorrectly classifying x by picking $t > 1$ values $r_1, \ldots, r_t$ independently from the range $\mathbb{Z}_n$, and computing $A(x, r_i)$ for all of them – in other words, by performing t independent iterations of the algorithm A on the same input x. If for any i we obtain $A(x, r_i) = 1$, we declare that x is in L, else we declare that x is not in L. By the independence of the trials, we are guaranteed that the probability of incorrectly classifying an input $x \in L$ (by declaring that it is not in L) is at most 2^{-t}.

Choosing t independent random numbers is expensive in that it requires $\Omega(t \log n)$ random bits. Suppose instead that we are only willing to use $O(\log n)$ random bits. In particular suppose that we wish to use only two independent samples from $\mathbb{Z}_n$. For a, b chosen independently from $\mathbb{Z}_n$, the naive usage of a and b as potential witnesses, i.e., computing $A(x,a)$ and $A(x,b)$, yields an upper bound of only $1/4$ on the probability of incorrect classification. Here is a better scheme: let $r_i = ai + b \bmod n$, and compute $A(x, r_i)$ for $1 \le i \le t$. As before, if for any i we obtain $A(x, r_i) = 1$, we declare that x is in L, else we declare that x is not in L. What is the probability of incorrectly classifying any input x? We show that this probability is much smaller than $1/4$.

We need to worry about the possibility of making error only in the case where the input x is in L. Our analysis will be insensitive to the actual values of r in $\mathbb{Z}_n$ which are witnesses for x; we will only rely on the fact that at least half the values of r are witnesses. Clearly $A(x, r_i)$ is a random variable over the probability space of pairs a and b chosen independently from $\mathbb{Z}_n$. By the result of Exercise 3.7, the random r_i's are pairwise independent and, therefore, so are the random variables $A(x, r_i)$, for $1 \le i \le t$. Let $Y = \sum_{i=1}^{t} A(x, r_i)$. Assuming that $x \in L$, $\mathbf{E}[Y] \ge t/2$ and $\sigma_Y^2 \le t/4$, or $\sigma_Y \le \sqrt{t}/2$. The probability that the pairwise independent iterations produce an incorrect classification corresponds to the event $\{Y = 0\}$, and

$$\mathbf{Pr}[Y = 0] \le \mathbf{Pr}[|Y - \mathbf{E}[Y]| \ge t/2].$$

By the Chebyshev inequality, the latter is at most $1/t$. Thus, the error probability is at most $1/t$, which is a considerable improvement over the error bound of $1/4$ achieved by the naive use of a and b. This improvement is sometimes referred to as *probability amplification.*

For a random variable X with expectation μ_X, we define the kth *central moment* to be $\mu_X^k = \mathbf{E}[(X - \mu_X)^k]$, if it exists (Appendix C). For example, the variance is the second central moment.

Exercise 3.9: The use of the variance of a random variable in bounding its deviation from its expectation is called *the second moment method*. In an analogous fashion, we can speak of the *kth moment method*: let k be even, and suppose we have a random variable X for which $\mu_X^k = \mathbf{E}[(X - \mu_X)^k]$ exists. Show that

$$\mathbf{Pr}[|X - \mu_X| > t\sqrt[k]{\mu_X^k}] \le \frac{1}{t^k}.$$

Why is the kth moment method difficult to invoke for odd values of k?

The second moment method is generally useful for a random variable X if σ_X is $o(\mu_X)$. In a manner similar to "two-point" sampling (the name comes from the independent choice of two points a and b from which the r_i are derived), one can speak of k-point sampling for $k > 2$. The reader is referred to Appendix C for a further discussion of k-wise independence.

3.5. The Stable Marriage Problem

Consider a society in which there are n men (denoted by capital letters A,B,C, ...) and n women (denoted by a,b,c...). A *marriage* M is a 1-1 correspondence between the men and the women. Assume a monogamous, heterosexual society. Each person has a preference list of the members of the opposite sex organized in a decreasing order of desirability. A marriage is said to be *unstable* if there exist two married couples X-x and Y-y such that X desires y more than x, and y desires X more than Y, implying that X-y will have a tendency to leave their current mates to marry each other. The pair X-y is said to be *dissatisfied* under this marriage. A marriage M in which there are no dissatisfied couples is called a *stable marriage.*

► **Example 3.1:**

For $n = 4$, consider the following preference lists.

$$\begin{array}{llll} A : abcd & B : bacd & C : adcb & D : dcab \\ a : ABCD & b : DCBA & c : ABCD & d : CDAB \end{array}$$

Consider the marriage M given by A-a, B-b, C-c, and D-d. Here C-d is a dissatisfied couple, implying that M is unstable. However, if C and d marry each other, and c and D marry each other, we obtain the stable marriage given by A-a, B-b, C-d, D-c.

The problem of finding stable marriages has several interesting applications, for example in matching medical graduates to residency positions in hospitals. It can be shown that for every choice of preference lists there exist at least one stable marriage. (Curiously enough, this is not the case in a homosexual, monogamous society with an even number of inhabitants.) We will prove this by presenting an algorithm to find a stable marriage. The naive approach of starting with an arbitrary marriage and trying to stabilize it by pairing up dissatisfied couples does not work.

Fortunately, an equally simple algorithm – the *Proposal Algorithm* – does the trick. The basic idea behind this algorithm can be summarized as "man proposes, woman disposes": each currently unattached man proposes to the most desirable woman on his list who has not already rejected him, and this woman then decides whether to accept or reject a proposal. The Proposal Algorithm is used by hospitals in North America in the match program that assigns medical graduates to residency positions.

More precisely, at any step, this algorithm will have a partial marriage. Assume that the men are numbered in some arbitrary manner. The lowest-numbered unmarried man X proposes to the most desirable woman on his list who has not already rejected him, call her x. The woman x will accept the proposal if she is currently unmarried, or if her current mate Y is less desirable to her than X (poor Y is jilted and reverts to the unmarried state). The algorithm repeats this process, terminating when every person has been married.

We show that this algorithm always terminates with a stable marriage. A woman once married will stay married during the course of the algorithm, although her mates may change with time. Furthermore, the desirability of her mates (in her view) can only improve with time. Thus at each step either a woman gets married for the first time, or an already married woman obtains a more desirable mate.

An unattached man always has at least one woman available that he can proposition. This is because every woman he has already proposed to is currently married, and if he runs out of women then all women are married – this cannot happen unless all men are married too. Since at each step the proposer will eliminate one woman on his list, and the total size of the lists is n^2, we conclude that the algorithm uses at most n^2 proposals.

We claim that the final marriage M is stable. Otherwise, let X-y be a dissatisfied pair, where in M they are paired as X-x and Y-y. Since X prefers y to x, he must have proposed to y before getting married to x. Since y either rejected X, or accepted him only to jilt him later, her mates thereafter (including Y) must be more desirable to her than X. Therefore, y must prefer Y to X, contradicting the assumption that y is dissatisfied.

Our interest here is in performing an average-case analysis of this algorithm. Thus we are considering a probabilistic analysis of a deterministic algorithm. We introduce this analysis here because it touches upon several tools that are important in the analysis of randomized algorithms.

For this average-case analysis, we assume that the men's lists are chosen independently and uniformly at random; the women's lists can be arbitrary but must be fixed in advance. Let the random variable T_P denote the number of proposals made during the execution of the Proposal Algorithm. It is clear that the running time of the algorithm is proportional to T_P. At first glance, it may appear that the distribution T_P is extremely difficult to analyze, owing to the various dependencies between the proposals. For instance, the choice of the proposer at any step is severely conditioned by the history of the process. The choice of the woman at each step also depends on the past proposals of the current proposer.

We present a very simple technique – the *Principle of Deferred Decisions* – for getting around such problems using the example of the card game called *Clock Solitaire.* In this game we start with a standard deck of 52 cards, which is assumed to be randomly shuffled. The pack is then divided into 13 piles of 4 cards each. Each pile is arbitrarily labeled with a distinct member of $\{A, 2, 3, \ldots, J, Q, K\}$. On the first move we draw a card from the pile labeled K. At each subsequent move, a card is drawn from the pile whose label is the face value of the card drawn at the previous move (the suits of the cards are ignored in this game). The game ends when an attempt is made to draw a card from an empty pile. We win the game if, on termination, all 52 cards have been drawn; in all other cases we lose the game.

Let us estimate the probability of winning the game. Observe that the game always terminates in an attempt to draw a card from the K pile: the last card drawn has to be a K. This is because there are 4 cards of each denomination, and except for the K pile, each pile initially has 4 cards.

A naive view of the probability space for this game considers all possible ways of dealing out the cards. Each point in this space corresponds to some partition of the 52 cards into 13 distinct piles, with an ordering defined on the 4 cards in each pile. Using this approach, computing the probability of a win would be a formidable task, since at each move of the game we introduce a new source of dependency.

We now examine a second probability space that better captures the dynamics of the game. The idea is to let the random choices unfold with the progress of the game, rather than fix the entire set of choices in advance. At each draw any unseen card is equally likely to appear. Thus, the process of playing this game is exactly equivalent to repeatedly drawing a card uniformly at random from a deck of 52 cards. A winning game corresponds to the situation where the first 51 cards drawn in this fashion contain exactly 3 Kings. The probability of the 52nd card drawn being a King is exactly $1/13$; this is also the probability of winning the game.

The idea of the Principle of Deferred Decisions is to not assume that the entire set of random choices is made in advance. Rather, at each step of the process we fix only the random choices that must be revealed to the algorithm.

The Principle of Deferred Decisions can be used to simplify the average-case analysis of the Proposal Algorithm as follows. We do not assume that the men have chosen their (random) preference list in advance. In fact, let us suppose that men do not know their lists to start with. Each time a man has to make a proposal, he picks a random woman from the set of women not already propositioned by him, and proceeds to propose to her. Clearly, this is equivalent to choosing the random preference lists prior to the execution of the algorithm.

The only dependency that remains is that the random choice of a woman at any step depends on the set of proposals made so far by the current proposer. We can eliminate even this dependency, albeit at the cost of modifying the behavior of the algorithm. Suppose that each time a man makes a proposal, he chooses a woman uniformly at random from the set of *all* n women, including those to whom he has already proposed. In other words, he forgets the fact that these women have already rejected him. Call this new algorithm the Amnesiac Algorithm.

How does the performance of the new algorithm relate to that of the original one? Every proposal a man makes to a woman who has already rejected him will be rejected again. Thus, the output produced by the Amnesiac Algorithm is exactly the same as that of the original Proposal Algorithm. The only difference is that there are some wasted proposals in the Amnesiac Algorithm. Let T_A denote the number of proposals made by the Amnesiac Algorithm. Clearly, T_A *stochastically dominates* T_P (Appendix C): for all m, $\mathbf{Pr}[T_A > m] \geq \mathbf{Pr}[T_P > m]$. Therefore, it suffices for an upper bound to analyze the distribution of T_A.

A benefit of analyzing T_A is that we need only count the total number of proposals made, without regard to the name of the proposer at each stage. This is because each proposal is independently made to one of the n women chosen uniformly at random. Moreover, the algorithm terminates with a stable marriage once all women have received at least one proposal each. As will become clear shortly, bounding the value of T_A is a special case of the Coupon Collector's

Problem described in the next section. The following theorem is implied by Theorem 3.8, a result about deviations in the Coupon Collector's Problem that we will prove below in Section 3.6.

Theorem 3.6: *For any constant $c \in \mathbb{R}$, and $m = n \ln n + cn$,*

$$\lim_{n\to\infty} \mathbf{Pr}[T_A > m] = 1 - e^{-e^{-c}}.$$

3.6. The Coupon Collector's Problem

In the coupon collector's problem, there are n types of coupons and at each trial a coupon is chosen at random. Each random coupon is equally likely to be of any of the n types, and the random choices of the coupons are mutually independent. Let m be the number of trials. The goal is to study the relationship between m and the probability of having collected at least one copy of each of the n types. The reader may wish to make the correspondence between this process and an occupancy problem (Section 3.1) in which m balls are randomly distributed in n bins. This process will arise again in the study of random walks (Chapter 6). In this section we provide an amazingly precise answer to this question, while illustrating some fundamental ideas in the analysis of stochastic processes of the type that arise in randomized algorithms.

3.6.1. An Elementary Analysis

Let X be a random variable defined to be the number of trials required to collect at least one of each type of coupon. We first determine the expected value of X. Let $C_1, C_2, \ldots, C_X$ denote the sequence of trials, where $C_i \in \{1, \ldots, n\}$ denotes the type of the coupon drawn in the ith trial. Call the ith trial C_i a *success* if the type C_i was not drawn in any of the first $i-1$ selections. Clearly C_1 and C_X are always successes.

We divide the sequence into *epochs*, where epoch i begins with the trial following the ith success and ends with the trial on which we obtain the $(i+1)$st success. Define the random variable X_i, for $0 \leq i \leq n-1$, to be the number of trials in the ith epoch, so that

$$X = \sum_{i=0}^{n-1} X_i.$$

Further, let p_i denote the probability of success on any trial of the ith epoch. This is the probability of drawing one of the $n-i$ remaining coupon types and so,

$$p_i = \frac{n-i}{n}.$$

The random variable X_i is geometrically distributed with parameter p_i (see

Appendix C). Thus, the expected value of X_i is $1/p_i$ and its variance is $(1-p_i)/p_i^2$.

By linearity of expectation,

$$\mathbf{E}[X] = \mathbf{E}[\sum_{i=0}^{n-1} X_i] = \sum_{i=0}^{n-1} \mathbf{E}[X_i] = \sum_{i=0}^{n-1} \frac{n}{n-i} = n\sum_{i=1}^{n} \frac{1}{i} = nH_n.$$

By Proposition B.4 the nth Harmonic number H_n is asymptotically equal to $\ln n + \Theta(1)$, implying that

$$\mathbf{E}[X] = n\ln n + \mathrm{O}(n).$$

Since the X_i's are independent, we can determine the variance of X using Proposition C.9.

$$\begin{aligned}
\sigma_X^2 &= \sum_{i=0}^{n-1} \sigma_{X_i}^2 \\
&= \sum_{i=0}^{n-1} \frac{ni}{(n-i)^2} \\
&= \sum_{i=1}^{n} \frac{n(n-i)}{i^2} \\
&= n^2\sum_{i=1}^{n} \frac{1}{i^2} - nH_n.
\end{aligned}$$

The sum $\sum_{i=1}^{n} 1/i^2$ converges to the constant $\pi^2/6$ for n approaching ∞; hence

$$\lim_{n\to\infty} \frac{\sigma_X^2}{n^2} = \frac{\pi^2}{6}.$$

Our next goal is to derive sharper estimates of the typical value of X. More precisely, we will show that the value of X is unlikely to deviate far from its expectation, or is *sharply concentrated around its expected value.* This entails bounding the tail probabilities of the distribution of X. The second moment method does not go far toward establishing such a result.

Exercise 3.10: Use the Chebyshev inequality to find an upper bound on the probability that $X > \beta n \ln n$, for a constant $\beta > 1$.

Let $\mathcal{E}_i^r$ denote the event that coupon type i is *not* collected in the first r trials. Using Proposition B.3 (Appendix B), we obtain that

$$\mathbf{Pr}[\mathcal{E}_i^r] = \left(1 - \frac{1}{n}\right)^r \le e^{-r/n}.$$

This bound is $n^{-\beta}$ for $r = \beta n \ln n$.

Using the fact that the probability of a union of events is always less than the sum of the probabilities of these events, we obtain for $r = \beta n \ln n$,

$$\mathbf{Pr}[X > r] = \mathbf{Pr}[\cup_{i=1}^{n} \mathcal{E}_i^r] \leq \sum_{i=1}^{n} \mathbf{Pr}[\mathcal{E}_i^r] \leq \sum_{i=1}^{n} n^{-\beta} = n^{-(\beta-1)}.$$

We now study the probability that X deviates from its expectation nH_n by the amount cn, for any real-valued constant c. We will see that this probability drops very quickly as we increase the absolute value of c.

3.6.2. The Poisson Heuristic

Before we show the sharp concentration result for X, the following heuristic argument will help to establish some intuition. The heuristic argument is based on the approximation of the binomial distribution by the Poisson distribution (see Appendix C for definitions of these distributions). The material in this section, although useful, is not an essential prerequisite for subsequent topics and may be omitted in the first reading.

Let N_i^r denote the number of times the coupon of type i is chosen during the first r trials; the event $\mathcal{E}_i^r$ is the same as the event $\{N_i^r = 0\}$. The random variable N_i^r has the binomial distribution with parameters r and $p = 1/n$ (see Appendix C). This means that the probability that $N_i^r = x$, for $0 \leq x \leq r$, is as follows:

$$\mathbf{Pr}[N_i^r = x] = \binom{r}{x} p^x (1-p)^{r-x}.$$

Let λ be a positive real number. A (non-negative integer) random variable Y has the Poisson distribution with parameter λ if for any non-negative integer y,

$$\mathbf{Pr}[Y = y] = \frac{\lambda^y e^{-\lambda}}{y!}.$$

For suitably small λ and as r approaches ∞, the Poisson distribution with parameter $\lambda = rp$ is a good approximation to the binomial distribution with parameters r and p. In the current setting, we can approximate the distribution of N_i^r by the Poisson distribution with parameter $\lambda = r/n$. We will ignore the fact that λ may not be "suitably small" and that there could be significant error in this approximation; after all, this is only intended to be a heuristic calculation. Using this approximation, we calculate the probability of the event $\mathcal{E}_i^r$ as follows:

$$\mathbf{Pr}[\mathcal{E}_i^r] = \mathbf{Pr}[N_i^r = 0] \approx \frac{\lambda^0 e^{-\lambda}}{0!} = e^{-r/n}. \tag{3.3}$$

The main benefit in using the Poisson approximation is that now we can claim that the events $\mathcal{E}_i^r$, for $1 \leq i \leq n$, are "almost independent," even though it is quite easy to see that there is indeed some dependence between these events. In particular, we make the following informal claim to complete the heuristic calculation.

Claim: For $1 \leq i \leq n$, and for any set of indices $\{j_1, \ldots, j_k\}$ not containing i,

$$\mathbf{Pr}[\mathcal{E}_i^r \mid \cap_{l=1}^k \mathcal{E}_{j_l}^r] \approx \mathbf{Pr}[\mathcal{E}_i^r].$$

PROOF: The proof follows from the following approximate calculations,

$$\begin{aligned}
\mathbf{Pr}[\mathcal{E}_i^r \mid \cap_{l=1}^k \mathcal{E}_{j_l}^r] &= \frac{\mathbf{Pr}[\mathcal{E}_i^r \cap (\cap_{l=1}^k \mathcal{E}_{j_l}^r)]}{\mathbf{Pr}[\cap_{l=1}^k \mathcal{E}_{j_l}^r]} \\
&= \frac{\left(1 - \frac{k+1}{n}\right)^r}{\left(1 - \frac{k}{n}\right)^r} \\
&\approx \frac{e^{-r(k+1)/n}}{e^{-rk/n}} \\
&= e^{-r/n}.
\end{aligned}$$

The first line follows from the definition of conditional expectation (Definition C.4), the second from an elementary probability calculation, and the third from Proposition B.3 (Appendix B). Since the last expression is the approximate value of $\mathbf{Pr}[\mathcal{E}_i^r]$, we obtain the desired result. □

If the approximation in (3.3) were exact, we would obtain that the events $\mathcal{E}_i^r$ are truly independent (Appendix C). In the following computation, we make the heuristic assumption of independence based on the approximation of (3.3). We then obtain that for $1 \leq i \leq n$, the probability that all coupon types are collected in the first m trials is given by:

$$\mathbf{Pr}[\neg(\cup_{i=1}^n \mathcal{E}_i^m)] = \mathbf{Pr}[\cap_{i=1}^n (\neg \mathcal{E}_i^m)] \approx (1 - e^{-m/n})^n \approx e^{-ne^{-m/n}}.$$

Let $m = n(\ln n + c)$ for any constant $c \in \mathbb{R}$. Then, by the preceding argument, we obtain that

$$\begin{aligned}
\mathbf{Pr}[X > m = n(\ln n + c)] &= \mathbf{Pr}[\cup_{i=1}^n \mathcal{E}_i^m] \\
&\approx \mathbf{Pr}[\cap_{i=1}^n (\neg \mathcal{E}_i^m)] \\
&= 1 - e^{-e^{-c}}.
\end{aligned}$$

Observe that this probability $e^{-e^{-c}}$ is close to 1 for large positive c, and is negligibly small for large negative c. Thus, the probability of having collected all n coupon types abruptly changes from nearly zero to almost one in a small interval centered around $n \ln n$. Of course, all this is contingent on our heuristic estimates being close to the true values. The power of this Poisson heuristic is that it gives a quick back-of-the-envelope type estimation of probabilistic quantities, which hopefully provides some insight into the true behavior of those quantities. As we will see in Section 3.6.3, a more rigorous but cumbersome argument can often be used to justify the conclusions obtained from such heuristic arguments.

3.6.3. A Sharp Threshold

We now convert the heuristic argument from the previous section into a rigorous (but significantly more complex) proof using the Boole-Bonferroni Inequalities (Proposition C.2). But first we prove the following technical lemma.

Lemma 3.7: *Let c be a real constant, and $m = n \ln n + cn$ for positive integer n. Then, for any fixed positive integer k,*

$$\lim_{n\to\infty} \binom{n}{k} \left(1 - \frac{k}{n}\right)^m = \frac{e^{-ck}}{k!}.$$

PROOF: Using Proposition B.3.2, we have that

$$e^{\frac{-km}{n}} \left(1 - \frac{k^2}{n}\right)^{\frac{m}{n}} \leq \left(1 - \frac{k}{n}\right)^m \leq e^{\frac{-km}{n}}.$$

Observe that $e^{-km/n} = n^{-k} e^{-ck}$. Further,

$$\lim_{n\to\infty} \left(1 - \frac{k^2}{n}\right)^{\frac{m}{n}} = 1$$

and (by Proposition B.2),

$$\lim_{n\to\infty} \binom{n}{k} \bigg/ \frac{n^k}{k!} = 1 .$$

Putting all this together yields the desired result. □

Theorem 3.8: *Let the random variable X denote the number of trials for collecting each of the n types of coupons. Then, for any constant $c \in \mathbb{R}$, and $m = n \ln n + cn$,*

$$\lim_{n\to\infty} \mathbf{Pr}[X > m] = 1 - e^{-e^{-c}}.$$

PROOF: We have that the event $\{X > m\} = \cup_{i=1}^{n} \mathcal{E}_i^m$. By the Principle of Inclusion-Exclusion,

$$\mathbf{Pr}[\cup_i \mathcal{E}_i^m] = \sum_{k=1}^{n} (-1)^{k+1} P_k^n$$

where

$$P_k^n \triangleq \sum_{1 \leq i_1 < i_2 < \cdots < i_k \leq n} \mathbf{Pr}[\cap_{j=1}^{k} \mathcal{E}_{i_j}^m].$$

Let $S_k^n = P_1^n - P_2^n + P_3^n - \cdots + (-1)^{k+1} P_k^n$ denote the partial sum formed by the first k terms of this series. By the Boole-Bonferroni inequalities (Proposition C.2), we have the bracketing property of the partial sums:

$$S_{2k}^n \leq \mathbf{Pr}[\cup_i \mathcal{E}_i^m] \leq S_{2k+1}^n.$$

By symmetry, all the k-wise intersections of the events $\mathcal{E}_i^m$ are equally likely. This implies that

$$P_k^n = \binom{n}{k} \mathbf{Pr}[\cap_{i=1}^k \mathcal{E}_i^m].$$

Moreover, the probability of the intersection of the k events $\mathcal{E}_1^m, \ldots, \mathcal{E}_k^m$ is the probability of not collecting any of the first k coupons in m trials, namely $(1 - k/n)^m$. Therefore

$$P_k^n = \binom{n}{k} \left(1 - \frac{k}{n}\right)^m.$$

For all positive integers k, define $P_k = e^{-ck}/k!$. By Lemma 3.7 we have that for each k

$$\lim_{n\to\infty} P_k^n = P_k.$$

Define the partial sums of the terms P_k as

$$S_k = \sum_{j=1}^{k} (-1)^{j+1} P_j = \sum_{j=1}^{k} (-1)^{j+1} \frac{e^{-cj}}{j!}.$$

Notice that the right-hand side consists precisely of the first k terms of the power series expansion of $f(c) = 1 - e^{-e^{-c}}$. We conclude that

$$\lim_{k\to\infty} S_k = f(c).$$

That is, for all $\epsilon > 0$, there exists $k^* > 0$ such that for any $k > k^*$,

$$|S_k - f(c)| < \epsilon.$$

Since $\lim_{n\to\infty} P_k^n = P_k$, it follows that $\lim_{n\to\infty} S_k^n = S_k$. Equivalently, for all $\epsilon > 0$ and k, when n is sufficiently large, $|S_k^n - S_k| < \epsilon$. Thus, for all $\epsilon > 0$, any fixed $k > k^*$, and n sufficiently large,

$$|S_k^n - S_k| < \epsilon \text{ and } |S_k - f(c)| < \epsilon,$$

which implies that

$$|S_k^n - f(c)| < 2\epsilon$$

and that

$$|S_{2k}^n - S_{2k+1}^n| < 4\epsilon.$$

Using the bracketing property of partial sums, we obtain that for any $\epsilon > 0$ and n sufficiently large,

$$|\mathbf{Pr}[\cup_i \mathcal{E}_i^m] - f(c)| < 4\epsilon.$$

This implies the desired result that

$$\lim_{n\to\infty} \mathbf{Pr}[\cup_i \mathcal{E}_i^m] = f(c) = 1 - e^{-e^{-c}}.$$

□

By this theorem, for any real constant c, we have

$$\lim_{n\to\infty} \mathbf{Pr}[X \leq n(\ln n - c)] = e^{-e^{c}}$$

and

$$\lim_{n\to\infty} \mathbf{Pr}[X \geq n(\ln n + c)] = 1 - e^{-e^{-c}}.$$

Thus, we obtain that

$$\lim_{n\to\infty} \mathbf{Pr}[n(\ln n - c) \leq X \leq n(\ln n + c)] = e^{-e^{-c}} - e^{-e^{c}}.$$

As the value of c is increased, it can be verified that this probability rapidly approaches 1. In other words, with extremely high probability, the number of trials for collecting all n coupon types lies in a small interval centered about its expected value. This result is *almost* like a deterministic result since it so sharply identifies the threshold value for collecting all coupons. We refer to such results as *sharp threshold* results.

Notes

Comprehensive treatises on occupancy problems are the books by Johnson and Kotz [222], and by Kolchin, Chistiakov, and Sevastianov [266]. However, most of the results in these books concern the behavior of the distributions of various random variables in the limit as n becomes large. (See also the various discussions of occupancy problems in the books by Feller [142, 143].) Generally, we will be concerned with statements resembling the ones in Section 3.1, involving asymptotic estimates on random variables and probabilities. We will return to such estimates for occupancy problems in Chapter 4. Recent work by Azar, Broder, Karlin, and Upfal [35] builds on the basic occupancy problem and points out many applications to computer science.

The history of tail inequalities such as the Chebyshev bound dates back to the early days of probability theory. Following Chebyshev's bound [394], Markov [293] observed that the same idea could be used with higher moments. Kolmogorov [267] went further and remarked that $\mathbf{Pr}[X \geq r] \leq \mathbf{E}[f(X)]/s$ for any function $f(X)$, provided that $\mathbf{E}[f(X)]$ exists and $f(x) \geq s > 0$ for all $x \geq r$. The latter idea was exploited by Bernstein and by Chernoff in a manner we will describe in Chapter 4.

Classic sources for deterministic selection algorithms are the papers of Blum, Floyd, Pratt, Rivest, and Tarjan [65], and of Schönhage, Paterson, and Pippenger [364]. The **LazySelect** algorithm presented here is a variant on one reported by Floyd and Rivest [151]. The algorithm described therein is a recursive algorithm, and does not sort after the first level of random sampling as we do. The lower bound of $2n$ for median selection is due to Bent and John [54].

The construction of pairwise independent random variables in Exercise 3.7 is given in Joffe [214]. Its application to the reduction of random bits used by abstract randomized algorithms is due to Chor and Goldreich [97]; Luby [282] presented this idea in the context of a concrete problem we will study in Chapter 12. The two-point sampling technique has been developed into a powerful technique for reducing the use of randomness, especially for the *derandomization* of algorithms (see the Notes section of Chapter 12).

The Proposal Algorithm for stable marriages is due to Gale and Shapley [161]. The book by Gusfield and Irving [188] provides a comprehensive treatment of results related to stable marriages. Our presentation of the average-case analysis of the Proposal Algorithm is drawn from Knuth's monograph [263]. The power and applicability of the Poisson heuristic is explored in great detail in the monograph by Aldous [12].

Problems

3.1 Consider an occupancy problem in which n balls are independently and uniformly distributed in n bins. Show that, for large n, the expected number of empty bins approaches n/e, where e is the base of the natural logarithm. What is the expected number of empty bins when m balls are thrown into n bins? (See Theorem 4.18.)

3.2 Suppose m balls are thrown into n bins. Give the best bound you can on m to ensure that the probability of there being a bin containing at least two balls is at least $1/2$.

3.3 A parallel computer consists of n processors and n memory modules. During a step, each processor sends a memory request to one of the memory modules. A memory module that receives either one or two requests can satisfy its request(s); modules that receive more than two requests will satisfy two requests and discard the rest.

(a) Assuming that each processor chooses a memory module independently and uniformly at random, what is the expected number of processors whose requests are satisfied? Use the approximation $(1-1/n)^n \approx 1/e$ if necessary.

(b) Repeat the computation for the case where each memory module can satisfy only one request during a step.

3.4 Consider the following experiment, which proceeds in a sequence of *rounds*. For the first round, we have n balls, which are thrown independently and uniformly at random into n bins. After round i, for $i \geq 1$, we discard every ball that fell into a bin by itself in round i. The remaining balls are retained for round $i+1$, in which they are thrown independently and uniformly at random into the n bins. Show that there is a constant c such that with probability $1-o(1)$, the number of rounds is at most $c \log\log n$.

3.5 Let X be a random variable with expectation μ_X and standard deviation σ_X.

(a) Show that for any $t \in \mathbb{R}^+$,

$$\mathbf{Pr}[X-\mu_X \geq t\sigma_X] \leq \frac{1}{1+t^2}.$$

This version of the Chebyshev inequality is sometimes referred to as the **Chebyshev-Cantelli bound**.

(b) Prove that

$$\mathbf{Pr}[|X-\mu_X| \geq t\sigma_X] \leq \frac{2}{1+t^2}.$$

Under what circumstances does this give a better bound than the Chebyshev inequality?

3.6 Let Y be a non-negative integer-valued random variable with positive expectation. Prove the following inequalities.

(a)

$$\mathbf{Pr}[Y=0] \leq \frac{\mathbf{E}[Y^2]-\mathbf{E}[Y]^2}{\mathbf{E}[Y]^2}$$

(b)

$$\frac{\mathbf{E}[Y]^2}{\mathbf{E}[Y^2]} \leq \mathbf{Pr}[Y \neq 0] \leq \mathbf{E}[Y]$$

(c) Explain why the second inequality always gives a stronger bound than the first inequality.

3.7 Let a and b be chosen independently and uniformly at random from $\mathbb{Z}_n = \{0, 1, 2, \ldots, n-1\}$, where n is a prime. Suppose we generate t pseudo-random numbers from $\mathbb{Z}_n$ by choosing $r_i = ai+b \bmod n$, for $1 \leq i \leq t$. For any $\epsilon \in [0, 1]$, show that there is a choice of the witness set $W \subset \mathbb{Z}_n$ such that $|W| \geq \epsilon n$ and the probability that *none* of the r_i's lie in the set W is at least $(1-\epsilon)^2/4t$.

3.8 Suggest a scheme for "four-point" sampling from the range $\mathbb{Z}_n$ where n is a prime. For $t < n$ samples $r_1, \ldots, r_t$ using this scheme, give an upper bound on the probability that all t attempts fail to discover a witness given $x \in L$ and compare this with the bound of $1/16$ that the naive use of four samples would yield. En route, derive an upper bound on the fourth central moment of the sum of four-way independent random variables.

3.9 (Due to D.R. Karger and R. Motwani [233].)
(a) Let S, T be two disjoint subsets of a universe U such that $|S| = |T| = n$. Suppose we select a random set $R \subseteq U$ by independently sampling each element of U with probability p. We say that the random sample R is *good* if the following two conditions hold: $R \cap S = \emptyset$ and $R \cap T \neq \emptyset$. Show that for $p = 1/n$, the probability that R is good is larger than some positive constant.

(b) Suppose now that the random set R is chosen by sampling the elements of U with only *pairwise* independence. Show that for a suitable choice of the value of p, the probability that R is good is larger than some positive constant.

3.10 The sharp threshold result in the coupon collector's problem does not imply that the probability of needing more than $cn \log n$ trials goes to zero at a doubly exponential rate if c were not a constant, but were allowed to grow with n. Let the probability of requiring more than $cn \log n$ trials be $p(c)$. For constant c, show that $1/p(c)$ can be bounded from above and below by polynomials in n.

3.11 Consider the extension of the coupon collector's problem to that of collecting at least k copies of each coupon type. Show that the sharp threshold for the number of selections required (denoted $X^{(k)}$) is centered at $n(\ln n + (k-1)\ln\ln n)$. In other words, for any positive integer k and constant $c \in \mathbb{R}$, prove that

$$\lim_{n\to\infty} \mathbf{Pr}[X^{(k)} > n(\ln n + (k-1)\ln\ln n + c)] = e^{-e^{-c}}.$$

3.12 Consider the following process related to the coupon collector problem. There are n bins and n players, and each player has an infinite supply of balls. The bins are all initially empty. We have a sequence of rounds: in each round, each player throws a ball into an empty bin chosen independently at random from all currently empty bins. Let the random variable Z be the number of rounds before every bin is non-empty. Determine the expected value of Z. What can you say about the tail of Z's distribution?

3.13 Let B be a random bipartite graph on two independent sets of vertices U and V, each with n vertices. For each pair of vertices $u \in U$ and $v \in V$, the probability that the edge between them is present is $p(n)$, and the presence of any edge is independent of all other edges. Let $p(n) = (\ln n + c)/n$ for some $c \in \mathbb{R}$.

(a) Show that the probability that B contains an isolated vertex is asymptotically equal to $e^{-2e^{-c}}$.

(b) Suggest and prove a generalization of this to random non-bipartite graphs.

3.14 (Due to R.M. Karp.) Consider a bin containing d balls chosen at random (without replacement) from a collection of n distinct balls. Without being able to see or count the balls in the bin, we would like to simulate random sampling *with replacement* from the original set of n balls. Our only access to the balls is that we can sample *without replacement* from the bin.

Consider the following strategy. Suppose that $k < d$ balls have been drawn from the bin so far. Flip a coin with the probability of HEADS being k/n. If HEADS appears, then pick one of the k previously drawn balls uniformly at random; otherwise, draw a random ball from the bin. Show that each choice is independently and uniformly distributed over the space of the n original balls. How many times can we repeat the sampling?

3.15 (Due to D. Angluin and L.G. Valiant [28].) Let B denote a random bipartite graph with n vertices in each of the vertex sets U and V. Each possible edge, independently, is present with probability $p(n)$. Consider the following algorithm for constructing a perfect matching (see Section 7.3) in such a random graph. Modify the Proposal Algorithm of Section 3.5 as follows. Each $u \in U$ can propose only to adjacent $v \in V$. A vertex $v \in V$ always accepts a proposal, and if a proposal causes a "divorce," then the newly divorced $u \in U$ is the next to propose. The sampling procedure outlined in Problem 3.14 helps implement the Principle of Deferred Decisions. How small can you make the value of $p(n)$ and still have the algorithm succeed with high probability? The following fact concerning the degree $d(v)$ of a vertex v in B proves useful:

$$\mathbf{Pr}[d(v) \leq (1-\beta)np] = \mathrm{O}\left(e^{-\beta^2 np/2}\right).$$

CHAPTER 4

Tail Inequalities

IN this chapter we present some general bounds on the tail of the distribution of the sum of independent random variables, with some extensions to the case of dependent or correlated random variables. These bounds are derived via the use of moment generating functions and result in "Chernoff-type" or "exponential" tail bounds. These Chernoff bounds are applied to the analysis of algorithms for global wiring in chips and routing in parallel communications networks. For applications in which the random variables of interest cannot be modeled as sums of independent random variables, martingales are a powerful probabilistic tool for bounding the divergence of a random variable from its expected value. We introduce the concept of conditional expectation as a random variable, and use this to develop a simplified definition of martingales. Using measure-theoretic ideas, we provide a more general description of martingales. Finally, we present an exponential tail bound for martingales and apply it to the analysis of an occupancy problem.

4.1. The Chernoff Bound

In Chapter 3 we initiated the study of techniques for bounding the probability that a random variable deviates far from its expectation. In this chapter we focus on techniques for obtaining considerably sharper bounds on such tail probabilities.

The random variables we will be most concerned with are sums of independent Bernoulli trials; for example, the outcomes of tosses of a coin. In designing and analyzing randomized algorithms in various settings, it is extremely useful to have an understanding of the behavior of this sum. Let $X_1, \ldots, X_n$ be independent Bernoulli trials such that, for $1 \leq i \leq n$, $\mathbf{Pr}[X_i = 1] = p$ and $\mathbf{Pr}[X_i = 0] = 1 - p$. Let $X = \sum_{i=1}^{n} X_i$; then X is said to have the *binomial distribution*. More generally, let $X_1, \ldots, X_n$ be independent coin tosses such that, for $1 \leq i \leq n$, $\mathbf{Pr}[X_i = 1] = p_i$ and $\mathbf{Pr}[X_i = 0] = 1 - p_i$. Such coin tosses are referred to as *Poisson trials.* Our discussion below will focus on the random variable $X = \sum_{i=1}^{n} X_i$, where the X_i are Poisson trials. Of course, all our bounds apply to the special case when the X_i are Bernoulli trials with identical probabilities, so that X has the binomial distribution.

We consider two questions regarding the deviation of X from its expectation $\mu = \sum_{i=1}^{n} p_i$. For a real number $\delta > 0$, we might ask "what is the probability that X exceeds $(1+\delta)\mu$?" We thus seek a bound on the *tail probability* of the sum of Poisson trials. An answer to this type of question is useful in *analyzing* an algorithm, showing that the chance it fails to achieve a certain performance is small. We face a different type of question in *designing* an algorithm: how large must δ be in order that the tail probability is less than a prescribed value ϵ?

Tight answers to such questions come from a technique known as the *Chernoff bound.* This technique proves to be extremely useful in designing and analyzing randomized algorithms. We focus on the Chernoff bound on the sum of independent Poisson trials.

For a random variable X, the quantity $\mathbf{E}[e^{tX}]$ is called the *moment generating function* of X. This is because $\mathbf{E}[e^{tX}]$ can be written as a power-series with terms of the form $t^k\mathbf{E}[X^k]/k!$, and $\mathbf{E}[X^k]$ is the kth *moment* of X for any positive integer k. The basic idea behind the Chernoff bound technique is to take the moment generating function of X and apply the Markov inequality to it. The sum of independent random variables appears in the exponent, and this turns into the product of random variables whose expectation we then bound.

Theorem 4.1: *Let $X_1, X_2, \ldots, X_n$ be independent Poisson trials such that, for $1 \le i \le n$, $\mathbf{Pr}[X_i = 1] = p_i$, where $0 < p_i < 1$. Then, for $X = \sum_{i=1}^{n} X_i$, $\mu = \mathbf{E}[X] = \sum_{i=1}^{n} p_i$, and any $\delta > 0$,*

$$\mathbf{Pr}[X > (1+\delta)\mu] < \left[\frac{e^\delta}{(1+\delta)^{(1+\delta)}}\right]^\mu. \tag{4.1}$$

PROOF: For any positive real t,

$$\mathbf{Pr}[X > (1+\delta)\mu] = \mathbf{Pr}[\exp(tX) > \exp(t(1+\delta)\mu)].$$

Applying the Markov inequality to the right-hand side, we have

$$\mathbf{Pr}[X > (1+\delta)\mu] < \frac{\mathbf{E}[\exp(tX)]}{\exp(t(1+\delta)\mu)}. \tag{4.2}$$

Notice that the inequality is strict: this stems from our assumption that the p_i are not all identically 0 or 1, so that X assumes more than one value. The reader may wish to recall the proof of the Markov inequality to see this.

We bound the right-hand side by observing that

$$\mathbf{E}[\exp(tX)] = \mathbf{E}[\exp(t\sum_{i=1}^{n} X_i)] = \mathbf{E}[\prod_{i=1}^{n}\exp(tX_i)].$$

Since the X_i are independent, the random variables $\exp(tX_i)$ are also independent. It follows that $\mathbf{E}[\prod_{i=1}^n \exp(tX_i)] = \prod_{i=1}^n \mathbf{E}[\exp(tX_i)]$. Using these facts in (4.2) gives

$$\mathbf{Pr}[X > (1+\delta)\mu] < \frac{\prod_{i=1}^n \mathbf{E}[\exp(tX_i)]}{\exp(t(1+\delta)\mu)}. \tag{4.3}$$

The random variable e^{tX_i} assumes the value e^t with probability p_i, and the value 1 with probability $1-p_i$. Computing $\mathbf{E}[e^{tX_i}]$ from these observations, we have that

$$\begin{aligned}\mathbf{Pr}[X > (1+\delta)\mu] &< \frac{\prod_{i=1}^n [p_i e^t + 1 - p_i]}{\exp(t(1+\delta)\mu)} \\ &= \frac{\prod_{i=1}^n [1 + p_i(e^t - 1)]}{\exp(t(1+\delta)\mu)}. \end{aligned} \tag{4.4}$$

Now we use the inequality $1 + x < e^x$ with $x = p_i(e^t - 1)$, to obtain

$$\begin{aligned}\mathbf{Pr}[X > (1+\delta)\mu] &< \frac{\prod_{i=1}^n \exp(p_i(e^t - 1))}{\exp(t(1+\delta)\mu)} \\ &= \frac{\exp(\sum_{i=1}^n p_i(e^t - 1))}{\exp(t(1+\delta)\mu)} \\ &= \frac{\exp((e^t - 1)\mu)}{\exp(t(1+\delta)\mu)}. \end{aligned} \tag{4.5}$$

Observe that all of the above has been proved for any positive real t; we are now free to choose a particular value for t that yields the best possible bound. For this, we differentiate the last expression with respect to t and set to zero; solving for t now yields $t = \ln(1+\delta)$, which is positive for $\delta > 0$. Substituting this value for t, we obtain our theorem. □

There were three main ingredients in the above proof:

1. We studied the random variable e^{tX} rather than X.
2. The expectation of the product of the e^{tX_i} turns into the product of their expectations owing to independence.
3. We pick a value of t to obtain the best possible upper bound – indeed, we choose a value of t that depends on the deviation δ.

These ingredients are generic and do not hinge on the particular case of the sum of Poisson trials. For example, Problem 4.4 is concerned with applying this technique to the sum of geometrically distributed random variables.

For succinctness in what follows, we define an upper tail bound function for the sum of Poisson trials.

► **Definition 4.1:** $F^+(\mu, \delta) \triangleq \left[e^\delta/(1+\delta)^{(1+\delta)}\right]^\mu$.

► **Example 4.1:** The Arkansas Aardvarks win each game they play with probability 1/3. Assuming that the outcomes of the games are independent, derive an upper bound on the probability that they have a winning season in a season lasting n games.

Let X_i be 1 if the Aardvarks win the ith game and 0 otherwise; let $Y_n = \sum_{i=1}^{n} X_i$. Applying Theorem 4.1 to Y_n, we find that $\mathbf{Pr}[Y_n > n/2] < F^+(n/3, 1/2) < (0.965)^n$. Thus, the probability that the Aardvarks have a winning season in n games is exponentially small in n, suggesting that the longer they play the more likely it is that their true colors show through.

The reader should verify that the term within the brackets in $F^+(\mu, \delta)$ is always strictly less than 1. Since the power μ is always positive, we will always get an upper bound that is less than 1.

The right-hand side of (4.1) is difficult to interpret, especially since we will require answers to questions such as "how large need δ be in order that $\mathbf{Pr}[X > (1+\delta)\mu]$ is at most 0.01?" We will presently work on simplifying it. But first, we consider deviations of X *below* its expectation μ.

Theorem 4.2: *Let $X_1, X_2, \ldots, X_n$ be independent Poisson trials such that, for $1 \le i \le n$, $\mathbf{Pr}[X_i = 1] = p_i$, where $0 < p_i < 1$. Then, for $X = \sum_{i=1}^{n} X_i$, $\mu = \mathbf{E}[X] = \sum_{i=1}^{n} p_i$, and $0 < \delta \le 1$,*

$$\mathbf{Pr}[X < (1-\delta)\mu] < \exp(-\mu\delta^2/2). \tag{4.6}$$

PROOF: The proof is very similar to the proof for the upper tail we saw in Theorem 4.1. As before,

$$\begin{aligned} \mathbf{Pr}[X < (1-\delta)\mu] &= \mathbf{Pr}[-X > -(1-\delta)\mu] \\ &= \mathbf{Pr}[\exp(-tX) > \exp(-t(1-\delta)\mu)], \end{aligned}$$

for any positive real t. Applying the Markov inequality and proceeding as in equations (4.2–4.3), we obtain that

$$\mathbf{Pr}[X < (1-\delta)\mu] < \frac{\prod_{i=1}^{n} \mathbf{E}[\exp(-tX_i)]}{\exp(-t(1-\delta)\mu)}.$$

Computing $\mathbf{E}[\exp(-tX_i)]$ and proceeding as in equations (4.4–4.5),

$$\mathbf{Pr}[X < (1-\delta)\mu] < \frac{\exp(\mu(e^{-t}-1))}{\exp(-t(1-\delta)\mu)}.$$

This time, we let $t = \ln(1/(1-\delta))$ to obtain that

$$\mathbf{Pr}[X < (1-\delta)\mu] < \left[\frac{e^{-\delta}}{(1-\delta)^{(1-\delta)}}\right]^{\mu}.$$

We simplify this by noting that for $\delta \in (0, 1]$,

$$(1-\delta)^{1-\delta} > \exp(-\delta + \delta^2/2),$$

using the McLaurin expansion for $\ln(1-\delta)$. This yields the desired result. □

We define the lower tail bound function for the sum of Poisson trials as follows.

► **Definition 4.2:** $F^-(\mu, \delta) \triangleq \exp\left(\frac{-\mu\delta^2}{2}\right)$.

It is immediate that $F^-(\mu, \delta)$ is always less than 1 for positive μ and δ. Note two differences between the proofs of Theorems 4.1 and 4.2. First, we directly apply the basic Chernoff technique to the random variable $-X$ rather than apply Theorem 4.1 to $Y = n - X$ (a plausible option, which leads, however, to a slightly weaker bound than the one derived below). Second, the form of the McLaurin expansion for $\ln(1 - \delta)$ allows us to obtain a "cleaner" closed form here, whereas the McLaurin expansion for $\ln(1 + \delta)$ did not permit this in Theorem 4.1.

► **Example 4.2:** The Arkansas Aardvarks hire a new coach, and critics revise their estimates of the probability of their winning each game to 0.75. What is the probability that the Aardvarks suffer a losing season assuming the critics are right and the outcomes of their games are independent of one another?

Setting up the random variable Y_n as before, we find that $\mathbf{Pr}[Y_n < n/2] < F^-(0.75n, 1/3)$, which evaluates to $< (0.9592)^n$. Thus, this probability is also exponentially small in n.

The bounds in Theorems 4.1 and 4.2 do not depend on n, but only on μ and δ. These bounds do not distinguish, for instance, between 1000 trials each with $p_i = 0.02$ and 100 each with $p_i = 0.2$, even though the distributions of X are different in the two cases. Thus, even if the actual tail probabilities are different in these cases, our estimates are the same in both cases.

We make the following definitions to facilitate our second kind of question, i.e.,"how large need δ be for $\mathbf{Pr}[X > (1 + \delta)\mu]$ to be less than ϵ?"

► **Definition 4.3:** For any positive μ and ϵ, $\Delta^+(\mu, \epsilon)$ is that value of δ that satisfies

$$F^+(\mu, \Delta^+(\mu, \epsilon)) = \epsilon. \tag{4.7}$$

Similarly, $\Delta^-(\mu, \epsilon)$ is that value of δ that satisfies

$$F^-(\mu, \Delta^-(\mu, \epsilon)) = \epsilon. \tag{4.8}$$

In other words, a deviation of $\delta = \Delta^+(\mu, \epsilon)$ suffices to keep $\mathbf{Pr}[X > (1 + \delta)\mu]$ below ϵ, irrespective of the values of n and the p_i's.

A nice feature of the bound in Theorem 4.2 is the convenient form of the right-hand side: it is easy to derive $\Delta^-(\mu, \epsilon)$ explicitly. Equating the right-hand side of (4.6) to ϵ yields

$$\Delta^-(\mu, \epsilon) = \sqrt{\frac{2 \ln 1/\epsilon}{\mu}}. \tag{4.9}$$

► **Example 4.3:** Suppose that $p_i = 0.75$. How large must δ be so that $\mathbf{Pr}[X < (1 - \delta)\mu]$ is less than n^{-5}? Using (4.9), we find that the value of δ that suffices for ϵ

to be less than n^{-5} is

$$\Delta^-(0.75n, n^{-5}) = \sqrt{\frac{10 \ln n}{0.75n}}.$$

Thus, to obtain a tail probability that is inversely polynomial in n, we need only go slightly away from the expectation – in this case out to $\delta = \sqrt{(13.333 \ln n)/n}$.

What if we wanted that $\mathbf{Pr}[X < (1-\delta)\mu]$ be less than $e^{-1.5n}$? Using (4.9), we find that for $\epsilon = e^{-1.5n}$,

$$\Delta^-(0.75n, e^{-1.5n}) = \sqrt{\frac{3n}{0.75n}} = 2,$$

which tells us nothing (for deviations below the expectation, values of δ bigger than 1 cannot occur).

We return to the simplification of (4.1) to obtain tractable estimates for $\Delta^+(\mu, \epsilon)$.

Exercise 4.1: Prove that

$$F^+(\mu, \delta) < [e/(1+\delta)]^{(1+\delta)\mu}. \tag{4.10}$$

Hence infer that if $\delta > 2e - 1$,

$$F^+(\mu, \delta) < 2^{-(1+\delta)\mu}.$$

Exercise 4.1 gives us a simple form for $F^+(\mu, \delta)$ when δ is "large." For such deviations, we have the bound

$$\Delta^+(\mu, \epsilon) < \frac{\log_2 1/\epsilon}{\mu} - 1. \tag{4.11}$$

We now present the following simplification of $F^+(\mu, \delta)$ for δ in a restricted range $(0, U]$. A pointer to the proof is given in the Notes section.

Theorem 4.3: *For* $0 < \delta \leq U$,

$$F^+(\mu, \delta) \leq \exp(-c(U)\mu\delta^2),$$

where $c(U) = [(1+U)\ln(1+U) - U]/U^2$.

For $U = 2e - 1$, this simplifies to $F^+(\mu, \delta) < \exp(-\mu\delta^2/4)$. Consequently, provided $\delta \leq 2e - 1$, we can use the estimate

$$\Delta^+(\mu, \epsilon) < \sqrt{\frac{4 \ln 1/\epsilon}{\mu}}. \tag{4.12}$$

Thus, between Theorem 4.3 and Exercise 4.1, we have bounds on $\Delta^+(\mu, \epsilon)$; however, we require some idea of the correct value of $\Delta^+(\mu, \epsilon)$ before deciding which of these forms to use. Moreover, the result of Exercise 4.1 may be slack for some values of μ and ϵ, as in the following example. This example uses Chernoff bounds to approach the occupancy problem considered in Section 3.1.

► **Example 4.4:** Consider throwing n balls uniformly and independently into n bins. Let the random variable Y_1 denote the number of balls that fall into the first bin. We wish to determine a quantity m such that $\mathbf{Pr}[Y_1 > m] \leq 1/n^2$.

Consider the Bernoulli trials indicating whether or not the ith ball falls into the first bin. Each of the p_i's is thus $1/n$. It follows that $\mu = 1$; the number m we seek is $1 + \Delta^+(1, 1/n^2)$. Guessing that $\Delta^+(1, 1/n^2)$ is larger than $2e$, we use the result in (4.1) to obtain $\Delta^+(1, 1/n^2) < 2\log_2 n - 1$.

Unfortunately, this is not the tightest possible answer in this case. Returning to (4.1), we can apply it with $\delta \approx (1.5 \ln n)/\ln\ln n$ and simplify to obtain $F^+(\mu, \delta)$ less than n^{-2}, so that our original estimate of $2\log_2 n - 1$ was asymptotically an overestimate.

A good rule of thumb from examples like this is: for ϵ of the order of n^{-c} (a value arising often in algorithmic applications), estimates such as (4.11) and (4.12) are satisfactory provided μ is $\Omega(\log n)$; when μ is smaller, we must return to (4.1) in order to obtain the tightest possible estimate.

► **Example 4.5 (Set Balancing):** This problem is known variously as *set-balancing*, or *two-coloring a family of vectors*. Given an $n \times n$ matrix A all of whose entries are 0 or 1, find a column vector $\boldsymbol{b} \in \{-1, +1\}^n$ minimizing $||A\boldsymbol{b}||_\infty$.

Consider the following algorithm for choosing $\boldsymbol{b}$: each entry of $\boldsymbol{b}$ is independently and equiprobably chosen from $\{-1, +1\}$. Note that this choice ignores the given matrix A. Clearly the inner product of any row of A with our randomly chosen $\boldsymbol{b}$ has expectation 0. We now study the deviation of this inner product from 0.

Consider the ith row of A. Applying (4.9), the probability that the inner product of this row with $\boldsymbol{b}$ is bounded by $-4\sqrt{n \ln n}$ is less than n^{-2}. An identical argument shows that the probability that the inner product of this row with $\boldsymbol{b}$ exceeds $4\sqrt{n \ln n}$ is less than n^{-2}. Thus, the probability that the absolute value of the inner product exceeds $4\sqrt{n \ln n}$ is less than $2n^{-2}$.

Let us say that the ith *bad event* occurs if the absolute value of the inner product of the ith row of A with $\boldsymbol{b}$ exceeds $4\sqrt{n \ln n}$. There are n possible bad events, one for each row, and the argument of the previous paragraph shows that the probability that any of them occurs is at most $2n^{-2}$. The probability of the union of the bad events is no more than the sum of their probabilities, which is $2/n$. In other words, with probability at least $1 - 2/n$, we find a vector $\boldsymbol{b}$ for which $||A\boldsymbol{b}||_\infty \leq 4\sqrt{n \ln n}$.

4.2. Routing in a Parallel Computer

Our first application of the Chernoff bound is another case where a randomized algorithm yields a performance that is *provably superior* to any deterministic algorithm. This application concerns a communication problem in a network of parallel processors.

We model a network of parallel processors by a directed graph on N nodes, each of which is a processing element. Edges in the graph represent communication links between processing elements. All communication between processors proceeds in a sequence of synchronous *steps*. Each link can carry a unit message, or *packet*, in a step. During a step, a processor can send at most one packet to each of its neighbors. Each processor has a unique identifying number, between 1 and N.

We consider the *permutation routing problem* on such a network. Each processor initially contains one packet destined for some processor in the network. Let v_i denote the packet originating at processor i; we denote its destination by $d(i)$. We consider the case when the $d(i)$'s, for $1 \leq i \leq N$, form a permutation of $\{1, \ldots, N\}$, i.e., every processor is the destination of exactly one packet. How many steps are necessary and sufficient to route an arbitrary permutation request $d(1)$, ..., $d(N)$? This special case is important in realizing abstract models of parallel computation (such as the PRAM model described in Chapter 12) by means of more feasible models.

A *route* for a packet is a sequence of edges it can follow from its source to its destination. An algorithm for the permutation routing problem must specify a route for each packet. In following a route, a packet may occasionally have to wait at an intermediate node because the next edge on its route is "busy" transmitting another packet. We assume that each node contains one queue for each edge leaving the node; the queue holds packets waiting to leave via that edge. A routing algorithm must also specify a *queueing discipline* for resolving conflicts between packets that simultaneously wish to follow the same edge out of a node.

We focus on a class of algorithms that are especially simple to implement in parallel computer hardware. An *oblivious algorithm* for the permutation routing problem satisfies the following property: the route followed by v_i depends on $d(i)$ alone, and not on $d(j)$ for any $j \neq i$. An oblivious algorithm specifies, for each pair $(i, d(i))$, a route between node i and node $d(i)$. Oblivious routing algorithms are attractive for their simplicity of implementation: the communication hardware at each node in the network can determine the next link on its route, simply by looking at the source and destination information carried by a packet. Often, the topology of the network makes this operation very simple. The communication hardware at a node does not have to compare the sources and destinations of different packets in its queues.

The following theorem gives a limit on the performance of deterministic oblivious algorithms; its proof is beyond the scope of this book (see the Notes section).

Theorem 4.4: *For any deterministic oblivious permutation routing algorithm on a network of N nodes each of out-degree d, there is an instance of permutation routing requiring $\Omega(\sqrt{N/d})$ steps.*

Consider the implications of this theorem for the case when the network is the *Boolean hypercube*, a popular network for parallel processing. The Boolean hypercube has $N = 2^n$ nodes connected in the following manner. Let $(i_0, \ldots, i_{n-1}) \in \{0,1\}^n$ be the (ordered) binary representation of i, i.e., $i = \sum_{j=0}^{n-1} i_j 2^j$. There is a link (a directed edge) from node i to node j if and only if $(i_0, \ldots, i_{n-1})$ and $(j_0, \ldots, j_{n-1})$ differ in exactly one position. Every node in the hypercube has $n = \log_2 N$ directed edges leaving it. Each edge incident on a node is associated with a distinct bit position in the node label, and traversing an edge corresponding to the position j will lead to a node whose label differs in exactly that bit position. Theorem 4.4 then tells us that for any deterministic oblivious routing algorithm on the hypercube, there is a permutation requiring $\Omega(\sqrt{N/n})$ steps.

We now establish a special case of the lower bound of Theorem 4.4 for the hypercube, showing that for a natural algorithm there is a natural permutation that results in poor performance. Given that the source and destination addresses are n-bit vectors, consider the following simple choice of route to send v_i from i to the node $\sigma(i)$: scan the bits of $\sigma(i)$ from left to right, and compare them with the address of the current location of v_i. Send v_i out of the current node along the edge corresponding to the left-most bit in which the current position and $\sigma(i)$ differ. Thus, in going from (1011) to (0000) in a 4-dimensional hypercube, the packet would pass through (0011) and then (0001) en route. This is referred to as the *bit-fixing* routing strategy for obvious reasons.

Exercise 4.2: Suppose that n is even. Consider the *transpose* permutation: writing i as the concatenation of two binary strings a_i and b_i each of length $n/2$, the destination of v_i is the concatenation of b_i and a_i. Show that the transpose permutation causes the bit-fixing strategy to take $\Omega(\sqrt{N/n})$ steps. Why is this permutation called a *transpose*?

We now study a randomized oblivious routing algorithm and show that its expected number of steps is considerably smaller than $\sqrt{N/n}$. This algorithm uses a simple two-phase scheme for permutation routing. Under this scheme, packet v_i executes the following two phases independently of all the other packets.

Phase 1: Pick a random intermediate destination $\sigma(i)$ from $\{1, \ldots, n\}$. Packet v_i travels to node $\sigma(i)$.

Phase 2: Packet v_i travels from $\sigma(i)$ on to its destination $d(i)$.

In each phase, each packet uses the bit-fixing strategy to determine its route.

Since each packet chooses its intermediate destination (in Phase 1) independently of the remaining packets, the scheme is oblivious. Because the $\sigma(i)$ are chosen independently at random, it may be that $\sigma(i) = \sigma(j)$ for $i \neq j$; thus σ is *not* a permutation. The choice of routes is now clear; it remains to specify the queueing discipline. For the above choice of routes, any of several queueing disciplines will in fact yield a result similar to Theorem 4.7 below. All that is required is that if at least one packet is ready to follow an edge e on a step, some packet follows e on that step. For concreteness, we adopt the following queueing discipline: each node maintains a queue for each outgoing edge, with packets leaving in FIFO (first in, first out) order. Ties occur only when two packets simultaneously arrive at a node and wish to leave by the same edge; these ties are broken arbitrarily. The reader should verify that any pair of packets may engage in such a tie at most once.

How many steps elapse before packet v_i reaches its destination? Let us first consider this question for Phase 1. Let ρ_i denote the route for v_i in Phase 1. The number of steps taken by v_i is equal to the length of ρ_i, which is at most n, plus the number of steps for which it is queued (delayed) at intermediate nodes in ρ_i. What is the delay encountered by packet v_i? To tackle this problem we require two additional facts; the first is a simple exercise.

Exercise 4.3: View each route in Phase 1 as a directed path in the hypercube from the source to the intermediate destination. Prove that once two routes separate, they do not rejoin.

We now establish an important step in the analysis. Like the statement in Exercise 4.3 above, it is a deterministic assertion that is independent of the randomization in our routing algorithm. In preparation for this step, the reader should first attempt the following exercise.

Exercise 4.4: Does the statement in Exercise 4.3 imply that for any two packets v_i and v_j, there is at most one queue q such that v_i and v_j are in the queue q at the same step?

Lemma 4.5: *Let the route of v_i follow the sequence of edges $\rho_i = (e_1, e_2, \ldots, e_k)$. Let S be the set of packets (other than v_i) whose routes pass through at least one of $\{e_1, e_2, \ldots, e_k\}$. Then, the delay incurred by v_i is at most $|S|$.*

PROOF: A packet in S is said to *leave* ρ_i at that time step at which it traverses an edge of ρ_i for the last time. If a packet is ready to follow edge e_j at time t, we define its *lag* at time t to be $t - j$. The lag of v_i is initially zero, and the delay incurred by v_i is its lag when it traverses e_k. We will show that each step at which the lag of v_i increases by one can be charged to a distinct member of S.

We argue that if the lag of v_i reaches $\ell + 1$, some packet in S leaves ρ_i with lag ℓ. When the lag of v_i increases from ℓ to $\ell + 1$, there must be at least one packet (from S) that wishes to traverse the same edge as v_i at that time step, since otherwise v_i would be permitted to traverse this edge and its lag would not increase. Thus, S contains at least one packet whose lag reaches the value ℓ.

Let t' be the last time step at which any packet in S has lag ℓ. Thus there is a packet v ready to follow edge $e_{j'}$ at t', such that $t' - j' = \ell$. We argue that some packet of S leaves ρ_i at t'; this establishes the lemma since by the result of Exercise 4.3, a packet that has left ρ_i will never again delay v_i.

Since v is ready to follow $e_{j'}$ at t', some packet ω (which may be v itself) in S follows $e_{j'}$ at t'. Now ω leaves ρ_i at t'; if not, some packet will follow $e_{j'+1}$ at step $t' + 1$ with lag still at ℓ, violating the maximality of t'. We charge to ω the increase in the lag of v_i from ℓ to $\ell + 1$; since ω leaves ρ_i, it will never be charged again. Thus, each member of S whose route intersects ρ_i is charged for at most one delay, establishing the lemma. □

Let the random variable $H_{ij} = 1$ if ρ_i and ρ_j share at least one edge, and 0 otherwise. It follows that the total delay incurred by v_i is at most $\sum_{j=1}^{N} H_{ij}$. Since the routes of the various packets are chosen independently at random, the H_{ij}'s are independent Poisson trials for $j \neq i$. Thus, to bound the delay of packet v_i from above using the Chernoff bound, it suffices to obtain an upper bound on $\sum_{j=1}^{N} H_{ij}$. To do this, we first bound $\mathbf{E}[\sum_{j=1}^{N} H_{ij}]$.

For an edge e in the hypercube, let the random variable $T(e)$ denote the number of routes that pass through e. Fix any route $\rho_i = (e_1, e_2, \ldots, e_k)$, with $k \leq n$. Then,

$$\sum_{j=1}^{N} H_{ij} \leq \sum_{l=1}^{k} T(e_l),$$

and therefore

$$\mathbf{E}[\sum_{j=1}^{N} H_{ij}] \leq \sum_{l=1}^{k} \mathbf{E}[T(e_l)]. \tag{4.13}$$

The following is an easy consequence of symmetry.

Exercise 4.5: Let e_l and e_m be any two edges in the hypercube. Prove that $\mathbf{E}[T(e_l)] = \mathbf{E}[T(e_m)]$. In other words, the expected number of routes passing through an edge is the same for all edges in the hypercube.

The expected length of ρ_j (number of edges traversed by v_j) is $n/2$ for all j, so that the expectation of the total route length summed over all the packets is $Nn/2$. The number of edges in the hypercube is Nn; by the result of Exercise 4.5, it follows that $\mathbf{E}[T(e)] = 1/2$ for all edges e. Using this in (4.13) gives

$$\mathbf{E}[\sum_{j=1}^{N} H_{ij}] \leq \frac{k}{2} \leq \frac{n}{2}.$$

By the Chernoff bound (the form in Exercise 4.1 is most convenient), the probability that $\sum_{j=1}^{N} H_{ij}$ exceeds $6n$ is less than 2^{-6n}. An important point: we apply the Chernoff bound to $\sum_{j=1}^{N} H_{ij}$ and *not* to $\sum_{l=1}^{k} T(e_l)$. We cannot apply the Chernoff bound to $\sum_{l=1}^{k} T(e_l)$ because the random variables $T(e_l)$ are *not* independent (and in fact are not Poisson trials). We use the quantity $\sum_{l=1}^{k} T(e_l)$ only to obtain an upper bound on $\mathbf{E}[\sum_{j=1}^{N} H_{ij}]$, and *then* apply the Chernoff bound to $\sum_{j=1}^{N} H_{ij}$, which is the sum of independent random variables.

Now $\sum_{j=1}^{N} H_{ij}$ is an upper bound on the delay incurred by v_i, so this delay exceeds $6n$ with probability less than 2^{-6n}. Since the total number of packets is $N = 2^n$, the probability that any of the N packets experiences a delay exceeding $6n$ is less than $2^n \times 2^{-6n} = 2^{-5n}$. Adding the length of the route to the delay gives $7n$ as the number of steps taken by v_i in Phase 1.

Theorem 4.6: *With probability at least $1 - 2^{-5n}$, every packet reaches its intermediate destination in Phase 1 in $7n$ or fewer steps.*

What happens to the packets in Phase 2? Observe that the routing scheme for Phase 2 can be viewed as the scheme for Phase 1 "run backwards." The same analysis then shows that with probability at least $1 - (1/32)^n$, every packet reaches its destination in $7n$ or fewer steps. The probability that any packet fails to reach its target in either phase is less than $2(1/32)^n$, which is less than $1/N$ for $n \geq 1$. Combining these facts, we have:

Theorem 4.7: *With probability at least $1 - (1/N)$, every packet reaches its destination in $14n$ or fewer steps.*

Note that we have bounded the delay of a packet in each phase by assuming it is delayed only by packets executing that phase. To avoid packets in Phase 1 delaying packets in Phase 2 and vice versa, rather than allow Phases 1 and 2 to proceed unchecked for the various packets, we make packets wait at their intermediate destinations until $7n$ steps have elapsed before beginning their Phase 2 travel.

An interesting feature of this scheme is that the distribution of the number of steps to completion is insensitive to the instance to be routed. Indeed, it is likely to take as long to route the identity permutation as any other "hard" permutation!

Exercise 4.6: Show that the expected number of steps within which all packets are delivered is less than $15n$.

Comparing the performance of the randomized algorithm with the negative result of Theorem 4.4, we find that our randomized oblivious algorithm is provably better in that it achieves an expected running time that no deterministic oblivious algorithm can achieve. In fact, any deterministic oblivious algorithm must have performance *exponentially* worse than that of our randomized oblivious algorithm.

4.3. A Wiring Problem

We now consider another application of the Chernoff bound. The problem is that of *global wiring in gate-arrays.* A gate-array is a two-dimensional $\sqrt{n} \times \sqrt{n}$ array of gates abutting each other, arranged at regularly spaced points in the plane. The gates are numbered from 1 through n. A logic circuit is implemented on such an array by connecting together some of the gates using wires. A *net* is a set of gates to be connected by a wire. Wires run over the array in "Manhattan" form, i.e., they run parallel to the axes of orientation of the gate-array. In Figure 4.1, n is 9, and we have 4 wires each of which connects a pair of gates. Each gate is represented as a square with thin lines defining the boundaries. Each net connects a pair of gates, and has the same number marking its end-points (i.e., the thick lines 1-1, 2-2, 3-3, and 4-4). Note that in some cases a gate contains the end-point of more than one net.

The wiring problem is the following: we are given a set of *nets*, each of which is a set of gates to be connected together (to form one electrical connection). Here we consider only the simplest case, where each net consists of two gates to be joined by a wire. We wish to specify for each net a physical path between the two gates in the net, subject to space constraints.

In practice, the wiring problem is usually accomplished in two sequential phases: *global wiring* and *detailed wiring.* In the global wiring phase, we only specify which gates a wire will pass over in connecting its end-points. Thus, in Figure 4.1, the global route for net 4-4 passes through the three gates in the right-most column of the array. This is followed by the detailed wiring phase, in which the exact positions of the wires along their routes are specified – in our example, we would specify that the wire for net 4-4 lies to the right of the wire for net 3-3 as it leaves the top-right gate, and so on. Here we only concern ourselves with the global wiring phase.

The boundary between adjacent gates in an array has a fixed physical dimension and can therefore accommodate only a prescribed maximum number of wires, say w. We wish to find an assignment of global routes to all the nets in the wiring problem, such that no more than w nets pass through any boundary. In Figure 4.1, the set of routes we have indicated is a feasible solution provided w is at least 2. It is not hard to see that in this instance, we cannot find a feasible global wiring of the wires if w were only 1 – four wires must leave the top row of gates, and we have only three boundaries through which they must all pass.

We will solve a somewhat harder optimization problem instead of the feasibility problem – for a boundary b between two gates in the array, let $w_S(b)$ denote the number of wires that pass through b in a solution S to the global wiring problem. Let $w_S = \max_b w_S(b)$ be the maximum number of wires through

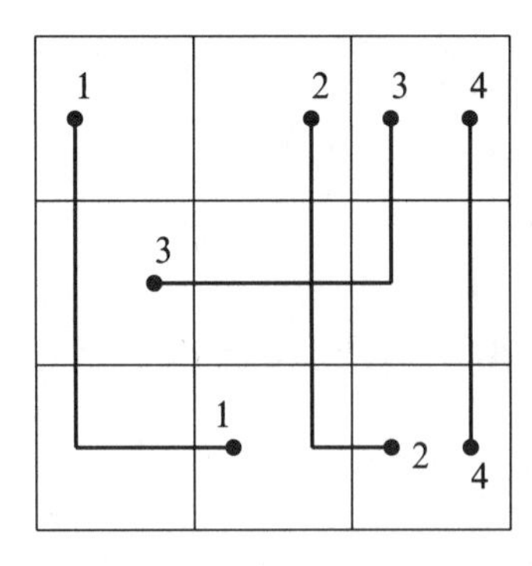

Figure 4.1: A gate-array with 9 gates.

any boundary in the solution S. If we can minimize w_S, we can surely decide the feasibility problem.

As a further simplification in our presentation, we assume that the global route for each net contains at most one 90° turn; we refer to such a route as a "one-bend" route. Thus, in joining the two end-points of a net, the wire will either first traverse the horizontal dimension and then the vertical dimension, or the other way around. In Figure 4.1, every net has been routed under this restriction. For net 4-4, which connects two gates in the same column of the array, we have only one choice under our restricted class of routes – to go right down the column; the reader should verify that the existence of such nets does not affect the following analysis. Our problem now becomes one of deciding, for each net, which of the two options to use.

This can be cast as a zero-one linear program as follows. For net i, we use two variables x_{i0} and x_{i1} to indicate which one of the two routes will be used for it. Thus, x_{i0} would be 1 if we chose the route that goes horizontally first, starting from the left end-point of net i, and 0 otherwise. For x_{i1} we adopt the opposite convention. In Figure 4.1, $x_{10} = 0$ and $x_{11} = 1$, whereas $x_{30} = 1$ and $x_{31} = 0$. For each boundary b in the array, let

$$T_{b0} = \{i \mid \text{net } i \text{ passes through } b \text{ if } x_{i0} = 1\}$$

and

$$T_{b1} = \{i \mid \text{net } i \text{ passes through } b \text{ if } x_{i1} = 1\}.$$

With these definitions, our integer program can be expressed as:

$$\text{minimize} \quad w$$

$$\text{where} \quad x_{i0}, x_{i1} \in \{0, 1\} \quad (\forall \text{ nets } i) \tag{4.14}$$

subject to

$$x_{i0} + x_{i1} = 1 \quad (\forall \text{ nets } i) \tag{4.15}$$

$$\sum_{i \in T_{b0}} x_{i0} + \sum_{i \in T_{b1}} x_{i1} \leq w \quad (\forall \text{ boundaries } b). \tag{4.16}$$

The constraint (4.15) ensures that a unique route is specified for every net. The constraint (4.16) specifies that at most w wires pass through any boundary b. The objective function seeks a solution of minimum w, with the zero-one constraint imposed. The optimum solution to this zero-one integer program gives the minimum w_S among the class of solutions allowing only one-bend routes. In general, allowing a less restrictive set of routes could result in a solution with a lower w_S.

Denote by w_O the value of the objective w in the optimum solution to (4.14–4.16). The general problem of zero-one linear programming is ***NP***-hard, and in fact even the particular class of zero-one linear programs (4.14–4.16) arising from our global wiring problem is known to be ***NP***-hard (i.e., our global wiring problem is ***NP***-hard). Thus we do not hope to compute w_O efficiently.

We solve instead the *linear program relaxation* of (4.14–4.16). This is a linear program in which the integrality constraint in (4.14) is replaced by the constraints $x_{i0}, x_{i1} \in [0, 1]$ for each i. In other words, we allow the x_{i0} and x_{i1} to assume real values between 0 and 1. This is a linear programming problem, and we know of several efficient methods for solving it (see Section 9.10). Let $\widehat{x}_{i0}$ and $\widehat{x}_{i1}$, for $1 \leq i \leq n$, be the solutions provided by the linear program, and let $\widehat{w}$ be the value of the objective function for this solution. Since the linear program is a relaxation of (4.14–4.16), it is clear that $w_O \geq \widehat{w}$. The $\widehat{x}_{i0}$'s and $\widehat{x}_{i1}$'s may be fractional values, and therefore may not constitute a feasible solution to our integer program. We must therefore "round" these fractional values to 0's and 1's to obtain a feasible global wiring; in doing so, we hope not to allow the objective w to drift too far from w_O.

We now describe a technique known as *randomized rounding* that rounds these fractional values to 0's and 1's. It finds a global wiring S with w_S provably not much larger than $\widehat{w}$, and thus w_O. Note that the fractional solutions $\widehat{x}_{i0}$ and $\widehat{x}_{i1}$ still satisfy the other constraints of the original integer program; in particular, $\widehat{x}_{i0} + \widehat{x}_{i1} = 1$ for each i. We will denote by $\overline{x}_{i0}$ the rounded value of $\widehat{x}_{i0}$, and define $\overline{x}_{i1}$ similarly.

Randomized rounding is the following process: independently for each i, set $\overline{x}_{i0}$ to 1 and $\overline{x}_{i1}$ to 0 with probability $\widehat{x}_{i0}$; otherwise set $\overline{x}_{i0}$ to 0 and $\overline{x}_{i1}$ to 1. Thus, for each i, $\mathbf{Pr}[\overline{x}_{i0} = 1] = \widehat{x}_{i0}$ and $\mathbf{Pr}[\overline{x}_{i1} = 1] = \widehat{x}_{i1}$. The idea of randomized rounding is to interpret the fractional solutions provided by the linear program as probabilities for the rounding process. Another interpretation is to imagine that the linear program, given the choice of two routes for wiring each net, routes the wire using two "fractional wires." Randomized rounding then picks one of these fractional wires, in proportion to its fraction. A nice property of randomized rounding is that if the fractional value of a variable is close to 0 (or 1), it is likely to be set to 0 (or 1).

Theorem 4.8: *Let ϵ be a real number such that $0 < \epsilon < 1$. Then with probability*

$1-\epsilon$, *the global wiring S produced by randomized rounding satisfies*

$$w_S \leq \widehat{w}(1+\Delta^+(\widehat{w},\epsilon/2n)) \leq w_O(1+\Delta^+(w_O,\epsilon/2n)).$$

PROOF: We establish that following the rounding process, with probability at least $1-\epsilon$, no boundary in the array has more than $\widehat{w}(1+\Delta^+(\widehat{w},\epsilon/2n))$ wires passing through it. We will do so by showing that for any particular boundary b, the probability that $w_S(b) > \widehat{w}(1+\Delta^+(\widehat{w},\epsilon/2n))$ is at most $\epsilon/2n$; then, since a $\sqrt{n}\times\sqrt{n}$ array contains fewer than $2n$ boundaries, we can sum this probability of failure over all the boundaries b to get an upper bound of ϵ on the failure probability.

Consider a boundary b; since the solutions of the linear program satisfy its constraints, we have

$$\sum_{i\in T_{b0}} \widehat{x}_{i0} + \sum_{i\in T_{b1}} \widehat{x}_{i1} \leq \widehat{w}. \tag{4.17}$$

The number of wires passing through b in the solution S is

$$w_S(b) = \sum_{i\in T_{b0}} \overline{x}_{i0} + \sum_{i\in T_{b1}} \overline{x}_{i1}. \tag{4.18}$$

But $\overline{x}_{i0}$ and $\overline{x}_{i1}$ are Poisson trials with probabilities $\widehat{x}_{i0}$ and $\widehat{x}_{i1}$, respectively. Further, $\overline{x}_{i0}$ and $\overline{x}_{i1}$ are each independent of $\overline{x}_{j0}$ and $\overline{x}_{j1}$ for $i \neq j$. Therefore, $w_S(b)$ is the sum of independent Poisson trials and, by (4.17) and (4.18),

$$\mathbf{E}[w_S(b)] = \sum_{i\in T_{b0}} \mathbf{E}[\overline{x}_{i0}] + \sum_{i\in T_{b1}} \mathbf{E}[\overline{x}_{i1}] = \sum_{i\in T_{b0}} \widehat{x}_{i0} + \sum_{i\in T_{b1}} \widehat{x}_{i1} \leq \widehat{w}.$$

Now, by the definition of $\Delta^+(\mu,\epsilon)$ in (4.7),

$$\mathbf{Pr}[w_S(b) > \widehat{w}(1+\Delta^+(\widehat{w},\epsilon/2n)] \leq \epsilon/2n,$$

and the theorem follows. □

Neither the theorem nor its proof makes any assumption on the value of ϵ – it can in fact be o(1), even n^{-c} for some constant c. Let us return to the guarantee provided by Theorem 4.8; how good is it? The answer depends on the value of w_O. Suppose we seek $\epsilon = 1/n$, so that $\epsilon/2n = 1/2n^2$. Then $w_S \leq w_O(1+\Delta^+(w_O,\epsilon/2n))$.

Consider first the case where $w_O = n^\gamma$, for some positive constant γ. We can use Theorem 4.3 to show that with probability $1-\epsilon$,

$$w_S \leq n^\gamma\left(1+\sqrt{\frac{4\ln 2n/\epsilon}{n^\gamma}}\right).$$

Thus, we find a solution with an additive term that is vanishingly small as n grows. Suppose, on the other hand, that $w_O = 20$. In this case, a calculation similar to that in Example 4.4 shows that w_S is $O((\log n)/\log\log n)$ with probability $1-1/n$. Randomized rounding is likely to perform well provided w_O is "not too small," and this appears to be the case in practice. When w_O is small (as in the latter case), we can in fact do substantially better than the $O((\log n)/\log\log n)$ guarantee provided by randomized rounding, as Exercise 4.7 below illustrates.

Exercise 4.7: Give a simple rounding procedure that obtains rounded solutions $\bar{x}$ from $\hat{x}$ so that $w_S \leq 2w_O$, where w_O is the optimum solution for our restricted class of one-bend routes.

We have focused on the quality of the solution produced in the probabilistic statement of Theorem 4.8. Our algorithm can be shown to run in time polynomial in the number of gates and nets in the instance. This is an example where we are interested in random variables other than running time of a randomized algorithm.

4.4. Martingales

Our discussion so far has centered on the sums of independent random variables. Frequently, it is necessary to consider the sum of random variables that are not totally independent. When relatively little knowledge of the random variables is available, we may resort to the Markov inequality or the Chebyshev inequality; in such cases, we cannot hope to show that a random variable is sharply concentrated about its expectation. There are, however, cases in which we can exploit additional structure in the random variables. An important case of such additional structure is that of *martingales*. (The material in this section, although useful, is not an essential prerequisite for subsequent chapters and may be omitted.)

Martingales originally referred to systems of betting in which a player increased his stake (usually by doubling) each time he lost a bet. Assuming unlimited capital, this system is guaranteed to eventually result in a net profit in any fair betting game; in the case of limited capital, it will eventually lead to net profit or total bankruptcy. It is no wonder that such systems have been outlawed in most casinos! Here we are interested in a far more general definition of martingales, which has proved to be very useful in showing that a random variable is sharply concentrated about its expectation. The following exposition concentrates on discrete martingales, as the continuous case seldom arises in computer science applications. The definition of martingales requires some exposure to the measure-theoretic underpinnings of probability theory, and we recommend a review of the material in Appendix C.

We begin by defining conditional distributions and expectations. Let X be a random variable and $\mathcal{E}$ any event that occurs with a non-zero probability. The *conditional density function* of X given $\mathcal{E}$ is given by $\mathbf{Pr}[X = x \mid \mathcal{E}]$. In particular, $\mathcal{E}$ can be the event that some other random variable Y takes on a specific value

y. Denoting the joint density function of X and Y by $p(x, y)$, we have

$$\mathbf{Pr}[X = x \mid Y = y] = \frac{p(x, y)}{\mathbf{Pr}[Y = y]} = \frac{p(x, y)}{\sum_x p(x, y)}$$

and

$$\mathbf{E}[X \mid Y = y] = \frac{\sum_x x p(x, y)}{\sum_x p(x, y)},$$

where $\mathbf{E}[X \mid Y = y]$ is the *conditional expectation* of X given that Y equals y. These definitions apply only for the values y for which $\mathbf{Pr}[Y = y] > 0$.

We can express the conditional expectation as a function of y, say $f(y)$. If the value of Y is not known, then the conditional expectation is itself a random variable. This is the random variable $f(Y)$.

► **Definition 4.4:** The random variable $\mathbf{E}[X \mid Y]$ is defined to be the random variable $f(Y)$ such that $f(y) = \mathbf{E}[X \mid Y = y]$.

Suppose that the random variables X and Y are defined over the probability space $(\Omega, \mathbb{F}, \mathbf{Pr})$. Consider the partition of Ω into the events $\{Y = y\}$ as y ranges over the subset of reals in which $\mathbf{Pr}[Y = y] > 0$. The function $f(y)$ is the average value of X over the various elementary events in the set $\{Y = y\}$. The random variable $\mathbf{E}[X \mid Y]$ takes on the value $f(y)$ when evaluated at some elementary outcome $\omega \in \{Y = y\}$. We can generalize this to define the random variable $\mathbf{E}[X \mid Y_1, \ldots, Y_r]$.

► **Example 4.6:** Consider independent throws of an unbiased 6-sided die. For $1 \leq i \leq 6$, let X_i denote the number of times the value i appears in n throws of the die. Consider the following conditional expectations:

$$\begin{aligned} \mathbf{E}[X_1 \mid X_2] &= \frac{n - X_2}{5}, \\ \mathbf{E}[X_1 \mid X_2, X_3] &= \frac{n - X_2 - X_3}{4}. \end{aligned}$$

These equations define the expected value of the random variable X_1 given the number of times 2 and 3 appear. Of course, the number of occurrences of 2 and 3 are themselves random variables, and so the expectation of X_1 is a random variable defined as a function of X_2 and X_3.

If we knew that there are α occurrences of 2, we can compute the expected value of X_1 as $(n-\alpha)/5$; given the further information that there are β occurrences of 3, we can compute the expected value of X_1 as $(n - \alpha - \beta)/4$. More succinctly,

$$\begin{aligned} \mathbf{E}[X_1 \mid X_2 = \alpha] &= \frac{n - \alpha}{5}, \\ \mathbf{E}[X_1 \mid X_2 = \alpha, X_3 = \beta] &= \frac{n - \alpha - \beta}{4}. \end{aligned}$$

We leave both the proofs of the following lemmas and their generalization to random variables such as $\mathbf{E}[X \mid Y_1, \ldots, Y_r]$ as an exercise.

Lemma 4.9: $\mathbf{E}[\mathbf{E}[X \mid Y]] = \mathbf{E}[X]$.

Lemma 4.10: $\mathbf{E}[Y \times \mathbf{E}[X \mid Y]] = \mathbf{E}[XY]$.

4.4.1. A Simple Definition

We start with a simplified definition of a martingale. No assumptions are made about the independence or the precise distributions of the random variables in this definition. In fact, this is just the reason why martingales are so powerful!

► **Definition 4.5:** A sequence of random variables X_0, X_1, ... is said to be a *martingale sequence* if for all $i > 0$,

$$\mathbf{E}[X_i \mid X_0, \ldots, X_{i-1}] = X_{i-1}.$$

Consider the example of a gambler who makes a sequence of bets. Her initial capital is X_0, and X_i represents the capital after the ith bet. Assume that the game is fair, so that the expected gain/loss from each bet is zero. We can then claim that the sequence X_0, X_1, ... forms a martingale. This is without the knowledge of the gambler's strategy; the gambler bets an arbitrary amount of money each time, and the amount bet may depend in any way upon the history (i.e., the previous results X_0, X_1, ..., X_{i-1}). The following lemma is an immediate consequence of Definition 4.5 and Lemma 4.9; it implies that the expected capital at any stage is exactly the initial amount X_0.

Lemma 4.11: *Let X_0, X_1, ... be a martingale sequence. Then, for all $i \geq 0$,* $\mathbf{E}[X_i] = \mathbf{E}[X_0]$.

An alternate view of the gambling example is provided by letting the random variable Y_i denote the net gain or loss from the ith bet. We can relate the sequences X_0, X_1, ... and Y_1, Y_2, ... as follows: $Y_i = X_i - X_{i-1}$ and $X_i = X_0 + \sum_{j=1}^{i} Y_j$. By fairness, regardless of the past history, the expected gain from each bet is zero, i.e., $\mathbf{E}[Y_i \mid Y_1, \ldots, Y_{i-1}] = 0$. Since the two views of the process are exactly equivalent, we make an alternate definition of a martingale.

► **Definition 4.6:** A sequence of random variables Y_1, Y_2, ... is said to be a *martingale difference sequence* if for all $i \geq 1$,

$$\mathbf{E}[Y_i \mid Y_1, \ldots, Y_{i-1}] = 0.$$

Of course, in a casino the games are known to be unfair to the gamblers. In that case, the sequence of capitals forms what is known as a super-martingale; from the point of view the casino, the situation is represented by what is called a sub-martingale.

► **Definition 4.7:** A sequence of random variables X_0, X_1, ... is said to be a *super-martingale* if for all i,

$$\mathbf{E}[X_i \mid X_0, \ldots, X_{i-1}] \leq X_{i-1}.$$

It is called a *sub-martingale* if for all i,

$$\mathbf{E}[X_i \mid X_0, \ldots, X_{i-1}] \geq X_{i-1}.$$

This definition can be adapted to a martingale difference sequence. Moreover, a super-martingale can be converted into a martingale by accounting for the expectation at each stage. In the case of a gambler playing an unfair game, suppose that the expected return on a bet of value 1 is the amount $1-\mu$. Assume that the gambler bets one dollar each time and gets a return of Y_i; let X_i be her net capital after the ith bet. Then the sequence Z_0, Z_1, ... forms a martingale, where

$$Z_i \stackrel{\Delta}{=} X_i + i\mu = X_0 + \sum_{j=1}^{i}(Y_j + \mu - 1).$$

A similar conversion can be performed for the sub-martingale corresponding to the casino's viewpoint.

Exercise 4.8 (Polya's Urn Scheme): Consider an urn that initially contains b black balls and w white balls. We perform a sequence of random selections from this urn, where at each step the chosen ball is replaced by c balls of the same color. Let X_i denote the fraction of black balls in the urn after the ith trial. Show that the sequence X_0, X_1, ... is a martingale.

Exercise 4.9 (Occupancy Problem): Suppose that m balls are thrown independently and uniformly at random into n bins. Let Z denote the number of bins that remain empty. Define time t to be the time at which exactly t balls have been thrown into the bins. For $0 \leq t \leq m$, define the random variable Z_t to be the *expectation at time* t of the number of bins that are *empty at time* m. The random variable Z_t depends on the placement of the first t balls, and is defined under the assumption that the remaining balls are placed at random. Show that the sequence of random variables $Z_0, \ldots, Z_m$ is a martingale, and that $Z_0 = \mathbf{E}[Z]$ and $Z_m = Z$.

Given our current description of a martingale, the latter exercise is non-trivial. In Section 4.4.2, we will develop a more general view of martingales that will reduce this exercise to a triviality.

4.4.2. A General Definition

Let us return to the example of the gambler discussed at the beginning of Section 4.4.1. Recall that X_t represents the gambler's capital at time t, i.e., after t bets have been placed. We observed that this sequence forms a martingale, and that $\mathbf{E}[X_i \mid X_0, \ldots, X_{i-1}] = X_{i-1}$. We would like to claim that this captures the fairness of the game in that, irrespective of the history and the gambler's strategy, the expected gain from each bet is exactly 0. However, this definition only says that the knowledge of the amounts won or lost in past bets does not help to predict the future. But what about other past information such as the exact set of cards dealt to various people, or the number of times a particular color or number shows up on the roulette table?

Specifically, suppose the gambler is playing roulette, and denote by Z_i the outcome on the roulette table during the ith bet; this random variable includes all information about the happenings on the roulette table, and not just the amount won or lost by this specific gambler. The gambler knows the value of Z_i and makes use of this knowledge in placing future bets. For example, if $Z_1, \ldots,$ Z_i indicate that the outcome on the table was always a red number, the gambler might then choose to bet on one of the red numbers the next time around. It is intuitively obvious that even this more refined knowledge of the past cannot help the gambler in the future, but the current definition of a martingale does not cater to the full generality of this intuition. The problem is that the conditioning is based on the amount of money lost or gained by the gambler from each bet, rather than the actual outcomes on the table. We would like a definition which gives

$$\mathbf{E}[X_i \mid Z_0, \ldots, Z_{i-1}] = X_{i-1}.$$

In fact, some authors define the notion of a martingale sequence X_0, X_1, ... with respect to a second sequence of random variables Z_0, Z_1, ... using precisely this equation.

Recall the definition of a σ-field $(\Omega, \mathbb{F})$ from Appendix C. In particular, we will consider only the probability spaces where the sample space Ω is a finite set and $\mathbb{F} = 2^{\Omega}$ contains all possible events in this sample space. Typically, we will assume that Ω is clear from the context and refer to $\mathbb{F}$ itself as a σ-field.

► **Definition 4.8:** Given the σ-field $(\Omega, \mathbb{F})$ with $\mathbb{F} = 2^{\Omega}$, a *filter* (sometimes also called a *filtration*) is a nested sequence $\mathbb{F}_0 \subseteq \mathbb{F}_1 \subseteq \cdots \subseteq \mathbb{F}_n$ of subsets of 2^{Ω} such that

1. $\mathbb{F}_0 = \{\emptyset, \Omega\}$
2. $\mathbb{F}_n = 2^{\Omega}$
3. for $0 \leq i \leq n$, $(\Omega, \mathbb{F}_i)$ is a σ-field

Let $\mathcal{E}_1$, $\mathcal{E}_2$, ... be any collection of events over the sample space Ω. The σ-field *generated* by these events is the minimal collection of subsets $\mathbb{F}$ that contains $\emptyset$ and each of $\mathcal{E}_1, \mathcal{E}_2, \ldots,$ and is closed under complement and union. If $\mathcal{E}_1$, $\mathcal{E}_2$, ... are disjoint events that partition Ω, then an event is in the generated σ-field $\mathbb{F}$ if and only if it can be expressed as the union of some subset of the events $\mathcal{E}_1$, $\mathcal{E}_2$, ...; we refer to the events $\mathcal{E}_1$, $\mathcal{E}_2$, ... as the elementary events in the σ-field $\mathbb{F}$.

An intuitive view of Definition 4.8 can now be obtained by associating with each $\mathbb{F}_i$ a partition of Ω into blocks B_1^i, B_2^i, ... such that the events B_j^i generate the σ-field $\mathbb{F}_i$. Furthermore, the partition associated with $\mathbb{F}_{i+1}$ is a refinement of the partition associated with $\mathbb{F}_i$, and $\mathbb{F}_0$ is generated by the trivial partition while $\mathbb{F}_n$ is generated by the partition of Ω into the singleton sets containing the sample points.

► **Example 4.7:** Consider a randomized algorithm $\mathcal{A}$ that uses a total of n random bits. The elementary events in the underlying sample space Ω are all possible 2^n choices of the n bits. For $0 \leq i \leq n$ and $w \in \{0,1\}^i$, let B_w denote the event that the first i random bits equal the bit string w. Let $\mathbb{F}_i$ be the σ-field generated by the partition of Ω into the blocks B_w, for $w \in \{0,1\}^i$. Then the sequence $\mathbb{F}_0$, $\mathbb{F}_1$, $\ldots$, $\mathbb{F}_n$ forms a filter. In the σ-field $\mathbb{F}_i$, the only valid events are the ones that depend on the values of the first i bits, and all such events are valid therein.

Recall that a random variable X over a probability space $(\Omega, \mathbb{F}, \mathbf{Pr})$ can be viewed as a function $X\colon \Omega \to \mathbb{R}$. In other words, given a sample $\omega \in \Omega$, the random variable takes on the value $X(\omega)$. Given a filter $\mathbb{F}_0, \ldots, \mathbb{F}_n$ with respect to this probability space, it is not clear that we can define the distribution of X relative to an arbitrary $\mathbb{F}_i$. This is because events of the type $\{X = x\}$ or $\{X \geq x\}$ may not exist in $\mathbb{F}_i$, although they will always be contained in the set $\mathbb{F}_n = \mathbb{F}$. We formalize this as follows.

► **Definition 4.9:** A random variable X is said to be $\mathbb{F}_i$*-measurable* if for each $x \in \mathbb{R}$, the event $\{X \leq x\}$ is contained in $\mathbb{F}_i$.

Since we are dealing only with the discrete case, the above definition could be made using the events $\{X = x\}$ rather than $\{X \leq x\}$.

► **Example 4.8:** Continuing with Example 4.7, consider the random variable X which is the parity of the n random bits used by algorithm $\mathcal{A}$. Clearly, X is $\mathbb{F}_i$-measurable only for $i = n$. On the other hand, let Y_j denote the number of ones in the first j random bits; then Y_j is $\mathbb{F}_i$-measurable for all $i \geq j$.

In general, a random variable X is $\mathbb{F}_i$-measurable if its value is constant over each block in the partition generating $\mathbb{F}_i$. Since the partitions generating the σ-fields in a filter are successively more refined, it follows that if X is $\mathbb{F}_i$-measurable, it is also $\mathbb{F}_j$-measurable for all $j \geq i$.

Suppose now that X is $\mathbb{F}_i$-measurable. What can we say about X with respect to the σ-field $\mathbb{F}_{i-1}$? An elementary event B in $\mathbb{F}_{i-1}$ is a block from its partition of Ω, and this is the union of some blocks $B_1, \ldots, B_r$ from the refined partition generating $\mathbb{F}_i$. Viewing X as a function over Ω, we know that X is constant over each of the blocks B_i, but is not necessarily so over B. However, the expected value of X is well-defined (and constant) over B. Thus, we can define $\mathbf{E}[X \mid \mathbb{F}_{i-1}]$ as the expected value of X conditioned on the events in $\mathbb{F}_{i-1}$. This conditional expectation is a random variable that can be viewed as a function into the reals from the blocks in the partition of $\mathbb{F}_{i-1}$. Moreover, this random variable is a constant if X is also $\mathbb{F}_{i-1}$-measurable. The converse is not always true; for example, when X is independent of the elementary events in $\mathbb{F}_{i-1}$, then $\mathbf{E}[X \mid \mathbb{F}_{i-1}]$ may be constant even though X is not $\mathbb{F}_{i-1}$-measurable.

There is nothing special about working with $\mathbb{F}_{i-1}$ in this discussion, and we can similarly define $\mathbf{E}[X \mid \mathbb{F}_j]$ for any j. The following is a general definition of conditional expectations.

► **Definition 4.10:** Let $(\Omega, \mathbb{F})$ be any σ-field, and Y any random variable that takes on distinct values on the elementary events in $\mathbb{F}$. Then $\mathbf{E}[X \mid \mathbb{F}] = \mathbf{E}[X \mid Y]$.

Notice that the conditional expectation $\mathbf{E}[X \mid Y]$ does not really depend on the precise value of Y on a specific elementary event. In fact, Y is merely an indicator of the elementary events in $\mathbb{F}$. Conversely, we can write $\mathbf{E}[X \mid Y] = \mathbf{E}[X \mid \sigma(Y)]$, where $\sigma(Y)$ is the σ-field generated by the events of the type $\{Y = y\}$, i.e., the smallest σ-field over which Y is measurable.

► **Example 4.9:** Consider the sample space Ω of all Americans, and let X be the random variable denoting the weight of a randomly chosen sample point. Consider the following filter with respect to Ω: $\mathbb{F}_0$ is the trivial σ-field; $\mathbb{F}_1$ is the σ-field generated by the partition of Ω into males and females; $\mathbb{F}_2$ is the σ-field generated by the refinement of the previous partition into sets corresponding to different heights; $\mathbb{F}_3$ is the further refinement of the partition based on age; and, $\mathbb{F}_4$ is the partition into singleton sets, each of which corresponds to an individual American.

Define $X_i = \mathbf{E}[X \mid \mathbb{F}_i]$, for $0 \leq i \leq 4$. Then $X_0 = \mathbf{E}[X]$ denotes the average weight of an American, X_1 is the average weight of Americans as a function of their sex, X_2 is the average weight of Americans as a function of their sex and height, and X_3 is the average weight of Americans as a function of their sex, height and age. Of course, $X_4 = X$ is the original random variable.

The "randomness" in these random variables results from the fact that a random American does not have a predetermined sex, weight, or age. For example, the sex of a random American is a random variable, and X_1 is a function of this random variable. Once the sex is known, the value of X_1 is completely determined.

► **Example 4.10:** Going back to Example 4.7, let T be the running time of the algorithm $\mathcal{A}$ on a specific input I. Clearly, T is a random variable whose value depends upon the specific values of the random bits used by $\mathcal{A}$. Observe that T is $\mathbb{F}_n$-measurable, but in general is not $\mathbb{F}_i$-measurable for any $i < n$.

Define the conditional expectation $T_i = \mathbf{E}[T \mid \mathbb{F}_i]$. Verify that $T_0 = \mathbf{E}[T]$ and that $T_n = T$. Also, T_i is a function of the values of the first i random bits denoting the expected running time for a random choice of the remaining $n - i$ bits. Given the value of the first i random bits, we may evaluate this random variable and obtain a constant. In fact, as will become clear shortly, the sequence $T_0, \ldots, T_n$ is a martingale.

We are now ready to give the more general definition of martingales.

► **Definition 4.11:** Let $(\Omega, \mathbb{F}, \mathbf{Pr})$ be a probability space with a filter $\mathbb{F}_0$, $\mathbb{F}_1$ .... Suppose that X_0, X_1, ... are random variables such that for all $i \geq 0$, X_i is $\mathbb{F}_i$-measurable. The sequence $X_0, \ldots, X_n$ is a martingale provided, for all $i \geq 0$,

$$\mathbf{E}[X_{i+1} \mid \mathbb{F}_i] = X_i.$$

As before, we can define martingale difference sequences using $Y_i = X_i - X_{i-1}$, and requiring that $\mathbf{E}[Y_{i+1} \mid \mathbb{F}_i] = 0$. We leave it as an exercise to verify that the definitions of Section 4.4.1 are special cases of Definition 4.11.

Suppose that X_0, X_1, ... is a martingale. Then it is intuitively clear that the sequence X_0, X_2, X_4, ... is also a martingale. This can be proved rigorously using the definition given above. The following theorem gives a general form of this result and the proof is left as Problem 4.18.

Theorem 4.12: *Any subsequence of a martingale is also a martingale (relative to the corresponding subsequence of the underlying filter).*

The following theorem gives us a way to construct a martingale sequence from any random variable. Martingales obtained in this manner are sometimes referred to as Doob martingales.

Theorem 4.13: *Let $(\Omega, \mathbb{F}, \mathbf{Pr})$ be a probability space, and let $\mathbb{F}_0, \ldots, \mathbb{F}_n$ be a filter with respect to it. Let X be any random variable over this probability space and define $X_i = \mathbf{E}[X \mid \mathbb{F}_i]$. Then, the sequence $X_0, \ldots, X_n$ is a martingale.*

The proof of this theorem is based on the following lemma, and these proofs are posed as Problems 4.19 and 4.20.

Lemma 4.14: *Let $(\Omega, \mathbb{F})$ and $(\Omega, \mathbb{G})$ be two σ-fields such that $\mathbb{F} \subset \mathbb{G}$. Then, for any random variable X, $\mathbf{E}[\mathbf{E}[X \mid \mathbb{G}] \mid \mathbb{F}] = \mathbf{E}[X \mid \mathbb{F}]$.*

► **Example 4.11:** Consider again the occupancy problem discussed in Exercise 4.9. There is an underlying filter $\mathbb{F}_0, \ldots, \mathbb{F}_n$ where $\mathbb{F}_t$ is the σ-field generated by the events corresponding to the placement of the first t balls. It then follows that the random variable Z_t equals $\mathbf{E}[Z \mid \mathbb{F}_t]$, and that the sequence $Z_0, \ldots, Z_m$ is a martingale.

► **Example 4.12 (Edge Exposure Martingale):** Let G be a random graph on the vertex set $V = \{1, \ldots, n\}$ obtained by independently choosing to include each possible edge with probability p. The underlying probability space is called $\mathcal{G}_{n,p}$. Arbitrarily label the $m = n(n-1)/2$ possible edges with the sequence 1, ..., m. For $1 \leq j \leq m$, define the indicator random variable I_j, which takes value 1 if edge j is present in G, and has value 0 otherwise. These indicator variables are independent and each takes value 1 with probability p.

Consider any real-valued function F defined over the space of all graphs, e.g., the clique number, which is defined as being the size of the largest complete subgraph. The *edge exposure martingale* is defined to be the sequence of random variables $X_0, \ldots, X_m$ such that

$$X_k = \mathbf{E}\left[F(G) \mid I_1, \ldots, I_k\right],$$

while $X_0 = \mathbf{E}[F(G)]$ and $X_m = F(G)$. The fact that this sequence of random variables is a (Doob) martingale is easy to verify – simply define the filter where $\mathbb{F}_k$ is the σ-field generated by the events corresponding to $I_1, \ldots, I_k$.

Exercise 4.10 (Vertex Exposure Martingale): In the same setting as in Example 4.12, we define a *vertex exposure martingale* as follows. For $1 \leq i \leq n$, let E_i be the set of all possible edges with both end-points in $\{1, \ldots, i\}$. Define Y_i as the (conditional) expectation of $F(G)$, conditioned by the knowledge of the indicator variables I_j for all $j \in E_i$. Show that the sequence $Y_0 = \mathbf{E}[F(G)]$, $Y_1, \ldots, Y_n$ forms a martingale.

At this point it is useful to review the intuition behind the above series of definitions. Recall the sequence $T_0, T_1, \ldots, T_n$ of conditional expectations of the running times defined in Example 4.10. This is a Doob martingale. We view the σ-field sequence $\mathbb{F}_0, \ldots, \mathbb{F}_n$ as representing the evolution of the algorithm, with each successive σ-field providing more information about the behavior of the algorithm (this information is determined by the values of the random bits given a fixed input). The random variables $T_0, \ldots, T_n$ represent the changing expectation of the running time as more information is revealed about the choice of the random bits. As we will see in the next section, if it can be shown that the absolute difference $|T_i - T_{i-1}|$ is suitably bounded, then the random variable T_n behaves like T_0 in the limit. In other words, the running time of the algorithm is sharply concentrated around its expected value provided that the choice of each individual random bit does not influence the behavior of the algorithm too dramatically. Similar arguments applied to the edge or vertex exposure martingales allow us to conclude that the value of a graph-theoretic function applied to a random graph is sharply concentrated around its expected value.

4.4.3. Martingale Tail Inequalities

In this section we present some inequalities for martingales that are reminiscent of the inequalities seen earlier for independent random variables. The reader may find it instructive to adapt these inequalities to the case of martingale difference sequences. The first inequality bears a resemblance to the Markov inequality.

Theorem 4.15 (Kolmogorov-Doob Inequality): *Let $X_0, X_1, \ldots$ be a martingale. Then, for any $\lambda > 0$,*

$$\mathbf{Pr}[\max_{0 \le i \le n} X_i \ge \lambda] \le \frac{\mathbf{E}[|X_n|]}{\lambda}.$$

The next bound is similar to the Chernoff bound for the sum of Poisson trials. Notice that X_0 equals $\mathbf{E}[X]$ in the case of a Doob martingale obtained from a random variable X, and so the following gives an exponentially small tail bound for X. It should also be noted that the tail bound does not require any knowledge of the expectation of X.

Theorem 4.16 (Azuma's Inequality): *Let $X_0, X_1, \ldots$ be a martingale sequence such that for each k,*

$$|X_k - X_{k-1}| \le c_k,$$

where c_k may depend on k. Then, for all $t \ge 0$ and any $\lambda > 0$,

$$\mathbf{Pr}[|X_t - X_0| \ge \lambda] \le 2\exp\left(-\frac{\lambda^2}{2\sum_{k=1}^t c_k^2}\right).$$

It is easy to see the connection between this bound and the Chernoff bound for the sum of Poisson trials. Let $Z_1, \ldots, Z_n$ be independent variables that take values 0 or 1 each with probability 1/2. The random variable $S = \sum_{i=1}^n Z_i$ has the binomial distribution with parameters n and $p = 1/2$. Define a martingale sequence $X_0, X_1, \ldots, X_n$ by setting $X_0 = \mathbf{E}[S]$, and, for $1 \le i \le n$, $X_i = \mathbf{E}[S \mid Z_1, \ldots, Z_i]$. It is clear that for $1 \le i \le n$, $|X_i - X_{i-1}| \le 1$, since fixing the value of any one variable Z_i can only affect the expected value of the sum S by at most 1. It follows that the probability that S deviates from its expected value $X_0 = \mathbf{E}[S] = n/2$ by more than λ is bounded by $2\exp(-\lambda^2/2n)$, a slightly weaker result than can be inferred from the Chernoff bound for binomial distributions.

The following is a useful corollary.

Corollary 4.17: *Let $X_0, X_1, \ldots$ be a martingale sequence such that for each k,*

$$|X_k - X_{k-1}| \le c$$

where c is independent of k. Then, for all $t \ge 0$ and any $\lambda > 0$,

$$\mathbf{Pr}[|X_t - X_0| \ge \lambda c\sqrt{t}] \le 2e^{-\lambda^2/2}.$$

The application of Azuma's inequality is sometimes called "the method of bounded differences." In applying this method to a martingale sequence, it is essential to set up the martingale in such a way as to guarantee the "bounded difference" property. We identify a general situation where this property is easily obtained.

► **Definition 4.12:** Let $f : \mathcal{D}_1 \times \cdots \times \mathcal{D}_n \rightarrow \mathbb{R}$ be a real-valued function with n arguments from possibly distinct domains. The function f is said to satisfy the *Lipschitz condition* if for any $x_1 \in \mathcal{D}_1, \ldots, x_n \in \mathcal{D}_n$, any $i \in \{1, \ldots, n\}$, and any $y_i \in \mathcal{D}_i$,

$$|f(x_1, \ldots, x_{i-1}, x_i, x_{i+1}, \ldots, x_n) - f(x_1, \ldots, x_{i-1}, y_i, x_{i+1}, \ldots, x_n)| \leq 1.$$

Basically, a function satisfies the Lipschitz condition if an arbitrary change in the value of any one argument does not change the value of the function by more than 1. Suppose we have a sequence of random variables $X_1, \ldots, X_n$, and a function $f(X_1, \ldots, X_n)$ defined over them such that f satisfies the Lipschitz condition. Define the Doob martingale sequence $Y_0, Y_1, \ldots, Y_n$ by setting $Y_0 = \mathbf{E}[f(X_1, \ldots, X_n)]$ and, for $1 \leq i \leq n$, $Y_i = \mathbf{E}[f(X_1, \ldots, X_n) \mid X_1, \ldots, X_i]$. It is easy to verify that the Lipschitz condition implies that for $1 \leq i \leq n$, $|Y_i - Y_{i-1}| \leq 1$. We can now employ the method of bounded differences. Of course, there is no particular reason to restrict the Lipschitz condition to absolute differences of 1, and we can appropriately generalize the definition to permit the exploitation of Azuma's inequality in its full generality.

The following exercise illustrates the power of the method of bounded differences.

Exercise 4.11: A legal coloring of a graph G with vertex set $V = \{1, \ldots, n\}$ is an assignment of colors (say, positive integers) to the vertices of the graph such that no two adjacent vertices receive the same color; the *chromatic number* of the graph G, denoted $\chi(G)$, is the minimum number of distinct colors needed for this purpose.

Consider a random graph G as defined in Example 4.12. Using the vertex exposure martingale from Exercise 4.10, employ the method of bounded differences to show that

$$\mathbf{Pr}[|\chi(G) - \mathbf{E}[\chi(G)]| > \lambda\sqrt{n}] \leq 2\exp(-\lambda^2/2).$$

Note that you will have to model the chromatic number as a function of n arguments, where the ith argument specifies the neighbors of vertex i from among the vertices $\{1, \ldots, i-1\}$, and then show that this satisfies the Lipschitz condition.

It may seem a bit surprising at first that such a sharp concentration result can be proved without even determining the expected value, but such is the power of martingale arguments.

4.4.4. Occupancy Revisited

We return to the occupancy problem and apply the martingale tail inequalities to it. We have m balls thrown independently and uniformly into n bins. Let Z denote the number of bins that remain empty. Our goal is to prove a sharp concentration result for Z.

Consider first the following easy application of the Lipschitz condition and the method of bounded differences. For $1 \leq i \leq m$, let the random variable X_i denote the bin chosen for the ith ball. We can view Z as a function $F(X_1, \ldots, X_m)$. It is easy to verify that this function satisfies the Lipschitz condition since moving any ball from one bin to another can change the number of empty bins by at most 1.

Exercise 4.12: Based on the Lipschitz condition deduced in the preceding paragraph, apply Corollary 4.17 to obtain that the probability that Z deviates from its expected value by more than λ is bounded by $2\exp(-\lambda^2/2m)$.

However, exploiting the full generality of Azuma's inequality allows us to derive a significantly stronger result for the case where $m \gg n$.

Theorem 4.18: *Let $r = m/n$, and Z be the number of empty bins when m balls are thrown randomly into n bins. Then,*

$$\mu = \mathbf{E}[Z] = n\left(1 - \frac{1}{n}\right)^m \sim ne^{-r}$$

and for $\lambda > 0$,

$$\mathbf{Pr}[|Z - \mu| \geq \lambda] \leq 2\exp\left(-\frac{\lambda^2(n - 1/2)}{n^2 - \mu^2}\right).$$

PROOF: The expected number of empty bins is studied in Problem 3.1. We concentrate here on proving the tail bound. Let time t refer to the point at which the first t balls have been thrown. Let $\mathbb{F}_t$ be the σ-field generated by the random choice of bins for the first t balls, i.e., the events corresponding to the state of the bins at time t. Let Z be the random variable denoting the number of empty bins at time m, and let $Z_t = \mathbf{E}[Z \mid \mathbb{F}_t]$ denote the conditional expectation of Z at time t. The random variables Z_0, Z_1, ..., Z_n form a martingale, with $Z_0 = \mathbf{E}[Z]$ and $Z_m = Z$.

Define $z(Y, t)$ as the expectation of Z given that Y bins are empty at time t. The probability that any of these bins does not receive a ball during the last $m - t$ time units is given by $(1 - 1/n)^{m-t}$. By linearity of expectations, we obtain that the number of these bins that remain empty at the end is given by

$$\begin{aligned} z(Y, t) &= \mathbf{E}[Z \mid Y \text{ bins are empty at time } t] \\ &= Y\left(1 - \frac{1}{n}\right)^{m-t}. \end{aligned}$$

Let the random variable Y_t denote the number of empty bins at time t. Then,

$$Z_{t-1} = z(Y_{t-1}, t-1) = Y_{t-1}\left(1 - \frac{1}{n}\right)^{m-t+1}.$$

Suppose we are at time $t - 1$ (i.e., in the σ-field $\mathbb{F}_{t-1}$.), so that the values of Y_{t-1} and Z_{t-1} are determined. At time t, there are two possibilities:

1. With probability $1 - Y_{t-1}/n$, the tth ball goes into a currently non-empty bin. Then, $Y_t = Y_{t-1}$, and

$$Z_t = z(Y_t, t) = z(Y_{t-1}, t) = Y_{t-1}\left(1 - \frac{1}{n}\right)^{m-t}.$$

2. With probability Y_{t-1}/n, the tth ball goes into a currently empty bin. Then, $Y_t = Y_{t-1} - 1$, and

$$Z_t = z(Y_t, t) = z(Y_{t-1} - 1, t) = (Y_{t-1} - 1)\left(1 - \frac{1}{n}\right)^{m-t}.$$

Let us now focus on the difference random variable $\Delta_t = Z_t - Z_{t-1}$. Corresponding to Z_t, the distribution of Δ_t (given the state at time $t-1$) can be characterized as follows.

1. With probability $1 - Y_{t-1}/n$, the value of Δ_t is

$$\begin{aligned}\delta_0 &= Y_{t-1}\left(1 - \frac{1}{n}\right)^{m-t} - Y_{t-1}\left(1 - \frac{1}{n}\right)^{m-t+1} \\ &= Y_{t-1}\left(1 - \frac{1}{n}\right)^{m-t}\left(1 - \left(1 - \frac{1}{n}\right)\right) \\ &= \frac{Y_{t-1}}{n}\left(1 - \frac{1}{n}\right)^{m-t}.\end{aligned}$$

2. With probability Y_{t-1}/n, the value of Δ_t is

$$\begin{aligned}\delta_1 &= (Y_{t-1} - 1)\left(1 - \frac{1}{n}\right)^{m-t} - Y_{t-1}\left(1 - \frac{1}{n}\right)^{m-t+1} \\ &= Y_{t-1}\left(1 - \frac{1}{n}\right)^{m-t}\left(1 - \left(1 - \frac{1}{n}\right)\right) - \left(1 - \frac{1}{n}\right)^{m-t} \\ &= -\left(1 - \frac{Y_{t-1}}{n}\right)\left(1 - \frac{1}{n}\right)^{m-t}.\end{aligned}$$

Observing that $0 \le Y_{t-1} \le n$, it follows that the value of the difference is bounded as follows:

$$-\left(1 - \frac{1}{n}\right)^{m-t} \le \Delta_t \le \left(1 - \frac{1}{n}\right)^{m-t}.$$

For $1 \le i \le m$, we set $c_t = \left(1 - \frac{1}{n}\right)^{m-t}$, and we have that $|Z_t - Z_{t-1}| \le c_t$. By a straightforward calculation,

$$\sum_{t=1}^{m} c_t^2 = \frac{1 - (1 - 1/n)^{2m}}{1 - (1 - 1/n)^2} = \frac{n^2 - \mu^2}{2n - 1}.$$

Invoking Azuma's inequality now gives the desired result. □

For large r, this tail bound is asymptotically equal to

$$2\exp\left(-\lambda^2/[n(1 - e^{-2r})]\right).$$

Compare this with a *heuristic* tail bound that can be obtained by using the fact that the distribution of Z approaches the normal distribution in the limit.

$$\mathbf{Pr}[|Z - \mu| \geq \lambda] \leq 2\exp\left(-\lambda^2 e^r / [2n(1 - e^{-r})]\right).$$

Notes

The general ideas behind the use of probability tail bounds derived from the moment generating function were presented by Chernoff [93]. The idea of using the moment-generating function to derive tail bounds is generally attributed to S. N. Bernstein [357]. The proof of Theorem 4.3 may be found in Raghavan's thesis [350]. Hoeffding [202] gives a similar bound that is insignificant unless $\mu\delta \gg n - \mu(1 + \delta)$. An alternative approach to proving these bounds in the setting of k-wise independent random variables is developed by Schmidt, Siegel, and Srinivasan [363]; they also provide general techniques for inferring Chernoff tail bounds for the sums of certain other types of correlated random variables. Janson [209] gives strong Chernoff-type bounds for the tail probabilites of the sum of Bernoulli variables that are either independent or negatively correlated. Hagerup and Rüb [189] give a detailed survey of Chernoff bounds on the tail of the binomial distribution.

Lower bounds on deterministic oblivious permutation routing such as Theorem 4.4 stemmed from work of Borodin and Hopcroft [75]; the form given here is an improvement due to Kaklamanis, Krizanc, and Tsantilas [225]. The power of randomization in solving the permutation routing problem was first demonstrated by Valiant [403]; his analysis was subsequently simplified by Valiant and Brebner [400]. Our presentation here is an adaptation of the latter analysis.

Notice that Valiant's scheme is an *oblivious randomized algorithm:* the route followed by a packet depends on its source, destination, and choice of random intermediate destination, but not on the sources, destinations, or choices of other packets. In Problem 4.11 below, we derive a result showing a limit on the performance of Valiant's scheme on an N-node, degree d network. In fact, such a lower bound has been shown for *any* randomized oblivious scheme by Borodin, Raghavan, Schieber, and Upfal [77], using the minimax principle of Section 2.2. In our model for parallel communication, we assumed that a node could transmit packets along all its links at each step. When the degree of a node is large, this assumption is unrealistic. Aleliunas [14] and Upfal [399] have addressed this problem by showing that there are bounded-degree networks for which Valiant's scheme routes any permutation in $O(\log N)$ steps with high probability.

The technique of solving a linear programming problem and then randomly rounding is due to Raghavan and Thompson [353]. Generalizations of the global wiring problem to more realistic settings and other details are also given in the paper by Raghavan and Thompson [353]. This technique has also been applied to the MAX-SAT problem in recent work of Goemans and Williamson [169]; we will explore this application in Section 5.2. Bertsimas and Vohra [58] explore randomized rounding in detail, applying the approach to unify approximation algorithms for a number of covering problems. Recent work of Goemans and Williamson [170], and Karger, Motwani, and Sudan [230], has extended the randomized rounding technique from linear programming relaxations to semi-definite programming relaxations, with applications to approximations for MAX-SAT and graph coloring. The idea here is to relax the requirement that the solutions be scalars, and instead allow them to be vectors in some high-dimensional space, thereby obtaining a polynomial-time solvable version of an ***NP***-hard problem; as before, randomization is then used to obtain approximate scalar solutions from the vector solutions. The article by Motwani, Naor, and Raghavan [314] surveys approximation algorithms for ***NP***-hard problems based on the randomized rounding of both linear programming and semi-definite programming relaxations of ***NP***-hard problems.

Several books on advanced probability theory cover martingales. Grimmett and Stirzaker [185] give an eminently readable description of martingale theory, as do Dubins and Savage [131]. The more measure-theoretic approach to martingales can be found in the books of Billingsley [61] and Feller [142, 143]. The reader seeking an in-depth understanding of martingales may refer to more advanced books such as those by Doob [129] and Hall and Heyde [191].

The tail inequality referred to as Azuma's inequality is due to Hoeffding [202] and Azuma [36]. The "method of bounded differences" has its origins in a paper by Maurey [300], and its various forms and applications are surveyed by McDiarmid [302]. The occupancy tail bound is due to Kamath, Motwani, Palem, and Spirakis [228], who provide a sequence of tail bounds for this problem. The classical results for occupancy problems can be found in the books by Johnson and Kotz [222] and Kolchin, Chistiakov, and Sevastianov [266]. While martingale arguments have been extremely useful for proving sharp concentration about expected values, it is only recently that they have attracted widespread attention in the computer science community, mainly due to the work of Shamir and Spencer [373] and Bollobás [70] on the chromatic number of random graphs; the book by Alon and Spencer [24] gives an excellent account of this work. Some notable successes in the application of martingales to computer science problems include: the work of McDiarmid and Reed [305] and Hayward and McDiarmid [198] on algorithms for building heaps; the results of McDiarmid and Hayward [304] on sharp concentration for quicksort; and the work of Aspnes and Waarts [34] on distributed algorithms for consensus.

Problems

4.1 Suppose you are given a biased coin that has $\mathbf{Pr}[\text{HEADS}] = p \geq a$, for some fixed a, without being given any other information about p. Devise a procedure for estimating p by a value $\tilde{p}$ such that you can guarantee that $\mathbf{Pr}[|p - \tilde{p}| > \epsilon p] < \delta$, for any choice of the constants $0 < a, \epsilon, \delta < 1$. Let N be the number of times you need to flip the biased coin to obtain the estimate. What is the smallest value of N for which you can still give this guarantee?

4.2 Let X be a random variable. Define the *kth factorial moment* of X, $\mathbf{E}[X^{\underline{k}}]$, as the expected value of $X^{\underline{k}} = X(X-1)\cdots(X-k+1)$. Let G_1 denote a random graph on n vertices where each edge independently is present with probability p, and G_2 denote a graph on n vertices that has m edges chosen uniformly at random. Let X_n denote the number of isolated vertices in G_1, and let Y_n be the number of isolated vertices in G_2. Consider the case $p = (\log n + c)/n$ and $m = n(\log n + c)/2$, for a real value c. Prove that $\mathbf{E}[X_n^{\underline{k}}]$ and $\mathbf{E}[Y_n^{\underline{k}}]$ are asymptotically equal to λ^k, where $\lambda = e^{-c}$.

4.3 For μ in the range $[1, \ln n]$, use (4.1) to obtain a closed-form upper bound for $\Delta^+(\mu, 1/n^2)$ (as a function of μ and n) that is within a constant factor of the best possible.

4.4 Let $X_1, X_2, \ldots, X_n$ be independent geometrically distributed random variables each having expectation 2 (each of the X_i is an independent experiment counting the number of tosses of an unbiased coin up to and including the first HEADS). Let $X = \sum_{i=1}^{n} X_i$ and δ be a positive real constant. Use moment generating functions and the Chernoff technique to derive the best upper bound you can on $\mathbf{Pr}[X > (1+\delta)(2n)]$.

4.5 The result of Theorem 4.2 bounds the probability of the sum of Poisson trials deviating far *below* its expectation. Use this to give a bound on the probability of the sum of independent geometric random variables deviating *above* its expectation, thus providing an alternative approach to that in Problem 4.4.

4.6 (**Hoeffding's Bound [202]**). Suppose $Y_1, \ldots, Y_n$ are independent Poisson trials such that $\mathbf{Pr}[Y_i = 1] = p_i$. Let $Y = \sum_{i=1}^{n} Y_i$, $\mu = \mathbf{E}[Y] = \sum_{i=1}^{n} p_i$ and $p = \mu/n$. Our goal is to show that from the standpoint of deviations from the mean, the worst case is when the p_i's are all equal. Let X be the sum of n independent Bernoulli trials each having probability p of assuming the value 1. Then, for any $a \geq \mu + 1$ and any $b \leq \mu - 1$, show that

$$\mathbf{Pr}[Y \geq a] \leq \mathbf{Pr}[X \geq a],$$

and

$$\mathbf{Pr}[Y \leq b] \leq \mathbf{Pr}[X \leq b].$$

4.7 (Due to W. Hoeffding [202].) This problem deals with a useful generalization of the Hoeffding bound in Problem 4.6.

(a) A function $f : \mathbb{R} \to \mathbb{R}$ is said to be *convex* if for any x_1, x_2 and $0 \leq \lambda \leq 1$, the following inequality is satisfied:

$$f(\lambda x_1 + (1-\lambda)x_2) \leq \lambda f(x_1) + (1-\lambda) f(x_2).$$

Show that the function $f(x) = e^{tx}$ is convex for any $t > 0$. What can you say when $t \leq 0$?

(b) Let Z be a random variable that assumes values in the interval $[0, 1]$, and let $p = \mathbf{E}[Z]$. Define the Bernoulli random variable X such that $\mathbf{Pr}[X = 1] = p$ and $\mathbf{Pr}[X = 0] = 1 - p$. Show that for any convex function f, $\mathbf{E}[f(Z)] \leq \mathbf{E}[f(X)]$.

(c) Let $Y_1, \ldots, Y_n$ be independent and identically distributed random variables over $[0, 1]$, and define $Y = \sum_{i=1}^{n} Y_i$. Using parts (a) and (b), derive upper and lower tail bounds for the random variable Y using the Chernoff bound technique. In particular, show that

$$\mathbf{Pr}[Y - \mathbf{E}[Y] > \delta] \leq \exp(-2\delta^2/n).$$

Remark: While the results in this problem hold for continuous random variables, they may be a bit easier to prove in the case where Z, $Y_1, \ldots, Y_n$ take on a discrete set of values in the interval $[0, 1]$. Also, it should be easy to generalize this to distributions defined over arbitrary intervals $[l, h]$. See also Problem 4.21.

4.8 Consider a ***BPP*** algorithm that has an error probability of $1/2 - 1/p(n)$, for some polynomially bounded function $p(n)$ of the input size n. Using the Chernoff bound on the tail of the binomial distribution, show that a polynomial number of independent repetitions of this algorithm suffice to reduce the error probability to $1/2^n$.

4.9 Consider now the following variant of the bit-fixing algorithm. Each packet randomly orders the bit positions in the label of its source, and then corrects the mismatched bits in that order. Show that there is a permutation for which with high probability this algorithm requires $2^{\Omega(n)}$ steps to complete the routing.

4.10 Suppose we run Valiant's scheme on an N-node network in which every node is of degree d; each packet first goes to a random destination chosen uniformly from all the nodes and then on to its final destination. Show that the expected number of steps for the completion of the first phase is

$$\Omega\left(\frac{\log N}{d \log \log N} + \frac{\log N}{\log d}\right).$$

4.11 The *lattice approximation problem* is an extension of the set-balancing problem (Example 4.5). As before, we are given an $n \times n$ matrix $\boldsymbol{A}$ all of whose entries are 0 or 1. In addition, we are given a column vector $\boldsymbol{p}$ with n entries, all of which are in the interval $[0, 1]$. We wish to find a column vector $\boldsymbol{q}$ with n entries, all of which are from the set $\{0, 1\}$, so as to minimize $||\boldsymbol{A}(\boldsymbol{p} - \boldsymbol{q})||_\infty$. We think of the vector $\boldsymbol{q}$ as an "integer approximation" to the given real vector $\boldsymbol{p}$, in the sense that $\boldsymbol{Aq}$ is close to $\boldsymbol{Ap}$ in every component. This has applications to approximating certain integer programs given solutions to their linear programming relaxations, along the lines of Section 4.3. Derive a bound on $||\boldsymbol{A}(\boldsymbol{p} - \boldsymbol{q})||_\infty$ assuming that $\boldsymbol{q}$ were derived from $\boldsymbol{p}$ using randomized rounding.

4.12 Consider the global wiring problem of Section 4.3. We wish to approximate the best possible solution *without* the restriction that only one-bend routes are used. Adapt the approach in Section 4.3 to devise an algorithm running in time polynomial in the number of gates and nets, achieving an approximation similar to that in Theorem 4.8.

4.13 The *set-cover* problem is the following: given sets $S_1, \ldots, S_n$ over a universe U, find the smallest set $T \subseteq U$ such that for $1 \leq i \leq n$, $T \cap S_i \neq \emptyset$. An alternative formulation of this problem is the following: given a 0-1 matrix $\boldsymbol{M}$, find a 0-1 column vector $\boldsymbol{c}$ such that the dot product of each row of $\boldsymbol{M}$ with $\boldsymbol{c}$ is positive while minimizing $||\boldsymbol{c}||_1$. The matrix $\boldsymbol{M}$ has n rows, and the ith row is the incidence vector of the set S_i.

Given a matrix $\boldsymbol{M}$, let $C(\boldsymbol{M})$ denote the size of the smallest set-cover for $\boldsymbol{M}$. Let n be the number of rows in $\boldsymbol{M}$. Show that we can adapt the technique of linear programming followed by randomized rounding to find a set-cover of size $O(\log n)$ times $C(\boldsymbol{M})$.

4.14 Show that the **RandQS** algorithm of Chapter 1 runs in time $O(n \log n)$ with high probability.

4.15 Redesign the parameters of the **LazySelect** algorithm of Chapter 3 and invoke the Chernoff bound to show that with high probability it finds the kth smallest of n elements in $n + k + \sqrt{n}\log^{O(1)} n$ steps, with probability $1 - o(1)$.

4.16 Prove Lemmas 4.9 and 4.10. Also, formulate and prove their generalizations to the case where the conditioning is done on more than one random variable. Finally, using these, prove Lemma 4.11.

4.17 Prove Theorem 4.12.

4.18 Prove Lemma 4.14.

4.19 Using Lemma 4.14, prove Theorem 4.13.

4.20 Derive the tail bounds described in Problem 4.4.7 (c) by applying Azuma's inequality (Corollary 4.17) to the Doob martingale sequence obtained from Y by setting $X_0 = \mathbf{E}[Y]$ and, for $1 \leq i \leq n$, $X_i = \mathbf{E}[Y \mid Y_1, \ldots, Y_i]$. How does this bound compare with the one obtained in Problem 4.7?

4.21 Prove Azuma's inequality (Theorem 4.16) for the case where $c_k = 1$ for all k. Note that this is the same as Corollary 4.17 with $c = 1$. Do you see how to generalize this to the case of arbitrary c_k's? (**Hint:** Concentrate on the upper tail bound, since the lower tail bound can be obtained by negating the random variables. Consider the martingale difference sequence Y_1, Y_2, ... obtained by setting $Y_i = X_i - X_{i-1}$, and note that $X_t = \sum_{i=1}^{t} Y_i$. You can essentially mimic the proof of Theorem 4.1, but be careful to use conditional expectations and the martingale property in going from the analog of equation (4.2) to that of equation (4.3). Since the random variables Y_i could have arbitrary distributions over the interval $[-1, 1]$, you will also have to make use of an argument similar to that in Problem 4.7.)

4.22 (Due to A. Kamath, R. Motwani, K. Palem, and P. Spirakis [228].) Consider again the issue of tail bounds on the number of empty bins studied in Theorem 4.18. In this setting, let I_i be the indicator variable whose value is 1 if and only if bin i is empty, and define $Z = \sum_{i=1}^{n} I_i$ as the number of empty bins. Define $p = \mathbf{E}[I_i] = (1 - 1/n)^m$, and let I_i' be mutually independent Bernoulli random variables that take value 1 with probability p and value 0 with probability $1 - p$; note that the sum $Y = \sum_{i=1}^{n} I_i'$ has the binomial distribution with parameters n and p.

(a) Show that for all $t \geq 0$, $\mathbf{E}[e^{tZ}] \leq \mathbf{E}[e^{tY}]$. Conclude that any Chernoff bound on the upper tail of Y's distribution also applies to the upper tail of Z's distribution, even though the Bernoulli variables I_i are not mutually independent. (The point is that their correlation is negative and only helps to reduce the tail probability.) How does the resulting bound on the upper tail of Z's distribution compare with the bound given in Theorem 4.18?

(b) Can you show that for all $t < 0$, $\mathbf{E}[e^{tZ}] \leq \mathbf{E}[e^{tY}]$? Repeat the exercise in part (a) for the lower tail.

CHAPTER 5

The Probabilistic Method

In this chapter we will study some basic principles of the *probabilistic method*, a combinatorial tool with many applications in computer science. This method is a powerful tool for demonstrating the existence of combinatorial objects. We introduce the basic idea through several examples drawn from earlier chapters, and follow that by a detailed study of the maximum satisfiability (MAX-SAT) problem. We then introduce the notion of expanding graphs and apply the probabilistic method to demonstrate their existence. These graphs have powerful properties that prove useful in later chapters, and we illustrate these properties via an application to probability amplification.

In certain cases, the probabilistic method can actually be used to demonstrate the existence of algorithms, rather than merely combinatorial objects. We illustrate this by showing the existence of efficient non-uniform algorithms for the problem of oblivious routing. We then present a particular result, the Lovász Local Lemma, which underlies the successful application of the probabilistic method in a number of settings. We apply this lemma to the problem of finding a satisfying truth assignment in an instance of the SAT problem where each variable occurs in a bounded number of clauses. While the probabilistic method usually yields only randomized or non-uniform deterministic algorithms, there are cases where a technique called the method of conditional probabilities can be used to devise a uniform, deterministic algorithm; we conclude the chapter with an exposition of this method for derandomization.

5.1. Overview of the Method

There are two recurrent ideas in the probabilistic method.

1. Any random variable assumes at least one value that is no smaller than its expectation, and at least one value that is no greater than its expectation. We know of many intuitive versions of this principle in real life – for instance, if we are told that the average annual income of theoretical computer scientists is

\$20,000, we know that there is at least one theoretical computer scientist whose income is \$20,000 or greater.

2. If an object chosen randomly from a universe satisfies a property with positive probability, then there must be an object in the universe that satisfies that property. For instance, if we were told that a ball chosen randomly from a bin is red with probability 1/3, then we know that the bin contains at least one red ball.

While these ideas may seem too obvious to be of much use, they turn out to give us a surprising amount of power. The power comes from our ability to recast counting arguments in the language of probability, and then bring to bear the tools of probability theory. In fact, we have already seen instances of the probabilistic method implicitly at work earlier in this book. Below we review some examples from earlier chapters, and then proceed to study some new techniques. This chapter is not meant to be a comprehensive guide to the probabilistic method in combinatorics, but rather a study of some ideas that have proved useful in randomized algorithms.

► **Example 5.1:** Theorem 1.2 asserts that for any set of n disjoint line segments in the plane, the expected size of the autopartition found by the **RandAuto** algorithm is $O(n \log n)$. From this we may conclude that for any set of n disjoint line segments in the plane, there *is always* an autopartition of size $O(n \log n)$. This follows directly from the fact that if we were to run the **RandAuto** algorithm, the random variable defined to be the size of the autopartition can assume a value that is no more than its expectation; thus, there is an autopartition of this size on any instance.

Our second example comes from the game tree evaluation problem of Section 2.1.

► **Example 5.2:** Any algorithm for game tree evaluation that produces the correct answer on every instance develops a *certificate of correctness:* for each instance, it can exhibit a set of leaves whose values together guarantee the value it declares is the correct answer. By Theorem 2.1, the expected number of leaves inspected by the algorithm of Section 2.1 on any instance of $T_{2,k}$ is at most $n^{0.793}$, where $n = 2^{2k}$. It follows that on any instance of $T_{2,k}$, there is a set of $n^{0.793}$ leaves whose values certify the value of the root for that instance. Note that we assert the existence of such a certificate with certainty, even though the technique used for establishing it was probabilistic. (Problem 5.2 describes a stronger version of this result.)

Our final example from an earlier chapter is the set-balancing problem described in Example 4.5.

► **Example 5.3:** We saw that for every $n \times n$ 0-1 matrix A, for a randomly chosen vector $\boldsymbol{b} \in \{-1, +1\}^n$, we have $||A\boldsymbol{b}||_\infty \leq 4\sqrt{n \ln n}$, with probability at least $1-2/n$.

From this we may conclude that for every such matrix A, there always *exists* a vector $b \in \{-1,+1\}^n$ such that $||Ab||_\infty \leq 4\sqrt{n \ln n}$.

The examples above show that the probabilistic method consists of two stages. First, we design a "thought experiment" in which a random process plays a role. In the case of set-balancing, for example, the thought experiment consists of independently and equiprobably assigning to each component of b either the value $+1$ or the value -1. The second part consists of analyzing the random experiment and then drawing a conclusion independent of the particular experiment.

Let us consider another example concerning the problem of finding a *large cut* in a graph. Given an undirected graph $G(V,E)$ with n vertices and m edges, we wish to partition the vertices of G into two sets A and B so as to maximize the number of edges (u,v) such that $u \in A$ and $v \in B$. This problem is sometimes referred to as the *max-cut* problem. The problem of finding an optimal max-cut is ***NP***-hard; in contrast, the min-cut problem studied in Section 1.1 has a polynomial time algorithm.

Theorem 5.1: *For any undirected graph $G(V,E)$ with n vertices and m edges, there is a partition of the vertex set V into two sets A and B such that*

$$|\{(u,v) \in E \mid u \in A \text{ and } v \in B\}| \geq m/2.$$

PROOF: Consider the following experiment. Each vertex of G is independently and equiprobably assigned to either A or B.

For an edge (u,v), the probability that its end-points are in different sets is $1/2$. By linearity of expectation, the expected number of edges with end-points in different sets is thus $m/2$. It follows that there must be a partition satisfying the theorem. □

We have viewed the process of partitioning the vertices of G as a thought experiment that yields the results mentioned. However, we could as well view it as a randomized algorithm. This would then require a further analysis bounding the probability that the algorithm fails to find a good partition on a given execution. The main difference between a thought experiment in the probabilistic method and a randomized algorithm is the end that each yields. When we use the probabilistic method, we are only concerned with showing that a combinatorial object exists; thus, we are content with showing that a favorable event occurs with non-zero probability. With a randomized algorithm, on the other hand, efficiency is an important consideration – we cannot tolerate a miniscule success probability. For instance, if we were only able to show that the experiment used in the proof of Theorem 5.1 succeeded with probability 2^{-n} in finding a cut of size $m/2$, we would be unable to derive from it an efficient randomized algorithm for finding a large cut. In this case however, the *expected* size of the cut is $m/2$ and so random partitioning can be viewed as an efficient randomized algorithm.

One of the questions we will deal with in this chapter and others is the following: having shown the existence of a combinatorial object using the probabilistic method, can we find the object efficiently? The answer to this general question varies widely. In some cases it is affirmative, and we have a deterministic polynomial-time algorithm that finds the combinatorial object whose existence is guaranteed by the probabilistic method. In others, we instead have a randomized polynomial-time algorithm that works with high probability. In yet others, we have a deterministic or randomized algorithm, but one that is non-uniform. And finally, we have instances where we know of no efficient algorithm for finding the object in question.

5.2. Maximum Satisfiability

We turn to the satisfiability problem defined in Section 1.5.2: given a set of m clauses in conjunctive normal form over n variables, decide whether there is a truth assignment for the n variables that satisfies all the clauses. We may assume without loss of generality that no clause contains both a literal and its complement, since such clauses are satisfied by *any* truth assignment. Consider the following optimization version of the satisfiability problem: rather than decide whether there is an assignment that satisfies *all* the clauses, we instead seek an assignment that *maximizes* the number of satisfied clauses. This problem, called the MAX-SAT problem, is known to be ***NP***-hard, but the following simple probabilistic argument shows that for any set of m clauses, there is an assignment to the input variables that satisfies at least $m/2$ clauses. Note that this is the best possible universal guarantee, since the instance may consist of the two clauses x and $\overline{x}$, in which case no better guarantee is possible.

Theorem 5.2: *For any set of m clauses, there is a truth assignment for the variables that satisfies at least $m/2$ clauses.*

PROOF: Suppose that each variable is set to TRUE or FALSE independently and equiprobably. For $1 \leq i \leq m$, let $Z_i = 1$ if the ith clause is satisfied and 0 otherwise. For any clause containing k literals, the probability that it is *not* satisfied by this random assignment is 2^{-k}, since this event takes place if and only if each literal gets a specific value, and the (distinct) literals in a clause are assigned independent values. This implies that the probability that a clause with k literals is satisfied is $1 - 2^{-k} \geq 1/2$, implying that $\mathbf{E}[Z_i] \geq 1/2$ for all i. The expected number of clauses satisfied by this random assignment is $\sum_{i=1}^{m} \mathbf{E}[Z_i] \geq m/2$. Thus, there exists at least one assignment of values to the variables for which $\sum_{i=1}^{m} Z_i \geq m/2$. □

Exercise 5.1: Consider the following weighted version of the MAX-SAT problem. Each clause has a positive real weight, and the goal is to maximize the sum of the weights of the satisfied clauses. Generalizing Theorem 5.2, show that there is a truth assignment that satisfies clauses the sum of whose of weights is at least half of the total clause weight.

This result holds regardless of whether the instance has a satisfying assignment. Let us continue with the MAX-SAT problem, in which our goal is to maximize the number of clauses that are satisfied. This problem being ***NP***-hard, we seek approximation algorithms. It turns out that variants of the probabilistic existence proof of Theorem 5.2 can actually be turned into approximation algorithms; we explore this theme for the remainder of this section.

Given an instance I, let $m_*(I)$ be the maximum number of clauses that can be satisfied, and let $m_A(I)$ be the number of clauses satisfied by an algorithm A. The *performance ratio* of an algorithm A is defined to be the infimum (over all instances I) of $m_A(I)/m_*(I)$. If A achieves a performance ratio of α, we call it an *α-approximation algorithm.* For a randomized algorithm A, the quantity $m_A(I)$ may be a random variable, in which case we replace $m_A(I)$ by $\mathbf{E}[m_A(I)]$ in the definition of the performance ratio. Note that unlike the satisfiability problem (in which we seek to satisfy all clauses), we may choose to leave some clauses unsatisfied in the MAX-SAT problem. Indeed this may be inevitable, for instance, as in the case of a set of contradictory clauses. Thus, our definition requires us to satisfy a number of clauses close to the best possible for the instance at hand, rather than satisfying all m clauses.

We now give a simple randomized algorithm that achieves a performance ratio of $3/4$. Before we begin, we observe that the proof of Theorem 5.2 actually yields a randomized $1/2$-approximation algorithm. In fact, we can say more: the procedure in the proof of Theorem 5.2 yields an algorithm whose performance guarantee is $1 - 2^{-k}$, provided every clause contains at least k literals.

It follows that we have a randomized $3/4$-approximation algorithm for instances of MAX-SAT in which every clause has at least two literals. It appears that the bottleneck for achieving a performance ratio of $3/4$ stems from clauses consisting of a single literal. We now give a second algorithm that performs especially well when there are many clauses consisting of single literals. We then argue that on any instance, one of the two algorithms will yield a randomized $3/4$-approximation. Thus, given an instance, we run both algorithms and take the better of the two solutions.

The algorithm we describe will not be entirely new to us: we have already encountered a variant in our study of the wiring problem in Section 4.3. The idea again is to formulate the problem as an integer linear program, solve the linear programming relaxation, and then to round using the randomized rounding technique of Section 4.3. With each clause C_j in the instance, we associate an indicator variable $z_j \in \{0,1\}$ in the integer linear program to indicate whether or not that clause is satisfied. For each variable x_i, we use an indicator variable y_i in the integer linear program to indicate the value assumed by that variable; thus $y_i = 1$ if the variable x_i is set TRUE, and $y_i = 0$ otherwise. Let C_j^+ be the set of indices of variables that appear in the uncomplemented form in clause C_j, and C_j^- be the set of indices of variables that appear in the complemented form in clause C_j. We may then formulate the MAX-SAT problem as follows: maximize

$$\sum_{j=1}^{m} z_j$$

where

$$y_i, z_j \in \{0, 1\} \ (\forall i \text{ and } j) \tag{5.1}$$

subject to

$$\sum_{i \in C_j^+} y_i + \sum_{i \in C_j^-} (1 - y_i) \geq z_j \ (\forall j). \tag{5.2}$$

The inequalities (5.1) ensure that a clause is deemed to be true (by assigning value 1 to its variable) only if at least one of the literals in that clause is assigned the value 1. Since $z_j = 1$ when clause C_j is satisfied, the objective function $\sum_j z_j$ counts the number of satisfied clauses. As in Section 4.3, we solve the relaxation linear program in which we relax the integrality constraints (5.2), i.e., we allow y_i and z_j to assume real values in the interval $[0, 1]$. Let $\widehat{y}_i$ be the value obtained for variable y_i by solving this linear program, and let $\widehat{z}_j$ be the value obtained for z_j. Clearly $\sum_j \widehat{z}_j$ is an upper bound on the number of clauses that can be satisfied in this instance. We first show that using randomized rounding, we obtain a truth assignment with which the expected number of clauses satisfied is at least $(1 - 1/e) \sum_j \widehat{z}_j$. This is already an improvement over the guarantee we get from Theorem 5.2; we will then show that for any instance, the number of clauses satisfied by the better of these two solutions is at least $(3/4) \sum_j \widehat{z}_j$.

For randomized rounding, each variable y_i is independently set to 1 (corresponding to x_i being set to TRUE) with probability $\widehat{y}_i$. For any positive integer k, let β_k denote $1 - (1 - 1/k)^k$. We will first show that for a clause C_j with k literals, the probability that it is satisfied by randomized rounding is at least $\beta_k \widehat{z}_j$. Noting that $\beta_k \geq 1 - 1/e$ for all positive integers k, and using linearity of expectation, we infer that the expected number of clauses satisfied by randomized rounding is at least $(1 - 1/e) \sum_j \widehat{z}_j$.

Lemma 5.3: *Let C_j be a clause with k literals. The probability that it is satisfied by randomized rounding is at least $\beta_k \widehat{z}_j$.*

PROOF: Since we are focusing on a single clause C_j, we may assume without loss of generality that all the variables contained in it appear in uncomplemented form. Moreover, we may assume that it is of the form $x_1 \vee \cdots \vee x_k$. By constraint (5.1) in the linear program,

$$\widehat{y}_1 + \cdots + \widehat{y}_k \geq \widehat{z}_j.$$

Clause C_j remains unsatisfied by randomized rounding only if every one of the variables y_i is rounded to 0. Since each variable is rounded independently, this occurs with probability $\prod_{i=1}^{k}(1 - \widehat{y}_i)$. It remains to show that

$$1 - \prod_{i=1}^{k}(1 - \widehat{y}_i) \geq \beta_k \widehat{z}_j.$$

The expression on the left is minimized when $\widehat{y}_i = \widehat{z}_j/k$ for all i. Therefore, it suffices to show that $1 - (1 - z/k)^k \geq \beta_k z$ for all positive integers k and $0 \leq z \leq 1$. Since $f(x) = 1 - (1 - x/k)^k$ is a concave function, to show that it is never less than a linear function $g(x)$ over the interval $[0, 1]$, it suffices to verify the inequality at the end-points $x = 0$ and $x = 1$ (see Problem 5.4). Applying this principle to the linear function $g(z) = \beta_k z$, the lemma follows. □

By Lemma 5.3 and from linearity of expectation we have:

Theorem 5.4: *Given an instance of MAX-SAT, the expected number of clauses satisfied by linear programming and randomized rounding is at least* $(1-1/e)$ *times the maximum number of clauses that can be satisfied on that instance.*

While Theorem 5.4 represents an improvement over Theorem 5.2, we will in fact be able to do even better. We have studied two randomized algorithms MAX-SAT: one that rounded each variable to 1 with probability 1/2, and a second that used the solutions to the linear program as a basis for randomized rounding. Figure 5.1 may help the reader appreciate the dependencies of these two algorithms on the clause length k.

k	$1 - 2^{-k}$	β_k
1	0.5	1.0
2	0.75	0.75
3	0.875	0.704
4	0.938	0.684
5	0.969	0.672

Figure 5.1: Performance of the two algorithms as a function of k.

We now argue that on any instance, one of the algorithms is a 3/4-approximation algorithm. Given any instance, we run both algorithms and choose the better solution. Let n_1 denote the expected number of clauses that are satisfied when each variable is independently set to 1 with probability 1/2 (corresponding to the procedure that yields Theorem 5.2). Let n_2 denote the expected number of clauses that are satisfied when we use the linear programming followed by randomized rounding (corresponding to Theorem 5.4).

Theorem 5.5:

$$\max\{n_1, n_2\} \geq \frac{3}{4} \sum_j \widehat{z}_j.$$

PROOF: It suffices to show that $(n_1 + n_2)/2 \geq (3/4) \sum_j \widehat{z}_j$. Letting S^k denote the set of clauses that contain k literals, we know that

$$n_1 = \sum_k \sum_{C_j \in S^k} (1 - 2^{-k}) \geq \sum_k \sum_{C_j \in S^k} (1 - 2^{-k})\widehat{z}_j. \tag{5.3}$$

By Lemma 5.3, we have

$$n_2 \geq \sum_k \sum_{C_j \in S^k} \beta_k \widehat{z}_j. \tag{5.4}$$

Thus

$$\frac{n_1 + n_2}{2} \geq \sum_k \sum_{C_j \in S^k} \frac{(1 - 2^{-k}) + \beta_k}{2} \widehat{z}_j.$$

An easy calculation shows that $(1 - 2^{-k}) + \beta_k \geq 3/2$ for all k, so that we have

$$\frac{n_1 + n_2}{2} \geq \frac{3}{4} \sum_k \sum_{C_j \in S^k} \widehat{z}_j = \frac{3}{4} \sum_j \widehat{z}_j.$$

□

5.3. Expanding Graphs

We now turn to a classic application of the probabilistic method, one that shows the existence of a class of graphs known as *expanding graphs*. Expanding graphs have found many uses in computer science and in telephone switching networks, and we will encounter them again in Chapters 6 and 11.

Intuitively, an expanding graph is a graph in which the number of neighbors of any set of vertices S is larger than some positive constant multiple of $|S|$. The following is a definition of a particular type of expanding graph called an OR-concentrator. It is important to keep in mind that several alternate definitions have been used in the literature; while they are similar in spirit, the precise definition varies (see for instance the slightly different definition used in Chapter 6). Recall that in a graph $G(V, E)$ for any set $S \subseteq V$, the set of neighbors of S is $\Gamma(S) = \{w \in V \mid \exists v \in S, (v, w) \in E\}$.

► **Definition 5.1:** An (n, d, α, c) OR-concentrator is a bipartite multigraph $G(L, R, E)$, with the independent sets of vertices L and R each of cardinality n, such that

1. Every vertex in L has degree at most d.
2. For any subset S of vertices from L such that $|S| \leq \alpha n$, there are at least $c|S|$ neighbors in R.

In most applications, it is desirable to have d as small as possible and c as large as possible. Of particular interest is the study of OR-concentrators in which α, c, and d are constants fixed independently of n, with $c > 1$. These are rather stringent requirements and it may seem quite surprising at first that such graphs can be constructed. Indeed, finding explicit constructions of such OR-concentrators is a non-trivial task, so we focus on the easier problem of demonstrating their existence. We will use the probabilistic method to show that a random graph chosen from a suitable probability space has a positive probability of being an $(n, 18, 1/3, 2)$ OR-concentrator. The particular constants in the proof are somewhat arbitrary, and the reader may easily adapt the proof to study other combinations of d, α, and c.

Theorem 5.6: *There is an integer n_0 such that for all $n > n_0$, there is an $(n, 18, 1/3, 2)$ OR-concentrator.*

PROOF: We give most of the proof in terms of general d, c, and α, pinning these constants down toward the end of the proof. Consider a random bipartite graph on the vertices in L and R, in which each vertex of L chooses its neighbors by sampling (with replacement) d vertices independently and uniformly from R. Since the sampling is with replacement, a vertex of L may choose a vertex in R more than once; we discard all but one copy of such multiple edges. Let $\mathcal{E}_s$ denote the event that a subset of s vertices of L has fewer than cs neighbors in R. We will first bound $\mathbf{Pr}[\mathcal{E}_s]$, and then sum $\mathbf{Pr}[\mathcal{E}_s]$ over the values of s no larger than αn to obtain an upper bound on the probability that the random graph fails to be an OR-concentrator with the parameters we seek.

Fix any subset $S \subseteq L$ of size s, and any subset $T \subseteq R$ of size cs. There are $\binom{n}{s}$ ways of choosing S, and $\binom{n}{cs}$ ways of choosing T. The probability that T contains all of the at most ds neighbors of the vertices in S is $(cs/n)^{ds}$. Thus, the probability of the event that all the ds edges emanating from some s vertices of L fall within any cs vertices of R is bounded as follows,

$$\mathbf{Pr}[\mathcal{E}_s] \leq \binom{n}{s}\binom{n}{cs}\left(\frac{cs}{n}\right)^{ds}.$$

Invoking the identity $\binom{n}{k} \leq (ne/k)^k$ from Proposition B.2 (Appendix B), we obtain

$$\begin{aligned}\mathbf{Pr}[\mathcal{E}_s] &\leq \left(\frac{ne}{s}\right)^s\left(\frac{ne}{cs}\right)^{cs}\left(\frac{cs}{n}\right)^{ds} \\ &= \left[\left(\frac{s}{n}\right)^{d-c-1} e^{1+c}c^{d-c}\right]^s.\end{aligned}$$

Simplifying for $\alpha = 1/3$ and using $s \leq \alpha n$, we have

$$\begin{aligned}\mathbf{Pr}[\mathcal{E}_s] &\leq \left[\left(\frac{1}{3}\right)^{d-c-1} e^{1+c}c^{d-c}\right]^s \\ &\leq \left[\left(\frac{c}{3}\right)^d (3e)^{c+1}\right]^s.\end{aligned}$$

Using $c = 2$ and $d = 18$, we have

$$\mathbf{Pr}[\mathcal{E}_s] \leq \left[\left(\frac{2}{3} \right)^{18} (3e)^3 \right]^s .$$

Let $r = (2/3)^{18}(3e)^3$, and note that $r < 1/2$. We obtain that

$$\sum_{s \geq 1} \mathbf{Pr}[\mathcal{E}_s] \leq \sum_{s \geq 1} r^s = \frac{r}{1 - r} < 1,$$

and the desired result follows. □

The reader may easily see that by bounding the probabilities $\mathbf{Pr}[\mathcal{E}_s]$ carefully, we can in fact show that our random graph has a fairly good (rather than merely non-zero) probability of being an $(n, 18, 1/3, 2)$ OR-concentrator. However, even if we were to generate a random graph and argue that it has a very high probability of being an OR-concentrator, we still do not know of an efficient way of verifying that the graph generated is indeed an OR-concentrator with the given parameters.

This is true of the verification of the expanding property of graphs for a variety of definitions of expansion, some of which we will encounter in Chapter 6. For instance, in Chapter 6 we will define and use a class of expanding graphs known as *expanders*. This indicates that the Monte Carlo algorithm implicit in the preceding discussion cannot be turned into a Las Vegas algorithm.

For many applications of expanding graphs, such a Monte Carlo guarantee is unacceptable – for instance, a telephone company may be uncomfortable that the network it plans to build may by chance be inadequate. Unfortunately, it is considerably harder to give a succinct "formula" or a deterministic algorithm that, given n, always generates such an expanding graph. We do have "explicit constructions" that will, given n, generate OR-concentrators with guaranteed bounds for d, α, and c; but these bounds are somewhat weaker than the bounds attainable using the probabilistic method (the Notes section contains more information on these).

This is another recurrent theme in the probabilistic method: whereas the existence proof can give strong (often the best possible) bounds for a combinatorial object, the version that can be constructed efficiently may be much weaker. We will see another instance of this in Section 5.5.

5.3.1. Probability Amplification

We now make use of an expanding bipartite graph to build on the idea of two-point sampling used in Section 3.4. Consider an $\boldsymbol{RP}$ algorithm A for deciding whether input strings x belong to a language L. Given x, A picks a random number r from the range $\mathbb{Z}_n = \{0, \ldots, n-1\}$, for a suitable choice of a prime n, and computes a binary value $A(x, r)$ with the following properties:

- If $x \in L$, then $A(x, r) = 1$ for at least half the possible values of r (we call these values of r the *witnesses* for x).

- If $x \notin L$, then $A(x,r) = 0$ for all possible choices of r.

By the two-point sampling approach of Section 3.4, we know that using $2\log n$ random bits to sample two numbers randomly from the range $\{0,\ldots,n-1\}$, we can achieve an error probability of less than $1/t$ in t (non-independent) trials of the algorithm A on a given input x. In this section, we will describe a way of achieving an error probability close to $1/n^{\log n}$ using only $\log^2 n$ random bits. The naive use of $\log^2 n$ bits to pick $\log n$ random numbers in the range $\{0,\ldots,n-1\}$ only yields a failure probability of $1/n$, so the scheme we will describe can be thought of as achieving "probability amplification."

We first establish the existence of an expanding graph that will serve our purpose, and then proceed to describe its application to amplifying randomness.

Theorem 5.7: *For n sufficiently large, there is a bipartite graph $G(L,R,E)$ with $|L| = n$, $|R| = 2^{\log^2 n}$ such that:*

1. *Every subset of $n/2$ vertices of L has at least $2^{\log^2 n} - n$ neighbors in R.*
2. *No vertex of R has more than $12\log^2 n$ neighbors.*

PROOF: Consider a random graph in which each vertex of L independently and uniformly chooses $d = 2^{\log^2 n}(4\log^2 n)/n$ neighbors in R. As before, the choices are made with replacement, i.e., a vertex of L may choose a vertex of R as neighbor more than once. We will show that this random graph violates each of the two properties with probability at most 1/2. It follows that with positive probability this random graph satisfies both properties, and we are done.

Following the reasoning in our proof of Theorem 5.6, the probability that there is a set of $n/2$ vertices in L having fewer than $2^{\log^2 n} - n$ neighbors in R is at most

$$\binom{n}{n/2}\binom{2^{\log^2 n}}{n}\left(1 - \frac{n}{2^{\log^2 n}}\right)^{dn/2}.$$

Using as before the upper bound for binomial coefficients from Proposition B.2 (Appendix B) together with the fact that $1 - n/2^{\log^2 n} \leq \exp(-n/2^{\log^2 n})$, it follows that the probability that property 1 is violated is (considerably) less than 1/2.

For property 2, we note that the expected number of neighbors for a vertex in R is $4\log^2 n$; the Chernoff bound (4.10) now shows that the probability of exceeding $12\log^2 n$ neighbors is less than $(e/3)^{12\log^2 n}$. Since R contains $2^{\log^2 n}$ vertices, this probability is small enough to guarantee that the probability that property 2 is violated at any vertex in R is also (considerably) less than 1/2. □

We return to probability amplification. Theorem 5.7 only guarantees the existence of a graph with the desired properties; in the sequel we will assume that we have an explicit graph with these properties. Of course, this graph has a super-polynomial number of vertices and it may not seem possible to perform polynomial-time computations based on its structure. However, we do not need an explicit representation of the graph; all we need is a polynomial-time neighborhood algorithm that can compute the neighbors of any given vertex in

R; we assume that the graph is represented by means of such a neighborhood algorithm. As we will see later, in Section 6.7, there do exist expanding graphs for which such neighborhood algorithms are known.

Given $\log^2 n$ bits of randomness, we use them to index a vertex in R, say v. Next, we use the neighborhood algorithm to identify the neighbors of v in L, which we denote $r_1, \ldots, r_k$. We then compute $A(x, r_i)$ for $1 \leq i \leq k$; note that $k \leq 12 \log^2 n$. If all k invocations of A return 0, we declare that x does not belong to L; else we declare that x does belong to L.

If $x \notin L$, our answer will be correct. But if $x \in L$, what is the probability that we fail to detect it using our procedure? The set of witnesses for x is a set of at least $n/2$ vertices of L. We err only if the vertex of R we choose is not a neighbor of any of the witnesses. By Theorem 5.7, the fraction of such vertices in R is at most $n/2^{\log^2 n}$, no matter how the witnesses are distributed in R. Thus using $\log^2 n$ random bits, we achieve a failure probability of at most $n/n^{\log n}$.

The reader may argue that the extra randomness we obtain is from the randomness "built into" the graph. However, we note that once we have built such a graph, it may be used over and over again for executions on arbitrarily many inputs x. More interestingly, it can be used on any ***RP*** algorithm, since the procedure works for any choice of $n/2$ witnesses in L. Thus the "one-time" randomness built into the graph serves as a reservoir that we can tap over and over again, for probability amplification. We know of no explicit construction for such graphs, nor do we know of an efficient procedure for verifying that a random graph has the properties we desire.

In Section 6.8 we will describe an alternative strategy for performing probability amplification without any of the drawbacks discussed above. Not surprisingly, this new scheme is also based on the use of expanders. But there we will use explicitly constructed expanders that have explicit polynomial time algorithms for determining the neighbors of a vertex.

5.4. Oblivious Routing Revisited

We turn now to another aspect of the probabilistic method. In the examples we have seen, the probabilistic method is used to prove the existence of a combinatorial object: an autopartition that is small, a vector $\boldsymbol{b}$ with certain properties in the case of set-balancing, or an expanding graph. The probabilistic method can also be used to design algorithms. We study one example here and will encounter other examples later in the book.

Let us return to the problem of oblivious permutation routing on the hypercube, studied in Section 4.2. In this section we focus on the number of random bits used by the randomized oblivious algorithm in Section 4.2. We first give a lower bound that suggests that the algorithm of Section 4.2 uses many more random bits than necessary. We then use the probabilistic method to show the existence of a randomized algorithm using (within a constant factor) the optimal number of random bits.

Comparing the performance of the randomized algorithm (the result of Exercise 4.6) with the negative result of Theorem 4.4, we find that our randomized oblivious algorithm achieves an expected running time that no deterministic oblivious algorithm can achieve. Given that randomness is absolutely necessary to beat the lower bound of $\sqrt{N/n}$ steps for deterministic oblivious algorithms (Theorem 4.4), we can ask the following question: how much randomness is actually needed to achieve an algorithm with an expected running time of $O(n)$?

We formulate the question more precisely. A randomized oblivious algorithm for permutation routing is a probability distribution on a set of deterministic oblivious routing algorithms. Each deterministic oblivious algorithm for an N-node network is a set of N^2 routes, one for each source–sink pair. Every randomized oblivious algorithm can be expressed as a pair of sets, $\{A_1, \ldots, A_R\}$ and $\{p_1, \ldots, p_R\}$, where each A_j is a deterministic oblivious algorithm and p_j is the probability that we use A_j on a run of the randomized algorithm. Naturally, $\sum_{j=1}^{R} p_j = 1$. For instance, in the randomized oblivious scheme of Section 4.2, each algorithm A_j is a set of possible routes of the form $i \to \sigma(i) \to d(i)$. There are N choices of $\sigma(i)$ for each i and $d(i)$.

Theorem 4.4 can be interpreted as follows: with zero bits, the expected running time of the algorithm is $\Omega(\sqrt{N/n})$. At the other extreme, the randomized algorithm of Section 4.2 has expected running time $O(n) = O(\log N)$ with Nn random bits; but are so many bits necessary?

Theorem 5.8: *Consider any randomized oblivious algorithm for permutation routing on the hypercube with $N = 2^n$ nodes. If this algorithm uses k random bits, then its expected running time is $\Omega(2^{-k}\sqrt{N/n})$.*

PROOF: We have observed that any randomized oblivious algorithm is a probability distribution on deterministic oblivious algorithms. Since only k random bits are used, at least one of these deterministic algorithms is chosen with probability at least 2^{-k}, on any execution. Denote this deterministic algorithm by A_1. By the lower bound of Theorem 4.4, there is an input that requires time $\Omega(\sqrt{N/n})$ on the deterministic algorithm A_1. Feed this input to the randomized algorithm; with probability 2^{-k}, the randomized algorithm chooses A_1 and takes time $\Omega(\sqrt{N/n})$. Thus, the expected running time of the randomized algorithm is $\Omega(2^{-k}\sqrt{N/n})$. □

Corollary 5.9: *Any randomized oblivious algorithm for permutation routing on the hypercube with $N = 2^n$ nodes must use $\Omega(n)$ random bits in order to achieve expected running time $O(n)$.*

The randomized oblivious algorithm of Section 4.2 uses about N times the number of bits of randomness deemed necessary by Corollary 5.9. Can we match this lower bound? The answer comes from an application of the probabilistic method.

Theorem 5.10: *For every n, there exists a randomized oblivious scheme for permutation routing on a hypercube with $N = 2^n$ nodes that uses $3n$ random bits and runs in expected time at most $15n$.*

PROOF: We will say that a set $\mathcal{B} = \{B_1, B_2, \ldots, B_t\}$ of deterministic oblivious permutation routing algorithms on the N-node hypercube is an *efficient N-scheme* if, for any instance, the expected number of steps using a randomly chosen algorithm from $\mathcal{B}$ is at most $15n$. To prove the theorem, we will show that for every $N = 2^n$, there is an efficient N-scheme for $t = N^3$.

The algorithm of Section 4.2 randomly chooses one of N^N possible deterministic algorithms on an execution: there are N sources, and we may choose from N possible intermediate destinations for each. Let us denote these N^N deterministic algorithms by A_j, for $1 \leq j \leq N^N$. On an N-node network, there are $N!$ distinct possible instances of permutation routing, one for each permutation on $\{1, \ldots, N\}$. For an instance π_i, $1 \leq i \leq N!$, call the deterministic algorithm A_j *good* if A_j routes π_i in $14n$ or fewer steps, and *bad* otherwise. By Theorem 4.7, for any particular instance π_i of the permutation routing problem, a fraction of at most $1/N$ of the algorithms A_j are bad. Which algorithms are bad may differ from instance to instance – we only know that for any particular instance π_i, at most $1/N$ of the A_j's are bad.

Consider now the following experiment: sample N^3 indices $i_1, i_2, \ldots, i_{N^3}$ independently and uniformly at random (for simplicity, with replacement) from the range $\{1, 2, \ldots, N^N\}$. We show that the set of deterministic algorithms $\mathcal{A} = \{A_{i_1}, \ldots, A_{i_{N^3}}\}$ is an efficient N-scheme with positive probability. From this, we will conclude that an efficient N-scheme exists for every $N = 2^n$.

For any instance π_i, a fraction of at most $1/N$ of the algorithms $A_1, \ldots, A_{N^N}$ is bad; thus the expected number of algorithms in $\mathcal{A}$ that are bad for π_i is at most $N^3(1/N) \leq N^2$. Let the indicator variable X_j be 1 if A_j is bad, for $1 \leq j \leq N^3$, and 0 otherwise. Thus $\mathbf{E}[\sum_j X_j] \leq N^2$. Since the X_j are independent Poisson trials, we may apply the Chernoff bound (the form in Exercise 4.1) to obtain that the probability that more than $2N^2$ of the algorithms in $\mathcal{A}$ are bad for π_i is less than $\exp(-N^2/4)$. Let $\mathcal{B}_i$ denote the *bad event* that more than $2N^2$ algorithms in $\mathcal{A}$ are bad for π_i. Then, for $n \geq 2$ (or $N \geq 4$),

$$\begin{aligned}
\mathbf{Pr}[\cup_{i=1}^{N!} \mathcal{B}_i] &\leq \sum_{i=1}^{N!} \mathbf{Pr}[\mathcal{B}_i] \\
&\leq N! \times \exp(-N^2/4) \\
&< 1,
\end{aligned}$$

where the last inequality follows from an application of Stirling's Formula (Proposition B.1, Appendix B).

Therefore, with positive probability, no more than $2N^2$ of the algorithms in $\mathcal{A}$ are bad for any instance π_i of permutation routing on the N-node hypercube. This means that there *exists* a subset of N^3 algorithms from $\{A_1, \ldots, A_{N^N}\}$ with the property that at most $2N^2$ algorithms in this subset are bad for any instance π_i; let us denote this subset by $\mathcal{B} = \{B_1, B_2, \ldots, B_{N^3}\}$.

It is easy to see now that $\mathcal{B}$ is an efficient N-scheme: on any instance π_i, a randomly chosen algorithm from $\mathcal{B}$ fails to route π_i within $14n$ steps with probability at most $2N^2/N^3 = 2/N$. By reasoning similar to that in Exercise 4.6, the expected number of steps using an algorithm randomly chosen from $\mathcal{B}$ is less than $15n$. □

We have used the probabilistic method to show the existence of a randomized algorithm meeting the lower bound of Corollary 5.9. It is important that the reader keep the two levels of randomization in the proof distinct – the first was to show probabilistically that a certain combinatorial object (the set $\mathcal{B}$) existed, and the second was to study the effect of choosing an algorithm at random from $\mathcal{B}$.

Does Theorem 5.10 settle the problem of designing a randomized algorithm for permutation routing using few random bits? It does not, for the following reason. The construction in the proof of Theorem 5.10 is not uniform: given N, we do not know how to obtain $\mathcal{B}$ efficiently. The reader is invited to draw a parallel between this result and that presented in Section 2.3.

5.5. The Lovász Local Lemma

The *Lovász Local Lemma* is a tool in the probabilistic method that has found many applications in extremal graph theory, in Ramsey theory, and in the theory of random graphs. Applications to algorithms and computer science have been fewer, so far, but it appears that this powerful technique will surely prove useful.

Suppose that we have n events, each of which occurs with probability at most $1/2$. In an instance of the probabilistic method, each of the n events may correspond to one of n ways in which the probabilistic experiment could fail. If the events were independent, we could then assert that with probability at least 2^{-n}, none of these events occurs. The Lovász Local Lemma generalizes this notion to the case where each of these events is independent of all but a small number of other events. In this section we give the lemma and apply it to show that any instance of SAT meeting certain conditions always has a satisfying assignment. We then give an algorithm that *finds* a satisfying assignment. Let $\mathcal{E}_i$, $1 \leq i \leq n$ be events in a probability space. Recall that $\mathcal{E}_i$ is mutually independent of a set S of events if $\mathbf{Pr}[\mathcal{E}_i | \bigcap_{j \in T} \mathcal{E}_j] = \mathbf{Pr}[\mathcal{E}_i]$, where T is any subset of events (or their complements) from S. The main device in establishing Lemma 5.11 below is a digraph we call the *dependency graph* G, in which there is a vertex representing each event $\mathcal{E}_i$. An event $\mathcal{E}_i$ is mutually independent of all other events $\mathcal{E}_j$ such that $(\mathcal{E}_i, \mathcal{E}_j)$ is not an edge of the graph. Before proceeding with the lemma, the reader may attempt the following exercise to better understand the notion of a dependency graph.

Exercise 5.2: Suppose that the events $\mathcal{E}_i$ are pairwise independent. What can you say about the structure of a dependency graph? Is the dependency graph always unique?

Lemma 5.11 (Lovász Local Lemma): *Let $G(V, E)$ be a dependency graph for events $\mathcal{E}_1, \ldots, \mathcal{E}_n$ in a probability space. Suppose that there exist $x_i \in [0, 1]$ for $1 \le i \le n$ such that*

$$\mathbf{Pr}[\mathcal{E}_i] \le x_i \prod_{(i,j)\in E} (1 - x_j).$$

Then

$$\mathbf{Pr}[\cap_{i=1}^n \overline{\mathcal{E}_i}] \ge \prod_{i=1}^n (1 - x_i).$$

PROOF: Let S denote a subset of the indices from $\{1, \ldots, n\}$. We first establish by induction on $k = |S|$ that for any S and for any i such that $i \notin S$,

$$\mathbf{Pr}[\mathcal{E}_i \mid \cap_{j\in S} \overline{\mathcal{E}}_j] \le x_i.$$

The base case, $S = \emptyset$, follows from our assumption on the probabilities $\mathbf{Pr}[\mathcal{E}_i]$. For the inductive step, we let $S_1 = \{j \in S : (i, j) \in E\}$, and let $S_2 = S \backslash S_1$. By the definition of conditional probability,

$$\mathbf{Pr}[\mathcal{E}_i \mid \cap_{j\in S} \overline{\mathcal{E}}_j] = \frac{\mathbf{Pr}[\mathcal{E}_i \cap (\cap_{j\in S_1} \overline{\mathcal{E}}_j) \mid \cap_{m\in S_2} \overline{\mathcal{E}}_m]}{\mathbf{Pr}[\cap_{j\in S_1} \overline{\mathcal{E}}_j \mid \cap_{m\in S_2} \overline{\mathcal{E}}_m]}. \tag{5.5}$$

We can bound the numerator of (5.5) from above as follows:

$$\begin{aligned} \mathbf{Pr}[\mathcal{E}_i \cap (\cap_{j\in S_1} \overline{\mathcal{E}}_j) \mid \cap_{m\in S_2} \overline{\mathcal{E}}_m] &\le \mathbf{Pr}[\mathcal{E}_i \mid \cap_{m\in S_2} \overline{\mathcal{E}}_m] \\ &= \mathbf{Pr}[\mathcal{E}_i] \\ &\le x_i \prod_{(i,j)\in E} (1 - x_j), \end{aligned}$$

since $\mathcal{E}_i$ is mutually independent of $\{\mathcal{E}_m : m \in S_2\}$. Also, we can bound the denominator from below as follows. Suppose that $S_1 = \{j_1, \ldots, j_r\}$. If $r = 0$, then the denominator is 1; for $r > 0$, we invoke the induction hypothesis:

$$\begin{aligned} \mathbf{Pr}[\overline{\mathcal{E}}_{j_1} \cap \cdots \cap \overline{\mathcal{E}}_{j_r} \mid \cap_{m\in S_2} \overline{\mathcal{E}}_m] &= (1 - \mathbf{Pr}[\mathcal{E}_{j_1} \mid \cap_{m\in S_2} \overline{\mathcal{E}}_m]) \\ &\quad \cdots (1 - \mathbf{Pr}[\mathcal{E}_{j_r} \mid \overline{\mathcal{E}}_{j_1} \cap \cdots \cap \overline{\mathcal{E}}_{j_{r-1}} \cap_{m\in S_2} \overline{\mathcal{E}}_m]) \\ &\ge (1 - x_{j_1}) \cdots (1 - x_{j_r}) \ge \prod_{(i,j)\in E} (1 - x_j). \end{aligned}$$

It follows that $\mathbf{Pr}[\mathcal{E}_i \mid \cap_{j\in S} \overline{\mathcal{E}}_j] \le x_i$. To complete the proof, we note that

$$\begin{aligned} \mathbf{Pr}[\cap_{i=1}^n \overline{\mathcal{E}_i}] &= (1 - \mathbf{Pr}[\mathcal{E}_1])(1 - \mathbf{Pr}[\mathcal{E}_2 \mid \overline{\mathcal{E}}_1]) \cdots (1 - \mathbf{Pr}[\mathcal{E}_n \mid \cap_{i=1}^{n-1} \overline{\mathcal{E}_i}]) \\ &\ge \prod_{i=1}^n (1 - x_i). \end{aligned}$$

□

Corollary 5.12: *Let $\mathcal{E}_1, \ldots, \mathcal{E}_n$ be events in a probability space, with* $\mathbf{Pr}[\mathcal{E}_i] \leq p$ *for all i. If each event is mutually independent of all other events except for at most d, and if $ep(d+1) \leq 1$, then* $\mathbf{Pr}[\cap_{i=1}^n \overline{\mathcal{E}}_i] > 0$.

We now apply Corollary 5.12 to show that an instance of SAT meeting certain conditions must have a satisfying truth assignment. Consider an instance of the k-CNF problem: we are given a CNF formula in which each clause contains k literals. This is also known as the k-SAT problem. Suppose further that each of the n variables appears (complemented or uncomplemented) in at most $2^{k/50}$ clauses. Let m denote the number of clauses.

Consider a random truth assignment of values to the variables, in which each variable is independently fixed to be 0 or 1 with probability $1/2$. For $1 \leq i \leq m$, let $\mathcal{E}_i$ denote the event that the ith clause is not satisfied by this random assignment. Since each clause contains k literals, we have $\mathbf{Pr}[\mathcal{E}_i] = 2^{-k}$, for $1 \leq i \leq m$. The event $\mathcal{E}_i$ that the ith clause is not satisfied is independent of all other events $\mathcal{E}_j$, except those j such that clause i and clause j share at least one variable. The number of clauses j that share a variable with a specific clause i cannot exceed the total number of clauses containing the variables that appear in clause i, and this is at most $k2^{k/50}$. We now apply Corollary 5.12 with $d = k2^{k/50}$, and conclude that with positive probability the random truth assignment satisfies all m clauses. Thus, there must be a satisfying truth assignment for any instance of SAT meeting these conditions.

Corollary 5.12 merely tells us that a random assignment is good with positive probability, but this probability may be miniscule. We may have to try the random assignment many times before we succeed in finding one that satisfies all m clauses. We now describe a Las Vegas randomized algorithm that runs in time polynomial in m (but not in k), yielding a satisfying truth assignment. From here on, the reader should think of k (and therefore $d = k2^{k/50}$) as a constant fixed independent of m, when we use the big-oh, O(), notation below.

Let G denote the dependency graph – each clause corresponds to a vertex of G, and two vertices are adjacent in G if the corresponding clauses share one or more variables. Note that if clause C_1 contains literal x_1, and clause C_2 contains literal $\overline{x}_1$, then the vertices C_1 and C_2 are adjacent. We know that every vertex of G has at most d neighbors.

At any point in the algorithm, some variables will have been fixed to 0 or 1, while others will remain unspecified as yet; initially, all variables are unspecified. The algorithm consists of two stages; the first stage will fix values for some of the variables and defer the rest to the second stage. In the first stage of the algorithm, we proceed sequentially through the variables, fixing each equiprobably to 0 or 1. We call a clause *dangerous* if both the following conditions hold:

1. $k/2$ literals of the clause C_i have been fixed.
2. C_i is not satisfied yet.

After fixing each variable, we identify any clause C_i that has turned dangerous.

For any dangerous clause, we defer its remaining unspecified variables to the second stage, skipping over them in the sequential random assignment. At the end of the first stage, we say that a clause has *survived* if it is not satisfied by the variables fixed in the first stage.

For the second stage we need only consider the variables that were unspecified at the end of the first stage, and the clauses that survived. A clause C_i can survive the first stage for one of two reasons:

1. It became dangerous, or
2. All variables corresponding to its unspecified literals were deferred because other clauses containing these variables (and, hence, adjacent to C_i) became dangerous.

Therefore, a clause C_i may survive as a result of any one of up to $d+1$ clauses becoming dangerous – C_i itself, and its d neighbors. Every clause that survived has at least $k/2$ unspecified variables.

Exercise 5.3: Apply Corollary 5.12 to show that there is a truth assignment of the deferred variables that satisfies all the surviving clauses. (Again, consider a random assignment.)

The second stage will find a truth assignment guaranteed by Exercise 5.3. The probability that any particular clause becomes dangerous during the first stage is at most $2^{-k/2}$, since exactly $k/2$ of its literals have their values fixed, and none of these random values satisfy the clause. This implies that the probability that a clause survives is at most $(d+1)2^{-k/2}$.

Consider the subgraph of G induced by the vertices corresponding to the surviving clauses. In Lemma 5.13 below, we will show that with high probability, all connected components of this induced subgraph of G have size $O(\log m)$. Notice that two surviving clauses from different connected components of this subgraph cannot share a deferred variable. Therefore, the deferred variables can be uniquely assigned to distinct connected components of the subgraph of G induced by the surviving clauses. For any particular connected component, the total number of deferred variables in its clauses must be $O(\log m)$; in time polynomial in m, we can enumerate the $2^{O(\log m)}$ truth assignments for these variables until we that one that satisfies all clauses in this component. The second stage consists of repeating this process independently for each connected component, giving a polynomial time algorithm for assigning values to the deferred variables so as to satisfy all surviving clauses.

Lemma 5.13: *With probability $1-o(1)$, all connected components of G induced by the clauses that survive the first stage have size at most $z \log m$, for a fixed constant z.*

PROOF: Consider a collection of clauses $C_1, \ldots, C_r$ such that every pair of these has distance at least 4 in G. Each clause C_i survives only if at least one of the

$d+1$ clauses at distance at most 1 from it turns dangerous during the first stage. For each C_i, let D_i be any one dangerous clause at distance at most 1 from it. Since the C_i's are at distance 4 from each other, the D_i's must be distinct.

There are at most $(d+1)^r$ possible ways of choosing the clauses $D_1, \ldots, D_r$. Since each of the clauses $D_1, \ldots, D_r$ is at distance at most 1 from some clause in the set $C_1, \ldots, C_r$, they must be at distance at least 2 from each other and hence have disjoint sets of variables. The probability that $D_1, \ldots, D_r$ all become dangerous is at most $2^{-rk/2}$. Thus, for a set of r clauses every pair of which is distance at least 4 apart in G, the probability that they all survive is at most

$$[(d+1)2^{-k/2}]^r. \tag{5.6}$$

We must bound the probability that some connected subgraph of G of size exceeding $z \log m$ survives. To this end we introduce a graph-theoretic device known as a *4-tree*. Call a subset T of clauses a 4-*tree* if the following two properties hold:

1. The distance in G between every pair of these clauses is at least 4.
2. If we form a new graph in which two clauses are adjacent if their distance in G is exactly 4, T is connected.

We first bound the number of 4-trees of size r and use this to bound the probability that a large 4-tree survives. By arguing that a large connected subgraph of G must contain a large 4-tree, we will finally conclude it is unlikely that a large connected subgraph survives.

Let us define a new graph G_4 as follows: there is a vertex for each clause, and two vertices are adjacent in G_4 if their distance in G is 4. Each vertex of G_4 has $O(d^4)$ neighbors. The number of 4-trees of size r in G is no more than the number of connected subgraphs in G_4 of size r. Problem 5.7 considers a general graph-theoretic bound on the number of connected subgraphs of a given size in a graph. The particular result from there that we now use is: the number of subgraphs of G_4 of size r is at most

$$amd^{8r} \tag{5.7}$$

for some constant a, and this is an upper bound on the number of 4-trees of size r in G. Multiplying (5.6) and (5.7), we conclude that the probability that *any* 4-tree of size larger than $b \log m$ survives the first round is $o(1)$, for a suitably large constant b.

What does this tell us about the probability that some connected subgraph of G of size exceeding $z \log m$ survives? For any connected subgraph in G there is a maximal 4-tree T, together with at most $3d^3 - 1$ other vertices within distance 3 of a vertex of T. Thus the size of this subgraph is at most $3|T|d^3$. We conclude that the probability of survival of any connected subgraph of size exceeding $3bd^3 \log m$ is $o(1)$. □

If the first stage results in a connected component larger than this bound, we repeat it; the expected number of repetitions is less than 2. Thus, we assume that we enter the second stage of the algorithm with every surviving connected component having size $O(\log m)$. The number of unspecified variables associated with each of these components is also $O(\log m)$, and in time polynomial in m we can find values for them that satisfy all the clauses. Since no variable is shared by two or more components, we can treat each component in isolation. Clearly the expected running time of this algorithm is polynomial in m.

Theorem 5.14: *The above algorithm finds a satisfying truth assignment for any instance of k-SAT containing m clauses in which each variable is contained in at most $2^{k/50}$ clauses, in expected time polynomial in m.*

It is worth noting that the constant 50 above can be strengthened somewhat; Problem 5.9 explores this further. The degree to which it can be strengthened depends on our aim: if we only wish to show that a satisfying truth assignment exists, we can obtain a better constant than if we actually want to show that the algorithm above will succeed in finding one. This is a feature of all known algorithms that, in polynomial time, find objects whose existence is guaranteed by the Lovász Local Lemma: the constants required for the algorithms are somewhat weaker than those for the corresponding existence proofs.

5.6. The Method of Conditional Probabilities

In Section 2.3 we saw that a randomized computation could sometimes be "derandomized." The derandomization in Section 2.3, however, led to a non-uniform deterministic algorithm. In this section, we will examine a technique that can derandomize certain randomized algorithms uniformly. We illustrate this method, known as the *method of conditional probabilities*, using the set-balancing problem of Example 4.5.

Recall the definition of the set-balancing problem: we are given an $n \times n$ matrix A all of whose entries are 0 or 1. We wish to find a column vector $\boldsymbol{b} \in \{-1, +1\}^n$, so as to minimize $||A\boldsymbol{b}||_\infty$. In Example 4.5, we used the following randomized algorithm: each entry of $\boldsymbol{b}$ is independently and equiprobably chosen from $\{-1, +1\}$. We argued that with probability at least $1 - 2/n$, this algorithm finds a vector $\boldsymbol{b}$ for which $||A\boldsymbol{b}||_\infty \leq 4\sqrt{n \ln n}$. We now describe the method of conditional probabilities, and use it to obtain a deterministic algorithm that finds a vector $\boldsymbol{b}$ for which $||A\boldsymbol{b}||_\infty \leq 4\sqrt{n \ln n}$.

Let us view the randomized algorithm as a computation tree. This tree is a complete binary tree of height n (there are $n + 1$ nodes on any root-leaf path, including the root and the leaf). The *level* of a node is its distance from the root. The computation begins at the root. Each node at the ith level is labeled by a distinct string from $\{-1, +1\}^i$, and corresponds to a setting of $b_1, \ldots, b_i$ in the obvious fashion. From any node whose level is less than n, the computation proceeds equiprobably to one of its children. If a node is labeled ℓ, its left child is labeled $\ell[-1]$ and its right child $\ell[+1]$, where $s[x]$ denotes the string that results when the bit x is appended to the string s. Each leaf of the tree is thus labeled by a distinct vector in $\{-1,+1\}^n$. An execution of the algorithm begins at the root and terminates on reaching a leaf. This process is a sequential view of the randomized algorithm of Example 4.5.

Call a leaf *good* if the vector $\boldsymbol{v}$ labeling it satisfies $||\boldsymbol{Av}||_\infty \leq 4\sqrt{n \ln n}$, and *bad* otherwise. From the argument of Example 4.5, we know that the randomized algorithm reaches a good leaf with probability at least $1 - 2/n$. For a node a in the tree, let $P(a)$ denote the probability that, starting from a, the randomized algorithm reaches a bad leaf. Thus $P(a)$ is the probability that the algorithm fails, conditional on its having reached the partial assignment that labels a. For the root r of the tree, we have $P(r) \leq 2/n < 1$ for $n > 2$.

Letting c and d denote the children of node a, we have

$$P(a) = \frac{P(c) + P(d)}{2}. \tag{5.8}$$

From (5.8), it follows that

$$\min\{P(c), P(d)\} \leq P(a).$$

In other words, every node has a child whose conditional probability of failure is no more than its own. This suggests the following deterministic algorithm for walking down the tree from r to a good leaf. Start from r; in general, from a node a, proceed to the child of a whose conditional probability of failure is no more than $P(a)$. Since $P(a) < 1$ when $a = r$, and never increases in the course of this walk, we arrive at a leaf ℓ for which $P(\ell) < 1$. But a leaf ℓ corresponds to a complete assignment to $\boldsymbol{b}$, so that its probability of being bad is either 0 or 1; since $P(\ell) < 1$, it must be the case that $P(\ell) = 0$. Thus this algorithm is guaranteed to arrive at a good leaf.

This scheme for derandomizing a randomized computation tree is quite general. Unfortunately, in most cases there is an obstacle to applying it: in order to choose which of the children (c or d) to proceed to from a node a, we must determine $P(c)$ and $P(d)$ (or at least determine which of them is smaller). We know of very few randomized algorithms for which this choice can be made efficiently. In the Notes section we will mention an approach to dealing with this problem. For the moment, we will tackle this problem for our set-balancing algorithm.

For $1 \leq i \leq n$, let us say that the ith *bad event*, denoted $\mathcal{E}_i$, occurs if the absolute value of the inner product of the ith row of $\boldsymbol{A}$ with $\boldsymbol{b}$ exceeds $4\sqrt{n \ln n}$. By the analysis of Example 4.5, we know that $\mathbf{Pr}[\mathcal{E}_i] \leq 2/n^2$, and so $\sum_i \mathbf{Pr}[\mathcal{E}_i] \leq 2/n$. For a node a in the computation tree, let $P(\mathcal{E}_i \mid a)$ denote the probability that $\mathcal{E}_i$ occurs conditional on the algorithm being at the intermediate stage a; clearly, $P(\mathcal{E}_i \mid r) \leq 2/n^2$. Let $\widehat{P}(a)$ denote $\sum_i P(\mathcal{E}_i \mid a)$; thus $P(a) \leq \widehat{P}(a)$ for all a. The deterministic algorithm now follows from the following three properties of $\widehat{P}(a)$. The first property has already been established; the reader may prove the other two in Problem 5.11.

1. $\widehat{P}(r) < 1$.
2. For any node a with children c and d,

$$\min\{\widehat{P}(c), \widehat{P}(d)\} \leq \widehat{P}(a).$$

3. For any node a, we can compute $\widehat{P}(a)$ in time polynomial in n.

The deterministic algorithm is clear: use the method of conditional probabilities as before, but with the value $\widehat{P}(a)$ instead of $P(a)$ at every step.

Theorem 5.15: *The algorithm based on the method of conditional probabilities determines a vector* $\boldsymbol{b}$ *such that* $||\boldsymbol{Ab}||_\infty \leq 4\sqrt{n \ln n}$, *in time polynomial in* n.

Notes

A comprehensive guide to the state of the art of the probabilistic method is the book by Alon and Spencer [24]. The books by Erdös and Spencer [139] and by Spencer [384] are quicker introductions to the field. The set-balancing problem has been widely studied, and the best known result is due to Spencer [383]: for every 0-1 matrix $\boldsymbol{A}$, there is a vector $\boldsymbol{b}$ such that $||\boldsymbol{Ab}||_\infty \leq 6\sqrt{n}$. It must be stressed that this result is existential, and there is no efficient (randomized or deterministic) algorithm known to find the vector whose existence is guaranteed by Spencer's result [383].

► **Research Problem 5.1:** Devise an efficient algorithm that for any 0-1 matrix $\boldsymbol{A}$ will find a vector $\boldsymbol{b}$ for which $||\boldsymbol{Ab}||_\infty$ is $o\left(\sqrt{n \ln n}\right)$.

The large cut example of Theorem 5.1 is taken from Luby [283]. The MAX-SAT problem is a classic problem in the theory of approximation algorithms. Johnson [219] gives a deterministic 1/2-approximation algorithm for the MAX-SAT problem that can be viewed as the derandomization (via the method of conditional probabilities) of the randomized algorithm in Theorem 5.2. Yannakakis [418] improved this result by presenting a deterministic 3/4-approximation algorithm. Our presentation in Section 5.2 is based on the work of Goemans and Williamson [169], who also describe how the algorithm may be made deterministic. In subsequent work [170], they have improved this using techniques from *semidefinite programming* to obtain a 0.878-approximation algorithm for instances of the MAX-SAT in which every clause has at most 2 literals (sometimes referred to as the MAX-2SAT problem). This implies an α-approximation algorithm for MAX-SAT, for a value of α that is slightly larger than 3/4. Improving on these bounds is an interesting challenge:

► **Research Problem 5.2:** Determine the largest value α for which there is a polynomial-time α-approximation algorithm for MAX-SAT.

Arora, Lund, Motwani, Sudan, and Szegedy [32] have shown that for a small constant $\epsilon > 0$ there is no polynomial time $(1 - \epsilon)$-approximation algorithm for MAX-3SAT, unless $\boldsymbol{P} = \boldsymbol{NP}$. Bellare and Sudan [50] have proved a similar result for ϵ close to 0.015

under a slightly weaker assumption than $P \neq NP$. These results carry over to other approximation problems, including the other versions of maximum satisfiability and the max-cut problem.

The history of expanding graphs can be traced to their origins in the construction of telephone networks. Cohen and Wigderson [108] provide a useful survey of the many different types of expanding graphs and their applications. Bien [59] also gives a good survey of the history of expanding graphs. The use of the probabilistic method for proving the existence of expanding graphs can be traced back to Pinsker [333]. The first explicit construction is due to Margulis [292]. Gabber and Galil [158] developed an explicit construction that we will use in Chapter 6. The probability amplification technique described in Section 5.3.1 is due to Sipser [378]. The use of expanding graphs for augmenting randomness is an idea that first appeared in work of Karp, Pippenger, and Sipser [248].

The number of bits used by an oblivious randomized permutation routing algorithm was studied by Peleg and Upfal [331]; they study a slightly more general question than that treated in Section 5.4. The following question remains open:

► **Research Problem 5.3:** Devise a uniform, randomized, oblivious scheme for permutation routing on the hypercube that uses $c_1 n$ bits of randomness and whose expected number of steps is $c_2 n$ on any instance of permutation routing on a hypercube with $N = 2^n$ nodes, for any constants c_1 and c_2.

The best known construction is due to Peleg and Upfal [331]: there is a uniform, randomized, oblivious scheme that uses $O(n^2)$ bits of randomness and runs in expected time $O(n)$.

The Lovász Local Lemma first appears in a paper by Erdös and Lovász [137]. Broder, Frieze, and Upfal have applied the Lovász Local Lemma to finding disjoint paths in expanders [84]. Leighton, Maggs, and Rao [272] have applied it to obtain an elegant result on packet routing, while Håstad, Leighton, and Newman have applied it to the probabilistic analysis of hypercubes with random faults [196]. The example of Section 5.5 is due to Beck [48]. A version of the algorithm that can be implemented as a "parallel algorithm" (see Chapter 12) is described by Alon [18].

The method of conditional probabilities is implicit in a paper of Erdös and Selfridge [138]. The connection to deterministic polynomial-time algorithms was developed by Spencer [384]. There are many applications for which we do not know how to compute the conditional probabilities that are compared at each step. One solution to this problem is the *method of pessimistic estimators* introduced by Raghavan [351]. The idea is to replace the conditional probability of failure at each stage by an efficiently computable estimate of the conditional probability. These papers [284, 351] demonstrate a number of algorithmic applications of the method of conditional probabilities. Chazelle and Friedman [91] have applied these tools to a number of problems in computational geometry. Berger and Rompel [55] and Motwani, Naor, and Naor [313] have applied a variant of the method of conditional probabilities to the derandomization of a variety of parallel algorithms.

Problems

5.1 (Due to J. Naor.) Let X be a random variable with expectation 0 such that moment generating function $\mathbf{E}[\exp(t|X|)]$ is finite for some $t > 0$. We can use the following two kinds of tail inequalities for X.

Chernoff Bound:

$$\mathbf{Pr}[|X| \geq \delta] \leq \min_{t \geq 0} \frac{\mathbf{E}[e^{t|X|}]}{e^{t\delta}}.$$

***k*th-Moment Bound:**

$$\mathbf{Pr}[|X| \geq \delta] \leq \frac{\mathbf{E}[|X|^k]}{\delta^k}.$$

(a) Show that for each δ, there exists a choice of k such that the kth-moment bound is stronger than the Chernoff bound. (**Hint:** Consider the Taylor expansion of the moment generating function and apply the probabilistic method.)

(b) Why would we still prefer the Chernoff bound to the (seemingly) stronger kth-moment bound?

5.2 In Example 5.2, we applied the probabilistic method to certificates for the value of a game tree in the setting of Section 2.1. We showed that for any instance of $T_{2,k}$ there is a set of $n^{0.793}$ leaves whose values certify the value of the root for that instance. Show that, in fact, for any instance of $T_{2,k}$, there is a set of $2^k = \sqrt{n}$ leaves whose values certify the value of the root for that instance.

5.3 Let G be a graph on n vertices, with $nd/2$ edges. Consider the following probabilistic experiment for finding an independent set in G. Delete each vertex of G (together with its incident edges) independently with probability $1 - 1/d$.

(a) Compute the expected number of vertices and edges that remain after the deletion process.

(b) From these, infer that there is an independent set with at least $n/2d$ vertices in any graph on n vertices with $nd/2$ edges.

(c) Let G be a 3-regular graph. Suppose that we wish to turn this probabilistic experiment into a randomized algorithm as follows. We delete each vertex independently with probability 2/3. For every edge that remains, delete one of its end-points. Derive an upper bound on the probability that this algorithm finds an independent set smaller than $n(1-\epsilon)/6$.

5.4 A function $f : \mathbb{R} \rightarrow \mathbb{R}$ is said to be *concave* if for any x_1, x_2 and $0 \leq \lambda \leq 1$, the following inequality is satisfied:

$$f(\lambda x_1 + (1-\lambda)x_2) \geq \lambda f(x_1) + (1-\lambda)f(x_2).$$

The reader may wish to compare this with the notion of convex functions defined in Problem 4.7.

(a) Suppose that f is a concave function and g is a linear function such that $g(0) \leq f(0)$ and $g(1) \leq f(1)$. Show that for any x in the interval $[0, 1]$, $g(x) \leq f(x)$.

(b) Show that the function $f(x) = 1 - (1 - x/k)^k$ is concave for any $k > 0$. What can you say when $k \leq 0$?

(c) Let $f(x) = 1 - (1 - x/k)^k$ and $g(x) = (1 - (1 - 1/k)^k)x$. Show that $f(x) \geq g(x)$ for positive k and $0 \leq x \leq 1$.

5.5 Use the probabilistic method to show that an expanding graph with the following properties exists for n sufficiently large:

- $|L| = |R| = n$.
- Every vertex in L has degree $n^{3/4}$, and every vertex in R has degree at most $3n^{3/4}$.
- Every subset of $n^{3/4}$ vertices in L has at least $n - n^{3/4}$ neighbors in R.

5.6 Suppose that you had access to the expanding graph described in Problem 5.5 for a certain value of n. Show that it can be used to run the **LazySelect** algorithm of Section 3.3 on any instance of size n, using $\log n$ random bits to choose the entire sample R. Show that the expected running time of this implementation is $O(n)$.

5.7 Let G be a d-regular graph on n vertices.

(a) Show that the number of connected subgraphs of G of size r is at most nd^{2r}.

(b) Suppose that each vertex of G is deleted independently with probability $1 - 1/2d^2$. Show that with probability $1 - n^{-\alpha}$, there is no surviving connected component of size exceeding $\log n$, for a suitable constant α.

5.8 Lemma 5.11 guarantees that with positive probability, none of the events $\mathcal{E}_i$ occurs. In this problem, we see how small this positive probability can be. Consider again the probabilistic experiment suggested in Problem 5.3 Let G be a $\sqrt{n}$-regular graph. Suppose that we delete vertices of G independently with probability $1 - 1/(3n^{1/4})$.

(a) Use Lemma 5.11 to make the (obvious) argument that with positive probability, an independent set remains after the deletion.

(b) Use the Chernoff bound to show that the probability that fewer than $n^{3/4}/6$ vertices survive is less than $\exp(-n^{3/4}/24)$.

(c) Now consider what happens when the above experiment is run on a $\sqrt{n}$-regular graph containing no independent set of size exceeding $\sqrt{n}$. What does this say about the positive probability in part (a)?

5.9 In Section 5.5, we assumed that a variable appears in at most $2^{k/50}$ clauses. Replace the constant 50 by the smallest constant you can for the following results:

(a) The existence proof using Corollary 5.12.

(b) The algorithm of Section 5.5.

5.10 (Due to J. Naor.) For a graph $G(V, E)$, and any $T \subseteq V$, define the cut function $c(T)$ as the number of edges in E which have exactly one end-point in T. For a suitably small function $f(n)$ and large enough even integer n, show that there exists a graph $G(V,E)$ with $|V| = n$ such that for *every* subset $T \subseteq V$ of size $n/2$,

$$\left| c(T) - \frac{n^2}{8} \right| \leq f(n).$$

How small can you make the function $f(n)$?

5.11 In this problem, we will complete establishing the properties of $\widehat{P}(a)$ leading to Theorem 5.15.

(a) Show that for a node a at the ith level of the computation tree, $\widehat{P}(a)$ is of the form $N(a)/2^{n-i}$, where $N(a)$ is a sum of binomial coefficients. Prove that for any node a with children c and d,

$$\min\{\widehat{P}(c), \widehat{P}(d)\} \leq \widehat{P}(a),$$

and that for any node a, we can compute $\widehat{P}(a)$ in time polynomial in n.

(b) Give an upper bound on the running time of the deterministic algorithm.

5.12 Show how the method of conditional probabilities can be applied to derandomize the **RandAuto** algorithm.

5.13 Consider the randomized algorithm implicitly described in the proof of Theorem 5.1, which finds a cut of expected size $m/2$ in a graph with m edges. Use the method of conditional probabilities to derandomize this algorithm and obtain a deterministic polynomial time algorithm that computes a cut of size at least $m/2$.

5.14 (Due to D.R. Karger and R. Motwani [233].) An *(n,m)-safe set instance* consists of a universe U of size n, a *safe* set $S \subseteq U$, and m target sets $T_1, \ldots, T_m \subseteq U$ such that

- $|S| = |T_1| = \cdots = |T_m|$,
- and, for $1 \leq i \leq m$, $S \cap T_i = \emptyset$.

An *isolator* for a safe set instance is a set $I \subseteq U$ that intersects all the target sets but not the safe set. An *(n,m)-universal isolating family* $\mathcal{F}$ is a collection of subsets of U such that $\mathcal{F}$ contains an isolator for *any* (n,m)-safe set instance.

Show that there exists a (n,m)-universal isolating family $\mathcal{F}$ such that $|\mathcal{F}|$ is polynomially bounded in n and m.

CHAPTER 6

Markov Chains and Random Walks

THE study of random walks on graphs is fascinating in its own right. In addition, it has a number of applications to the design and analysis of randomized algorithms. This chapter will be devoted to studying random walks on graphs, and to some of their algorithmic applications. We start by describing a simple algorithm for the 2-SAT problem, and analyze it by studying the properties of random walks on the line. Following a brief treatment of the basics of Markov chains, we consider random walks on undirected graphs. It is shown that there is a strong connection between random walks and the theory of electric networks. Random walks are then applied to the problem of determining the connectivity of graphs. Next, we turn to the study of random walks on expander graphs. We define a class of expanders and use algebraic graph theory to characterize their properties. Finally, we illustrate the special properties of random walks on expanders via an application to probability amplification.

Let $G = (V, E)$ be a connected, undirected graph with n vertices and m edges. For a vertex $v \in V$, $\Gamma(v)$ denotes the set of neighbors of v in G. A *random walk* on G is the following process, which occurs in a sequence of discrete *steps*: starting at a vertex v_0, we proceed at the first step to a randomly chosen neighbor of v_0. This may be thought of as choosing a random edge incident on v_0 and walking along it to a vertex $v_1 \in \Gamma(v_0)$. At the second step, we proceed to a randomly chosen neighbor of v_1, and so on. Unless otherwise stated, "randomly chosen neighbor" will mean a neighbor chosen uniformly at random; the choice at each step is independent of all previous choices.

Here are some typical questions about the simple random walk that we study: what is the expected number of steps to get from vertex u to another vertex v? Starting from a given vertex u, what is the expected number of steps to visit *every* vertex in the graph?

Exercise 6.1: Let G be the complete graph K_n on n vertices. Let u and v be two vertices in G. Prove that:

1. The expected number of steps in a simple random walk that begins at u and ends upon first reaching v is $n-1$.

2. The expected number of steps to visit all the vertices in G starting from u is $(n-1)H_{n-1}$, where $H_{n-1} = \sum_{j=1}^{n-1} 1/j$ is the Harmonic number.

Is the random walk on K_n exactly the same process as coupon collection with $n-1$ coupons?

6.1. A 2-SAT Example

Recall that the k-SAT problem is the special case of the SAT problem in which each clause in the input formula contains exactly k literals. We seek an assignment of (Boolean) values to the variables such that all the clauses are satisfied, or an assurance that no such assignment exists. While the k-SAT problem is ***NP***-hard for $k \geq 3$, it is solvable in polynomial time for $k = 1$ or $k = 2$. In this section we present a simple polynomial-time (Monte Carlo) algorithm for solving the 2-SAT problem.

Suppose we start with an arbitrary assignment of values to the literals. As long as there is a clause that is unsatisfied, we modify the current assignment as follows: we choose an arbitrary unsatisfied clause, and pick one of the (two) literals in it uniformly at random; the new assignment is obtained by complementing the value of the chosen literal. After each such step, we check to see if there exists an unsatisfied clause under the current assignment; if not, the algorithm terminates successfully with a satisfying assignment. If there is a satisfying assignment for this instance, how long does it take for this process to discover it?

Given an instance with a satisfying assignment, let us fix our attention on a particular satisfying assignment A, and refer to the values assigned by A to the literals as the "correct values." Let n be the number of variables in an instance. The progress of this algorithm can be represented by a particle moving between the integers $\{0, 1, \ldots, n\}$ on the real line. The position of the particle indicates how many variables in the current solution have the correct values. At each iteration, we complement the current value of one of the literals of some unsatisfied clause, so that the particle's position changes by 1 at each step. In particular, a particle currently at position i, for $0 < i < n$, can only move to positions $i-1$ or $i+1$. A particle at location 0 can only move to 1, and the process terminates when the particle reaches position n, although it may terminate at some other position with a satisfying assignment other than A.

The crucial observation is the following: in an unsatisfied clause, at least one of the two literals has an incorrect value. With probability at least $1/2$ we increase (by one) the number of variables having their correct values. The motion of the particle thus resembles a random walk on the line.

The reader may relate this process to a familiar gambling experience (see also Section 4.4). A gambler goes to a casino with n dollars. At each step he bets \$1, and loses it with probability at least 1/2. If he wins, his bet of \$1 is returned to him, and in addition he is given \$1. The gambler must quit when his capital is reduced to 0. Note the similarity to the process in the previous paragraph, with the coordinates on the line reversed.

The random walk on the line is one of the most extensively studied stochastic processes. Using the tools developed in this chapter, we will be able to prove:

Theorem 6.1: *The expected number of steps for the above 2-SAT algorithm to find a satisfying assignment is* $O(n^2)$.

Exercise 6.2: Using Theorem 6.1, devise a one-sided error Monte Carlo algorithm for the 2-SAT problem. This algorithm should run in polynomial time, always return UNSATISFIABLE for unsatisfiable formulas, and with high probability it should return a satisfying truth assignment for satisfiable formulas.

6.2. Markov Chains

Although we can deal with some of the questions concerning random walks using basic probability theory (as in Exercise 6.1), they are more conveniently studied using an abstraction known as a *Markov chain.* A Markov chain $\mathcal{M}$ is a discrete-time stochastic process defined over a set of states S in terms of a matrix $\boldsymbol{P}$ of *transition probabilities.* The set S is either finite or countably infinite. The transition probability matrix $\boldsymbol{P}$ has one row and one column for each state in S. The Markov chain is in one state at any time, making state-transitions at discrete time-steps $t = 1, 2, \ldots$. The entry P_{ij} in the transition probability matrix is the probability that the next state will be j, given that the current state is i. Thus, for all i, $j \in S$, we have $0 \leq P_{ij} \leq 1$, and $\sum_j P_{ij} = 1$.

An important property of a Markov chain is the *memorylessness property*: the future behavior of a Markov chain depends only on its current state, and not on how it arrived at the present state. This follows from the observation that the transition probabilities P_{ij} depend only on the current state i. We will denote by X_t the state of the Markov chain at time t; thus, the sequence $\{X_t\}$ specifies the history or the evolution of the Markov chain. The memorylessness property can be stated more formally as follows:

$$\mathbf{Pr}[X_{t+1} = j \mid X_0 = i_0, X_1 = i_1, \ldots, X_t = i] = \mathbf{Pr}[X_{t+1} = j \mid X_t = i] = P_{ij}.$$

A Markov chain (indeed, a random walk) need not have a prespecified initial state; in general, its initial state X_0 is permitted to be chosen according to some probability distribution over S. Of course, an initial probability distribution includes as a special case the deterministic specification that the initial state X_0 be i. Given a distribution for the initial state X_0, we have a probability distribution for the history $\{X_t\}$.

For states $i, j \in S$, define the *t-step transition probability* as $P_{ij}^{(t)} = \mathbf{Pr}[X_t = j \mid X_0 = i]$. Given an initial state $X_0 = i$, the probability that the first transition *into* state j occurs at time t is denoted by $r_{ij}^{(t)}$ and is given by

$$r_{ij}^{(t)} = \mathbf{Pr}[X_t = j, \text{and, for } 1 \leq s \leq t-1, X_s \neq j \mid X_0 = i].$$

Also, for $X_0 = i$, the probability that there is a visit to (transition into) state j at some time $t > 0$ is denoted by f_{ij}, and is given by

$$f_{ij} = \sum_{t>0} r_{ij}^{(t)}.$$

Finally, the expected number of time steps to reach state j starting from state i is denoted by h_{ij} and is given by

$$h_{ij} = \sum_{t>0} t\, r_{ij}^{(t)}.$$

For $f_{ij} = 1$, and $h_{ij} = \infty$ otherwise.

► **Definition 6.1:** A state i for which $f_{ii} < 1$ (and hence $h_{ii} = \infty$) is said to be *transient*, and one for which $f_{ii} = 1$ is said to be *persistent*. Those persistent states i for which $h_{ii} = \infty$ are said to be *null persistent* and those for which $h_{ii} \neq \infty$ are said to be *non-null persistent*.

We restrict our attention to finite Markov chains, i.e., Markov chains whose states are finite in number. We claim that every state in such a Markov chain is either transient or non-null persistent. We define the underlying directed graph of a Markov chain as follows: there is one vertex in the graph for each state of the Markov chain; and there is an edge directed from vertex i to vertex j if and only if $P_{ij} > 0$.

► **Definition 6.2:** A *strong component* of a directed graph G is a maximal subgraph C of G such that for any pair of vertices i and j in the vertex set of C, there is a directed path from i to j, as well as a directed path from j to i.

► **Definition 6.3:** A strong component C is said to be a *final strong component* if there is no edge going from a vertex in C to a vertex not in C.

In a finite Markov chain, starting from any vertex in a strong component C, there is a non-zero probability of reaching any other vertex in the same strong component in a finite number of steps. If C is a final strong component, this probability is 1 since the Markov chain can never leave the component C once it enters it. It follows that a state is persistent if and only if it lies in a final strong component.

► **Definition 6.4:** A Markov chain is said to be *irreducible* whenever its underlying graph consists of a single strong component.

The unique strong component in an irreducible Markov chain must be final, and hence all states are persistent.

► **Definition 6.5:** Define $\boldsymbol{q}^{(t)} = (q_1^{(t)}, q_2^{(t)}, \ldots, q_n^{(t)})$, the *state probability vector* (also called the *distribution of the chain at time t*), to be the *row* vector whose ith component is the probability that the chain is in state i at time t.

Henceforth, whenever we mention a probability distribution on the states of a Markov chain, we mean such a vector. It is easy to check that $\boldsymbol{q}^{(t+1)} = \boldsymbol{q}^{(t)}\boldsymbol{P}$, so we have by induction that $\boldsymbol{q}^{(t)} = \boldsymbol{q}^{(0)}\boldsymbol{P}^t$. It follows that a Markov chain's behavior for all time is specified by its initial distribution $\boldsymbol{q}^{(0)}$ and its transition matrix $\boldsymbol{P}$.

Some remarks about our notation are in order. Throughout this chapter, when multiplying a probability vector $\boldsymbol{q}$ with a transition probability matrix $\boldsymbol{P}$, we will use $\boldsymbol{qP}$ instead of $\boldsymbol{Pq}$ since the correct interpretation is that the entry P_{ij} represents the probability of going from state i to state j, and that the entry q_i is the probability of being in state i. For notational convenience, we interpret a probability vector as a row vector whenever it premultiplies a matrix in this fashion.

► **Definition 6.6:** A *stationary distribution* for the Markov chain with transition matrix $\boldsymbol{P}$ is a probability distribution $\boldsymbol{\pi}$ such that $\boldsymbol{\pi} = \boldsymbol{\pi P}$.

Intuitively, if the Markov chain is in the stationary distribution at step t, it remains in the stationary distribution at step $t+1$. Thus the stationary distribution is thought of as a description of the steady-state behavior of the Markov chain.

► **Definition 6.7:** The *periodicity* of a state i is the maximum integer T for which there exists an initial distribution $\boldsymbol{q}^{(0)}$ and positive integer a such that, for all t, if at time t we have $q_i^{(t)} > 0$, then t belongs to the arithmetic progression $\{a + Ti \mid i \geq 0\}$. A state is said to be *periodic* if it has periodicity greater than 1, and is said to be *aperiodic* otherwise. A Markov chain in which every state is aperiodic is known as an *aperiodic Markov chain.*

Consider a Markov chain in which the underlying graph is bipartite. It follows that every state is periodic with periodicity at least 2. As we will see later, this is really the only possible source of periodicity in Markov chains obtained from random walks. Periodic Markov chains cause complications (for example, they do not converge to the stationary distribution), but we will show that there is a simple trick for dealing with this source of periodicity.

► **Definition 6.8:** An *ergodic* state is one that is aperiodic and non-null persistent.

► **Definition 6.9:** An *ergodic* Markov chain is one in which all states are ergodic.

The following basic theorem on Markov chains may be found in most texts on stochastic processes.

Theorem 6.2 (Fundamental Theorem of Markov Chains): *Any irreducible, finite, and aperiodic Markov chain has the following properties.*

1. *All states are ergodic.*
2. *There is a unique stationary distribution* $\boldsymbol{\pi}$ *such that, for* $1 \leq i \leq n$, $\pi_i > 0$.
3. *For* $1 \leq i \leq n$, $f_{ii} = 1$ *and* $h_{ii} = 1/\pi_i$.
4. *Let* $N(i, t)$ *be the number of times the Markov chain visits state* i *in* t *steps. Then,*

$$\lim_{t \to \infty} \frac{N(i, t)}{t} = \pi_i.$$

6.3. Random Walks on Graphs

Let $G = (V, E)$ be a connected, non-bipartite, undirected graph where $|V| = n$ and $|E| = m$. It induces a Markov chain M_G as follows: the states of the M_G are the vertices of G, and for any two vertices $u, v \in V$,

$$P_{uv} = \begin{cases} \frac{1}{d(u)} & \text{if } (u, v) \in E \\ 0 & \text{otherwise,} \end{cases}$$

where $d(w)$ is the degree of vertex w. Because G is connected, M_G is irreducible. For a connected, undirected graph G, the periodicity of the states in M_G is the greatest common divisor (gcd) of the length of all closed walks in G, where a closed walk is any walk that starts and ends at the same vertex. As G is undirected, there are closed walks of length 2 that traverse the same edge twice in succession. Further, since G is non-bipartite it has odd cycles that give closed walks of odd length. It follows that the gcd of the closed walks is 1, and hence M_G is aperiodic. Noting that G is finite, Theorem 6.2 now implies that M_G has a unique stationary distribution $\boldsymbol{\pi}$.

Lemma 6.3: *For all* $v \in V$, $\pi_v = d(v)/2m$.

PROOF: From the fact that $\boldsymbol{\pi}$ is stationary and the definition of P_{uv} we get the following system of equations

$$\pi_v = [\boldsymbol{\pi} \boldsymbol{P}]_v = \sum_{u\ v} \pi_u P_{uv}, \text{ for all } v.$$

This system admits the solution $\pi_u = d(u)/2m$ which, by Theorem 6.2, is the unique stationary distribution. □

As a direct consequence of Theorem 6.2 and Lemma 6.3, we obtain the following lemma.

Lemma 6.4: *For all $v \in V$, $h_{vv} = 1/\pi_v = 2m/d(v)$.*

► **Definition 6.10:** The *hitting time* h_{uv} (sometimes called the *mean first passage time*) is the expected number of steps in a random walk that starts at u and ends upon first reaching v.

► **Definition 6.11:** We define C_{uv}, the *commute time* between u and v, to be $C_{uv} = h_{uv} + h_{vu} = C_{vu}$. This is the expected time for a random walk starting at u to return to u after at least one visit to v.

► **Definition 6.12:** Let $C_u(G)$ denote the expected length of a walk that starts at u and ends upon visiting every vertex in G at least once. The *cover time* of G, denoted $C(G)$, is defined by $C(G) = \max_u C_u(G)$.

► **Example 6.1:** A graph that tells us a great deal about the behavior of random walks is the n-vertex *lollipop graph* L_n (Figure 6.1). This graph consists of a clique on $n/2$ vertices, and a path on the remaining vertices. There is a vertex u in the clique to which the path is attached; let v denote the other end of the path.

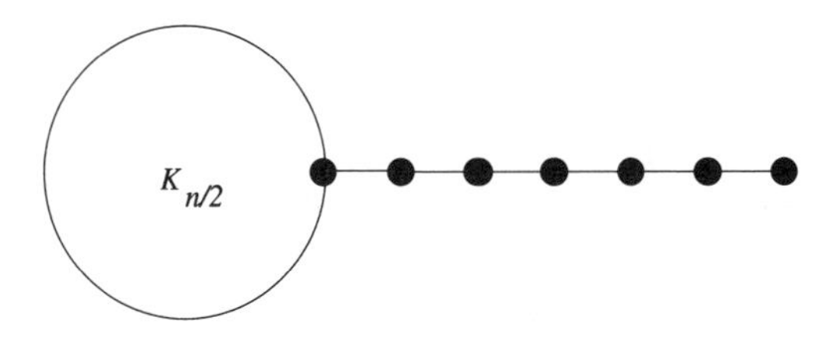

Figure 6.1: The lollipop graph L_n.

By elementary probability (or using methods for studying random walks that we will encounter shortly), it turns out that in L_n, h_{uv} is $\Theta(n^3)$, whereas h_{vu} is $\Theta(n^2)$. Thus, in general, $h_{uv} \neq h_{vu}$, and the asymptotic difference (as in this case) can be as much as a factor of n.

Another misconception that L_n dispels is that "adding more edges should help reduce the cover time $C(G)$." This is false, because L_n has cover time $\Theta(n^3)$; on the other hand, it can be built by adding edges to a chain on n vertices, which can be shown to have cover time $\Theta(n^2)$. In turn, the complete graph K_n can be built by adding edges to L_n, and the cover time of K_n is $\Theta(n \log n)$. Thus the cover time of a graph is not monotone in the number of edges.

The following lemma establishes an important property of the commute time across an edge and will prove useful in Section 6.5 below.

Lemma 6.5: *For any edge* $(u, v) \in E$, $h_{uv} + h_{vu} \leq 2m$.

PROOF: The proof considers a new Markov chain defined on the edges of G. The current state is defined to be the pair composed of the edge most recently traversed in the random walk, together with the direction of this traversal; equivalently, replacing each undirected edge by two oppositely directed edges, the directed edges form the state space. There are $2m$ states in this new Markov chain. The transition matrix Q for this Markov chain has non-zero entry

$$Q_{(u,v),(v,w)} = P_{vw} = 1/d(v),$$

corresponding to an edge (v, w). This matrix is *doubly stochastic*, meaning that not only do the rows sum to one (as in every Markov chain), but the columns sum to one as well. To see this, fix a (directed) edge (v, w) and observe that the column sum corresponding to this state is given by

$$\begin{aligned} \sum_{x \in V, y \in \Gamma(x)} Q_{(x,y),(v,w)} &= \sum_{u \in \Gamma(v)} Q_{(u,v),(v,w)} \\ &= \sum_{u \in \Gamma(v)} P_{vw} \\ &= d(v) \times \frac{1}{d(v)} \\ &= 1. \end{aligned}$$

Noting the result in Problem 6.6, it follows that the uniform distribution on the edges is stationary for this Markov chain, so the stationary probability of each directed edge is $1/2m$. By part (3) of Theorem 6.2, we can conclude that the expected time between successive traversals of the directed edge (v, u) is $2m$.

Consider now $h_{uv} + h_{vu}$, and interpret this as the expected time for a walk starting from vertex u to visit vertex v and return to u. Conditioned on the event that the initial entry into u was via the directed edge (v, u), we conclude that the expected time to go from there to v and then to u along (v, u) is $2m$. The memorylessness property of a Markov chain now allows us to remove the conditioning: since the sequence of transitions from u onward is independent of the fact that we arrived at u along (v, u) at the start of the commute, the expected time back to u is at most $2m$. □

We emphasize that the result in Lemma 6.5 is valid only for vertices u and v that are connected by an edge in G.

6.4. Electrical Networks

Many random variables associated with the simple random walk on an undirected graph are studied conveniently using the tools and the language of electrical network theory. Our focus here will be on characterizing h_{uv} and C_{uv} in terms of properties of the graph G. We begin with a review of some basics of resistive electrical networks.

A *resistive electrical network* is an undirected graph; each edge has associated with it a positive real *branch resistance.* The flow of current in such networks is governed by two rules: *Kirchhoff's Law* and *Ohm's Law.* Kirchhoff's Law stipulates that the sum of the currents entering a node in the network equals the sum of the currents leaving it. Ohm's Law states that the voltage across a resistance equals the product of the resistance and the current through it.

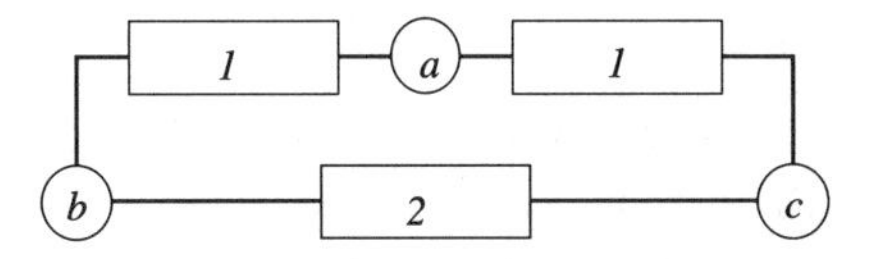

Figure 6.2: A resistive electrical network. Each rectangle signifies a branch resistance.

Consider the simple example in Figure 6.2. If a current of one ampere were injected into node b and removed from node c in this network, a simple calculation using Kirchhoff's Law and Ohm's Law yields the following: half an ampere of current flows along the branch bc, and the other half ampere through branch ba and onto ac. The voltage difference between c and b is one volt, while the voltage difference between c and a (and between a and b) is half a volt.

One final notion we need is that of the *effective resistance* between two nodes in a network. The effective resistance between two nodes u and v is the voltage difference between u and v when one ampere is injected into u and removed from v; equivalently, one ampere could be injected into v and removed from u. The effective resistance between u and v is always at most the branch resistance between u and v and can be much less, as we shall see. This distinction between branch and the effective resistances is important. In the example in Figure 6.2, for instance, the effective resistance between b and c is 1, whereas the branch resistance is 2.

Given an undirected graph G, let $\mathcal{N}(G)$ be the electrical network defined as follows: it has a node for each vertex in V; for every edge in E, it has a one ohm resistance between the corresponding nodes in $\mathcal{N}(G)$. For two vertices $u, v \in V$, R_{uv} denotes the effective resistance between the corresponding nodes in $\mathcal{N}(G)$. The following theorem establishes a close relation between commute times for the simple random walk on G and effective resistances in the electrical network $\mathcal{N}(G)$.

Theorem 6.6: *For any two vertices u and v in G, the commute time $C_{uv} = 2mR_{uv}$.*

PROOF: For a vertex x in G, let $\Gamma(x)$ denote the set of vertices in V that are adjacent to x, and let $d(x)$ denote its degree $|\Gamma(x)|$. Let ϕ_{uv} denote the voltage at u in $\mathcal{N}(G)$ with respect to v, if $d(x)$ amperes of current are injected into each node $x \in V$, and $2m$ amperes are removed from v. We will first prove that for all $u \in V$,

$$h_{uv} = \phi_{uv}. \tag{6.1}$$

Using Kirchhoff's Law and Ohm's Law, we obtain that for all $u \in V \setminus \{v\})$,

$$d(u) = \sum_{w \in \Gamma(u)} (\phi_{uv} - \phi_{wv}). \tag{6.2}$$

By the definition of expectation, for all $u \in V \setminus \{v\})$,

$$h_{uv} = \sum_{w \in \Gamma(u)} \frac{1}{d(u)} (1 + h_{wv}). \tag{6.3}$$

Equations (6.2) and (6.3) are both linear systems with unique solutions; furthermore, they are identical if we identify ϕ_{uv} in (6.2) with h_{uv} in (6.3). This proves (6.1). To complete the proof of the theorem, we note that h_{vu} is the voltage ϕ_{vu} at v in $\mathcal{N}(G)$ measured with respect to u, when currents are injected into all nodes and removed from u. Changing signs, ϕ_{vu} is now the voltage at u relative to v when current is injected at u, and removed from all other nodes. Since resistive networks are linear, we can determine C_{uv} by super-posing (taking care with the sign!) the networks on which ϕ_{uv} and ϕ_{vu} are measured. Currents at all nodes except u and v cancel, resulting in C_{uv} being the voltage between u and v when $\sum_{w \in V} d(w) = 2m$ amperes are injected into u and removed from v, which yields the theorem by Ohm's Law. □

Exercise 6.3: Verify all the hitting times claimed in Example 6.1 using the ideas in the above proof.

Exercise 6.4: Consider a random walk on the integer points $1, 2, \ldots, n$, starting at 1. If the walk is at 1, it always proceeds to 2 at the next step; when the walk is at a point $i > 1$, it proceeds at the next step to $i - 1$ or to $i + 1$ with equal probability. Show that the expected number of steps that elapse before the walk first reaches n is $(n - 1)^2$.

Exercise 6.5: Prove Theorem 6.1. Why does the bound of $O(n^2)$ steps hold only for finding *some* satisfying assignment, rather than the specified assignment A? What happens if each clause has 3 literals rather than 2?

The effective resistance between two nodes u and v is at most the length of the shortest path between them in G. This observation yields an alternative proof of Lemma 6.5. The length of the shortest path between any two vertices of G is at most the diameter of G. We thus have the following corollary, which by Example 6.1 is asymptotically tight.

Corollary 6.7: *In any n-vertex graph, and for all vertices u and v,*

$$C_{uv} < n^3.$$

6.5. Cover Times

We are now ready to prove a classic theorem on the cover time of the simple random walk on G.

Theorem 6.8: $\mathcal{C}(G) \leq 2m(n-1)$.

PROOF: Let T be any spanning tree of G. There is a traversal of T, visiting vertices $v_0, v_1, \ldots, v_{2n-2} = v_0$ that traverses each edge of T exactly once in each direction. Further, every vertex of G appears at least once in the sequence $v_0, v_1, \ldots, v_{2n-2}$. Consider a random walk that starts at v_0 and terminates upon returning to v_0, having visited the vertices $v_1, v_2, \ldots$ in the order prescribed by the traversal. Since this walk has visited every vertex in G, an upper bound on the expected length of this walk is an upper bound on $\mathcal{C}_{v_0}(G)$. Now

$$\mathcal{C}_{v_0}(G) \leq \sum_{j=0}^{2n-3} h_{v_j, v_{j+1}} = \sum_{(u,w) \in T} C_{uw}.$$

Since the vertices v_j, v_{j+1} are adjacent for all j, we have by Lemma 6.5 that

$$C_{v_j, v_{j+1}} \leq 2m.$$

Since there are $n-1$ edges in T, $\mathcal{C}_{v_0}(G) \leq 2m(n-1)$. But this upper bound holds no matter which vertex of G we designate to be the starting point v_0 in the traversal; therefore $\mathcal{C}(G) \leq 2m(n-1)$. □

Note that Theorem 6.8 gives (asymptotically) the right answer for the lollipop graph: $\mathcal{C}(L_n)$ is $\Theta(n^3)$. On the other hand, it gives the same $O(n^3)$ upper bound for the complete graph K_n, whereas we have already seen (Exercise 6.1) that $\mathcal{C}(K_n)$ is $\Theta(n \log n)$. Theorem 6.8 can be slack for some graphs: in the proof, we measure the time for the vertices of G be visited in one specific order. In fact, we can often refine the upper bound on cover time as follows.

Let $R(G) = \max_{u,v \in V} R_{uv}$; we call R the *resistance of* G. The resistance of a graph characterizes its cover time fairly tightly:

Theorem 6.9: $mR(G) \leq \mathcal{C}(G) \leq 2e^3 mR(G) \ln n + n$.

PROOF: The proof of the lower bound follows from the fact that there exist vertices u, v such that $R(G) = R_{uv}$ and $\max(h_{uv}, h_{vu}) \geq C_{uv}/2$; the bound then follows from Theorem 6.6.

For the upper bound, we will show that the probability that all the vertices are not visited within $2e^3mR(G)\ln n$ steps is at most $1/n^2$; this, together with Corollary 6.7 will yield the result.

Divide the random walk of length $2e^3mR(G)\ln n$ into $\ln n$ *epochs* each of length $2e^3mR(G)$. For any vertex v, the hitting time h_{uv} is at most $2mR(G)$, regardless of the vertex u at which an epoch starts. By the Markov inequality, the probability that v is not visited during any single epoch is at most $1/e^3$. Thus, the probability that v is not visited during any of the $\ln n$ epochs is at most $1/n^3$. Summing this probability over the n choices of the vertex v, the probability that *any* vertex is not visited within $2e^3mR(G)\ln n$ steps is at most $1/n^2$. When this happens (there is a vertex that has not been visited within $2e^3mR(G)\ln n$ steps), we continue the walk until all vertices are visited, and n^3 steps suffice for this (by Corollary 6.7). Thus the expected total time is at most

$$2e^3mR(G)\ln n + (1/n^2)n^3 = 2e^3mR(G)\ln n + n.$$

□

The bounds in Theorem 6.9 cannot in general be improved; the upper bound is tight (within constant factors) for the complete graph (Problem 6.10 below) and the lower bound is tight for the chain on n vertices.

There are also graphs for which neither bound of Theorem 6.9 is tight. Note that Theorem 6.9 gives an estimate for the cover time that is tight to within a factor of $\log n$. This is because effective resistances in a graph (and therefore the resistance of the graph, $R(G)$) can be computed efficiently using matrix inversions. Note also that neither Theorem 6.8 nor 6.9 is universally superior; we have already seen that for the complete graph K_n, Theorem 6.8 gives a loose upper bound. For the lollipop graph L_n, Theorem 6.9 gives an upper bound of $O(n^3 \log n)$, which is an overestimate by a factor of $\log n$.

Often, we are interested not so much in determining the cover time of a single graph, as in bounding the cover times of a family of graphs. A simple fact that is of great use in bounding effective resistances in electrical networks is following *Rayleigh's Short-cut Principle:*

Effective resistance is never raised by lowering the resistance on an edge (e.g., by "shorting" two nodes together), and is never lowered by raising the resistance on an edge (e.g., by "cutting" it). Similarly, resistance is never lowered by "cutting" a node, leaving each incident edge attached to only one of the two resulting halves of the node.

A second useful fact about effective resistances in an electrical network is that they obey the triangle inequality. As one very simple application of these facts, observe that in a graph with minimum degree d, $R \geq 1/d$: short all vertices except the one of minimum degree. Another simple application is the following lemma.

Lemma 6.10: *Suppose that g contains p edge-disjoint paths of length at most ℓ from s to t. Then, $R_{st} \leq \ell/p$.*

6.6. Graph Connectivity

We are now ready for our first algorithmic application of random walks. Two vertices in an undirected graph G are said to be *connected* if there exists a path between them. A *connected component* of G is a (maximal) subset of vertices in which every pair of vertices is connected.

6.6.1. Undirected Graphs

The *undirected s–t* connectivity (USTCON) problem is the following: given an undirected graph G and two vertices s and t in G, decide whether s and t are in the same connected component. The USTCON problem is important in the study of space-bounded complexity classes and is a natural abstraction of a number of graph search problems. It is easy to see that a standard graph search algorithm such as depth-first search solves the problem in $O(m)$ steps. In doing so, the algorithm keeps track of all the vertices of G that the search has visited and, therefore, uses workspace at least linear in n.

A probabilistic log-space Turing machine for a language L is a probabilistic Turing machine using space $O(\log n)$ on instances of size n, and running in time polynomial in n. We say that a language (equivalently, a decision problem) A is in ***RLP*** if there exists a probabilistic log-space Turing machine M such that on any input x,

$$\mathbf{Pr}[M \text{ accepts } x] \begin{cases} \geq 1/2 & x \in A \\ 0 & x \notin A. \end{cases} \tag{6.4}$$

Here space $O(\log n)$ refers to the workspace of the Turing machine; the input is given on a read-only tape, and the only storage available to it with write-access is a log-space tape.

Theorem 6.11: *USTCON $\in$ **RLP**.*

PROOF: The log-space probabilistic Turing machine simulates a (simple) random walk of length $2n^3$ through the input graph, starting from s. If it encounters the vertex t in the course of this walk, it outputs YES; otherwise it outputs NO. Clearly the machine will never output YES on an instance of USTCON in which s and t are not in the same connected component. What is the probability that it outputs NO when it should have said YES?

By Theorem 6.6, $h_{st} \leq n^3$. By the Markov inequality, if t is in the same component of G as s, the probability that it is not visited in a random walk of $2n^3$ steps starting from s is at most $1/2$. The Turing machine uses its workspace to count up to $2n^3$, and to keep track of its position in the graph during the walk; both of these require only space $O(\log n)$. □

We have thus seen a uniform, randomized algorithm for deciding USTCON in log-space and polynomial time. This randomized algorithm can also be made deterministic while still using logarithmic space, albeit non-uniformly. We consider a specific class of non-uniform, deterministic log-space algorithms for USTCON known as *universal traversal sequences.* We focus on n-vertex graphs that are *regular of degree d* – every vertex has degree d – throughout our discussion of universal traversal sequences. Such a graph is said to be *labeled* if, at each vertex in the graph, each of the d edges incident on that vertex has a unique (integer) label in $\{1,\ldots,d\}$. There is no requirement that an edge receive the same label at both end-points. Figure 6.3 gives an example of a labeled 3-vertex, 2-regular graph. Note that the edge joining vertices a and b has different labels at its end-points.

Any sequence of symbols $\sigma = (\sigma_1, \sigma_2, \ldots)$ from $\{1,\ldots,d\}$ together with a *starting vertex* v in a labeled graph describes a walk through the graph in the following natural fashion. The walk begins at v, and at its first step walks along the edge incident on v whose label is σ_1. It now arrives at another vertex, say u, and leaves by the edge whose label is σ_2, and so on. For example, in Figure 6.3, if the starting vertex were a and σ were $(1,2,1,1,2)$, the walk would proceed to visit the vertices b, a, b, c, a. On the other hand, if the starting vertex were b, the same sequence σ visits the vertices c, a, b, c, a.

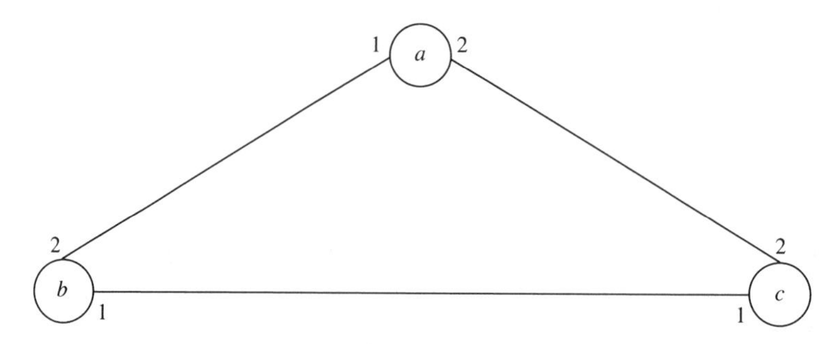

Figure 6.3: A labeled 3-vertex, 2-regular graph.

A sequence σ is said to *traverse* a labeled graph G if the walk it prescribes visits every vertex of G regardless of the starting vertex. The reader may verify that the sequence $(1,2,2)$ traverses the labeled graph in Figure 6.3, and that no shorter sequence does so. A sequence σ is said to be *universal traversal sequence* for a class of labeled graphs if it traverses every labeled graph in the class. By this we mean every labeling of every graph in the class, and for every starting vertex.

A universal traversal sequence whose length is polynomial in n can be used by a deterministic log-space Turing machine to decide instances of USTCON of size n as follows. The sequence is stored in the finite-state control of the Turing machine and is used to traverse G starting from s on an instance of USTCON. (However, in order for it to be a uniform log-space algorithm, the universal traversal sequence should be constructible by the log-space Turing machine, rather than be encoded in the machine's finite-state control.)

Let $\mathcal{G}$ be a family of connected labeled d-regular graphs on n vertices. Each member of each graph counts as a distinct member of $\mathcal{G}$. Let $U(\mathcal{G})$ denote the length of the shortest universal traversal sequence for all the labeled graphs in $\mathcal{G}$. Let $R(\mathcal{G})$ denote the maximum resistance between any pair of vertices in any graph in $\mathcal{G}$.

Theorem 6.12: $U(\mathcal{G}) \leq 5mR(\mathcal{G})\log_2(n|\mathcal{G}|)$.

PROOF: Given a labeled graph $G \in \mathcal{G}$, let v be a vertex of G. Consider a random walk of length $5mR(\mathcal{G})\log_2(n|\mathcal{G}|)$, divided into $\log_2(n|\mathcal{G}|)$ "epochs" each of length $5mR(\mathcal{G})$. The probability that the walk fails to visit v in any epoch is at most $2/5$ by Theorem 6.6 and Markov's inequality, regardless of the vertex of G at which the epoch began. The probability that v is not visited during any of the $\log_2(n|\mathcal{G}|)$ epochs is thus at most $(n|\mathcal{G}|)^{-c}$ for a value of $c > 1$. Summing this probability over all n choices of the vertex v and all $|\mathcal{G}|$ choices of the labeled graph G, the probability that the random walk (sequence) fails to be universal is less than one. Thus there is a sequence of this length that is universal for the class $\mathcal{G}$. □

The constant 5 in Theorem 6.12 can be improved slightly. Let $U(d, n)$ denote the length of the shortest universal traversal sequence for connected, labeled, n-vertex, d-regular graphs.

Exercise 6.6: Show that the number of labeled n-vertex graphs that are d-regular is $(nd)^{O(nd)}$.

Putting together Theorem 6.12 and the result of Exercise 6.6, we have:

Corollary 6.13: $U(d, n) = O\left(n^3 d \log n\right)$.

PROOF: The diameter of every connected n-vertex, d-regular graph is $O(n/d)$ and so, therefore, is its resistance. The number of edges $m = nd/2$.

The result now follows from Exercise 6.6 and Theorem 6.12. □

This suggests that there is a deterministic log-space Turing machine that decides USTCON on n-vertex, d-regular graphs. Unfortunately, all we have given here is a proof (by the probabilistic method) that such a universal traversal sequence *exists*, and thus a non-uniform deterministic log-space machine. We do not know how to construct such a sequence by a deterministic log-space machine; in fact, we do not in general know how to do this even with a polynomial-time machine.

6.6.2. Directed Graphs

Are the techniques of Section 6.6.1 applicable to s–t connectivity (STCON) in directed graphs? There is certainly no immediate way of using the results on random walks, since the cover time of the random walk may no longer be finite. For instance, a directed graph may contain vertices with no outgoing edges, so that a random walk may get trapped at such a vertex. What if we were to perform a random walk from the vertex s, and to jump back to s whenever we are stuck at such a vertex? We will use a variant of this idea to give a Monte Carlo algorithm that decides s–t connectivity in directed graphs using space $O(\log n)$. The running time of this algorithm may be large – its expectation may be of the order of n^n. The algorithm has one-sided error: whenever it terminates and outputs YES, it is correct, but when it outputs NO, it is wrong with some probability.

As before, let the edges leaving a vertex v be labeled $1, 2, \ldots, d(v)$. Thus any path in the graph can be associated with a string whose symbols are drawn from $\{1, 2, \ldots, n-1\}$, as in our discussion of universal traversal sequences. If we could begin at s and enumerate the walks corresponding to all such strings of length $n-1$, we would be assured of discovering a path from s to t if one existed. The number of such strings being of the order of n^n, we would require $\Omega(n \log n)$ space to maintain a counter that could index these strings. Since we only wish to use $O(\log n)$ space, we use randomization to achieve this reduction in space.

The algorithm consists of repeatedly executing the following two steps until either step results in termination.

1. Starting at s, simulate a random walk of $n-1$ steps. Each step consists of choosing an edge leaving the current vertex uniformly at random. If t is reached, output YES and stop. If the walk reaches a vertex with no outgoing edge, or a vertex other than t after $n-1$ steps, return to s. This step can be implemented using $O(\log n)$ bits of memory.

2. Flip $\log n^n$ unbiased coins. If they all come up HEADS, halt and output NO. This can be implemented by a counter that keeps track of the number of coins that have been flipped. The number of bits required in this counter is $\log(\log n^n)$, which is $O(\log n)$, as required.

We wish to bound the probability of terminating and erroneously outputting NO when in fact there is a path from s to t. Since the number of distinct walks from s is at most n^n, the probability of discovering an s–t path on a trial (in Step 1) is at least n^{-n}. The probability of terminating in Step 2 on a trial is the probability that all the coins come up HEADS, and this is n^{-n}. Thus on each trial, the algorithm terminates successfully with probability at least n^{-n}, and erroneously with probability at most $(1 - n^{-n})n^{-n} \leq n^{-n}$. Let p_w denote the probability of outputting YES on termination; then we have

$$p_w \geq n^{-n} + (1 - 2n^{-n})p_w,$$

where the first term on the right-hand side denotes the probability of succeeding on the very first trial, while the second term denotes success thereafter. Solving, we have $p_w \geq 1/2$.

Theorem 6.14: *The above algorithm will, given an instance of STCON,*

1. *Always output* NO *if there is no path from s to t.*
2. *Output* YES *with probability at least* $1/2$ *if there is a path from s to t.*

The algorithm uses space $O(\log n)$.

Exercise 6.7: Derive a bound on the expected running time of the above algorithm.

6.7. Expanders and Rapidly Mixing Random Walks

In previous sections of this chapter, we have focused on the expected lengths of random walks. In this section, we study a different aspect of random walks. We know by Theorem 6.2 that the probability vector of the random walk eventually converges to the stationary distribution whenever one exists. We now study the rate at which the probability vector approaches this stationary distribution. This study will yield useful applications here and in Chapter 11.

In particular, we will focus our attention on random walks on a special class of graphs called *expanders*. An expander (see also Section 5.3) is a graph in which the neighborhood of any set of vertices S is large relative to the size of S. Since the expansion property cannot be destroyed by the addition of edges to the graph, a complete graph is the best possible expander. However, in most applications we require sparse expander graphs; ideally, the graph should have a linear number of edges, and in fact be of bounded degree. Henceforth, we will use the term *expander* to refer to bounded-degree graphs with the desired expansion properties; a formal definition appears below in Section 6.7.1.

In Section 5.3 we saw that a sparse random graph is quite likely to be an expanding graph. We also noted there that giving an explicit construction of an expander is a much harder problem. That this is a non-trivial task is supported by the fact that the problem of deciding whether a graph is an expander is known to be co-$\mathcal{NP}$-complete. The bottleneck appears to be that we need to verify the expansion of an exponentially large number of subsets of vertices. Happily for us, there exists a *partial* characterization of expanders using the machinery of algebraic graph theory. The power of these algebraic methods lies in their ability to simultaneously describe the properties of all possible subsets of vertices, although some precision is lost in the process. This leads to a proof that certain explicitly specified graphs are expanders.

After studying this algebraic characterization, we turn to random walks on expanders. An important property of random walks on expanders is that they are *rapidly mixing:* the corresponding Markov chain will quickly converge to its stationary distribution regardless of the starting state. The major result of this section determines just how quickly this convergence occurs.

6.7.1. Expanders and Eigenvalues

This section assumes knowledge of elementary linear algebra, and the reader may wish to review the material in Appendix B before proceeding further. Recall that in a multigraph there can be more than one undirected edge between any pair of vertices. The discussion in this section is more easily stated in terms of multigraphs, and we allow all graphs under consideration to have multiple edges. A multigraph may also have self-loops at vertices.

Consider an undirected (multi)graph $G(V, E)$ with n vertices. The adjacency matrix $A(G)$ of G is the $n \times n$ symmetric matrix where $A_{ij} = A_{ji}$ is the number of edges between the vertices v_i and v_j. When G is bipartite, we assume that it has two independent sets of vertices $X = \{v_1, \ldots, v_{n/2}\}$ and $Y = \{v_{n/2+1}, \ldots, v_n\}$. Observe that in this case the adjacency matrix can be decomposed into four blocks of equal size as shown below, where 0 denotes the all-zeros matrix and $\boldsymbol{B}$ encodes the edges between X and Y.

$$A(G) = \begin{bmatrix} 0 & \boldsymbol{B} \\ \boldsymbol{B}^T & 0 \end{bmatrix}.$$

Since $A(G)$ is symmetric, even if the eigenvalues $\lambda_1 \geq \lambda_2 \geq \cdots \geq \lambda_n$ are not necessarily all distinct, we can fix corresponding eigenvectors $\boldsymbol{e}_1, \ldots, \boldsymbol{e}_n$ that form an orthonormal basis.

We state without proof the following basic result from algebraic graph theory; pointers may be found in the Notes section; the reader is asked to verify some parts of this theorem in Problems 6.20–6.23.

Theorem 6.15 (Fundamental Theorem of Algebraic Graph Theory): *Let $G(V, E)$ be an n-vertex, undirected (multi)graph with maximum degree d. Then, under the canonical labeling of eigenvalues λ_i and orthonormal eigenvectors $\boldsymbol{e}_i$ for the matrix $A(G)$,*

1. *If G is connected, then $\lambda_2 < \lambda_1$.*
2. *For $1 \leq i \leq n$, $|\lambda_i| \leq d$.*
3. *d is an eigenvalue if and only if G is regular.*
4. *If G is d-regular, then the eigenvalue $\lambda_1 = d$ has the eigenvector $\boldsymbol{e}_1 = \frac{1}{\sqrt{n}}(1, 1, 1, \ldots, 1)$.*
5. *The graph G is bipartite if and only if for every eigenvalue λ there is an eigenvalue $-\lambda$ of the same multiplicity.*
6. *Suppose that G is connected. Then, G is bipartite if and only if $-\lambda_1$ is an eigenvalue.*

7. *If G is d-regular and bipartite, $\lambda_n = -d$ and $\mathbf{e}_n = \frac{1}{\sqrt{n}}(1,\ldots,1,-1,\ldots,-1)$ (the last $n/2$ entries in $\mathbf{e}_n$ are -1).*

If G consists of more than one connected component, Theorem 6.15 can be applied independently to each connected component. For convenience, in the sequel we will restrict our attention to studying the eigenvalue properties only for graphs that are *connected, bipartite,* and *regular.* For a d-regular graph G, $A(G)$ is a symmetric matrix with all row and column sums equal to d.

What does all this have to do with expanders? Consider the algebraic characterization of connectedness in terms of a separation between the first and the second eigenvalues. Note also that a graph is connected if and only if every set of vertices S has at least one neighbor outside of S. We can view the expansion property as a stronger version of this connectivity condition. Might it not be the case that the property of being an expander is equivalent to having a *strong* separation between these two eigenvalues? It turns out that this is close to the truth. But first we formally define an expander; while the usual definition of an expander requires a graph of maximum degree d, we prefer to work with d-regular graphs.

► **Definition 6.13:** An *(n,d,c)-expander* is a d-regular bipartite (multi)graph $G(X,Y,E)$ with $|X| = |Y| = n/2$ such that for any $S \subseteq X$,

$$|\Gamma(S)| \geq \left(1 + c\left(1 - \frac{2|S|}{n}\right)\right)|S|.$$

As we remarked above, it will be convenient to assume that any expander under consideration is connected. In most applications, it is desirable to have d as small as possible and c as large as possible. In particular, we would like d to be bounded and c to be a constant greater than 0. Much as in Section 5.3, it is possible to give a probabilistic proof of existence of expanders for suitable values of n, d, and c by showing that a random graph chosen from an appropriate probability space is likely to be an expander. Several explicit constructions of such expanders are also known, but we describe only the so-called Gabber-Galil expanders.

For a positive integer m, let $n = 2m^2$. Each vertex in X is given a distinct label consisting of a pair (a,b) for a, $b \in \mathbb{Z}_m$; the vertices in Y are labeled similarly. A vertex labeled (x,y) in X has edges going to the vertices in Y whose labels are: (x,y), $(x,x+y)$, $(x,x+y+1)$, $(x+y,y)$, and $(x+y+1,y)$. The addition is done modulo m. Each of these linear functions is a permutation and defines a perfect matching between X and Y. The graph is 5-regular, and it can be shown that the expansion factor for this graph is $\alpha = (2-\sqrt{3})/4$, giving us a family of $(n,5,\alpha)$-expanders. We can obtain $(n,7,2\alpha)$-expanders using instead the following seven linear functions modulo m: (x,y), $(x,2x+y)$, $(x,2x+y+1)$, $(x,2x+y+2)$, $(x+2y,y)$, $(x+2y+1,y)$, and $(x+2y+2,y)$. The proof of the expansion property is beyond the scope of this book. Note that both graphs have multiple edges but these could be removed without affecting the expansion properties.

Usually, explicit construction of expanders such as these is required to specify an *(n, d, c)-expander family*. This means that the construction must provide an infinite sequence of graphs G_1, G_2, ..., such that the number of vertices in these graphs forms a strictly increasing sequence. Since the choice of the number m is arbitrary, the Gabber-Galil expander definition is easily seen to specify such a family of graphs.

As we saw in Section 5.3.1, in some applications we have to use expanders with a super-polynomial number of vertices. This presents problems when we are trying to perform some polynomial-time computation based on the structure of such graphs. However, we do not need to explicitly represent the Gabber-Galil expanders. It is easy to see that there is a polynomial-time neighborhood algorithm that can compute the neighbors of any given vertex in R; we can implicitly represent the graph by means of this neighborhood algorithm.

Finally, we note the following theorems, which make explicit the connection between the expansion properties of graphs and their eigenvalues. A pointer to their proofs is given in the Notes section. The proofs of these theorems are somewhat complicated and involve numerous calculations and estimates, but below we derive a closely related result (Theorem 6.19) that captures much of the intuition behind their proofs.

Theorem 6.16: *If G is an (n, d, c)-expander, then $A(G)$ has*

$$|\lambda_2| \leq d - \frac{c^2}{1024 + 2c^2}.$$

Theorem 6.17: *If $A(G)$ has $|\lambda_2| \leq d - \epsilon$, then G is an (n, d, c)-expander with*

$$c \geq \frac{2d\epsilon - \epsilon^2}{d^2}.$$

Since the largest eigenvalue λ_1 is exactly d, this gives a (partial) characterization of the expansion factor c in terms of the gap between the absolute values of the first and second eigenvalues.

Exercise 6.8: Given an (n, d, c)-expander, Theorem 6.16 yields a bound on λ_2; if we were now to use this bound on λ_2 in Theorem 6.17, what bound on c do we obtain and how does it compare with the value c that we started with?

Note that we are assuming that the expanders are connected; otherwise, $\lambda_2 = \lambda_1$ and we will have to use the eigenvalue of the second-largest absolute value to play the role of λ_2. It should be easy to see that relaxing this assumption makes no essential difference to the following discussion, but does make the notation more cumbersome.

We now give a result related to Theorems 6.16 and 6.17 to motivate the intuition behind these theorems. For a d-regular graph $G = (V, E)$, define

$$\text{split}(G) = \min_{\emptyset \subset X \subset V} \frac{|e(X, V \setminus X)|}{|X||V \setminus X|},$$

where $e(A, B)$ denotes the multiset of edges of G between subsets of vertices A, B. We now relate split(G) to λ_2, the second eigenvalue of the adjacency matrix $A(G)$ of G. First, we give a technical lemma concerning λ_2.

Lemma 6.18:

$$\lambda_2 = \max\{2 \sum_{(i,j) \in E} x_i x_j\},$$

where the max *is taken over vectors* $\boldsymbol{x}$ *such that* $||\boldsymbol{x}|| = 1$ *and* $\sum_{i=1}^{n} x_i = 0$.

The proof of Lemma 6.18 follows from the *Courant-Fisher equalities* established in Problem 6.19.

Theorem 6.19: *If G is d-regular, then*

$$\text{split}(G) \geq \frac{d - \lambda_2}{n}.$$

PROOF: Let $W \subset V$ and $|W| = k$. Define the vector $\boldsymbol{x}$ by

$$x_i = \begin{cases} \sqrt{\frac{n-k}{nk}} & \text{if } i \in W; \\ -\sqrt{\frac{k}{n(n-k)}} & \text{if } i \notin W. \end{cases}$$

Then $\boldsymbol{e}_1^T \boldsymbol{x} = 0$ and $||\boldsymbol{x}|| = 1$. By Lemma 6.18,

$$\lambda_2 \geq 2 \sum_{(i,j) \in E} x_i x_j = d - \sum_{(i,j) \in E} (x_i - x_j)^2 \tag{6.5}$$

$$= d - |e(W, V \setminus W)| \left(\sqrt{\frac{n-k}{nk}} + \sqrt{\frac{k}{n(n-k)}} \right)^2 \tag{6.6}$$

$$= d - \frac{n\, |e(W, V \setminus W)|}{k(n-k)}. \tag{6.7}$$

The result now follows from the definition of split(G). □

Corollary 6.20: *If G is d-regular then for any* $W \subset V$,

$$|W \cup \Gamma(W)| \geq [1 + (1 - \lambda_2/d)/2]\, |W|.$$

In the applications of expanders discussed here, we are primarily concerned with the eigenvalue separation for the adjacency matrix and we do not explicitly employ the expansion property itself. In fact, we could very well have defined expanders in terms of the eigenvalue separation, but the expansion property does serve to provide some intuition behind the algebraic machinery.

6.7.2. Random Walks on Expanders

Consider the simple random walk on an (n, d, c)-expander G. Since we permit multigraphs in the definition of expanders, it is necessary to generalize the definition of the random walk, as follows: at each step, the random walk proceeds along a randomly chosen edge among those incident on the current vertex v; thus, if there are k edges from v to w, then the probability that the random walk goes from v to w is $k/d(v)$. For an (n, d, c)-expander G, this corresponds to a Markov chain with the probability transition matrix $\boldsymbol{P} = \boldsymbol{A}(G)/d$.

Simple algebra shows that the eigenvalues of $\boldsymbol{P}$ are given by λ_i/d, and the corresponding eigenvectors remain unchanged. Notice that now all the eigenvalues lie between 1 and -1, and the gap between the first and second eigenvalue is reduced by a factor of d. A technical problem is that the random walk on such a bipartite graph results in a periodic Markov chain. We use a standard trick to get around this problem: reduce all transition probabilities by a factor of 2, and add a self-loop of probability 1/2 at each vertex. Observe that the new Markov chain still has G as its underlying graph, but the transition probability matrix $\boldsymbol{Q} = (\boldsymbol{I} + \boldsymbol{P})/2$ now has a stationary distribution.

Let the eigenvalues of $\boldsymbol{Q}$ be $\lambda_1', \ldots, \lambda_n'$. Since the identity matrix has all its eigenvalues equal to 1, it can be verified (see Problem 6.26) that the eigenvalues of $\boldsymbol{Q}$ are given by

$$\lambda_i' = \frac{(1 + \lambda_i/d)}{2}.$$

Thus, $1 = \lambda_1' \geq \lambda_2' \geq \cdots \geq \lambda_n' = 0$ and, assuming that $\lambda_2 = d - \epsilon$, we have that $\lambda_2' = 1 - \epsilon/2d$. The eigenvectors of $\boldsymbol{Q}$ can be chosen to form an orthonormal basis since it is a symmetric matrix. In fact, the first eigenvector $\boldsymbol{e}_1'$ is the same as that of $\boldsymbol{A}$, i.e., $\frac{1}{\sqrt{n}}(1, 1, 1, \ldots, 1)$.

Exercise 6.9: Verify that Q is a doubly stochastic matrix. Using the result from Problem 6.6, conclude that for the transition matrix Q, the stationary distribution is necessarily the uniform distribution.

We show that the Markov chain defined by $\boldsymbol{Q}$ is "rapidly mixing" in the following sense. Starting from any initial distribution, the Markov chain converges to its stationary distribution in a small number of steps. To make this notion more precise, we first define measure of convergence to the stationary distribution.

► **Definition 6.14:** Let $\boldsymbol{q}^{(t)}$ denote the state probability vector of a Markov chain defined by $\boldsymbol{Q}$ at time $t \geq 0$, given any initial distribution $\boldsymbol{q}^{(0)}$. Let $\boldsymbol{\pi}$ denote the stationary distribution of $\boldsymbol{Q}$. The *relative pointwise distance* (r.p.d.) of the Markov chain at time t is a measure of deviation from the limit and is defined as

$$\Delta(t) = \max_i \frac{|q_i^{(t)} - \pi_i|}{\pi_i}.$$

Intuitively, the change in Δ with t measures the rate of convergence to the stationary distribution, independent of the choice of the initial distribution $\boldsymbol{q}^{(0)}$. There are several types of distance functions defined in the literature for measuring the difference between two probability distributions; in Problem 6.24, we explore the connections between the relative pairwise distance and these other measures.

The next theorem shows that the relative pointwise distance for the random walk on an expander converges to zero at an exponential rate.

Theorem 6.21: *Let $\boldsymbol{Q}$ be the transition matrix of the aperiodic random walk on a (n, d, c)-expander G with $\lambda_2 \leq d - \epsilon$. Then, for any initial distribution $\boldsymbol{q}^{(0)}$, the relative pointwise distance is bounded as follows:*

$$\Delta(t) \leq n^{1.5}(\lambda_2')^t \leq n^{1.5}\left(1 - \frac{\epsilon}{2d}\right)^t.$$

PROOF: We know that the distribution of the Markov chain at time t is given by the following equation:

$$\boldsymbol{q}^{(t)} = \boldsymbol{q}^{(0)}\boldsymbol{Q}^t. \tag{6.8}$$

Now the eigenvectors of $\boldsymbol{Q}$ are chosen to form an orthonormal basis for $\mathbb{R}^n$. This implies that we can write $\boldsymbol{q}^{(0)}$ as a linear combination of those vectors, as follows:

$$\boldsymbol{q}^{(0)} = \sum_{i=1}^{n} c_i \boldsymbol{e}_i. \tag{6.9}$$

Combining (6.8) and (6.9), we obtain

$$\boldsymbol{q}^{(t)} = \sum_{i=1}^{n} c_i \boldsymbol{e}_i \boldsymbol{Q}^t = \sum_{i=1}^{n} c_i (\lambda_i')^t \boldsymbol{e}_i.$$

Let $\mathcal{L} \subset \mathbb{R}^n$ be the vector space spanned by the first eigenvector $\boldsymbol{e}_1$. This space contains all scalar multiples of the all-ones vector; the orthogonal space $\mathcal{L}^\perp$ contains all linear combinations of the remaining $n-1$ eigenvectors. Then $\boldsymbol{q}^{(0)} = \boldsymbol{x} + \boldsymbol{y}$ for some $\boldsymbol{x} \in \mathcal{L}$ and $\boldsymbol{y} \in \mathcal{L}^\perp$; in fact, $\boldsymbol{x} = c_1\boldsymbol{e}_1$ and $\boldsymbol{y} = \sum_{i=2}^{n} c_i \boldsymbol{e}_i$. Since $\boldsymbol{x}$ and $\boldsymbol{y}$ are orthogonal, the Pythagoras Inequality (Proposition B.8) implies that $||\boldsymbol{x}|| \leq ||\boldsymbol{q}^{(0)}||$ and $||\boldsymbol{y}|| \leq ||\boldsymbol{q}^{(0)}||$.

Since $\lambda_1' = 1$, $\boldsymbol{x}\boldsymbol{Q} = \boldsymbol{x}$ and we can write

$$\boldsymbol{q}^{(t)} = \boldsymbol{q}^{(0)}\boldsymbol{Q}^t = (\boldsymbol{x} + \boldsymbol{y})\boldsymbol{Q}^t = \boldsymbol{x} + \sum_{i=2}^{n} c_i (\lambda_i')^t \boldsymbol{e}_i.$$

We now obtain the following bounds on the L_1-norm of $\boldsymbol{q}^{(t)} - \boldsymbol{x}$.

$$
\begin{aligned}
\|\boldsymbol{q}^{(t)} - \boldsymbol{x}\|_1 &\leq \sqrt{n}\|\boldsymbol{q}^{(t)} - \boldsymbol{x}\| && (6.10)\\
&= \sqrt{n}\|\sum_{i=2}^{n} c_i(\lambda_i)^t \boldsymbol{e}_i\| && (6.11)\\
&= \sqrt{n}\sqrt{\sum_{i=2}^{n} c_i^2(\lambda_i)^{2t}} && (6.12)\\
&\leq \sqrt{n}\sqrt{\sum_{i=2}^{n} c_i^2(\lambda_2')^{2t}} && (6.13)\\
&\leq \sqrt{n}(\lambda_2')^t\sqrt{\sum_{i=2}^{n} c_i^2} && (6.14)\\
&\leq \sqrt{n}(\lambda_2')^t\|\boldsymbol{y}\| && (6.15)\\
&\leq \sqrt{n}(\lambda_2')^t\|\boldsymbol{q}^{(0)}\|. && (6.16)
\end{aligned}
$$

The inequality (6.13) relies on the fact that λ_2' has the second largest absolute value; the inequality (6.15) follows from the fact that $\boldsymbol{y} = \sum_{i=2}^{n} c_i \boldsymbol{e}_i$; the inequality (6.16) is a consequence of the Pythagoras Inequality. Since $\boldsymbol{q}^{(0)}$ is a probability distribution, its components are all non-negative and sum to 1; thus, by Proposition B.10, $\|\boldsymbol{q}^{(0)}\| \leq \|\boldsymbol{q}^{(0)}\|_1 = 1$. We obtain that

$$\|\boldsymbol{q}^{(t)} - \boldsymbol{x}\|_1 \leq \sqrt{n}(\lambda_2')^t.$$

By Problem 6.6 we know that for any doubly stochastic matrix, the stationary distribution $\boldsymbol{\pi}$ must be uniform. Since $\lambda_2' < 1$, we know that as t increases, $\|\boldsymbol{q}^{(t)} - \boldsymbol{x}\|$ goes to 0 and $\boldsymbol{q}^{(t)}$ converges to $\boldsymbol{x}$. We conclude that $\boldsymbol{x} = \boldsymbol{\pi}$, and that

$$\|\boldsymbol{q}^{(t)} - \boldsymbol{\pi}\|_1 \leq \sqrt{n}(\lambda_2')^t.$$

The relative pointwise distance can now be bounded as follows.

$$
\begin{aligned}
\Delta(t) &= \max_i \frac{|q_i^{(t)} - \pi_i|}{\pi_i}\\
&= n \times \max_i |q_i^{(t)} - \pi_i|\\
&\leq n \times \|\boldsymbol{q}^{(t)} - \boldsymbol{\pi}\|_1\\
&\leq n \times \sqrt{n}(\lambda_2')^t\\
&= n^{1.5}(\lambda_2')^t.
\end{aligned}
$$

□

Exercise 6.10: For any $0 < \delta < 1$, let $T(\delta)$ denote the time at which the relative pointwise distance of the random walk defined by $\boldsymbol{Q}$ first falls below δ. Show that

$$T(\delta) \leq \frac{\log n^{1.5}/\delta}{-\log \lambda_2'}.$$

By this exercise, to get a relative pointwise distance that is bounded from above by an inverse polynomial in n, it suffices to run the random walk for only a logarithmic number of steps. Notice that this is the best possible bound since the length of the random walk must be at least the diameter of the expander. Since our expander has bounded degree, it has diameter $\Omega(\log n)$.

6.8. Probability Amplification by Random Walks on Expanders

Recall the 2-point sampling scheme of Section 3.4. Given an ***RP*** algorithm, which uses n random bits to obtain a probability of error $1/2$, this scheme reduced the probability of error to $O(1/t)$ while using only $2n$ random bits and t trials of the algorithm. Even using k-point sampling for $k > 2$, there is no hope of achieving a probability of error that is exponentially small in the number of trials, without using a significantly larger number of random bits. Also, in Section 5.3 we saw that expander-type graphs could be used to achieve a stronger probability amplification, but several important issues remained unresolved in that discussion and in any case that scheme did not provide the desired exponentially small error probability with a small number of random bits. Here we present a related technique that achieves the desired exponential behavior, even in the case of ***BPP*** algorithms, and without any of the drawbacks of the earlier scheme based on expanders. The version of this technique that establishes the same result for ***RP*** algorithms is slightly easier to analyze (see Problem 6.29).

Without loss of generality, we modify the standard definition of ***BPP*** such that the probability of error is $1/100$; clearly, this can be achieved via $O(1)$ independent iterations of an algorithm meeting only the standard definition.

► **Definition 6.15:** The class ***BPP*** consists of all languages L that have a randomized polynomial-time algorithm $\mathcal{A}$ such that for any $x \in \Sigma^*$, given a suitably long random string r,

- $x \in L \Rightarrow \mathbf{Pr}[\mathcal{A}(x,r) \text{ rejects}] \leq \frac{1}{100}$.
- $x \notin L \Rightarrow \mathbf{Pr}[\mathcal{A}(x,r) \text{ accepts}] \leq \frac{1}{100}$.

Fix an input x, and consider a ***BPP*** algorithm $\mathcal{A}$ that uses n random bits on inputs of length $|x|$. Suppose we choose k independent n-bit random strings r_1, $\ldots$, r_k, and compute $\mathcal{A}(x,r_1), \ldots, \mathcal{A}(x,r_k)$. By the Chernoff bound, the probability that the majority of these outputs is incorrect is $1/2^{\Omega(k)}$. Thus, we have made the error probability exponentially small in k using nk random bits. The *probability amplification* problem is that of achieving this error probability while using the minimum possible number of random bits. What is the minimum number of random bits required for the exponentially small error probability of $1/2^k$?

Consider forming a crude estimate as follows. Imagine that a single execution of the algorithm consumes n random bits and delivers one bit as the result of execution (i.e., a decision on the membership of x in L); it appears plausible that $n-1$ random bits remain available for future executions of the algorithm. Intuitively then, we should not have to use more than $n+k-1$ random bits for k repeated executions of the algorithm. The following scheme comes very close to realizing this intuition, using as it does $n+\mathrm{O}(k)$ bits.

Consider the $(N, 7, 2\alpha)$-expander family described in Section 6.7.1. We assume that n is odd; otherwise we can increase by one the number of random bits used by $\mathcal{A}$ by throwing in a dummy random bit. Choose $m = 2^{(n-1)/2}$ and $N = 2m^2 = 2^n$, and label each vertex with a distinct sequence of bits from $\{0,1\}^n$. Let $\boldsymbol{A}$ be the adjacency matrix of the resulting expander. Let $\boldsymbol{Q} = (\boldsymbol{I} + \boldsymbol{A}/7)/2$ be the probability transition matrix of the ergodic Markov chain obtained by performing a random walk on this graph, with a self-loop of probability 1/2 at each vertex. We assume that the random walk starts at a uniformly chosen initial vertex. Denote by X_0, X_1, ... the states of the resulting Markov chain. Note that each X_i corresponds to a particular setting of the random bits used by $\mathcal{A}$.

Choose a positive integer β such that $\lambda_2^\beta \leq 1/10$, where λ_i is the ith largest eigenvalue of $\boldsymbol{Q}$. Since the graph is an expander, λ_2 is bounded away from 1 and we are guaranteed that a value of β that is O(1) will suffice.

Given the output from the random walk process described above, the probability amplification scheme works as follows. For $0 \leq i \leq 7k$, let $r_i = X_{i\beta}$. Run the algorithm $\mathcal{A}(x, *)$ using these $7k$ different choices of random inputs. Declare the majority of these $7k$ YES/NO decisions to be the final decision; for convenience, we assume that k is odd. We will show that the resulting decision is wrong with probability at most $1/2^k$. Note that the total number of random bits used is $n + \mathrm{O}(k)$: we need n bits to choose the starting vertex of the random walk, and 4 bits for each of the $7k\beta$ subsequent steps of the random walk. Also, the locally defined neighborhood structure of the Gabber-Galil expander has the crucial advantage that we do not need to explicitly construct the entire graph, whose size is exponential in n (the number of random bits given to $\mathcal{A}$). In particular, given the index for any vertex in the expander, it is possible to compute the indices of the neighboring vertices in time polynomial in the *length* of the index, i.e., n. This suffices for the purposes of obtaining a polynomial time implementation of each step of the random walk.

The intuition behind this scheme is as follows. We know that the random walk on an expander is rapidly mixing. In other words, given any starting vertex, after a small number of steps we expect the random walk to be at a uniformly distributed vertex independent of the choice of the initial vertex. We can view the above process as using the composition of $7k$ different random walks, each generating a different random string r_i. The catch here is that each of these smaller random walks has length $\beta = \mathrm{O}(1)$, whereas we would require $\Theta(\log N)$ steps to get close to the stationary distribution. On the other hand, we choose the initial vertex according to the stationary distribution, and this should work in our favor.

Let us denote the probability distribution vector for $r_i = X_{i\beta}$ as $\boldsymbol{p}^{(i)}$. Define $\boldsymbol{B} = \boldsymbol{Q}^{\beta}$; this is the transition matrix for the Markov chain corresponding to the sequence of r_i's. We have that $\boldsymbol{p}^{(i)} = \boldsymbol{p}^{(0)}\boldsymbol{B}^i$, where $\boldsymbol{p}^{(0)}$ is the uniform distribution that we start with.

Let $\mathcal{W}$ denote the set of witnesses for the input x. In other words, $\mathcal{W} = \{r \in \{0,1\}^n \mid \mathcal{A}(x,r) \text{ is correct}\}$. We are guaranteed that $|\mathcal{W}| \geq 0.99N$. The set of non-witnesses has cardinality $|\overline{\mathcal{W}}| \leq 0.01N$. We define the 0-1 $N \times N$ diagonal matrix $\boldsymbol{W}$ such that $W_{ii} = 1$ if and only if the ith vertex corresponds to a string that is a witness for x; similarly, the 0-1 $N \times N$ diagonal matrix $\overline{\boldsymbol{W}} = \boldsymbol{I} - \boldsymbol{W}$. The reader is invited to verify that $||\boldsymbol{p}^{(i)}\boldsymbol{W}||_1$ and $||\boldsymbol{p}^{(i)}\overline{\boldsymbol{W}}||_1$ are the probabilities that r_i is a witness or a non-witness, respectively. This is because the linear transformation $\boldsymbol{W}$ zeros out the entries corresponding to the non-witnesses, leaving the others untouched; the transformation $\overline{\boldsymbol{W}}$ does the converse.

Consider the sequence of strings $r_1, \ldots, r_{7k}$. Let the event sequence of matrices $\mathcal{S} = (\boldsymbol{S}_1, \ldots, \boldsymbol{S}_{7k}) \in \{\boldsymbol{W}, \overline{\boldsymbol{W}}\}^{7k}$ be such that $\boldsymbol{S}_i = \boldsymbol{W}$ if and only if $r_i \in \mathcal{W}$. Thus, $\mathcal{S}$ encodes the pattern of errors in the various executions of the algorithm. The following lemma is a direct consequence of these definitions.

Lemma 6.22: *For any fixed event sequence $\mathcal{S}$,*

$$\mathbf{Pr}[\mathcal{S} \text{ occurs}] = ||\boldsymbol{p}^{(0)}(\boldsymbol{BS}_1)(\boldsymbol{BS}_2)\cdots(\boldsymbol{BS}_{7k-1})(\boldsymbol{BS}_{7k})||_1.$$

The proof of the next lemma is deferred for the moment.

Lemma 6.23: *For all vectors $\boldsymbol{p} \in \mathbb{R}^N$,*

1. $||\boldsymbol{pBW}|| \leq ||\boldsymbol{p}||$,
2. $||\boldsymbol{pB\overline{W}}|| \leq \frac{1}{5}||\boldsymbol{p}||$.

We now prove that this probability amplification scheme gives the desired error probability, and then we complete the analysis by giving the proof of Lemma 6.23.

Theorem 6.24: *The probability that the majority of the outputs $\mathcal{A}(x,r_1)$, ..., $\mathcal{A}(x,r_{7k})$ is incorrect is at most $1/2^k$.*

PROOF: Note that the majority of the outputs is incorrect only if the event sequence $\mathcal{S}$ has more than half of its elements equal to $\overline{\boldsymbol{W}}$. Fix any particular $\mathcal{S}$ whose elements contain a majority of $\overline{\boldsymbol{W}}$'s, say $\kappa \geq 7k/2$ of them. By Lemma 6.22,

$$\begin{aligned}\mathbf{Pr}[\mathcal{S} \text{ occurs}] &= ||\boldsymbol{p}^{(0)}(\boldsymbol{BS}_1)(\boldsymbol{BS}_2)\cdots(\boldsymbol{BS}_{7k-1})(\boldsymbol{BS}_{7k})||_1 && (6.17)\\ &\leq \sqrt{N}||\boldsymbol{p}^{(0)}(\boldsymbol{BS}_1)(\boldsymbol{BS}_2)\cdots(\boldsymbol{BS}_{7k-1})(\boldsymbol{BS}_{7k})|| && (6.18)\end{aligned}$$

$$\leq \sqrt{N}\left(\frac{1}{5}\right)^{\kappa} ||\boldsymbol{p}^{(0)}|| \tag{6.19}$$

$$\leq \sqrt{N}\left(\frac{1}{5}\right)^{7k/2} ||\boldsymbol{p}^{(0)}||, \tag{6.20}$$

where the inequality (6.19) follows from a repeated application of Lemma 6.23. Since we chose $\boldsymbol{p}^{(0)}$ to be uniform on the N vertices, it is clear that its L_2 norm is exactly $1/\sqrt{N}$. Finally, using the overestimate that the number of sequences $\mathcal{S}$ with a majority of $\overline{\boldsymbol{W}}$'s is at most 2^{7k}, we obtain

$$\begin{aligned}
\mathbf{Pr}[\text{Majority vote is incorrect}] &\leq 2^{7k}\sqrt{N}\left(\frac{1}{5}\right)^{7k/2} ||\boldsymbol{p}^{(0)}|| \\
&\leq \frac{2^{7k}}{5^{7k/2}} \\
&\leq \frac{1}{2^k}.
\end{aligned}$$

□

We complete the analysis by giving the proof of Lemma 6.23.

Proof of Lemma 6.23: Recall that the eigenvalues of $\boldsymbol{Q}$ are all in the interval $[0,1]$ with $\lambda_1 = 1$ and $\lambda_2^\beta \leq 1/10$. Let $\boldsymbol{e}_1, \ldots, \boldsymbol{e}_N$ be an orthonormal set of eigenvectors corresponding to these eigenvalues. The vector $\boldsymbol{p}$ can be expressed as a linear combination of the eigenvectors, say $\sum_{i=1}^N c_i\boldsymbol{e}_i$; further, $||\boldsymbol{p}|| = \sqrt{\sum_{i=1}^N c_i^2}$.

To prove the first part of the lemma, note that $||\boldsymbol{pBW}|| \leq ||\boldsymbol{pB}||$. This is because the diagonal matrix $\boldsymbol{W}$ only has 1's on its diagonal, and this can zero out only some of the components of the vector $\boldsymbol{pB}$, thereby decreasing its L_2-norm. Moreover, $\boldsymbol{pB} = \sum_{i=1}^N c_i\boldsymbol{e}_i\boldsymbol{B} = \sum_{i=1}^N c_i\lambda_i^\beta \boldsymbol{e}_i$. We thus have

$$||\boldsymbol{pBW}|| \leq ||\boldsymbol{pB}|| = ||\sum_{i=1}^N c_i\lambda_i^\beta \boldsymbol{e}_i|| = \sqrt{\sum_{i=1}^N c_i^2\lambda_i^{2\beta}} \leq \sqrt{\sum_{i=1}^N c_i^2}.$$

The last inequality makes use of the fact that each λ_i lies in $[0,1]$. Since the last expression is $||\boldsymbol{p}||$, we obtain the desired result.

Consider now the second part of the lemma. Let us decompose $\boldsymbol{p} = \boldsymbol{x} + \boldsymbol{y}$, where $\boldsymbol{x} = c_1\boldsymbol{e}_1$ and $\boldsymbol{y} = \sum_{i=2}^N c_i\boldsymbol{e}_i$. By the Pythagoras Inequality (Proposition B.8), $||\boldsymbol{x}|| \leq ||\boldsymbol{p}||$ and $||\boldsymbol{y}|| \leq ||\boldsymbol{p}||$. We first derive independent inequalities for $\boldsymbol{x}$ and $\boldsymbol{y}$.

Observe that $\boldsymbol{xB} = c_1\boldsymbol{e}_1\boldsymbol{B} = c_1\lambda_1\boldsymbol{e}_1 = \boldsymbol{x}$, since $\lambda_1 = 1$. We claim that $||\boldsymbol{x}\overline{\boldsymbol{W}}|| \leq ||\boldsymbol{x}||/10$. Recall that $\overline{\boldsymbol{W}}$ is a 0-1 diagonal matrix, where the fraction of non-zero entries on its diagonal is no more than $1/100$. Therefore, it zeros out all but a $1/100$ fraction of the entries in $\boldsymbol{x}$. Moreover, $\boldsymbol{x}$ is a scalar multiple of the all-ones vector, and so all its components are equal. Reducing all but a $1/100$ fraction of its components to 0 will reduce its L_2-norm by a factor of $\sqrt{100}$. Thus, $||\boldsymbol{xB}\overline{\boldsymbol{W}}|| = ||\boldsymbol{x}\overline{\boldsymbol{W}}|| \leq ||\boldsymbol{x}||/10$.

A similar inequality can be obtained for $\boldsymbol{y}$ as follows. Observe that $\boldsymbol{yB} = \sum_{i=2}^{N} c_i \boldsymbol{e}_i \boldsymbol{B} = \sum_{i=2}^{N} c_i \lambda_i^{\beta} \boldsymbol{e}_i$. It is also clear that $||\boldsymbol{y}\boldsymbol{B}\overline{\boldsymbol{W}}|| \leq ||\boldsymbol{yB}||$ since $\overline{\boldsymbol{W}}$ is only zeroing out some entries in the vector $\boldsymbol{yB}$. Since $\lambda_2^{\beta} \leq 1/10$ corresponds to the second largest eigenvalue,

$$||\boldsymbol{y}\boldsymbol{B}\overline{\boldsymbol{W}}|| \leq \sqrt{\sum_{i=2}^{N} c_i^2 \lambda_i^{2\beta}} \leq \lambda_2^{\beta} \sqrt{\sum_{i=2}^{N} c_i^2} \leq \frac{1}{10} ||\boldsymbol{y}||.$$

Using the triangle inequality, we obtain that

$$||\boldsymbol{p}\boldsymbol{B}\overline{\boldsymbol{W}}|| \leq ||\boldsymbol{x}\boldsymbol{B}\overline{\boldsymbol{W}}|| + ||\boldsymbol{y}\boldsymbol{B}\overline{\boldsymbol{W}}|| \leq \frac{1}{10} \left(||\boldsymbol{x}|| + ||\boldsymbol{y}||\right).$$

Finally, applying inequalities $||\boldsymbol{x}|| \leq ||\boldsymbol{p}||$ and $||\boldsymbol{y}|| \leq ||\boldsymbol{p}||$, we obtain the desired bound.

Notes

Aldous [13] is a comprehensive source for random walks on graphs, as well as some advanced algorithmic applications that are beyond the scope of this book. The 2-SAT algorithm of Section 6.1 is due to Papadimitriou [325]. McDiarmid [303] has independently given a number of applications of this technique to coloring the vertices of a hypergraph. An excellent source for basic Markov chain theory is the book by Kemeny, Snell, and Knapp [253]. The relationship of random walks to electrical networks has been known for over a century. Doyle and Snell [130] demonstrate many interesting relations between random walks in graphs and electrical networks. Their work deals with finite as well as infinite graphs and highlights many tools from electrical network analysis that are useful in the study of random walks. Theorem 6.6 is due to Chandra, Raghavan, Ruzzo, Smolensky, and Tiwari [89]. Tetali [396] gives an interesting refinement and generalization of Theorem 6.6. The Short-cut Principle is due to Rayleigh and is described in [130, 301].

Theorem 6.8 is due to Aleliunas, Karp, Lipton, Lovász, and Rackoff [15] and builds on work of Göbel and Jagers [168]. A version of Theorem 6.12 is derived in [15]; our presentation follows Chandra, Raghavan, Ruzzo, Smolensky, and Tiwari [89]. Our proof of Theorem 6.9 is also taken from [89], although in fact a stronger version of Theorem 6.9 appears earlier in work of Matthews [296, 297]. Matthews gives an elegant approach to proving upper and lower bounds on cover times in terms of hitting times (this is the subject of Problem 6.9). Broder and Karlin [85] give a number of relations between the cover time of a graph and the second-largest eigenvalue of its adjacency matrix. Undirected *s–t* connectivity is a natural abstraction of many graph search procedures; in addition, it has applications to complexity theory [276]. Borodin, Cook, Dymond, Ruzzo, and Tompa [74] have given a Las Vegas algorithm for USTCON whose running time is polynomial in *n*. The idea of using a probabilistic counter for a space-efficient algorithm for directed *s–t* connectivity is due to Gill [166]. The reader may refer to the paper by Borodin, Ruzzo, and Tompa [78] for further material on universal traversal sequences.

The books by Biggs [60] and Cvetkovic, Doob, and Sachs [117] provide comprehensive treatments of algebraic graph theory. The article by Bien [59] surveys the definitions, properties, and explicit construction of expanders, as well as many other related classes types of graphs such as magnifiers. The co-*NP*-completeness of the problem of verifying the expansion property in graphs is due to Blum, Karp, Vornberger, Papadimitriou, and Yannakakis [67]. The explicit expanders introduced in Section 6.7.1 are due to Gabber and Galil [158]. Theorems 6.16 and 6.17 are due to Alon [17] and are extensions of the earlier work of Tanner [389] and Alon and Milman [23]. Theorem 6.19 is due to Donath and Hoffman [128], and Corollary 6.20 due to Alon and Milman [23]. The rapid mixing property of random walks was first exploited by Ajtai, Komlós, and Szemerédi [9] in a complexity-theoretic setting. The result on probability amplification is due to Cohen and Wigderson [108], and independently due to Impagliazzo and Zuckerman [205]. The former paper is also a good source for the known results on expanders and their applications. Gillman [167] bounds the probability that in a random walk on an expander, the frequency of visits to any subset of vertices deviates substantially from the sum of the stationary probabilities of those vertices. Dinwoodie [126] provides further results along these lines.

Problems

6.1 Consider a random walk on the infinite line. At each step, the position of the particle is one of the integer points. At the next step, it moves to one of the two neighboring points equiprobably. Show that the expected distance of the particle from the origin after n steps is $\Theta(\sqrt{n})$.

6.2 Consider the randomized algorithm for 2-SAT discussed in Section 6.1. Show that the analysis is tight, in that there exist satisfiable 2-SAT formulas with n variables such that the expected time for this algorithm to find a satisfying truth assignment is $\Omega(n^2)$.

6.3 Consider a *1-dimensional random walk with a reflecting barrier*, which is defined as follows. For each natural number i, there is a state i. At state 0, with probability 1 the walk will move to state 1. At every other state $i > 0$, the walk will move to state $i+1$ with probability p and to state $i-1$ with probability $1-p$. Prove the following for the resulting Markov chain:

(a) For $p > \frac{1}{2}$, each state is transient.

(b) For $p = \frac{1}{2}$, each state is null persistent.

(c) For $p < \frac{1}{2}$, each state is non-null persistent.

6.4 Consider a Markov chain with the states $0, 1, \ldots, N$. This Markov chain induces a sequence of random variables $X_0, X_1, \ldots$, each of which takes an integer value between 0 and N, i.e. X_t is the state at time t. Suppose this sequence of random variables forms a martingale.

(a) A state q is said to be an *absorbing* state if the transition probability $P_{qq} = 1$. Identify all the absorbing states and the transient states of this Markov chain.

(b) Given that the initial state of this Markov chain is i, compute the probability of being absorbed into each of the absorbing states.

6.5 (Due to C.J.H. McDiarmid [303].) Let G be a 3-colorable graph. Consider the following algorithm for coloring the vertices of G with 2 colors so that no triangle of G is monochromatic. The algorithm begins with an arbitrary 2-coloring of G. While there is a monochromatic triangle in G, it chooses one such triangle, and changes the color of a randomly chosen vertex of that triangle. Derive an upper bound on the expected number of such recoloring steps before the algorithm finds a 2-coloring with the desired property.

6.6 An $n \times n$ matrix $\boldsymbol{P}$ is said to be *stochastic* if all its entries are non-negative and for each row i, $\sum_j P_{ij} = 1$. It is said to be *doubly stochastic* if, in addition, $\sum_i P_{ij} = 1$.

(a) Show that for any stochastic matrix P, there exists an n-dimensional vector π with non-negative entries such that $\sum_i \pi_i = 1$ and $\pi\boldsymbol{P} = \pi$.

(b) Suppose that the transition probability matrix $\boldsymbol{P}$ for a Markov chain is doubly stochastic. Show that the stationary distribution for this Markov chain is necessarily the uniform distribution.

6.7 Consider a random walk on a graph whose edges have positive real *costs*: the interpretation of these costs is that every time the random walk traverses an edge (ij), it incurs a given cost $c_{ij} > 0$; $c_{ij} = c_{ji}$, and $c_{ii} = 0$. Consider the random walk on a graph G with m edges that have such costs associated with them, with transition probabilities

$$P_{ij} = \frac{1/c_{ij}}{\sum_k 1/c_{ik}}.$$

Let S_{uv} denote the expected total cost incurred by a walk that begins at vertex u and terminates upon returning to u after having visited v at least once. Show that

$$S_{uv} = 2mR_{uv},$$

where R_{uv} is the effective resistance between node u and node v in an electrical network whose underlying graph is G, and where the branch resistance between i and j is c_{ij}.

6.8 In a connected graph G, an edge is called a *bridge* if the removal of the edge disconnects the graph. Let G be a connected graph with n vertices and m edges. Let (u, v) be any edge in G. For the simple random walk on G, show that

$$h_{uv} + h_{vu} = 2m$$

if and only if the edge (u, v) is a bridge.

6.9 (Due to P.C. Matthews [296, 297].) The goal of this problem is to derive a cleaner version of Theorem 6.9. Consider a random permutation of the vertices of a connected graph G, and let J_i denote the ith vertex in this permutation. For $1 \le k \le n$, define $F_k = \max_{i \le k} T_{J_i}$ to be the time by which all of $\{J_1, J_2, \ldots, J_k\}$ have been visited (in some order). Let L_k be the last of the vertices in $\{J_1, J_2, \ldots, J_k\}$ to be visited. Let $\delta(ij)$ be the delta function, defined to be 1 if $i = j$ and 0 otherwise.

(a) Show that conditioned on the sequence of vertices visited until time F_{k-1}, and for a fixed set $\{J_1, J_2, \ldots, J_k\}$,

$$\mathbf{E}[F_k - F_{k-1}] = \delta(L_k J_k) h_{L_{k-1} J_k}.$$

(b) Hence infer that

$$\mathbf{E}[F_k - F_{k-1}] \geq \min_{i \neq j} h_{ij} \mathbf{Pr}[L_k = J_k].$$

(c) Now use the fact that the J_i are randomly ordered to show that

$$\min_u \mathcal{C}_u(G) \geq H_{n-1} \min_{i \neq j} h_{ij}.$$

(d) Repeat the above arguments to obtain an upper bound on cover time:

$$\max \mathcal{C}_u(G) \leq H_{n-1} \max_{i \neq j} h_{ij}.$$

6.10 By showing that the resistance of the complete graph K_n is $\Theta(1/n)$, show that the upper bound of Theorem 6.9 cannot be improved in general.

6.11 Let G be a regular graph with every vertex having degree d. Show that C_G is $O(n^2 \log n)$.

Remark: This shows that regular graphs have lower cover times than graphs that have large disparities in their vertex-degrees (such as the lollipop graph, which had $\mathcal{C}_{L_n}(G)$ as large as $\Theta(n^3)$). In fact, using a more careful argument, Kahn, Linial, Nisan, and Saks [224] show that for every regular graph, C_G is $O(n^2)$.

6.12 The result in Problem 6.11 can be improved for dense regular graphs. Let G be a regular graph with every vertex having degree $\geq 2n/3$. Show that C_G is $O(n \log n)$. Complement this upper bound by showing that for $d < n/2$ such that $d+1$ divides n, there exists a d-regular graph whose cover time is $\Omega(n^2)$. Derive an upper bound on $U(d, n)$ for $d \geq 2n/3$.

6.13 Consider the *two-dimensional mesh:* a graph in which each vertex is a point with integer coordinates in the plane, all coordinates being in the interval $[1, n^{1/2}]$. An edge connects two vertices if they differ in one coordinate by 1. Show that the maximum commute time in this graph is $\Theta(n \log n)$.

6.14 Consider next the *three-dimensional mesh:* a graph in which each vertex is thought of as a point with integer coordinates in three dimensions, all coordinates being in the interval $[1, n^{1/3}]$. Show that the cover time for this graph is $O(n \log n)$. Derive upper bounds for the lengths of the universal traversal sequences for labeled two-dimensional and three-dimensional meshes.

6.15 (a) Show that for $n = 3$ and $d = 2$, there exists a universal traversal sequence $U(d, n)$ of length 3.

(b) What is the smallest UTS you can construct for the case $n = 4$ and $d = 2$?

6.16 Show that the expected time for a random walk to visit every vertex of a strongly connected directed graph is not bounded above by any polynomial function of n, the number of vertices. In other words, construct a directed graph that is strongly connected and where the expected cover time is super-polynomial.

6.17 Show that any probabilistic, log-space, polynomial-time Turing machine can be simulated by a deterministic, non-uniform, log-space, polynomial-time Turing machine. (**Hint:** Use the ideas of Section 2.3.)

6.18 (Due to D. Zuckerman [424].) Let $G(V, E)$ be a graph with n vertices such that for some constant $\alpha > 0$ and every set $S \subseteq V$ with $n/2$ vertices,

$$|\{w \in V \mid \exists v \in S,\ (v, w) \in E\}| \geq \frac{n}{2} + \alpha n.$$

For any positive integer k, let $W_1, \ldots, W_k$ be subsets of V of size at least $(1-\alpha)n$ each. Show that there exists a path $(v_1, \ldots, v_k)$ in G such that, for $1 \leq i \leq k$, $v_i \in W_i$.

6.19 (Courant-Fisher equalities.) Let A be an $n \times n$ symmetric matrix with real entries, and let $\boldsymbol{e}_1$ denote the eigenvector corresponding to the first eigenvalue λ_1. Show that
(1) $\lambda_1 = \max\{\boldsymbol{x}^T A \boldsymbol{x}\}$, where the max is taken over $\boldsymbol{x}$ such that $||\boldsymbol{x}|| = 1$.
(2) $\lambda_n = \min\{\boldsymbol{x}^T A \boldsymbol{x}\}$, where the min is taken over $\boldsymbol{x}$ such that $||\boldsymbol{x}|| = 1$.
(3) $\lambda_2 = \max\{\boldsymbol{x}^T A \boldsymbol{x}\}$, where the max is taken over $\boldsymbol{x}$ such that $||\boldsymbol{x}|| = 1$ and $\boldsymbol{x}^T \boldsymbol{e}_1 = 0$.

6.20 Let $G(V, E)$ be a connected, d-regular, undirected (multi)graph with n vertices. Show that for the adjacency matrix $A(G)$, $\lambda_1 = d$ and $\boldsymbol{e}_1 = \frac{1}{\sqrt{n}}(1, 1, 1, \ldots, 1)$.

6.21 Let $G(V, E)$ be a connected, d-regular, undirected (multi)graph. Show that for the adjacency matrix $A(G)$, each eigenvalue λ_i has absolute value bounded by d.

6.22 Show that a connected graph G with maximum eigenvalue λ_1 is bipartite if and only if $-\lambda_1$ is also an eigenvalue.

6.23 Show that a graph G is bipartite if and only if for every eigenvalue λ, there is an eigenvalue $-\lambda$ of the same multiplicity.

6.24 Consider the setting of Definition 6.14 and the following measures of deviation from the limit. Let S denote the set of states of the Markov chain under consideration. The *total variation distance* is defined as

$$\overline{\Delta}(t) = \max_{T \subseteq S} |\sum_{i \in T} q_i^{(t)} - \sum_{i \in T} \pi_i|.$$

(a) Define the L_1 distance as

$$||\boldsymbol{q}^{(t)} - \boldsymbol{\pi}||_1 = \sum_i |q_i^{(t)} - \pi_i|.$$

Determine the relation between the L_1 distance and the total variation distance.

(b) Suppose that the relative pointwise distance is bounded by ϵ at time t. Give the tightest bound you can on the total variation distance at time t.

(c) Suppose that the total variation distance at time t is bounded by ϵ. What can you say about the relative pointwise distance at time t?

6.25 Does Theorem 6.21 hold true if the relative pointwise distance is replaced by the total variation distance defined in Problem 6.24?

6.26 Let G be d-regular, and define the matrix $\boldsymbol{Q} = (\boldsymbol{I} + \boldsymbol{A}(G)/d)/2$. Show that if the ith eigenvalue of $\boldsymbol{A}(G)$ is λ_i, then the ith eigenvalue of $\boldsymbol{Q}$ equals $(1 + \lambda_i/d)/2$.

6.27 (Due to N. Alon and F.R.K. Chung [20].) Let $G(X, Y, E)$ be a d-regular, connected, bipartite (multi)graph. Show that for any sets $S \subseteq X$ and $T \subseteq Y$, the number of edges connecting S and T is at least

$$\frac{\lambda_1 |S||T|}{n} - \lambda_2 \sqrt{|S||T|}.$$

(**Hint:** Consider the adjacency matrix of G premultiplied by the characteristic vector of S, and postmultiplied by the characteristic vector of T. (The characteristic vector of S is a vector of dimension equal to the cardinality of S, with a 1 in every position corresponding to a member of S, and 0 everywhere else.)

Remark: Note that in a random d-regular graph, the expected number of edges from S to T is $d|S||T|/n$, which is $\lambda_1|S||T|/n$. This result can be viewed as bounding the deviation from the behavior of a random graph in terms of the eigenvalue λ_2, thereby adding to the intuition that an expander "looks" like a random graph.

6.28 (Due to M. Ajtai, J. Komlós, and E. Szemerédi [9].) Let G be an (n, d, c)-expander. Show that there exist constants β, $\delta > 0$ such that for any "bad" set of vertices B of cardinality at most δn, the following property holds: the probability that, starting from a vertex chosen uniformly at random, a random walk of length ℓ does not visit any vertex outside of B is at most $\exp(-\delta \ell)$. Exactly what properties of G are essential for your proof of this fact?

6.29 Using the result in Problem 6.28, obtain a probability amplification result for ***RP*** algorithms similar to that obtained in Section 6.8 for ***BPP*** algorithms.

Remark: While it is an easy consequence of the result for ***BPP*** algorithms, this problem requires you to derive a direct proof based only on the property stated in Problem 6.28.

CHAPTER 7

Algebraic Techniques

SOME of the most notable results in theoretical computer science, particularly in complexity theory, have involved a non-trivial use of algebraic techniques combined with randomization. In this chapter we describe some basic randomization techniques with an underlying algebraic flavor. We begin by describing Freivalds' technique for the verification of identities involving matrices, polynomials, and integers. We describe how this generalizes to the Schwartz-Zippel technique for identities involving multivariate polynomials, and we illustrate this technique by applying it to the problem of detecting the existence of perfect matchings in graphs. Then we present a related technique that leads to an efficient randomized algorithm for pattern matching in strings. We conclude with some complexity-theoretic applications of the techniques introduced here. In particular, we define interactive proof systems and demonstrate such systems for the graph non-isomorphism problem and the problem of counting the number of satisfying truth assignments for a Boolean formula. We then refine this concept into that of an efficiently verifiable proof and demonstrate such proofs for the satisfiability problem. We indicate how these concepts have led to a completely different view of classical complexity classes, as well as the new results obtained via the resulting insight into the structure of these classes.

Most of these techniques and their applications involve (sometimes indirectly) a *fingerprinting* mechanism, which can be described as follows. Consider the problem of deciding the equality of two elements x and y drawn from a large universe U. Under any "reasonable" model of computation, testing the equality of x and y then has a deterministic complexity of at least $\log |U|$. An alternative approach is to pick a random mapping from U into a significantly smaller universe V in such a way that there is a good chance that x and y are identical if and only if their images are identical. The images of x and y are their *fingerprints*, and their equality can be verified in $\log |V|$ time by comparing the fingerprints.

Throughout this chapter we will be working over some unspecified field $\mathbb{F}$. Part of the reason we do not explicitly specify the underlying field is that typically the randomization will involve uniform sampling from a finite subset of the field; in such cases, we do not have to worry about whether the field is finite or not. The reader may find it helpful to think of $\mathbb{F}$ as the field $\mathbb{Q}$ of the rational numbers; when we restrict ourselves to finite fields, it may be useful to assume that $\mathbb{F}$ is $\mathbb{Z}_p$, the field of integers modulo some prime number p. We will use the unit-cost RAM model from Section 1.5.1 to measure the running time of an algorithm over the field $\mathbb{F}$. In this model each field operation (addition, subtraction, multiplication, division, comparison, or choosing a random element) takes unit time, provided the operand magnitude is polynomially related to the input size. For example, over the field of rationals we will assume that operations involving $O(\log n)$-bit numbers take unit time. This is not completely realistic as arithmetic operations are significantly more expensive in practice. However, in most applications described below this small additional factor in the running time is inconsequential, and we would get essentially the same result in the more expensive model.

7.1. Fingerprinting and Freivalds' Technique

We illustrate fingerprinting by describing a technique for verifying matrix multiplication. The fastest known algorithm for matrix multiplication runs in time $O(n^{2.376})$, which improves significantly on the obvious $O(n^3)$ time algorithm but has the disadvantage of being extremely complicated. Suppose we are given an implementation of this algorithm and would like to verify its correctness. Since program verification is a difficult task, a reasonable goal might be to verify the correctness of the output produced on specific executions of the algorithm. (Such verification on specific inputs has been studied in the theory of *program checking*.) In other words, given $n \times n$ matrices $\boldsymbol{A}$, $\boldsymbol{B}$, and $\boldsymbol{C}$ over the field $\mathbb{F}$, we would like to verify that $\boldsymbol{AB} = \boldsymbol{C}$. We cannot afford to use a simpler but slower algorithm for matrix multiplication to verify the output $\boldsymbol{C}$, as this would defeat the purpose of using the fast matrix multiplication algorithm. Moreover, we would like to use the fact that we do not have to compute $\boldsymbol{C}$; rather, our task is to verify that this product is indeed $\boldsymbol{C}$. The following technique, known as *Freivalds' technique*, provides an elegant solution. It gives an $O(n^2)$ time randomized algorithm with a bounded error probability.

The randomized algorithm first chooses a random vector $\boldsymbol{r} \in \{0,1\}^n$; each component of $\boldsymbol{r}$ is chosen independently and uniformly at random from 0 and 1, the additive and multiplicative identities of the field $\mathbb{F}$. We can compute $\boldsymbol{x} = \boldsymbol{Br}$, $\boldsymbol{y} = \boldsymbol{Ax} = \boldsymbol{ABr}$, and $\boldsymbol{z} = \boldsymbol{Cr}$ in $O(n^2)$ time; clearly, if $\boldsymbol{AB} = \boldsymbol{C}$ then $\boldsymbol{y} = \boldsymbol{z}$. We now show that for $\boldsymbol{AB} \neq \boldsymbol{C}$, the probability that $\boldsymbol{y} \neq \boldsymbol{z}$ is at least $1/2$. The algorithm errs only if $\boldsymbol{AB} \neq \boldsymbol{C}$ but $\boldsymbol{y}$ and $\boldsymbol{z}$ turn out to be equal.

Theorem 7.1: *Let $\boldsymbol{A}$, $\boldsymbol{B}$, and $\boldsymbol{C}$ be $n \times n$ matrices over $\mathbb{F}$ such that $\boldsymbol{AB} \neq \boldsymbol{C}$. Then for $\boldsymbol{r}$ chosen uniformly at random from $\{0,1\}^n$, $\mathbf{Pr}[\boldsymbol{ABr} = \boldsymbol{Cr}] \leq 1/2$.*

PROOF: Let $D = AB - C$; we know that D is not the all-zeroes matrix. We wish to bound the probability that $y = z$, or, equivalently, the probability that $Dr = 0$. Without loss of generality, we may assume that the first row in D has a non-zero entry, and that all the non-zero entries in that row precede the zero entries. Let d be the vector consisting of the entries from the first row in D, and assume that the first $k > 0$ entries in d are non-zero. We concentrate on the probability that the inner product of d and r is non-zero; since the first entry in Dr is exactly $d^T r$, this yields a lower bound on the probability that $y \neq z$.

Now, the inner product $d^T r = 0$ if and only if

$$r_1 = \frac{-\sum_{i=2}^{k} d_i r_i}{d_1}. \tag{7.1}$$

We invoke the Principle of Deferred Decisions (Section 3.5) and assume that all the other random entries in r are chosen before r_1. Then the right-hand side of (7.1) is fixed at some value $v \in \mathbb{F}$. Since r_1 is uniformly distributed over a set of size 2, the probability that it equals v cannot exceed 1/2. □

Exercise 7.1: Verify that there is nothing magical about choosing r to have only entries drawn from $\{0, 1\}$. In fact, any two elements of $\mathbb{F}$ may be used instead.

Thus, in $O(n^2)$ time we have reduced the matrix product verification problem to that of verifying the equality of two vectors, and the latter can be done in $O(n)$ time. This gives an overall running time of $O(n^2)$ for this Monte Carlo procedure. The probability of error can be reduced to $1/2^k$ by performing k independent iterations. The following exercise gives an alternative approach to reducing the probability of error.

Exercise 7.2: Suppose that each component of r is chosen uniformly and independently from some subset $\mathbb{S} \subseteq \mathbb{F}$. Show that the probability of error in the verification procedure is no more than $1/|\mathbb{S}|$. Compare the usefulness of the two different methods for reducing the error probability.

Freivalds' technique is applicable to verifying any matrix identity $X = Y$. Of course, if X and Y are explicitly provided, just comparing their entries takes only $O(n^2)$ time. But as in the case of matrix multiplication, there are situations where computing X explicitly is expensive (or even infeasible, as we will see in Section 7.8), whereas computing Xr is easy.

7.2. Verifying Polynomial Identities

Freivalds' technique is fairly general in that it can be applied to the verification of several different kinds of identities. In this section we show that it also applies to the verification of identities involving polynomials. Two polynomials $P(x)$ and $Q(x)$ are said to be equal if they have the same coefficients for corresponding powers of x. Verifying identities of integers, or, in general, strings over any fixed alphabet, is a special case since we can represent any string of length n as a polynomial of degree n. This is achieved by treating the kth element in the string as the coefficient of the kth power of a symbolic variable.

We first consider the *polynomial product verification* problem: given polynomials $P_1(x)$, $P_2(x)$, $P_3(x) \in \mathbb{F}[x]$, verify that $P_1(x) \times P_2(x) = P_3(x)$. Assume that the polynomials $P_1(x)$ and $P_2(x)$ are of degree at most n; then $P_3(x)$ cannot have degree exceeding $2n$. Polynomials of degree n can be multiplied in $O(n \log n)$ time using Fast Fourier Transforms, whereas the evaluation of a polynomial at a fixed point requires $O(n)$ time.

The basic idea underlying the randomized algorithm for polynomial product verification is similar in spirit to the algorithm for matrices. Let $\mathbb{S} \subseteq \mathbb{F}$ be a set of size at least $2n+1$. Pick $r \in \mathbb{S}$ uniformly at random and evaluate $P_1(r)$, $P_2(r)$, and $P_3(r)$ in $O(n)$ time. The polynomial identity $P_1(x)P_2(x) = P_3(x)$ is declared correct unless $P_1(r)P_2(r) \neq P_3(r)$. This algorithm errs only when the polynomial identity is false but the evaluation of the polynomials at r fails to detect this.

Define the polynomial $Q(x) = P_1(x)P_2(x) - P_3(x)$ of degree $2n$. We say that a polynomial P is *identically zero*, or $P \equiv 0$, if all of its coefficients are zero. Clearly, $Q(x)$ is identically zero if and only if the polynomial product is correct. We complete the analysis of the randomized verification algorithm by showing that if $Q(x) \not\equiv 0$, then with high probability $Q(r) = P_1(r)P_2(r) - P_3(r) \neq 0$. Elementary algebra tells us that Q can have at most $2n$ distinct roots. Hence, unless $Q \equiv 0$, not more that $2n$ different choices of $r \in \mathbb{S}$ will have $Q(r) = 0$. Thus, the probability of error is at most $2n/|\mathbb{S}|$. This probability can be reduced by either using independent iterations of the entire algorithm or by choosing a sufficiently large set $\mathbb{S}$.

In the case where $\mathbb{F}$ is an infinite field (such as the reals), the error probability can be reduced to 0 by choosing r uniformly from the entire field $\mathbb{F}$. Unfortunately, this requires an infinite number of random bits! We could also use a deterministic version of this algorithm where each choice of $r \in \mathbb{S}$ is tried once. But this requires $2n+1$ different evaluations of each polynomial, and the best algorithm for this requires $\Theta(n \log^2 n)$ time, which is more than the time required to actually multiply $P_1(x)$ and $P_2(x)$.

This verification procedure is not restricted to polynomial product verification. It is a generic procedure for testing any polynomial identity of the form $P_1(x) = P_2(x)$, by transforming it into the identity $Q(x) = P_1(x) - P_2(x) \equiv 0$. Obviously, if the polynomials P_1 and P_2 are explicitly provided, we can perform this task deterministically in $O(n)$ time by comparing corresponding coefficients. The randomized algorithm will take as long to just evaluate the polynomials at a random point. However, the verification procedure pays off in situations where the polynomials are provided implicitly, such as when we have only a "black box" for computing the polynomial, with no means of accessing its coefficients. There are also situations where the polynomials are provided in a form where computing the actual coefficients is exceedingly expensive. One example is provided by the following problem concerning the determinant of a symbolic matrix; in fact, this problem will turn out to be the same as that of verifying a polynomial identity involving *multivariate* polynomials, necessitating a generalization of the idea used for univariate polynomials.

Let $\boldsymbol{M}$ be an $n \times n$ matrix. The determinant of $\boldsymbol{M}$ is defined by

$$\det(\boldsymbol{M}) = \sum_{\pi \in \mathbb{S}_n} \operatorname{sgn}(\pi) \prod_{i=1}^{n} M_{i,\pi(i)}, \tag{7.2}$$

where $\mathbb{S}_n$ is the symmetric group of permutations of size n, and $\operatorname{sgn}(\pi)$ is the sign of the permutation π. Recall that $\operatorname{sgn}(\pi) = (-1)^t$, where t is the number of pairwise element exchanges required to transform the identity permutation into π. Although the determinant has $n!$ terms, it can be evaluated in polynomial time given explicit values for the matrix entries M_{ij}.

► **Definition 7.1:** The *Vandermonde matrix* $\boldsymbol{M}(x_1, \ldots, x_n)$ is defined in terms of the indeterminates $x_1, \ldots, x_n$ such that $M_{ij} = x_i^{j-1}$, that is

$$\boldsymbol{M} = \begin{pmatrix} 1 & x_1 & x_1^2 & \ldots & x_1^{n-1} \\ 1 & x_2 & x_2^2 & \ldots & x_2^{n-1} \\ & & & \cdot & \\ & & & \cdot & \\ & & & \cdot & \\ 1 & x_n & x_n^2 & \ldots & x_n^{n-1} \end{pmatrix}.$$

Vandermonde's identity states that for this matrix $\boldsymbol{M}$, $\det(\boldsymbol{M}) = \prod_{j<i}(x_i - x_j)$. Suppose that we did not have a proof of this identity and would like to verify it efficiently. Computing the determinant of this symbolic matrix is prohibitively expensive since it has $n!$ terms. Instead, we will formulate this as the problem of verifying that the polynomial $Q(x_1, \ldots, x_n) = \det(\boldsymbol{M}) - \prod_{j<i}(x_i - x_j)$ is identically zero. Drawing upon our experience with Freivalds' technique, it seems natural to substitute random values for each x_i and check whether $Q \equiv 0$. The polynomial Q is easy to evaluate at a specific point since the determinant can be computed in polynomial time for specified values of the variables $x_1, \ldots, x_n$.

We formalize this intuition by extending the analysis of Freivalds' technique for univariate polynomial identity verification to the multivariate case. In a multivariate polynomial $Q(x_1, \ldots, x_n)$, the degree of any term is the sum of the exponents of the variables, and the *total degree* of Q is the maximum of the degrees of its terms.

Theorem 7.2 (Schwartz-Zippel Theorem): *Let $Q(x_1, \ldots, x_n) \in \mathbb{F}[x_1, \ldots, x_n]$ be a multivariate polynomial of total degree d. Fix any finite set $\mathbb{S} \subseteq \mathbb{F}$, and let $r_1, \ldots, r_n$ be chosen independently and uniformly at random from $\mathbb{S}$. Then*

$$\mathbf{Pr}[Q(r_1, \ldots, r_n) = 0 \mid Q(x_1, \ldots, x_n) \not\equiv 0] \leq \frac{d}{|\mathbb{S}|}.$$

PROOF: The proof is by induction on the number of variables n. The base case $n = 1$ involves a univariate polynomial $Q(x_1)$ of degree d, and by the preceding discussion we already know that for $Q(x_1) \not\equiv 0$, the probability that $Q(r_1) = 0$ is at most $d/|\mathbf{S}|$. Assume now that the induction hypothesis is true for a multivariate polynomial with up to $n-1$ variables, for $n > 1$.

Consider the polynomial $Q(x_1, \ldots, x_n)$, and factor out the variable x_1 to obtain

$$Q(x_1, \ldots, x_n) = \sum_{i=0}^{k} x_1^i Q_i(x_2, \ldots, x_n),$$

where $k \leq d$ is the largest exponent of x_1 in Q. (Assume that x_1 affects Q, so that $k > 0$). The coefficient of x_1^k, $Q_k(x_2, \ldots, x_n)$, is not identically zero by our choice of k. Since the total degree of Q_k is at most $d - k$, the induction hypothesis implies that the probability that $Q_k(r_2, \ldots, r_n) = 0$ is at most $(d-k)/|\mathbf{S}|$.

Suppose that $Q_k(r_2, \ldots, r_n) \neq 0$. Consider the following univariate polynomial:

$$q(x_1) = Q(x_1, r_2, r_3, \ldots, r_n) = \sum_{i=0}^{k} x_1^i Q_i(r_2, \ldots, r_n).$$

The polynomial $q(x_1)$ has degree k, and it is not identically zero since the coefficient of x_1^k is $Q_k(r_2, \ldots, r_n)$. The base case now implies that the probability that $q(r_1) = Q(r_1, r_2, \ldots, r_n)$ evaluates to 0 is at most $k/|\mathbf{S}|$.

Thus, we have shown the following two inequalities.

$$\begin{aligned} \mathbf{Pr}[Q_k(r_2, \ldots, r_n) = 0] &\leq \frac{d-k}{|\mathbf{S}|}; \\ \mathbf{Pr}[Q(r_1, r_2, \ldots, r_n) = 0 \mid Q_k(r_2, \ldots, r_n) \neq 0] &\leq \frac{k}{|\mathbf{S}|}. \end{aligned}$$

Invoking the result in Exercise 7.3, we find that the probability that $Q(r_1, r_2, \ldots, r_n) = 0$ is no more than the sum of these two probabilities, which is $d/|\mathbf{S}|$. This completes the induction. □

Exercise 7.3: Show that for any two events $\mathcal{E}_1$ and $\mathcal{E}_2$,

$$\mathbf{Pr}[\mathcal{E}_1] \leq \mathbf{Pr}[\mathcal{E}_1 \mid \overline{\mathcal{E}}_2] + \mathbf{Pr}[\mathcal{E}_2].$$

The randomized verification procedure for polynomials has one potential problem. In the case of infinite fields, the intermediate results in the evaluation of the polynomial could involve enormous values. This problem can be avoided in the case of integers by performing all the computations modulo a small random prime number, without adversely affecting the error probability. We will return to this issue in Example 7.1.

As suggested in Problem 7.1, Theorem 7.2 can be viewed as a generalization of Freivalds' technique from Section 7.1. A generalized version of this theorem is described in Problem 7.6.

7.3. Perfect Matchings in Graphs

We illustrate the power of the techniques of Section 7.2 by giving a fascinating application. Consider a bipartite graph $G(U, V, E)$ with the independent sets of vertices $U = \{u_1, \ldots, u_n\}$ and $V = \{v_1, \ldots, v_n\}$. A *matching* is a collection of edges $M \subseteq E$ such that each vertex occurs at most once in M. A *perfect matching* is a matching of size n. Each perfect matching M in G can be viewed as a permutation from U into V. More precisely, the perfect matchings in G can be put into a one-to-one correspondence with the permutations in $\mathbb{S}_n$, where the matching corresponding to a permutation $\pi \in \mathbb{S}_n$ is given by the pairs $(u_i, v_{\pi(i)})$, for $1 \leq i \leq n$. The following theorem draws a connection between determinants and matchings.

Theorem 7.3 (Edmonds' Theorem): *Let A be the $n \times n$ matrix obtained from $G(U, V, E)$ as follows:*

$$A_{ij} = \begin{cases} x_{ij} & (u_i, v_j) \in E \\ 0 & (u_i, v_j) \notin E \end{cases}.$$

Define the multivariate polynomial $Q(x_{11}, x_{12}, \ldots, x_{nn})$ as being equal to $\det(A)$. *Then, G has a perfect matching if and only if $Q \not\equiv 0$.*

Remark: The matrix of indeterminates is sometimes referred to as the *Edmonds matrix* of a bipartite graph. We do not explicitly specify the underlying field because any field will do for the purposes of this theorem.

PROOF: The determinant of A is given by

$$\det(A) = \sum_{\pi \in \mathbb{S}_n} \operatorname{sgn}(\pi) A_{1,\pi(1)} A_{2,\pi(2)} \ldots A_{n,\pi(n)}.$$

Since each indeterminate x_{ij} occurs at most once in A, there can be no cancellation of the terms in the summation. Therefore the determinant is not identically zero if and only if there is a permutation π for which the corresponding term in the summation is non-zero. The latter happens if and only if each of the entries $A_{i,\pi(i)}$, for $1 \leq i \leq n$, is non-zero. This is equivalent to having a perfect matching (the one corresponding to π) in G. □

We can now construct a simple randomized test for the existence of perfect matchings. Using the algorithm from Section 7.2, we can determine whether the determinant is identically zero or not. The time required is dominated by the cost of computing a determinant, which is essentially that of multiplying two matrices. As it turns out, there are algorithms for *constructing* a maximum matching in a graph in time $O(m\sqrt{n})$, where $m = |E|$. Since the time to compute the determinant exceeds $m\sqrt{n}$ for small m, the payoff in using this randomized decision procedure is marginal at best. However, we will see later (in Section 12.4) that this decision procedure is essential for devising a fast *parallel* algorithm for computing a maximum matching in a graph. In Problem 7.8 we will see that this technique also applies to the case of non-bipartite graphs.

7.4. Verifying Equality of Strings

We have seen that the idea of fingerprinting is useful in verifying identities of algebraic objects. In this section we introduce a different form of fingerprinting, motivated by the problem of testing the equality of two strings. As mentioned earlier, the string equality verification problem can be reduced to that of verifying polynomial identities. However, the new type of fingerprint introduced here has important benefits when extended to the pattern matching problem discussed later in Section 7.6.

Suppose that Alice maintains a large database of information. Bob maintains a second copy of the database. Periodically, they must compare their databases for consistency. Because transmission between Alice and Bob is expensive, they would like to discover the presence of an inconsistency without transmitting the entire database between them. Denote Alice's data by the sequence of bits $(a_1, \ldots, a_n)$, and Bob's by the sequence $(b_1, \ldots, b_n)$. It is clear that any deterministic consistency check that transmits fewer than n bits will fail if an adversary could decide which bits of either database to modify. We describe a randomized strategy that detects an inconsistency with high probability while transmitting far fewer than n bits of information.

We use the following simple fingerprint mechanism. Interpret the data as n-bit integers a and b, by defining $a = \sum_{i=1}^{n} a_i 2^{i-1}$ and $b = \sum_{i=1}^{n} b_i 2^{i-1}$. Define the fingerprint function $F_p(x) = x \bmod p$ for a prime p. Then Alice can transmit $F_p(a)$ to Bob, who in turn can compare this with $F_p(b)$. The hope is that if $a \neq b$, then it will also be the case that $F_p(a) \neq F_p(b)$. The number of bits to be transmitted is $O(\log p)$, which will be much smaller than n for a small prime p. This strategy can be easily foiled by an adversary for any fixed choice of p since, for any p and b, there exist many choices of a for which $a \equiv b \pmod{p}$. We get around this problem by choosing p at random.

For any number k, let $\pi(k)$ be the number of distinct primes less k. A well-known result in number theory is the Prime Number Theorem, which states that $\pi(k)$ is asymptotically $k/\ln k$. Consider now the non-negative integer $c = |a - b|$. The fingerprint defined above fails only when $c \neq 0$ and p divides c. How many primes can divide c? Define $N = 2^n$; we know that $c \leq N$.

Lemma 7.4: *The number of distinct prime divisors of any number less than 2^n is at most n.*

PROOF: Each prime number is greater than 1. If N has more than t distinct prime divisors, then $N \geq 2^t$. □

Choose a threshold τ that is larger than $n = \log N$. The number of primes smaller than τ is $\pi(\tau) \sim \tau / \ln \tau$. Of these, at most n can be divisors of c and cause our fingerprint function to fail. Therefore, we pick a random prime p smaller than τ for defining F_p. The number of bits of communication is $O(\log \tau)$. Choose

$\tau = tn \log tn$, for large t. The following theorem is immediate. The probability is taken over the random choice of p.

Theorem 7.5: $\Pr[F_p(a) = F_p(b) \mid a \neq b] \leq \frac{n}{\pi(\tau)} = \mathrm{O}\left(\frac{1}{t}\right)$.

Thus, we get an error probability of at most $\mathrm{O}(1/t)$, and the number of bits to be transmitted is $\mathrm{O}(\log t + \log n)$. Choosing $t = n$ gives us an excellent strategy for this problem. We remark that the task of picking a random prime is non-trivial, primarily because verifying the primality of a number is difficult. Some algorithms for this purpose will be presented in Chapter 14.

► **Example 7.1:** This integer equality verification technique can be used to solve the problem alluded to at the end of Section 7.2. In verifying that a multivariate polynomial $Q(x_1, \ldots, x_n)$ is identically zero, we evaluate the polynomial at a random point. The problem is that the intermediate values arising in the evaluation of $q = Q(r_1, \ldots, r_n)$ could be extremely large. Of course, we do not really wish to compute q; our goal is to merely verify that $q = 0$. By the preceding discussion, it suffices to verify that $q \bmod p = 0$ for some small random prime p.

But how can we possibly hope to perform the verification without evaluating q explicitly? The trick is to use arithmetic modulo p while evaluating $Q(r_1, \ldots, r_n)$ and thereby obtain the residue of q modulo p directly, rather than first computing q and then reducing it modulo p. The intermediate values are all smaller than p, and p itself is chosen to be a small random prime. By Theorem 7.5, the probability of error does not increase significantly for a suitable choice of t.

7.5. A Comparison of Fingerprinting Techniques

It is useful at this point to compare the two types of fingerprinting techniques that we have seen so far. Suppose that we wish to verify the equality of two strings or vectors $\boldsymbol{a} = (a_1, \ldots, a_n)$ and $\boldsymbol{b} = (b_1, \ldots, b_n)$ with each component drawn from a finite alphabet Σ. We can encode the alphabet symbols using the set of numbers $\Gamma = \{0, 1, \ldots, k-1\}$, where $k = |\Sigma|$. It is then possible to view the two strings as the polynomials $A(z) = \sum_{i=0}^{n-1} a_i z^i$ and $B(z) = \sum_{i=0}^{n-1} b_i z^i$, each of which has integer coefficients and degree at most n. Clearly, the two vectors are identical if and only if the two polynomials are identical.

The fingerprinting technique of Sections 7.1 and 7.2 can be summarized as follows. Fix a prime number p greater than both $2n$ and k. View the polynomials $A(z)$ and $B(z)$ as polynomials over the field $\mathbb{Z}_p$. By our choice of p, the set Γ is contained in this field and arithmetic modulo p will not render identical any two non-identical polynomials. The fingerprint of the two polynomials is obtained by choosing a random element $r \in \mathbb{Z}_p$ and substituting it for the symbolic variable z. If $\boldsymbol{a} = \boldsymbol{b}$, then the two polynomials are identical and the fingerprint will also be identical; on the other hand, when $\boldsymbol{a} \neq \boldsymbol{b}$, the two polynomials are distinct and the probability that their fingerprints turn out to be the same is at most n/p, and this is bounded by $1/2$ for our choice of p. For $k = 2$ and $p = O(n)$, this can be viewed as reducing the problem of comparing n-bit numbers to that of comparing $O(\log n)$-bit numbers.

The fingerprinting technique from Section 7.4 is in some sense a dual of the first technique. In this approach, we fix $z = 2$ and choose a random prime q of a reasonably small magnitude. The fingerprints are obtained by evaluating $A(2)$ and $B(2)$ over the field $\mathbb{Z}_q$. Thus, instead of fixing the field and evaluating at a random point in the field, the second type of fingerprint is obtained by fixing the point of evaluation and choosing a random field over which the evaluation is to be performed. By our analysis in Section 7.4, this also reduces the problem of comparing n-bit numbers to that of comparing $(\log n)$-bit numbers. However, as we will see in the next section, there are certain applications where the second type of fingerprinting proves to be more useful.

A third version of the fingerprinting approach works as follows. Assume that $k = 2$, and interpret the bit vectors $\boldsymbol{a}$ and $\boldsymbol{b}$ as the n-bit integers a and b. Fix a prime number $p > 2^n$. Choose a random polynomial $P(z)$ over the field $\mathbb{Z}_p$, and obtain the fingerprints by evaluating this polynomial at the integers a and b, performing all arithmetic over the field $\mathbb{Z}_p$, and then reducing the resulting values modulo a number of magnitude close to $\log n$. This is the main idea behind the construction of the so-called *universal hash functions* discussed in Section 8.4.

7.6. Pattern Matching

Consider now the problem of pattern matching in strings. A *text* is a string $X = x_1x_2\dots x_n$ and a *pattern* is a string $Y = y_1y_2\dots y_m$, both over a fixed finite alphabet Σ, such that $m \leq n$. Without loss of generality, we restrict ourselves to the case $\Sigma = \{0,1\}$. The pattern occurs in the text if there is a $j \in \{1,2,\dots,n-m+1\}$ such that for $1 \leq i \leq m$, $x_{j+i-1} = y_i$. The pattern matching problem is that of finding an occurrence (if any) of a given pattern in the text. This problem can be trivially solved in $O(nm)$ time by trying for a match at all possible locations i; moreover, there are deterministic algorithms that achieve the best possible running time of $O(n+m)$.

We describe a Monte Carlo algorithm that also achieves a running time of $O(n+m)$; later, we will convert this into a Las Vegas algorithm. This algorithm is interesting despite the existence of linear-time deterministic algorithms because it is significantly simpler, has a "real-time" implementation (this is explained below), and generalizes to the problem of pattern matching in two-dimensional strings (or matrices).

Define the string $X(j) = x_jx_{j+1}\dots x_{j+m-1}$ as the sub-string of length m in X that starts at position j. A match occurs if there is a choice of j, for $1 \leq j \leq n-m+1$, for which $Y = X(j)$. We make the solution unique by requiring that the algorithm find the smallest value of j such that $X(j) = Y$.

The brute-force $O(nm)$ time algorithm compares Y with each of the strings $X(j)$. Our randomized algorithm will choose a fingerprint function F and compare $F(Y)$ with each of the fingerprints $F(X(j))$. An error occurs if $F(Y) = F(X(j))$ but $Y \neq X(j)$. We would like to choose a function F that has a small probability of error and can be efficiently computed.

In fact, we use the same fingerprint function as in Section 7.4: for any string $Z \in \{0,1\}^m$, interpret Z as an m-bit integer and define $F_p(Z) = Z \bmod p$. Assume that p is chosen uniformly at random from the set of primes smaller than a threshold τ. Suppose that we interpret the strings Y and $X(j)$ as m-bit integers, and compare their fingerprints $F_p(Y)$ and $F_p(X(j))$ instead of trying to match each symbol in the two strings. The only possible error is that we get identical fingerprints when $Y \neq X(j)$. By Theorem 7.5, we bound the probability of such a *false match* as follows:

$$\mathbf{Pr}[F_p(Y) = F_p(X(j)) \mid Y \neq X(j)] \leq \frac{m}{\pi(\tau)} = O\left(\frac{m \log \tau}{\tau}\right).$$

Then, the probability that a false match occurs for any of the at most n values of j is $O((nm \log \tau)/\tau)$. We choose $\tau = n^2 m \log n^2 m$, and this gives

$$\mathbf{Pr}[\text{a false match occurs}] = O\left(\frac{1}{n}\right).$$

The Monte Carlo version of this algorithm simply compares the fingerprints of all $X(j)$ to that of Y, and outputs the first j for which a match occurs; the Las Vegas version will be described below. We first show that the running time of this algorithm is as claimed. For $1 \leq j \leq n - m + 1$,

$$X(j+1) = 2\left[X(j) - 2^{m-1} x_j\right] + x_{j+m}.$$

From this we obtain the recurrence

$$F_p(X(j+1)) = 2\left[F_p(X(j)) - 2^{m-1} x_j\right] + x_{j+m} \bmod p.$$

It is now clear that given the fingerprint of $X(j)$, the incremental cost of computing the fingerprint of $X(j+1)$ is $O(1)$ field operations. In fact, there is no need to use the more expensive operations of multiplication and division, because each x_j is 0 or 1. Thus, the total time required for this algorithm is $O(n+m)$ even under the more stringent log-cost RAM model. This efficient incremental update property is the main motivation for using the second form of fingerprinting; the reader may verify that more complex computations would be required if the first form of fingerprinting was used instead (see Section 7.5).

Theorem 7.6: *The Monte Carlo algorithm for pattern matching requires* $O(n+m)$ *time and has a probability of error* $O(1/n)$.

It is easy to convert this into a Las Vegas algorithm. Whenever a match occurs between the fingerprints of Y and some $X(j)$, we compare the strings Y and $X(j)$ in $O(m)$ time. If this is a false match, we detect it and abandon the whole process in favor of using the brute-force $O(nm)$ time algorithm. The new algorithm does not make any errors and has expected running time $O((n+m)(1-1/n)+nm(1/n))$, which works out to be $O(n+m)$. An alternative Las Vegas version of this algorithm restarts the entire algorithm with a new random choice of p whenever a false match is detected. In the latter approach, the probability of having to restart more than t times is bounded by $1/n^t$. This leads to a very small variance in the running time. In contrast, the first approach has a relatively high probability of being forced to use the $O(nm)$ time algorithm, and hence has a high variance in the running time.

An alternative fingerprint function with a similar behavior is described in Problem 7.12. In Problem 7.13 it is required to show that this algorithm extends to the case of two-dimensional pattern matching.

The method for computing the fingerprints of the various $X(j)$'s will work well in on-line or real-time settings where the string X is provided incrementally, possibly a bit at a time. This feature is also useful when the text is extremely large and cannot be completely stored in the primary memory of a machine.

Exercise 7.4: Consider the fingerprint function used for polynomial identities and adapt it to the problem of testing string equality. Why is this not a good choice of a fingerprint for the pattern matching problem?

7.7. Interactive Proof Systems

We have seen the power of combining randomization and algebra in devising fingerprinting techniques with applications to efficient verification of simple identities involving objects such as matrices, polynomials, and strings. We have also seen that the basic idea used in the verification of the equality of two strings x and y could be taken a step further and be used for the efficient detection of a pattern y in a string x. How far can we push this approach?

Suppose, for example, the string x represents a graph G, and the "pattern" y represents some graph property P. Can we then use the ideas developed here for efficient "pattern matching" in terms of verifying the property P in G? More specifically, suppose that the pattern y corresponds to the property of *not* being an expanding graph. The problem of verifying this property belongs to $\mathcal{NP}$ and so there exist short proofs of non-expansion. Moreover, given such a proof, it is possible to efficiently verify its correctness. Thus, the pattern matching task can be efficiently performed provided the pattern y includes a "proof" of this fact, i.e., a description of a set of vertices in G that do not have too many neighbors. In this context, efficiency means time polynomial in the length of the inputs, and this requires that the proof itself be of polynomial length.

Suppose instead the pattern matching task corresponds to the verification of the property of being an expander. As we mentioned earlier (Section 6.7), this problem is co-***NP***-complete and it is quite unlikely that there is a polynomial length proof of this property. Intuitively, verifying the expander property requires checking almost all subsets of the vertices. But could it be that it is possible to verify such proofs efficiently, even though their length is not polynomially bounded? At an intuitive level this seems impossible, since we at least have to read a proof completely to verify its correctness. Quite surprisingly, however, we will show how the use of randomization combined with elementary algebra allows us to efficiently verify an exponential length proof of such co-***NP*** properties, provided the proof itself is written in a specific format. In fact, there are more profound complexity-theoretic results that can be obtained using randomized algebraic techniques. In this section and the next, we will describe some aspects of these complexity-theoretic results.

7.7.1. Verifying Graph Non-Isomorphism

Let us start by considering the problem of graph isomorphism. Informally, two graphs are isomorphic if they have exactly the same structure. We make a formal definition for the case of labeled graphs.

► **Definition 7.2:** Let $G_1(V, E_1)$ and $G_2(V, E_2)$ be two graphs on the same set of labeled vertices $V = \{1, \ldots, n\}$. The two graphs are said to be *isomorphic* if there exists a permutation $\pi \in \mathbb{S}_n$ such that an edge $(i, j) \in E_1$ if and only if the edge $(\pi(i), \pi(j)) \in E_2$; the permutation π is referred to as an *isomorphism* from G_1 to G_2. Two graphs are *non-isomorphic* if there does not exist any isomorphism from one graph to the other.

Consider the *graph isomorphism (GI) problem*: given two graphs G_1 and G_2, decide whether they are isomorphic to each other. This problem lies in ***NP*** since it is possible to "guess" an isomorphism and verify that it maps edges correctly. That is, there is a short proof of isomorphism (the description of a permutation π), and its validity can be verified efficiently. It is believed that GI does not belong to ***P***, and yet there is no proof that this problem is ***NP***-complete. In fact, there is strong evidence that this problem is *not* ***NP***-complete, making it one of the few natural problems believed to have this property. This evidence is derived from results closely related to those discussed in this section.

The complementary problem, *graph non-isomorphism* (GNI), is that of verifying that G_1 and G_2 are non-isomorphic. By definition, this problem lies in co-***NP***. Unlike the case of isomorphism, there is no known short proof of non-isomorphism, and it appears that verifying non-isomorphism will essentially require checking that none of the $n!$ permutations provides an isomorphism from G_1 to G_2. However, as we show next, using a more active "prover" instead of a passive "proof" together with randomization in the verification process leads to an efficient scheme for verifying non-isomorphism.

The model that we adopt is the following. A verifier V that can perform any randomized polynomial-time computation is attempting to verify that two graphs G_1 and G_2 are non-isomorphic. The verifier can enlist the help of a prover P, which is an all-powerful adversarial entity whose goal is to convince the verifier that G_1 is not isomorphic to G_2, *even if the two graphs are indeed isomorphic.* The prover's computational power is not constrained in any way; in particular, it is not restricted to polynomial-time computations, and it knows precisely the strategy employed by the verifier V. The only limitation on the prover is that it does not have access to the random bits used by V in the course of its computations, except as revealed in the information communicated to it by V.

The interaction between the two entities can be viewed as being composed of a sequence of rounds of communication, where in each round V poses a question to P, and P responds with a possibly maliciously chosen incorrect answer. Upon termination, V decides to either accept that G_1 is not isomorphic to G_2, or else reject the prover's answers as being incorrect or unconvincing. A *protocol* is the specification of a randomized polynomial-time algorithm for V such that: when G_1 and G_2 are non-isomorphic, it is possible for a prover P to convince V to accept; and when G_1 and G_2 are isomorphic, even a malicious prover cannot respond so as to persuade V to accept with probability more than $1/2$ (say).

It turns out that the following simple protocol suffices. In the description of the protocol, $\sigma(G)$ denotes the graph isomorphic to G that is obtained by applying the permutation σ to the labels of the vertices in G.

Verifier V:

- picks index $i \in \{1, 2\}$ and permutation $\sigma \in \mathbb{S}_n$, both uniformly at random;
- computes $H = \sigma(G_i)$;
- specifies H to the prover P and asks for an index j such that H is isomorphic to G_j;

Prover P: responds with an index j;

Verifier V: if $j = i$ then it accepts that G_1 and G_2 are non-isomorphic, else it rejects.

Fix any two graphs $G_1(V, E_1)$ and $G_2(V, E_2)$. Consider the execution of this protocol with prover P following an adversarial strategy as discussed earlier. The following theorem shows that if the verifier V follows this protocol, then it achieves the desired result.

Theorem 7.7: *If G_1 and G_2 are non-isomorphic, an honest prover P can ensure that V will accept; otherwise, for any (possibly maliciously dishonest) prover P', the probability that V accepts is $1/2$.*

PROOF: Consider first the case where the two graphs are non-isomorphic. Suppose that V used $i = 1$. Then, H and G_1 are isomorphic, while H and G_2 are non-isomorphic since G_1 is not isomorphic to G_2. An honest prover can use its unbounded power to determine that H is isomorphic to G_1 but not to G_2.

Therefore, the prover can ensure that it sends back $j = 1$, thereby persuading V to accept. A similar argument applies in the case when $i = 2$.

Suppose now that G_1 and G_2 are isomorphic to each other. The graph H must then be isomorphic to both G_1 and G_2. Let σ_1 denote an isomorphism from G_1 to H, and σ_2 an isomorphism from G_2 to H. Given that the verifier follows the protocol,

$$\mathbf{Pr}[\sigma = \sigma_1 \text{ and } i = 1 \mid V \text{ specifies } H] = \mathbf{Pr}[\sigma = \sigma_2 \text{ and } i = 2 \mid V \text{ specifies } H].$$

The prover does not know the value of i or the permutation σ used to determine H from G_i. We claim that even knowing H and regardless of its strategy for choosing j, the probability that $j = i$ is exactly $1/2$. It follows that the probability that V accepts is $1/2$.

To verify the claim, we invoke the Principle of Deferred Decisions (Section 3.5) as follows: assume that the verifier first decides upon H, using it to obtain the value of j from the prover, and only then does it decide upon the choice of i and σ. This is equivalent to assuming that V chooses H at random from the uniform distribution on the space of all graphs isomorphic to G_1 and G_2. Then, after it has forced the prover to commit to the value of j, it makes a random choice of i and determines the isomorphism σ_i from G_i to H. Of course, this would require V to solve the GI problem efficiently, which is not believed to be possible for any randomized polynomial-time algorithm. But the point is that as far as the prover is concerned, it cannot distinguish between the two types of verifiers and we postulate the existence of a "deferring" verifier only for the purposes of our analysis. We assume that this verifier is still honest in that it chooses i at random even though it already knows the value of j; this is because the verifier just wants to ensure that it does not get cheated by the prover, and it does not gain anything by cheating itself. □

Exercise 7.5: Verify that independent iterations of this protocol can be used to reduce the probability that the verifier accepts erroneously. Argue that the prover does not gain additional power to cheat as the iterations proceed.

7.7.2. The Class *IP* and #3SAT

We now formalize the notion of an *interactive proof system* used in Section 7.7.1. Given any language L over an alphabet Σ, an interactive proof system for L consists of a verifier V and prover P such that: the verifier V can perform any randomized polynomial-time computation and can communicate with the prover P in an attempt to verify that an input x belongs to L; the prover P can perform arbitrary computations but does not have access to the random bits used by V. Typically, we use the symbol P to denote an *honest* prover that always provides truthful responses to the queries posed by V. Let $V(x, P')$ be the outcome (acceptance or rejection) of the computation performed by the verifier given an input string x and communicating with a prover P', where P' denotes a prover that does not necessarily behave in the manner expected of the honest prover P. We define a complexity class consisting of all languages (or decision problems) that have interactive proof systems such as the one demonstrated for graph non-isomorphism.

► **Definition 7.3:** The class ***IP*** consists of all languages L that have an interactive proof system (P, V) with a randomized polynomial-time verifier V and an honest prover P such that for any $x \in \Sigma^*$,

- $x \in L \Rightarrow$ for the honest prover P, $\mathbf{Pr}[V(x,P) \text{ accepts}] = 1$.
- $x \notin L \Rightarrow$ for all provers P', $\mathbf{Pr}[V(x,P') \text{ accepts}] \leq \frac{1}{2}$.

We have already shown that GNI $\in$ ***IP***, and it is not very hard to verify that GI $\in$ ***IP***. As we will see shortly, this is not a coincidence since both ***NP*** and co-***NP*** are contained in ***IP***. Intuitively, ***IP*** can be viewed as a generalization of ***NP*** obtained by permitting randomization. It turns out that ***IP*** = ***PSPACE***, the class of languages whose membership can be decided using only a polynomial amount of space (see Problems 7.16–7.17). While it is relatively easy to argue that ***IP*** $\subseteq$ ***PSPACE***, the proof of ***PSPACE*** $\subseteq$ ***IP*** turns out to be more difficult, and this is where randomized algebraic techniques prove to be useful. We illustrate some of the key ideas behind the latter proof by showing that the problem of verifying the number of satisfying truth assignments for a 3-CNF Boolean formula lies in ***IP***.

Let $X_1, \ldots, X_n$ be Boolean variables whose values can be either TRUE or FALSE. A 3-CNF formula $F(X_1, \ldots, X_n)$ is the conjunction of a collection of *clauses* C_1, $\ldots$, C_m, where each clause is a disjunction of three literals L_{i1}, L_{i2}, and L_{i3}. Recall that a formula F is said to be satisfiable if there exists an assignment of values to its variables that results in F evaluating to TRUE, and then the assignment is called a satisfying truth assignment.

In the 3-SAT problem, we are required to determine whether a given 3-CNF formula F is satisfiable. We are interested in a counting version of this problem called #3SAT: given a 3-CNF formula F and an integer s, verify that the number of distinct satisfying truth assignments for F is s. We will establish the following theorem.

Theorem 7.8: *#3SAT* $\in$ ***IP***.

What are the implications of this result? Recall that the 3-SAT problem is ***NP***-complete, which implies that if 3-SAT $\in$ ***P***, then any $L \in$ ***NP*** is also in ***P*** (and therefore ***P*** = ***NP***). The 3-SAT problem is a special case of #3SAT and this means that 3-SAT $\in$ ***IP***. The following exercise then implies that ***NP*** $\subseteq$ ***IP***.

Exercise 7.6: Let L_1 and L_2 be two languages such that $L_2 \in \boldsymbol{IP}$ and there is a polynomial reduction from L_1 to L_2. Show that $L_1 \in \boldsymbol{IP}$.

This is not very interesting since it is easy to argue directly that $\boldsymbol{NP} \subseteq \boldsymbol{IP}$ (see Problem 7.14). However, consider now the special case of #3SAT where $s = 0$. This is the problem of deciding that a 3-CNF formula is *not* satisfiable; since verifying unsatisfiability is a co-$\boldsymbol{NP}$-complete problem, it follows that co-$\boldsymbol{NP} \subseteq \boldsymbol{IP}$. This is much more interesting since it is not immediately obvious from the definition of $\boldsymbol{IP}$ that it contains co-$\boldsymbol{NP}$. Actually, #3SAT is complete for a class of problems called $\#\boldsymbol{P}$, which is defined formally in Chapter 11. It follows that $\#\boldsymbol{P} \subseteq \boldsymbol{IP}$. It is known that $\#\boldsymbol{P} \subseteq \boldsymbol{PSPACE}$, and so we are proving a weaker result than $\boldsymbol{IP} = \boldsymbol{PSPACE}$. We choose to focus on this weaker result since it introduces some of the key ideas involving randomization that are used in the proof of $\boldsymbol{IP} = \boldsymbol{PSPACE}$. Problems 7.16–7.17.

7.7.3. Arithmetization of Satisfiability

A key step in the proof of #3SAT $\in \boldsymbol{IP}$ is the conversion of the Boolean formula F into an algebraic formula. This process is called the *arithmetization* of a Boolean formula. Let us view any truth assignment A for the variables in F as an n-dimensional vector over the integers. More precisely, we represent it by a vector $\boldsymbol{a} = (a_1, a_2, \ldots, a_n)$ such that

$$a_i = \begin{cases} 1 & \text{if } A_i = \text{TRUE} \\ 0 & \text{if } A_i = \text{FALSE} \end{cases}.$$

At the same time, we convert F into a polynomial with the variables $x_1, \ldots, x_n$, as follows. Any literal L_{ij} is turned into a linear polynomial l_{ij} by replacing a Boolean variable X_j by $1 - x_j$, and a negated variable $\overline{X}_j$ by x_j. A clause $C_i = L_{i1} \vee L_{i2} \vee L_{i3}$ is replaced by a degree 3 polynomial $c_i = 1 - l_{i1} l_{i2} l_{i3}$. Finally, the Boolean formula $F(X_1, \ldots, X_n)$ is represented by the following polynomial of degree $3m$:

$$f(x_1, \ldots, x_n) = \prod_{i=1}^{m} c_i = \prod_{i=1}^{m} (1 - l_{i1} l_{i2} l_{i3}).$$

► **Example 7.2:** Consider the 3-CNF Boolean formula

$$F(X_1, X_2, X_3, X_4) = (X_1 \vee X_2 \vee \overline{X}_3) \wedge (\overline{X}_1 \vee \overline{X}_3 \vee X_4).$$

Then, the arithmetization of F yields the following polynomial of degree 6:

$$f(x_1, x_2, x_3, x_4) = (1 - (1 - x_1)(1 - x_2)x_3) \times (1 - x_1 x_3 (1 - x_4)).$$

Exercise 7.7: Show that there is no essential difference between the Boolean formula F and its arithmetization f: let A be any truth assignment, and a the corresponding vector over $\{0, 1\}$. Show that $F(A_1, \ldots, A_n) = \text{TRUE}$ if and only if $f(a_1, \ldots, a_n) = 1$.

Let $\#F$ denote the number of satisfying truth assignments for F, and define

$$\#f = \sum_{x_1=0}^{1} \sum_{x_2=0}^{1} \cdots \sum_{x_n=0}^{1} f(x_1, \ldots, x_n).$$

Since $\#F = \#f$, the problem of verifying that $\#F = s$ is the same as the problem of verifying that $\#f = s$.

It will be convenient to work over a finite field and so we treat the polynomial f as a polynomial over the field $\mathbb{Z}_p$, for some prime p. Since the value of $\#f$ cannot exceed the total number of truth assignments, this restriction to a finite field will not affect the value of $\#f$ provided we choose $p > 2^n$. By Bertrand's Postulate, there is a prime p such that $2^n \leq p < 2^{n+1}$ and we can use any such prime number. A technical issue is that there is no known polynomial time algorithm for finding such a prime. But this issue can be easily handled in the setting of an interactive proof system. The verifier asks the prover to specify such a prime p, and to prevent cheating it also asks for a proof of the primality of p. As we will see in Section 14.6, there exist polynomial length "certificates of primality" that can be verified in polynomial time, and the all-powerful prover can easily provide such a certificate of primality along with the value of p.

The following notation will be useful in describing the interactive proof system. For any polynomial $f(x_1, \ldots, x_n)$, and for $0 \leq i \leq n$, define the partial sum polynomials

$$f_i(x_1, \ldots, x_i) = \sum_{x_{i+1}=0}^{1} \cdots \sum_{x_n=0}^{1} f(x_1, \ldots, x_n).$$

The proof of the following set of properties for the partial sum polynomial is left as Problem 7.15.

Lemma 7.9: *The partial sum polynomials have the following properties:*

1. $f_0 = \#f$.
2. $f_n(x_1, \ldots, x_n) = f(x_1, \ldots, x_n)$.
3. *for* $1 \leq i \leq n$, $f_{i-1}(x_1, \ldots, x_{i-1}) = f_i(x_1, \ldots, x_{i-1}, 0) + f_i(x_1, \ldots, x_{i-1}, 1)$.

7.7.4. The Interactive Proof System for #3SAT

We now provide an interactive proof system that takes as input a polynomial $f(x_1, \ldots, x_n)$ over $\mathbb{Z}_p$ and an integer $s \in \mathbb{Z}_p$, and verifies that $\#f = s$. Since a verifier can easily compute the arithmetization of a 3-CNF formula F, this suffices to show that $\#3\text{SAT} \in \boldsymbol{IP}$.

The basic step in the interactive proof system is for the verifier to ask the prover for the description of the polynomial $f_1(z)$, where z is a symbolic variable. Suppose that the prover responds with a polynomial $g(z)$, which may or may not be the desired polynomial $f_1(z)$. Assuming that the prover does not cheat, it must be the case that $\#f = g(0) + g(1)$. Therefore, the verifier compares s with $g(0) + g(1)$ and rejects if the two are unequal. It must now verify that g is indeed the same polynomial as f_1.

Of course, we have to concern ourselves only with the case where $\#f \neq s$, since it will be clear that an honest prover can always make the verifier accept by providing correct responses. Now, since $g(0)+g(1) = s$, and $\#f = f_1(0)+f_1(1) \neq s$, it follows that $g(z) \neq f_1(z)$. The verifier's goal is to make sure that the polynomial equality $g(z) = f_1(z)$ is satisfied, so as to ensure that it catches a prover that is attempting to cheat by sending a polynomial $g(z) \neq f_1(z)$ such that $g(0) + g(1) = s$. The only problem is that while the verifier knows $g(z)$ explicitly, the polynomial $f_1(z)$ is only implicitly defined by the equation

$$f_1(z) = \sum_{x_2=0}^{1} \cdots \sum_{x_n=0}^{1} f(z, x_2, \ldots, x_n).$$

Computing $f_1(z)$ explicitly from this equation would require super-polynomial time. But this is precisely the kind of situation where we use the technique described in Section 7.2 for verifying polynomial identities.

The verifier chooses an element $r \in \mathbb{Z}_p$ uniformly at random, evaluates $s' = g(r)$, and asks the prover to show that $s' = f_1(r)$. Again, we are only interested in analyzing the case where $g(z) \neq f_1(z)$. Of course, it is still possible that $s' = f_1(r)$. In this case, the prover will succeed in cheating the verifier as it will be able to pass all subsequent tests (described next). But this "error" happens with a small probability since the polynomials in question are of low degree; in particular, the error probability is given by

$$\mathbf{Pr}[g(r) = f_1(r) \mid g(z) \neq f_1(z)] \leq \frac{3m}{p},$$

as the degree of these polynomials is at most $3m$.

Assuming that this error does not occur, the verifier has a value $s' \neq f_1(r)$, and the subsequent interaction is geared toward detecting this fact. The verifier now asks the prover to show that $s' = f_1(r)$, or equivalently that

$$s' = \sum_{x_2=0}^{1} \cdots \sum_{x_n=0}^{1} f'(x_2, \ldots, x_n),$$

where we define the polynomial $f'(x_2, \ldots, x_n) = f(r, x_2, \ldots, x_n)$. This is exactly the original verification problem all over again, but the crucial point is that the number of variables has been reduced to $n - 1$ from n.

The verifier can perform this new verification by recursively running the same protocol, and the recursion bottoms out with the problem of verifying the equality of two degree 0 polynomials, which is a trivial task. The probability of error accumulates over the various stages of recursion, but since the number of stages is n, we can bound the overall error probability by $3mn/p$. Recall that p was chosen to be larger than 2^n and so the error probability is small. The net running time of the verifier is bounded by a polynomial in n and m.

7.8. *PCP* and Efficient Proof Verification

We continue with our excursion into complexity theory and describe the application of Freivalds' technique to a problem in proof verification. In Section 1.5.2, we defined ***NP*** in terms of the verification of proofs by deterministic polynomial-time verifiers. In Section 7.7, we replaced the notion of a proof with that of an active prover and, in addition, we permitted the verifier to use randomness in the verification process. A natural question to ask is: what is the additional power of a polynomial-time verifier working with proofs (as opposed to provers) when they use randomization? It turns out that the answer to this question again involves the use of algebraic methods together with randomization.

Before addressing the question posed above, it is important to understand the difference between a proof and a prover. A prover is active in the sense that it can cheat in an adaptive and online manner by using its knowledge of the earlier queries from the verifier to decide upon its responses to subsequent queries. A proof, on the other hand, is passive and non-adaptive. We can view the proof as being written down by an adversarial prover that knows the particular input x being tested for membership in L, as well as a description of the protocol that will be followed by the verifier. The prover can attempt to use this knowledge to write down a fallacious proof of x's membership in L, even though it is the case that $x \notin L$. In effect, a proof is a predetermined set of responses to all possible questions that could be asked by a specific verifier when its random bits are as yet undetermined. The crucial difference is that unlike the responses of an online prover, a proof cannot change in response to the questions posed by the verifier, and thus it cannot adapt to even the partial information of the verifier's random bits that can be inferred from the questions themselves. Since a prover can simulate an offline proof, a prover has more power to cheat and, conversely, a verifier working with a proof has more power than a verifier working with a prover.

We modify the definition of ***IP*** to that of ***PCP*** (for Probabilistically Checkable Proofs), the only difference being that the prover is replaced by a proof. By the preceding discussion, this is a possibly wider class of languages than ***IP***. We define ***PCP*** as the class of all languages whose proofs of membership can be verified by a randomized polynomial time verifier V with random access to a proof, i.e., the verifier can query arbitrary bits in the proof by specifying their indices or positions.

► **Definition 7.4:** The class ***PCP*** consists of all languages L that have a randomized polynomial-time verifier V such that for any $x \in \Sigma^*$,

- $x \in L \Rightarrow$ there exists a proof Π, such that $\mathbf{Pr}[V(x,\Pi)$ accepts$] = 1$.
- $x \notin L \Rightarrow$ for all purported proofs Π, $\mathbf{Pr}[V(x,\Pi)$ accepts$] \leq \frac{1}{2}$.

When $x \notin L$, all purported proofs Π must be erroneous, and the verifier is required to spot an error with high probability.

We would like to point out that an equivalent definition of ***PCP*** is in terms of a multi-prover interactive proof system where the verifier has access to two or more provers, and the provers are not allowed to communicate with each other once the verifier starts the interaction with the provers (see Problem 7.18). It has been shown that the class ***PCP*** is the same as ***NEXP*** (non-deterministic exponential time), clearly a superset of ***NP***. Our interest is in a restricted version of ***PCP*** where we account for the use of randomness and the number of bits in the proof examined by the verifier.

► **Definition 7.5:** The class $\boldsymbol{PCP}[r(n), q(n)]$ consists of all languages $L \in \boldsymbol{PCP}$ that have a randomized polynomial-time verifier V which, on inputs of length n, uses $O(r(n))$ random bits and examines $O(q(n))$ bits of a purported proof Π.

Let $poly(n)$ denote a function of n that is polynomially bounded. It follows that $\boldsymbol{P} = \boldsymbol{PCP}[0,0]$, $\boldsymbol{NP} = \boldsymbol{PCP}[0, poly(n)]$ and co-$\boldsymbol{RP} = \boldsymbol{PCP}[poly(n), 0]$ (see Problem 7.19). Our goal is to establish the following result, which is far less obvious; the rest of this section is devoted to the proof of this theorem.

Theorem 7.10: $\boldsymbol{NP} \subseteq \boldsymbol{PCP}[poly(n), 1]$.

It is possible to improve Theorem 7.10 by reducing $r(n)$ to a logarithmic function of n, but we omit the rather intricate proof of the stronger version. This result is quite amazing in the sense that it requires a proof that can be verified by examining only $O(1)$ of its bits, regardless of the length of the input. The power of Theorem 7.10 can be fully appreciated by noting that it may be applied to the verification of the (suitably encoded) proof of any mathematical statement.

7.8.1. Arithmetization Revisited

To prove Theorem 7.10, it suffices to show that the ***NP***-complete problem 3-SAT belongs to $\boldsymbol{PCP}[poly(n), 1]$. A proof of the satisfiability of a 3-SAT formula F is easy to construct: write down the satisfying truth assignment $A = (A_1, A_2, \ldots, A_n) \in \{\text{TRUE}, \text{FALSE}\}^n$ for the variables in F. A verifier can substitute these values into the definition of F and verify that it evaluates to TRUE. Unfortunately, this requires that the verifier access all n bits of the proof. If the verifier were to access only a small number of bits in the proof, that would not give sufficient information to decide whether the truth assignment would satisfy F. We will get around this problem by requiring that the proof

Π be a very redundant encoding of A, much like an error-correcting code. To do this, we convert this Boolean formula F into an algebraic formula using an arithmetization similar to that in Section 7.7.

This time we convert F into a polynomial over the field $\mathbb{Z}_2$, as follows. A clause $C_i = L_{i1} \vee L_{i2} \vee L_{i3}$ is replaced by a polynomial c_i of degree 3 obtained by replacing any Boolean variable X_j by $(1 - x_j)$, any negated variable $\overline{X}_j$ by x_j, and the Boolean operation $\vee$ by the field operation of multiplication. For example, when $C_i = X_{i1} \vee \overline{X}_{i2} \vee X_{i3}$, we obtain the term $c_i = (1 - x_{i1})x_{i2}(1 - x_{i3})$; for notational convenience, we omit the dependence of c_i on the variables by writing c_i instead of $c_i(x_{i1}, x_{i2}, x_{i3})$. The assignment A causes C_i to evaluate to TRUE if and only if the corresponding vector $\boldsymbol{a}$ causes c_i to evaluate to 0. We replace the Boolean operator $\wedge$ by the field operation of addition. The arithmetization of F is now given by the degree 3 multivariate polynomial $f(x_1, \ldots, x_n) = \sum_{i=1}^{m} c_i$. It is important to keep in mind that all additions and multiplications are performed modulo 2.

The reason we choose this different arithmetization is that it yields a polynomial of degree 3 instead of $3m$, and this is important for reducing the number of random bits and queries used by the verifier. The problem with this arithmetization is that the polynomial f does not correspond exactly to the Boolean formula F, as indicated in the following exercise.

Exercise 7.8: Let A be a truth assignment for F, and $\boldsymbol{a}$ the corresponding integer vector. Show that if $F(A) = \text{TRUE}$, then $f(\boldsymbol{a}) = 0$. Show also that the converse need not be true, i.e., $f(\boldsymbol{a})$ could evaluate to 0 even though $F(A) \neq \text{TRUE}$.

To get around this problem we use a variant of Freivalds' technique: choose a random vector $\boldsymbol{r} = (r_1, \ldots, r_m)$ uniformly at random from $\mathbb{Z}_2^m$, and redefine f to be

$$f(x_1, \ldots, x_n) = \sum_{i=1}^{m} r_i c_i.$$

The proof of the following lemma is very similar to the argument used in the proof of Theorem 7.1.

Lemma 7.11: *If* $F(A) = \text{FALSE}$, *then* $\mathbf{Pr}[f(\boldsymbol{a}) = 0] = 1/2$.

Thus, with sufficiently high probability (which can be further boosted by repeating the entire verification protocol several times), the polynomial f has a root (in $\mathbb{Z}_2^n$) if and only if the Boolean formula F is satisfiable. We concentrate on the verification of the existence of a root of a multivariate degree 3 polynomial over $\mathbb{Z}_2$. More precisely, we seek a verifier V such that: if f has a root, there exists a proof that will convince the verifier; if not, any proof will deceive the verifier with probability at most $1/2$.

7.8.2. A Proof of Satisfiability

In this section we describe a proof of satisfiability (actually of the existence of a root for f) that the verifier would expect to see in the case when the formula is satisfiable. Later we will see how the verifier can efficiently look for errors or fallacies in the proof.

► **Definition 7.6:** Given an n-dimensional vector $\boldsymbol{x}$ and an m-dimensional vector $\boldsymbol{y}$, their *outer product* $\boldsymbol{z} = \boldsymbol{x} \circ \boldsymbol{y}$ is an $n \times m$ matrix $\boldsymbol{z}$ such that $z_{ij} = x_i y_j$.

We will sometimes view the matrix $\boldsymbol{z}$ as an (nm)-dimensional vector by writing it in a row-major form; this should be clear from the context. (The row-major form of a matrix is obtained by concatenating its rows in the order of increasing row indices.) Consider the vector $\boldsymbol{a}$ of the assignment of values to the variables in f. Define $\boldsymbol{b} = \boldsymbol{a} \circ \boldsymbol{a}$ and $\boldsymbol{c} = \boldsymbol{a} \circ \boldsymbol{b}$, where the second definition views $\boldsymbol{b}$ as a vector; then, $b_{ij} = a_i a_j$ and $c_{ijk} = a_i b_{jk} = a_i a_j a_k$. The vectors $\boldsymbol{a}$, $\boldsymbol{b}$, and $\boldsymbol{c}$ will be used to define three linear functions over $\mathbb{Z}_2$ as follows.

$$
\begin{aligned}
G_a(z_1, \ldots, z_n) &= \sum_{i=1}^{n} a_i z_i, \\
G_b(z_{11}, \ldots, z_{nn}) &= \sum_{i,j} b_{ij} z_{ij} = \sum_{i=1}^{n} \sum_{j=1}^{n} a_i a_j z_{ij}, \\
G_c(z_{111}, \ldots, z_{nnn}) &= \sum_{i,j,k} c_{ijk} z_{ijk} = \sum_{i=1}^{n} \sum_{j=1}^{n} \sum_{k=1}^{n} a_i a_j a_k z_{ijk}.
\end{aligned}
$$

Note that $G_a : \mathbb{Z}_2^n \to \mathbb{Z}_2$, $G_b : \mathbb{Z}_2^{n^2} \to \mathbb{Z}_2$, and $G_c : \mathbb{Z}_2^{n^3} \to \mathbb{Z}_2$. These functions allow us to compute the sum of a subset of the entries in $\boldsymbol{a}$, $\boldsymbol{b}$, or $\boldsymbol{c}$, by encoding the subset into a characteristic vector, which is then used as an assignment to the variables.

The coefficients of the terms in any polynomial over $\mathbb{Z}_2$ must be either 0 or 1. Applying this fact to the degree 3 polynomial f, we can assume that it is of the form

$$
f(\boldsymbol{x}) = \alpha + \sum_{i \in S_1} x_i + \sum_{(i,j) \in S_2} x_i x_j + \sum_{(i,j,k) \in S_3} x_i x_j x_k,
$$

where α is a fixed element of $\mathbb{Z}_2$, S_1 is a set of indices, S_2 is a set of pairs of indices, and S_3 is a set of triples of indices. For a fixed assignment $\boldsymbol{a}$, this expression can be simplified into the following.

$$
\begin{aligned}
f(\boldsymbol{a}) &= \alpha + \sum_{i \in S_1} a_i + \sum_{(i,j) \in S_2} a_i a_j + \sum_{(i,j,k) \in S_3} a_i a_j a_k \\
&= \alpha + G_a(\chi_{S_1}) + G_b(\chi_{S_2}) + G_c(\chi_{S_3}).
\end{aligned}
$$

Here χ_S denotes the characteristic vector of a set S, i.e., the ith component of χ_S is 1 if and only if the ith element of the universe belongs to S. Our definition of the linear functions G_a, G_b, and G_c is such that the three sums can be determined by evaluating each of these functions at a single point.

The desired proof Π of the existence of a root of f consists of the values of G_a, G_b, and G_c at all points in their respective domains. Thus, the verifier V can determine the value of $f(\boldsymbol{a})$ by examining three bits, one each to determine the values of $G_a(\chi_{S_1})$, $G_b(\chi_{S_2})$, and $G_c(\chi_{S_3})$. This would solve the proof verification problem with $r(n) = O(n^3)$ random bits and and $q(n) = 3$ in the case of correct proofs. But the whole point is to be able to deal with erroneous proofs. What if the function f does not have a root but an adversary chooses some functions G_a, G_b, and G_c that result in the verifier being deceived with high probability? Of course, the adversary has to fix the proof by writing down G_a, G_b, and G_c before the verifier chooses the random bits $\boldsymbol{r}$ used to obtain f from F, but this may not prevent the adversary from cheating successfully. In fact, the adversary may not even choose G_a, G_b, and G_c to be linear functions. For example, if they are random functions, the probability of acceptance of an incorrect proof is $1/2$.

7.8.3. The Verification

We now complete the argument by showing how the verifier can test a proof that is purported to be correct and in the form described above. There are two properties of the proof that the verifier would like to ensure. First, that the functions G_a, G_b, and G_c are linear functions. Second, they should all be determined by the same vector $\boldsymbol{a}$. Given these two properties, the strategy described above will work. The constraint is that V is allowed to expend only polynomially many random bits and to examine only a constant number of bits in the proof to achieve this goal. In the verification procedure described below, there are several sub-verifications to be performed. Each of these will be shown to succeed with a constant probability, where failure means that the verifier fails to detect a particular type of error in the proof. We will not compute these probabilities explicitly; it suffices to observe that they can all be made smaller than any fixed constant by repeating the sub-verification O(1) times. Since the whole process makes only O(1) probes into the proof, the number of sub-verifications is also bounded and the total probability of error is no more than the sum of the error probabilities at each stage. We can thus guarantee that the overall probability of error is bounded away from $1/2$.

► **Definition 7.7:** Let $f, g : \mathcal{I} \rightarrow \mathcal{O}$ be two functions with identical finite domain and finite range. Their distance $\Delta(f, g)$ is defined as

$$\Delta(f, g) = \mathbf{Pr}[f(x) \neq g(x)],$$

where x is chosen uniformly at random from $\mathcal{I}$.

In other words, the distance between these functions is the fraction of the domain in which they take on different values.

▶ **Definition 7.8:** For $0 \le \delta \le 1$, the functions f and g are said to be *δ-close* if $\Delta(f, g) \le \delta$.

A linear function $f(x) : \mathbb{Z}_2 \to \mathbb{Z}_2$ is one that can be expressed as $f(x) = ax+b$, for some choice of the coefficients a, $b \in \mathbb{Z}_2$. For historical reasons, in the rest of this section we will abuse terminology somewhat by defining linear functions to be those functions that can be expressed as $f(x) = ax$. It can be shown that a univariate function $f(x) : \mathbb{Z}_2 \to \mathbb{Z}_2$ is linear if and only if for all a and b, $f(a) + f(b) = f(a + b)$. In the case of multivariate functions $f(\boldsymbol{x}) : \mathbb{Z}_2^n \to \mathbb{Z}_2$, we say that f is linear if it is of the form $\sum_{i=1}^{n} a_i x_i$. Again, it can be shown that f is linear if and only if for all $\boldsymbol{a}$ and $\boldsymbol{b}$, $f(\boldsymbol{a}) + f(\boldsymbol{b}) = f(\boldsymbol{a} + \boldsymbol{b})$ (see Problem 7.22). We define a *nearly linear* function as one that satisfies this property for random choices of $\boldsymbol{a}$ and $\boldsymbol{b}$ with probability bounded away from zero.

The following lemma is intuitively obvious, but the proof is non-trivial. We outline the proof in Problem 7.24.

Lemma 7.12: *Fix any δ such that $0 \le \delta < 1/3$. Suppose that $\widehat{G} : \mathbb{Z}_2^n \to \mathbb{Z}_2$ is a function such that for $\boldsymbol{x}$ and $\boldsymbol{y}$ chosen independently and uniformly at random from $\mathbb{Z}_2^n$,*

$$\mathbf{Pr}[\widehat{G}(\boldsymbol{x}) + \widehat{G}(\boldsymbol{y}) = \widehat{G}(\boldsymbol{x} + \boldsymbol{y})] \ge 1 - \frac{\delta}{2}.$$

Then, there exists a linear *function $G : \mathbb{Z}_2^n \to \mathbb{Z}_2$ such that G and $\widehat{G}$ are δ-close.*

Essentially, this lemma says that if $\widehat{G}$ satisfies the linearity condition on most pairs of points, then modifying its value at a few points will make it a linear function.

Suppose now that the proof Π contains the values of three arbitrary (possibly non-linear) functions $\widehat{G}_a$, $\widehat{G}_b$, and $\widehat{G}_c$. The verifier uses the lemma to ensure that they are all nearly linear and can then assume that the δ-close linear functions G_a, G_b, and G_c are actually presented in the proof. We illustrate this for the case of $\widehat{G}_a$. Suppose the verifier V chooses $\boldsymbol{x}$ and $\boldsymbol{y}$ uniformly at random from $\mathbb{Z}_2^n$. Then it probes the proof and verifies that $\widehat{G}_a(\boldsymbol{x}) + \widehat{G}_a(\boldsymbol{y}) = \widehat{G}_a(\boldsymbol{x} + \boldsymbol{y})$. If this test fails, the entire proof can be rejected since it is clear that $\widehat{G}_a$ is not a linear function. When the function passes this test, however, it is not guaranteed that it is indeed a linear function. But with high probability, the function $\widehat{G}_a$ satisfies the above lemma and is nearly linear. Repeating this test boosts the probability of spotting a function that is not δ-close to a linear function.

At this point, V knows that with high probability, each of the three functions in the proof is δ-close to some linear function. In fact, the verifier can now evaluate these linear functions at arbitrary points via the following *self-correction* mechanism. Suppose that the verifier needs to compute $G_a(\boldsymbol{z})$ for an arbitrary $\boldsymbol{z} \in \mathbb{Z}_2^n$, while using the values of the function $\widehat{G}_a$. It chooses $\boldsymbol{x} \in \mathbb{Z}_2^n$ uniformly at random, and evaluates $G_a(\boldsymbol{z}) = \widehat{G}_a(\boldsymbol{z} - \boldsymbol{x}) + \widehat{G}_a(\boldsymbol{x})$. Since $\widehat{G}_a$ is δ-close to G_a, evaluating it at random points gives us the value of G_a at those points with probability $1 - \delta$. Even though the random points $z - x$ and x are highly correlated, the probability that they are both evaluated correctly is at least $1-2\delta$. This can be repeated for independent choices of x to reduce the probability of error below any desired constant. We may now assume that V can evaluate the linear functions G_a, G_b, and G_c at O(1) points each, with the error probability being smaller than any desired constant. Thus, we may as well assume that the proof contains the correct values of G_a, G_b, and G_c at all points.

Of course, the functions G_a, G_b, and G_c could be linear but not related in the desired fashion. Suppose V could verify that these functions are determined by some coefficients $\boldsymbol{a}$, $\boldsymbol{b}$, and $\boldsymbol{c}$ such that $\boldsymbol{b} = \boldsymbol{a} \circ \boldsymbol{a}$ and $\boldsymbol{c} = \boldsymbol{a} \circ \boldsymbol{b}$, with a small probability of error. Then it is possible to verify the existence of a root for f as described earlier. Let us now concentrate on verifying the outer product property.

The following lemma can be proved in a manner similar to Theorem 7.1.

Lemma 7.13: *Let $\boldsymbol{r}$, $\boldsymbol{s} \in \mathbb{Z}_2^n$ be chosen independently and uniformly at random. Suppose that $\boldsymbol{b} \neq \boldsymbol{a} \circ \boldsymbol{a}$, then*

$$\mathbf{Pr}[\boldsymbol{r}^T(\boldsymbol{a} \circ \boldsymbol{a})\boldsymbol{s} \neq \boldsymbol{r}^T\boldsymbol{b}\boldsymbol{s}] \geq \frac{1}{4}.$$

Note that $\boldsymbol{a} \circ \boldsymbol{a}$ and $\boldsymbol{b}$ are now being interpreted as $n \times n$ matrices, and we are applying Freivalds' matrix identity verification technique to determine whether $(\boldsymbol{a} \circ \boldsymbol{a})\boldsymbol{s} = \boldsymbol{b}\boldsymbol{s}$. To verify the equality of these two vectors, we merely apply the technique once more by taking the inner product with the random vector $\boldsymbol{r}$.

This test of the outer product construction can be performed with access to the functions G_a and G_b by observing that $\boldsymbol{a}^T\boldsymbol{s} = G_a(\boldsymbol{s})$, $\boldsymbol{r}^T\boldsymbol{a} = G_a(\boldsymbol{r})$, and $\boldsymbol{r}^T\boldsymbol{b}\boldsymbol{s} = G_b(\boldsymbol{r} \circ \boldsymbol{s})$; thus, V merely confirms that $G_a(\boldsymbol{r})G_a(\boldsymbol{s}) = G_b(\boldsymbol{r} \circ \boldsymbol{s})$. This requires only three probes into the proof. A similar test will verify that $\boldsymbol{c} = \boldsymbol{a} \circ \boldsymbol{b}$.

Finally, we invite the reader to check that the total number of probes into the proof is O(1). In making any probe, the only use of randomness is in the choice of the point at which the function is being evaluated, and each of these uses $O(n^3)$ random bits. We conclude by pointing out that the length of the proof is enormous, being $2^{\Theta(n^3)}$. As we remarked earlier, this proof verification process can be improved such that the length of the proof reduces to a polynomial in n and the number of random bits reduces to a logarithmic function of n, while still preserving the property that only O(1) bits of the proof need to be examined.

Notes

The notion of program checking alluded to in Section 7.1 is due to Blum and Kannan [66]. The technique for verifying matrix and univariate polynomial multiplication is due to Freivalds [157]. More efficient versions of this test (in terms of the number of random bits used) have been devised by Naor and Naor [319], with further improvements by Kimbrel and Sinha [254]. Blum, Chandra, and Wegman [64] have applied Freivalds' technique to obtain an ***RP*** algorithm for deciding the equivalence of *free Boolean graphs,* also known as ordered Boolean decision diagrams (see Problem 7.3). The generalization to multivariate polynomial identities has been rediscovered many times. Although it is usually attributed to the independent and simultaneous articles by Schwartz [367] and Zippel [422], essentially the same result appears in an article by DeMillo and Lipton [123] on the testing of algebraic programs.

The fast matrix multiplication algorithm, running in $O(n^{2.376})$ time, is due to Coppersmith and Winograd [113]. The book by Aho, Hopcroft, and Ullman [5] is a good source for deterministic algorithms for problems involving polynomials and matrices, and most of the basic results assumed in this chapter can be found therein. Zippel's book [423] provides comprehensive coverage of randomized and deterministic algorithms for computations with polynomials. For general information on prime numbers, in particular Bertrand's Postulate and the Prime Number Theorem, the reader may refer to the books on number theory mentioned in the Notes section of Chapter 14.

Tutte [398] first pointed out the close connection between matchings in graphs and matrix determinants, as described in Problem 7.8. The simpler relation between bipartite matchings and matrix determinants was given by Edmonds [134], who also showed that the size of the maximum matching equals the rank of the matrix (see Problem 7.7). The application of the randomized polynomial identity verifier to the problem of matchings in graphs was first pointed out by Lovász [280], who also established a tight relation between the matrix rank and the size of the maximum matching (see Problem 7.9 for a simpler proof). These ideas were applied to the construction of simple algorithm for maximum matchings by Rabin and Vazirani [348, 349]. Although their randomized algorithms for matchings are simple and elegant, they are slower than the deterministic $O(m\sqrt{n})$ time algorithms for bipartite matchings due to Hopcroft and Karp [203], and for non-bipartite matchings due to Micali and Vazirani [308, 406]; the bound for bipartite matchings has been marginally improved to $O(n^{2.5}/\log n)$ by Feder and Motwani [140]. As we shall see in Chapter 12, this algebraic view of matchings and the algorithmic ideas of Rabin and Vazirani have had considerable influence on the development of efficient parallel algorithms for matchings.

The discussion on randomized pattern matching algorithms is based on the work of Karp and Rabin [249]. The deterministic linear time algorithms for pattern matching mentioned above are due to Knuth, Morris, and Pratt [262] and to Boyer and Moore [82].

The survey articles by Babai [39, 40], Goldreich [174, 175], and Johnson [217, 218] give excellent and comprehensive accounts of results in the area of interactive proof systems and proof verification. The protocol for graph non-isomorphism is due to Goldreich, Micali, and Wigderson [176]. The concept of an interactive proof system was introduced by Goldwasser, Micali, and Rackoff [179]. Their motivation was derived from cryptography, and with this application in mind they defined a special type of interactive proof system called a *zero-knowledge interactive proof system* in which the prover would like to prevent the verifier from gaining any useful information while participating in the protocol. Around the same time, Babai [38] introduced the notion of Arthur-Merlin games which are essentially the same as interactive proof systems, the key difference being that the prover (Merlin) has access to the random bits of the verifier (Arthur). Babai's definition was motivated by the desire to classify the complexity of certain group-theoretic problems. A related concept is that of "games against nature" introduced by Papadimitriou [324]. The evidence that graph isomorphism is unlikely to be ***NP***-complete is obtained by combining the results of Boppana, Håstad, and

Zachos [72] with those of Goldreich, Micali, and Wigderson [176] and Schöning [365]; the details are beyond the scope of this book; we refer the reader to Johnson [217] for an overview of this argument.

The result that #3SAT is in ***IP*** is originally due to Lund, Fortnow, Karloff, and Nisan [288]. The proof presented here also includes ideas from Babai and Fortnow [41] and Shamir [372]. In showing that ***IP*** = ***PSPACE***, the easy direction that ***IP*** ⊆ ***PSPACE*** follows from the work of Papadimitriou [324], while the more difficult proof of ***PSPACE*** ⊆ ***IP*** was devised by Shamir [372] based on the techniques used by Lund, Fortnow, Karloff, and Nisan [288] (see Problems 7.16–7.17). The techniques used in these results were inspired by the ideas used in program checking by Blum, Luby, and Rubinfeld [68] and Lipton [277], as well as the idea of representing Boolean formulas as polynomials in the work of Beaver and Feigenbaum [47]. The generalization of ***IP*** to ***MIP***, via the introduction of multiple provers, is due to Ben-Or, Goldwasser, Kilian, and Wigderson [53]. Fortnow, Rompel, and Sipser [153] showed that ***MIP*** ⊆ ***NEXP***, while the more difficult direction ***NEXP*** ⊆ ***MIP*** was established by Babai, Fortnow, and Lund [43].

The complexity class ***PCP*** was defined by Arora, and Safra [33] based on a notion implicit in the work of Feige, Goldwasser, Lovász, Safra, and Szegedy [141]. Efficiently and probabilistically checkable proofs are sometimes also referred to as *transparent proofs* – a terminology introduced earlier by Babai, Fortnow, Levin, and Szegedy [42]. These concepts are variants of the probabilistic oracle machines introduced by Fortnow, Rompel, and Sipser [153] as an alternate view of multiprover systems. Refer to the survey articles cited above for a more thorough discussion of proof systems and the evolution of the current definitions.

Theorem 7.10 is due to Arora, Lund, Motwani, Sudan, and Szegedy [32]; they also established that ***NP*** ⊆ ***PCP***$[\log n, 1]$, combining ideas from various articles mentioned above. The theses by Sudan [388] and Arora [31] contains more complete expositions of the latter result. An important motivation for this work on the ***PCP*** model was to derive the hardness of approximation results for problems such as cliques in graphs [141] and MAX-SAT [32] (see the Notes section of Chapter 5). Lemma 7.12 is originally due to Blum, Luby, and Rubinfeld [68]. The version we state here can be inferred from the results of Rubinfeld [360] and Gemmell, Lipton, Rubinfeld, Sudan, and Wigderson [165].

Problems

7.1 In this problem we will see that Theorem 7.1 is actually just a special case of Theorem 7.2. In the setting of Theorem 7.1, construct a multivariate polynomial Q such that $Q \equiv 0$ if and only if $\boldsymbol{AB} = \boldsymbol{C}$, and then apply Theorem 7.2 to derive result in Theorem 7.1.

7.2 Two rooted trees T_1 and T_2 are said to be *isomorphic* if there exists a one-to-one onto mapping f from the vertices of T_1 to those of T_2 satisfying the following condition: for each internal vertex v of T_1 with the children $v_1, \ldots, v_k$, the vertex $f(v)$ has as children exactly the vertices $f(v_1), \ldots, f(v_k)$. Observe that no ordering is assumed on the children of any internal vertex. Devise an efficient randomized algorithm for testing the isomorphism of rooted trees and analyze its performance. (**Hint:** Associate a polynomial P_v with each vertex v in a tree T. The polynomials are defined recursively, the base case being that the leaf vertices all have $P = x_0$. An internal vertex v of height h with the children $v_1, \ldots, v_k$ has its polynomial defined to be

$$(x_h - P_{v_1})(x_h - P_{v_2}) \cdots (x_h - P_{v_k}).$$

Note that there is exactly one indeterminate for each level in the tree.)

Remark: There is a linear time *deterministic* algorithm for this problem based on a similar approach. Refer to Aho, Hopcroft and Ullman [5].

7.3 (Due to M. Blum, A.K. Chandra, and M.N. Wegman [64].) A labeled directed acyclic graph $G(V, E)$ may be used to represent a Boolean function of n variables $x_1, \ldots, x_n$, as follows. One vertex of V is the *start* vertex, and another the *finish* vertex. Every vertex has out-degree zero or two; if two edges leave a vertex, one must be labeled with a variable and the other by the complement of this variable. Such a graph is said to be *free* if there is at most one occurrence of every variable – complemented or not – on any (directed) path of G. The Boolean function represented by such a graph is the sum of all product terms, where each product term is a product of all the variables on a path from the start vertex to the finish vertex.

Devise a randomized algorithm that, given two free graphs, decides whether they represent the same Boolean function. If the functions are different, the algorithm should output NO; otherwise, it should output YES with probability at least 1/2.

7.4 (Due to R.J. Lipton [277]; see also M. Blum and S. Kannan [66].) Consider the problem of deciding whether two integer *multisets* S_1 and S_2 are identical in the sense that each integer occurs the same number of times in both sets. This problem can be solved by sorting the two sets in $O(n \log n)$ time, where n is the cardinality of the multisets. Suggest a way of representing this as a problem involving a verification of a polynomial identity, and thereby obtain an efficient randomized algorithm. Discuss the relative merits of the two algorithms, keeping in mind issues such as the model of computation and the size of the integers being operated upon. (See also Problem 8.20.)

7.5 (Due to J. Naor.) Two $n \times n$ matrices $\boldsymbol{A}$ and $\boldsymbol{B}$ over a field $\mathbb{Z}_2$ are said to be *similar* if there exists a non-singular matrix $\boldsymbol{T}$ such that $\boldsymbol{T}\boldsymbol{A}\boldsymbol{T}^{-1} = \boldsymbol{B}$. Devise a randomized algorithm for testing the similarity of the matrices $\boldsymbol{A}$ and $\boldsymbol{B}$. (**Hint:** View the entries in T as a collection of variables, and from the definition of similarity, obtain a homogeneous set of linear equations that these variables must satisfy. Any solution T must be a linear combination of the basic solutions to this family of equations. Apply the randomized techniques from this chapter to determining whether there exists a linear combination of the basic solutions that yields a non-singular matrix T.)

7.6 Let $Q(x_1, x_2, \ldots, x_n)$ be a multivariate polynomial over a field $\mathbb{Z}_2$ with the degree sequence $(d_1, d_2, \ldots, d_n)$. A degree sequence is defined as follows: let d_1 be the maximum exponent of x_1 in Q, and $Q_1(x_2, \ldots, x_n)$ be the coefficient of $x_1^{d_1}$ in Q; then, let d_2 be the maximum exponent of x_2 in Q_1, and $Q_2(x_3, \ldots, x_n)$ be the coefficient of $x_2^{d_2}$ in Q_1; and, so on.

Let $\mathbb{S}_1, \mathbb{S}_2, \ldots, \mathbb{S}_n \subseteq \mathbb{Z}_2$ be arbitrary subsets. For $r_i \in \mathbb{S}_i$ chosen independently and uniformly at random, show that

$$\mathbf{Pr}[Q(r_1, r_2, \ldots, r_n) = 0 \mid Q \not\equiv 0] \leq \left(\frac{d_1}{|\mathbb{S}_1|} + \frac{d_2}{|\mathbb{S}_2|} + \cdots + \frac{d_n}{|\mathbb{S}_n|} \right).$$

7.7 (Due to J. Edmonds [134].) Let $G(U, V, E)$ be a bipartite graph, and let $\mathcal{A}$ be the corresponding matrix of indeterminates as defined in Section 7.3. Show that the size of a maximum matching in G is exactly equal to the rank of the matrix $\mathcal{A}$.

7.8 **(Tutte's Theorem [398])** In this problem we generalize Theorem 7.3 to the case of an arbitrary (possibly non-bipartite) graph $G(V, E)$ where $V = \{v_1, \ldots, v_n\}$. A skew-symmetric matrix $\mathcal{A}$ is defined to be a matrix in which for all i and j, $A_{ij} = -A_{ji}$. Let $\mathcal{A}$ be the $n \times n$ skew-symmetric matrix obtained from $G(V, E)$ as follows. A distinct indeterminate x_{ij} is associated with the edge (v_i, v_j), where $i < j$, and the corresponding matrix entries are given by $A_{ij} = x_{ij}$ and $A_{ji} = -x_{ij}$; more succinctly,

$$A_{ij} = \begin{cases} x_{ij} & (v_i, v_j) \in E \text{ and } i < j \\ -x_{ji} & (v_i, v_j) \in E \text{ and } i > j \\ 0 & \text{otherwise} \end{cases}.$$

This matrix is called the *Tutte matrix* of the graph G. Define the multivariate polynomial $Q(x_{11}, x_{12}, \ldots, x_{nn})$ as being equal to $\det(\mathcal{A})$. Show that G has a perfect matching if and only if $Q \not\equiv 0$.

7.9 (Due to M.O. Rabin and V.V. Vazirani [348, 349].) Consider the Tutte matrix of a (non-bipartite) graph $G(V, E)$ defined in Problem 7.8. Show that the rank of the Tutte matrix of G is twice the size of a maximum matching in G.

Hint: Let $\mathcal{A}$ be an $n \times n$ skew-symmetric matrix of rank r. For any two sets S, $T \subset \{1, \ldots, n\}$, denote by $\mathcal{A}_{ST}$ the sub-matrix of $\mathcal{A}$ obtained by including only the rows with indices in S and columns with indices in T. Then, for any two sets S, $T \subset \{1, \ldots, n\}$ of size r,

$$\det(\mathcal{A}_{SS}) \times \det(\mathcal{A}_{TT}) = \det(\mathcal{A}_{ST}) \times \det(\mathcal{A}_{TS}).$$

7.10 Given a randomized algorithm for testing the existence of a perfect matching in a graph G, describe how you would actually construct such a matching. Assuming that you use the randomized testing algorithm from Problem 7.8, compare the running time of your approach with that of the best known deterministic algorithm perfect matching mentioned in the Notes section.

7.11 Given a randomized algorithm for testing the existence of a *perfect* matching in a graph, describe how we can use this to construct a *maximum* matching in a graph G.

7.12 (Due to R.M. Karp and M.O. Rabin [249].) In this problem we will use a different fingerprinting technique to solve the pattern matching problem. The idea is to map any bit string s into a 2×2 matrix $M(s)$, as follows.

- For the empty string ϵ, $M(\epsilon) = \begin{bmatrix} 1 & 0 \\ 0 & 1 \end{bmatrix}$.

- $M(0) = \begin{bmatrix} 1 & 0 \\ 1 & 1 \end{bmatrix}$.
- $M(1) = \begin{bmatrix} 1 & 1 \\ 0 & 1 \end{bmatrix}$.
- For non-empty strings x and y, $M(xy) = M(x) \times M(y)$.

Show that this fingerprint function has the following properties.

1. $M(x)$ is well-defined for all $x \in \{0, 1\}^*$.
2. $M(x) = M(y) \Rightarrow x = y$.
3. For $x \in \{0, 1\}^n$, the entries in $M(x)$ are bounded by Fibonacci number F_n (see Appendix B).

By considering the matrices $M(x)$ modulo a suitable prime p, show how you would perform efficient randomized pattern matching. Explain how you would implement this as a "real-time" algorithm.

7.13 (Due to R.M. Karp and M.O. Rabin [249].) Consider the two-dimensional version of the pattern matching problem. The text is an $n \times n$ matrix X, and the pattern is an $m \times m$ matrix Y. A pattern match occurs if Y appears as a (contiguous) sub-matrix of X. To apply the randomized algorithm described above, we convert the matrix Y into an m^2-bit vector using the row-major format. The possible occurrences of Y in X are the m^2-bit vectors $X(j)$ obtained by taking all $(n-m+1)^2$ sub-matrices of X in a row-major form. It is clear that the earlier algorithm can now be applied to this scenario. Analyze the error probability in this case, and explain how the fingerprints of each $X(j)$ can be computed at a small incremental cost.

7.14 Prove the following relations directly from the definition of ***IP***, i.e., without invoking any results regarding ***IP*** stated in this chapter.

(a) Show that $NP \subseteq IP$.

(b) Show that if the definition of ***IP*** is modified to require that the probability of error be zero, then the resulting complexity class would be exactly the class ***NP***.

(c) Show that co-$RP \subseteq IP$.

7.15 Prove Lemma 7.9.

7.16 (Due to C.H. Papadimitriou [324].) Let ***PSPACE*** be the class of all languages whose membership can be decided using space polynomial in the input size, with no explicit constraint on the running time. Show that $IP \subseteq PSPACE$.

7.17 (Due to A. Shamir [372].) A *quantified Boolean formula* (QBF) is a Boolean formula Φ of the form

$$(Q_1 x_1)(Q_2 x_2) \cdots (Q_n x_n) F(x_1, x_2, \ldots, x_n),$$

where each x_i is a Boolean variable, each Q_i is either the universal ($\forall$) or the existential ($\exists$) quantifier, and F is quantifier-free Boolean formula. It is known that QBF is ***PSPACE***-complete. By devising an interactive proof system for QBF, show that $PSPACE \subseteq IP$.

Hint: The following is a brief sketch of a reformulation of Shamir's proof as presented by A. Shen. The first step is to arithmetize the QBF formula Φ. For any Boolean expression G, possibly a single Boolean variable or a quantified formula, construct an integer polynomial $\widehat{G}$ using the following rules recursively: replace TRUE by 1 and FALSE by 0; replace Boolean variables x_i by arithmetic variables $\widehat{x_i}$; replace $P \wedge Q$ by $\widehat{P} \times \widehat{Q}$; replace the negation of an expression P by $1 - \widehat{P}$; replace $P \vee Q$ by $\overline{\overline{P} \wedge \overline{Q}}$ and apply the previous two rules; replace $(\forall x_i)P(x_i)$ by $\widehat{P}(0) \times \widehat{P}(1)$; and, replace $(\exists x_i)P(x_i)$ by $\widehat{P}(0) + \widehat{P}(1) - (\widehat{P}(0) \times \widehat{P}(1))$. Apply the ideas used in devising an interactive proof system for the arithmetized version of #3SAT to the problem of verifying the value of the arithmetized version, $\widehat{\Phi}$, of the QBF formula Φ. One serious problem in the case of QBF is that the intermediate polynomials need not be of a small degree, primarily to the arithmetization of the the quantifiers. To handle this problem, assume that the arithmetization of the sequence of quantifiers $\mathcal{Q}_1, \ldots, \mathcal{Q}_n$ is interleaved with the application of the following *reduce* operation: for each (integer) variable $\widehat{x_i}$, replace any non-zero power $\widehat{x_i}^k$ by $\widehat{x_i}$. Argue that in the case where we assign only the values 0 or 1 to each $\widehat{x_i}$, the reduce operation does not change the value of the resulting polynomial.

Remark: Combining this result with that of Problem 7.17, we conclude that ***IP*** = ***PSPACE***. It is known that ***PSPACE*** is closed under complementation, and so it follows that ***IP*** = co-***IP***.

7.18 (Due to L. Fortnow, J. Rompel, and M. Sipser [153].) Define the complexity class ***MIP*** as the generalization of ***IP*** where the verifier has access to two provers and the provers are not allowed to communicate with each other once the verifier starts executing. Show that ***MIP*** = ***PCP***.

7.19 Prove the following relations directly from the definition of ***PCP***, i.e., without invoking any results regarding ***PCP*** stated in this chapter.

(a) Show that ***P*** = ***PCP***$[0, 0]$.

(b) Show that ***NP*** = ***PCP***$[0, poly(n)]$.

(c) Show that co-***RP*** = ***PCP***$[poly(n), 0]$.

7.20 (Due to S. Arora and S. Safra [33].) Show that ***PCP***$[\log n, 1] \subseteq$ ***NP***.

7.21 Prove Lemma 7.11.

7.22 Consider a multivariate function $f(\boldsymbol{x}) : \mathbb{Z}_2^n \rightarrow \mathbb{Z}_2$. Show that f is linear if and only if for all $\boldsymbol{a}$ and $\boldsymbol{b}$, $f(\boldsymbol{a}) + f(\boldsymbol{b}) = f(\boldsymbol{a} + \boldsymbol{b})$.

7.23 This problem is concerned with some properties of the distance measure defined in Definition 7.7.

(a) Show that the distance measure Δ satisfies the triangle inequality: for all functions $f, g, h : \mathcal{I} \rightarrow \mathcal{O}$,

$$\Delta(f, h) \leq \Delta(f, g) + \Delta(g, h).$$

(b) For a class of functions $F = \{f : \mathcal{I} \rightarrow \mathcal{O}\}$, define $\Delta_{min}(F)$ as the minimum distance between any two functions in F. Show that for any function g (not necessarily in F), there is at most one function from F at distance $\Delta_{min}(F)/2$ or less.

(c) Suppose that F is the set of all linear functions from $\mathbb{Z}_2^n$ to $\mathbb{Z}_2$. What is $\Delta_{min}(F)$?

7.24 (Due to M. Blum, M. Luby, and R. Rubinfeld [68].) Prove Lemma 7.12 using the following sketch of a proof due to D. Coppersmith. Define the function G such that for each $\boldsymbol{x}$,

$$G(\boldsymbol{x}) = \text{majority}_{\boldsymbol{y}}[\widehat{G}(\boldsymbol{x}+\boldsymbol{y}) - \widehat{G}(\boldsymbol{y})],$$

where the "majority" denotes the value occurring most often over all choices of $\boldsymbol{y}$, breaking ties arbitrarily.

(a) Show that for all $\boldsymbol{x}$, and for $\boldsymbol{y}$ chosen uniformly at random,

$$\mathbf{Pr}[G(\boldsymbol{x}) = \widehat{G}(\boldsymbol{x}+\boldsymbol{y}) - \widehat{G}(\boldsymbol{y})] \geq 1 - \delta.$$

(b) Show that the functions G and $\widehat{G}$ are δ-close.

(c) Show that G is a linear function.

(d) Show that G is uniquely defined.

7.25 Prove Lemma 7.13.

7.26 Appropriately generalizing Lemma 7.13, describe how the verifier can check that $\boldsymbol{c} = \boldsymbol{a} \circ \boldsymbol{b}$.

PART TWO
Applications

CHAPTER 8

Data Structures

The fundamental data-structuring problem is that of maintaining sets of items drawn from an ordered universe so as to efficiently support search queries, update operations, and operations involving entire sets. This chapter begins by identifying some drawbacks in traditional approaches to data structuring using either balanced search trees or self-adjusting search trees. We then describe simple and elegant solutions to these problems using randomization.

8.1. The Fundamental Data-structuring Problem

Consider the fundamental data-structuring problem: we are required to maintain a collection $\{S_1, S_2, \ldots\}$ of sets of items so as to efficiently support certain types of queries and operations. Each item i is an arbitrary record indexed by a key $k(i)$ drawn from a totally ordered universe $\mathcal{U}$. We assume that each item belongs to a unique set and that the keys are all distinct. The operations to be supported are:

MAKESET(S): create a new (empty) set S.
INSERT(i, S): insert item i into the set S.
DELETE(k, S): delete the item indexed by the key value k from the set S.
FIND(k, S): return the item indexed by the key value k in the set S.
JOIN(S_1, i, S_2): replace the sets S_1 and S_2 by the new set $S = S_1 \cup \{i\} \cup S_2$, where

- for all items $j \in S_1$, $k(j) < k(i)$,
- for all items $j \in S_2$, $k(j) > k(i)$.

PASTE(S_1, S_2): replace the sets S_1 and S_2 by the new set $S = S_1 \cup S_2$, where for all items $i \in S_1$ and $j \in S_2$, $k(i) < k(j)$.
SPLIT(k, S): replace the set S by the new sets S_1 and S_2 where

- $S_1 = \{j \in S \mid k(j) < k\}$,
- $S_2 = \{j \in S \mid k(j) > k\}$.

Since it is clear that the structure of the record constituting an item i is irrelevant, we will not distinguish between an item and its key. For example, we will refer to the INSERT operation as INSERT(k, S) and omit all references to the actual item indexed by the key value k. It should be clear that a solution that works when the items consist only of their key values will generalize to more complex record structures. We will refer to the FIND operation as a search, and the INSERT and DELETE operations as an update.

A standard solution to this problem is to represent the set S as a binary search tree. Recall that in a binary search tree the keys are stored at the nodes of a binary tree, and the assignment of keys to nodes must satisfy the following *search tree property*: at a node containing a key value k, the left sub-tree contains only key values smaller than k and the right sub-tree contains only key values larger than k. The keys associated with the nodes in a binary tree are said to be in a *symmetric order* if the search tree property is satisfied. It will be convenient to assume that any node v in a binary search tree contains three pointers in addition to the key value: $L(v)$ points to the left child of v, $R(v)$ points to the right child of v, and $P(v)$ points to the parent of v.

We will assume that the binary search trees we deal with are *endogenous*, in that all key values are stored at internal nodes, and all leaf nodes are empty. This will ensure that the trees are *full*, which means that every non-leaf (internal) node has exactly two children. The pointers $L(v)$ and $R(v)$ are NIL pointers if and only if v is a leaf node, and the pointer $P(v)$ is a NIL pointer if and only if v is the root. In pictorial representations, we will use circles for internal nodes, rectangles for leaf nodes (although usually these are not explicitly specified), and triangles for sub-trees whose internal structure is not relevant (see Figure 8.1). While it is not essential to introduce the dummy leaf nodes or to ensure endogenousness, this does help to simplify the description of the implementation of the various operations.

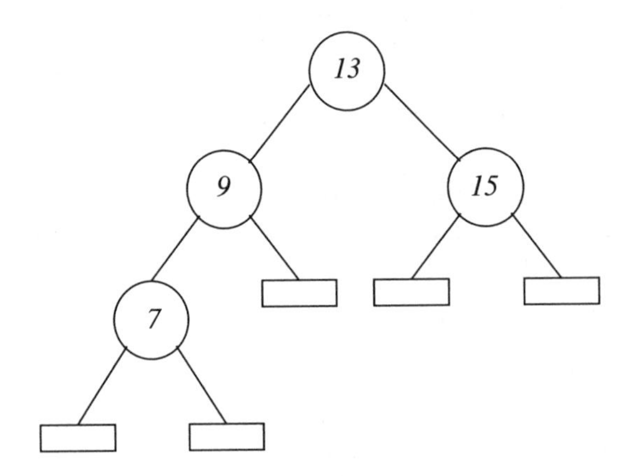

Figure 8.1: A full, endogenous binary search tree for the set of keys $\{7, 9, 13, 15\}$.

Exercise 8.1: In the implementation of a binary search tree described above, we are using three pointers per node. Show that it is possible to reduce this to two pointers per node such that the children and the parent of any node can be accessed by following at most two pointers.

Let us now briefly review the standard implementation of the operations using the binary search tree representation. The operation MAKESET(S) is trivial – simply initialize an empty tree for the set S. To perform a FIND(k, S) is also easy and requires just the standard binary search process. To implement INSERT(k, S), perform FIND(k, S) and, if the value k is not found, insert k into the (empty) leaf node where the search terminates with failure. The operation JOIN(S_1, k, S_2) can be performed by creating a new node containing the key k, and making it the root of a new tree with the trees representing S_1 and S_2 as its left and right sub-trees, respectively. It is easy to handle DELETE(k, S) if the node v containing k (which can be located by a FIND(k, S)) has a leaf as one of its two children. For example, if the right child of v is a leaf, then replace v by $L(v)$ as the child of $P(v)$. If neither of the children is a leaf, then let k' be the key value that is the predecessor of k in the set S; clearly, k' must be at the node arrived at by starting at $L(v)$ and doing FIND($\infty, L(v)$). Now, we can delete the node containing k' since its right child is a leaf, and replace the key value k by k' in the node v, preserving the search tree property. The operation PASTE(S_1, S_2) can be implemented by first deleting the largest key value, say k, from S_1 and then applying JOIN(S_1, k, S_2). Notice that k can be found by doing a FIND(∞, S_1). Finally, doing a SPLIT(k, S) is easy if k is at the root of S; simply do the reverse of the steps employed in JOIN(S_1, k, S_2). When k is not at the root, we can make use of rotations to move it to the root as described in Exercise 8.2.

Each operation can be performed in time proportional to the height(s) of the tree(s), although some operations like JOIN can be performed in constant time. Ideally, the height of a tree would be logarithmic in the size n of the set it represents. Unfortunately, it is easy to devise a sequence of INSERT operations that creates a tree of height linear in n. Several strategies have been devised to handle this problem, usually involving balancing operations to ensure that the tree has height $O(\log n)$. The most commonly used strategy is to perform *rotations* during the update operations so as to ensure that all leaves remain within a distance $O(\log n)$ of the root. In Figure 8.2, we illustrate the two basic types of rotations that are needed.

Each type of rotation moves a node together with one of its sub-trees closer to the root (and some others away from the root), while preserving the search tree property. We will not discuss the specific details of implementing balanced trees using rotations.

Exercise 8.2: Devise a strategy for moving any specified node of a binary search tree to the root using rotations, while preserving the search tree property.

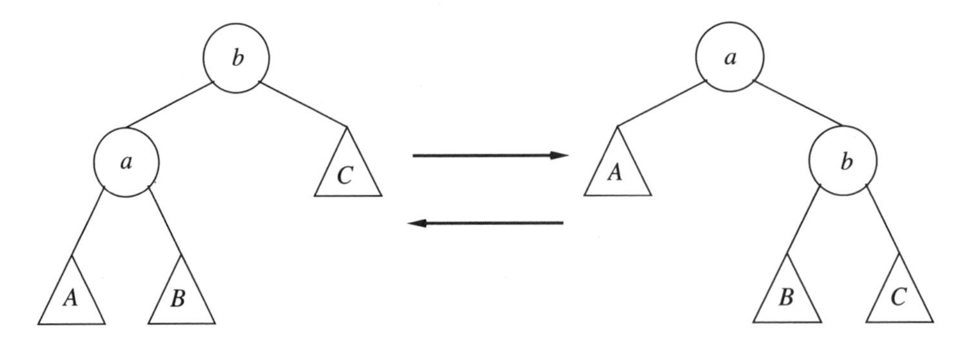

Figure 8.2: The basic rotations.

A balanced search tree guarantees a worst-case time bound of $O(\log n)$ for each of the operations described above. There is an inherent logarithmic lower bound on the number of comparisons required for searching in an ordered list; this lower bound generalizes to randomized searching. Some of the other operations (for example, DELETE) are at least as hard as the FIND operations, and so the lower bound applies to them also. This means that a balanced binary search tree is optimal, at least with respect to the comparison-based model of computation (see Section 8.4 for a further discussion on this issue).

A different strategy, called *splaying*, is used in "self-adjusting" search trees to guarantee an *amortized* time bound of $O(\log n)$; the splay operation moves a specified node to the root via a sequence of rotations. Amortization is the partitioning of the total cost of a sequence of operations among the individual operations in that sequence; thus, an amortized time bound can be viewed as the average cost of the operations in a sequence.

The idea behind self-adjusting trees is to use a particular implementation of the splay operation to move to the root a node accessed by a FIND operation. If a node is accessed often enough, it will remain close to the root and will not contribute much to the total running time; an infrequently accessed node cannot contribute much to the total running time in any case. While these self-adjusting trees guarantee only amortized logarithmic time per operation, they have the advantage of being relatively simple to implement and do not require explicit balance information to be stored at nodes. Furthermore, splay trees can be shown to be optimal with respect to arbitrary access frequencies for the items being stored; in fact, they achieve this optimality without having any explicit information about the access frequencies.

Although self-adjusting trees provide optimal (amortized) solutions to the fundamental data structuring problem, they suffer from some drawbacks. First of all, they restructure the entire tree not only during updates but also while performing simple search operations. This extensive restructuring can cause a significant slowdown in practice in caching and paging environments. Moreover, during any given operation splay trees may perform a linear number of rotations. This is particularly inefficient in implementing higher dimensional search trees common in computational geometry. The reason is that there are secondary data structures associated with each node of these higher dimensional trees, and the secondary data structure at any node depends on the set of keys stored in the sub-tree rooted at that node. Since the entire secondary data structure has to be recomputed during each rotation, the cost of performing a single rotation could increase from a constant to some super-linear function of the sub-tree size. Finally, by its very nature, an amortized time bound leads to the unsatisfying situation where we do not have the guarantee that every operation will run quickly; instead, we obtain bounds only on the total cost of the operations.

We describe an elegant and efficient randomized alternative to the balanced tree and self-adjusting tree, called *treaps*. Treaps achieve essentially the same time bounds in the *expected* sense, do not require any explicit balance information, and the expected number of rotations performed is small for each operation. They have the further advantage of being extremely simple to implement. We also describe an alternative (but closely related) randomized data structure called *skip lists* with similar benefits. Next, we consider the possibility of circumventing the logarithmic lower bound on searching in some interesting special cases. We show that using hash tables, we can guarantee that the expected time required for a search can be made O(1). In the process, we introduce the notion of universal hash functions, which have found numerous applications outside the domain of data structures. Finally, we focus on the version of the data structuring problem without any update operations and provide a hashing scheme that has worst-case search time O(1).

8.2. Random Treaps

A (full, endogenous) binary tree whose nodes have key values associated with them is a binary search tree if the key values are in the symmetric order. If the key values decrease monotonically along any root–leaf path, we call the structure a *heap* and say that the keys are stored in a *heap order*.

Consider a binary tree where each node v contains a pair of values: a key $k(v)$ as well as a *priority* $p(v)$. We call this structure a *treap* if it is a binary search tree with respect to the key values and, simultaneously, a heap with respect to the priorities. More precisely, consider a set of items $S = \{(k_1, p_1), \ldots, (k_n, p_n)\}$ such that the key value of item i is k_i, and its priority is p_i. Assume that the key values and the priorities are drawn from (possibly different) totally ordered universes and that all key values and priorities are distinct. A treap for S will ensure that the k_i's are stored in symmetric order, while the p_i's are stored in heap order. The reader may verify that for the set

$$\{(2,13),(4,26),(6,19),(7,30),(9,14),(11,27),(12,22)\}$$

the tree shown in Figure 8.3 is a valid treap.

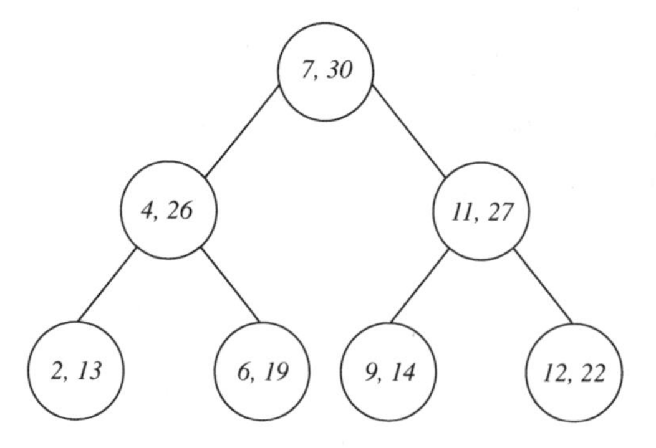

Figure 8.3: A treap.

It is not immediately obvious that any such set has a valid treap but, as we show in the following theorem, there exists a *unique* treap for any set of key-priority pairs.

Theorem 8.1: *Let $S = \{(k_1, p_1), \ldots, (k_n, p_n)\}$ be any set of key-priority pairs such that the keys and the priorities are distinct. Then, there exists a unique treap $T(S)$ for it.*

PROOF: Our proof is constructive, and the construction is recursive. It is obvious that the theorem is true for $n = 0$ and for $n = 1$. Suppose now that $n \geq 2$, and assume that (k_1, p_1) has the highest priority in S. Then, a treap for S can be constructed by putting item 1 at the root of $T(S)$. A treap for the items in S of key value smaller than k_1 can be constructed recursively, and this is stored as the left sub-tree of item 1. Similarly, a treap for the items of key value larger than k_1 is constructed recursively and becomes the right sub-tree of item 1. It is also fairly easy to see that any treap for S must have this decomposition at the root. □

The shape of the tree underlying the treap is determined by the relative priorities of the key values, and any particular shape can be obtained by choosing the priorities suitably. To solve the fundamental data structuring problem, we must somehow pick a good set of priorities for the items being stored and then implement the various operations as described below.

We implement a MAKESET(S) or a FIND(k, S) operation exactly as before. The update operation INSERT(k, S) is implemented by starting as before and doing a FIND(k, S) and inserting k at the empty leaf node where the search terminates with failure. While this maintains the binary search tree property, it will violate the heap order property if the priority of the key k is higher than that of its parent. However, a rotation of k will maintain the heap property at all nodes, except that the order of the node containing k and its parent is now reversed. Thus, we can restore the heap order by using rotations to move k towards the root until its priority value is smaller than that of its parent. A DELETE(k, S) operation is exactly the reverse of an insertion: rotate the node containing k

downward until both its children are leaves, and then simply discard the node. The choice of the rotation (left or right) at each stage depends on the relative order of the priorities of the children of the node being deleted. It is easy to verify that the DELETE operation can be implemented such that it preserves the treap property.

We implement a JOIN(S_1, k, S_2) operation as before, and the resulting structure is a treap provided the priority of k is higher than that of any item in S_1 or S_2. If the new root (containing k) violates the heap order, we simply rotate that node downward until each of the two children of the node has a smaller priority or is a leaf. A PASTE(S_1, S_2) operation can be implemented exactly as in the case of binary search trees. Finally, a SPLIT(k, S) operation can be implemented easily by first deleting k from S, and then inserting it into S with a priority of ∞. Clearly, the node containing k is the root of the new tree and its sub-trees S_1 and S_2 constitute the desired partition of S. These trees can be easily extracted.

Exercise 8.3: The JOIN, PASTE, and SPLIT operations are implemented in terms of the INSERT and DELETE operations. Show how the INSERT and DELETE operations can be implemented in terms of JOIN, PASTE, and SPLIT, and how the latter can be implemented directly.

Clearly, we need only analyze the performance of the FIND, INSERT, and DELETE operations. It is easy to verify that these take time proportional to the depth of the tree representing the treap. However, a slightly stronger statement can be made about the number of rotations required during a DELETE, and by symmetry, during an INSERT operation. Define the *left spine* of a tree as the path obtained by starting at the root and repeatedly moving to the left child until a leaf is reached; the *right spine* is defined similarly.

Exercise 8.4: Show that the number of rotations during a DELETE operation on a node v is equal to the sum of the lengths of the left spine of the right sub-tree and the right spine of the left sub-tree of v.

Before we analyze the running times of the various operations, we must specify how the priorities are chosen for any given key. The idea is to create a *random treap* by choosing the priorities p_i independently from some probability distribution $\mathcal{D}$. The only restriction on the choice of $\mathcal{D}$ is that it should ensure that with probability 1 the priorities are all distinct; in general, it suffices to use any continuous distribution such as the uniform distribution $\mathcal{U}[0, 1]$ on the real interval $[0, 1]$. The priority of an item is chosen at random from $\mathcal{D}$ when the item is first inserted into a set, and the priority for this item remains fixed until it is deleted; moreover, if the item is re-inserted after a deletion, a completely new random priority is assigned to it. The following technicality arises: in our model of computation, we cannot sample a continuous distribution. However, for simplicity of presentation, we temporarily assume in this section that such sampling from a continuous distribution is permissible. Later, in Problem 8.12, we show that treaps can in fact be implemented in our model of computation using only a finite number of random bits.

The ordering of the priorities associated with the various items is completely uncorrelated with the ordering of their key values, ensuring that the tree underlying the treap will remain balanced and have expected depth $O(\log n)$. The choice of the priorities is an implementation detail that is kept hidden, so that an adversary cannot request a sequence of operations that is likely to cause the tree to be unbalanced. The formal verification of this intuition uses the analysis of a set of probabilistic games called *Mulmuley games*, which are described in the next section.

8.2.1. Mulmuley Games

Mulmuley games are useful abstractions of processes underlying the behavior of certain geometric algorithms. We use this abstraction here only for pedagogical purposes; a more direct analysis is possible.

The cast of characters in these games is:

- a set $\mathcal{P} = \{P_1, \ldots, P_p\}$ of *players*;
- a set $\mathcal{S} = \{S_1, \ldots, S_s\}$ of *stoppers*;
- a set $\mathcal{T} = \{T_1, \ldots, T_t\}$ of *triggers*;
- a set $\mathcal{B} = \{B_1, \ldots, B_b\}$ of *bystanders*.

The set $\mathcal{P} \cup \mathcal{S}$ is drawn from a totally ordered universe and all players are smaller than all stoppers: for all i and j, $P_i < S_j$. We assume that the sets are pairwise disjoint. Depending upon the set of active characters, we formulate four different games, with each game being more general than the previous one. Before we describe and analyze the games, it will be useful to list an important property of the Harmonic numbers.

Exercise 8.5: Let $H_k = \sum_{i=1}^{k} 1/i$ denote the kth Harmonic number. Show that $\sum_{k=1}^{n} H_k = (n+1)H_{n+1} - (n+1)$.

Recall that $H_k = \ln k + O(1)$ (Proposition B.4).

Game A. This game starts with the initial set of characters $\mathcal{X} = \mathcal{P} \cup \mathcal{B}$. The game proceeds by repeatedly sampling from $\mathcal{X}$ *without* replacement, until the set $\mathcal{X}$ becomes empty. Each sample is a character chosen uniformly at random from the remaining pool in $\mathcal{X}$. Let the random variable V denote the number of samples in which a player P_i is chosen such that P_i is larger than all previously chosen players. We define the value of the game A_p to be $\mathbf{E}[V]$.

Lemma 8.2: *For all $p \geq 0$, $A_p = H_p$.*

PROOF: Assume that the set of players is ordered as $P_1 > P_2 > \cdots > P_p$. The key observation is that the bystanders are irrelevant to the game: the value of the game is not influenced by the number of bystanders. Thus, we can assume that the initial number of bystanders $b = 0$. Conditional upon the first random sample being a particular player P_i, the expected value of the game is $1 + A_{i-1}$. This is because the players $P_{i+1}, \ldots, P_p$ cannot contribute to the game any more and are effectively reduced to being bystanders. Since i is uniformly distributed over the set $\{1, \ldots, p\}$, we obtain the following recurrence.

$$A_p = \sum_{i=1}^{p} \frac{1 + A_{i-1}}{p} = 1 + \sum_{i=1}^{p} \frac{A_{i-1}}{p}. \tag{8.1}$$

Upon rearrangement, using the fact that $A_0 = 0$, we obtain that $\sum_{i=1}^{p-1} A_i = pA_p - p$. Now, by the property of the Harmonic numbers described in Exercise 8.5, it is easy to see that the Harmonic numbers are the solution to (8.1). □

Game C. In this game, the initial pool is given by $\mathcal{X} = \mathcal{P} \cup \mathcal{B} \cup \mathcal{S}$. The process is exactly the same as that in Game A, treating the stoppers as players as well. The only difference is that the game stops when a stopper is chosen for the first time. Note that since all players are smaller than all stoppers, we will always get a contribution of 1 to the game value from the first stopper. The value of the game is $C_p^s = \mathbf{E}[V + 1] = 1 + \mathbf{E}[V]$, where V is defined exactly as in Game A.

Lemma 8.3: *For all $p, s \geq 0$, $C_p^s = 1 + H_{s+p} - H_s$.*

PROOF: As before, we assume that the set of players is ordered as $P_1 > P_2 > \cdots > P_p$ and that the number of bystanders is 0. Now, if the first sample is a stopper then the game value is 1, and if the first sample is a player P_i then the game value is $1 + C_{i-1}^s$. Noting that the probability of the first event is $s/(s + p)$ and that of the second event is $1/(s + p)$, we obtain the following recurrence:

$$C_p^s = \left(\frac{s}{s+p} \times 1\right) + \left(\frac{1}{s+p} \times \sum_{i=1}^{p}(1 + C_{i-1}^s)\right).$$

Upon rearrangement, using the fact that $C_0^s = 1$, we obtain that

$$C_p^s = \frac{s+p+1}{s+p} + \frac{\sum_{i=1}^{p-1} C_i^s}{s+p}$$

which is equivalent to

$$\sum_{i=1}^{p-1} C_i^s = (s+p)C_p^s - (s+p+1).$$

Once again, using Exercise 8.5 it can be verified that the solution to the recurrence is given by $C_p^s = 1 + H_{s+p} - H_s$. □

Games D and E. Games D and E are similar to Games A and C, the only difference being that their initial pools consist of $\mathcal{X} = \mathcal{P} \cup \mathcal{B} \cup \mathcal{T}$ and $\mathcal{X} = \mathcal{P} \cup \mathcal{B} \cup \mathcal{S} \cup \mathcal{T}$, respectively. The role of the triggers is that the counting process begins only after the first trigger has been chosen. More precisely, a player or a stopper contributes to V only if it is sampled after a trigger and before any stopper, and if it is larger than all previously chosen players. Letting D_p^t and $E_p^{s,t}$ denote the expected values of the two games, the following lemmas can be proved as before.

Lemma 8.4: *For all* $p, t \geq 0$, $D_p^t = H_p + H_t - H_{p+t}$.

Lemma 8.5: *For all* $p, s, t \geq 0$, $E_p^{s,t} = \dfrac{t}{s+t} + (H_{s+p} - H_s) - (H_{s+p+t} - H_{s+t})$.

The proofs of these lemmas are left as problems.

8.2.2. Analysis of Treaps

In order to apply the games described above to the analysis of the performance of random treaps, it will be useful to identify an important property of random treaps – the *memoryless* property. Consider a random treap obtained by inserting the elements of a set S into an initially empty treap. Since the random priorities for the elements of S are chosen independently, we can assume that the priorities are chosen before the insertion process is initiated. Once the priorities have been fixed, Theorem 8.1 implies that the treap T is uniquely determined. This implies that the order in which the elements are inserted does not affect the structure of the tree. Thus, without loss of generality, we can assume that the elements of set S are inserted into T in the order of decreasing priority. An advantage of this view is that it implies that all insertions take place at the leaves and no rotations are required to ensure the heap order on the priorities.

Exercise 8.6: Using the memoryless property, derive a connection between the structure of a treap and the behavior of the Quicksort algorithm (see Chapter 1).

Define the *depth* of a node in a treap as its distance from the root. The following lemma establishes that the expected depth of the element of rank k in S is $O(\log k + \log(n - k + 1))$, which is always $O(\log n)$.

Lemma 8.6: *Let T be a random treap for a set S of size n. For an element $x \in S$ having rank k,*

$$\mathbf{E}[depth(x)] = H_k + H_{n-k+1} - 1.$$

PROOF: Define the sets $S^- = \{y \in S \mid y \leq x\}$ and $S^+ = \{y \in S \mid y \geq x\}$. Since x has rank k, it follows that $|S^-| = k$ and $|S^+| = n - k + 1$. Denote by $Q_x \subseteq S$

the set of elements that are stored at nodes on the path from the root of T to the node containing x, i.e., the ancestors of x. Let Q_x^- denote $S^- \cap Q_x$. We will establish that $\mathbf{E}[|Q_x^-|] = H_k$. By symmetry, it follows that the expected size of $Q_x^+ = S^+ \cap Q_x$ is H_{n-k+1}. This will imply that the expected length of the path from the root to x is $H_k + H_{n-k+1} - 1$, since $Q_x^- \cap Q_x^+ = \{x\}$.

Consider any ancestor $y \in Q_x^-$ of the node x. By the memoryless assumption, y must have been inserted prior to x, and the priorities must satisfy the inequality $p_y > p_x$. Since $y < x$, it must be the case that x lies in the right sub-tree of y. In fact, we claim that all elements z such that $y < z < x$ lie in the right sub-tree of y. Consider the searches for the elements x, y, and z in T. Clearly, the searches for x and y will follow the path from the root to the node containing y. But then there cannot be any node on this path whose value is between y and x. This implies that the search for every element whose value lies between y and x must follow the path from the root to y, and in fact go into the right sub-tree of y. We conclude that y is an ancestor of every node containing an element of value between y and x. By our assumption about the order of insertion, this implies that every element whose value lies between y and x must have been inserted after y, and hence is of lower priority than y.

The preceding argument establishes that an element $y \in S^-$ is an ancestor of x, or a member of Q_x^-, if and only if it was the largest element of S^- in the treap at the time of its insertion. Since the order of insertion is determined by the order of the priorities, and the latter is uniformly distributed, the order of insertion can be viewed as being determined by uniform sampling without replacement from the pool S. We can now claim that the distribution of $|Q_x^-|$ is the same as that of the value of Game A when $\mathcal{P} = S^-$ and $\mathcal{B} = S \backslash S^-$. Since $|S^-| = k$, the expected size of $|Q_x^-| = H_k$. □

Exercise 8.7: Obtain an alternate proof of Lemma 8.6 by using the analysis of Game C when x is a stopper, $\mathcal{P} = S^- \backslash \{x\}$, and $\mathcal{B} = S^+ \backslash \{x\}$.

The next lemma helps us bound the expected number of rotations required during an update operation (see Exercise 8.4). For any element x in a treap, let L_x denote the length of the left spine of the right sub-tree of x, and R_x the length of the right spine of the left sub-tree of x.

Lemma 8.7: *Let T be a random treap for a set S of size n. For an element $x \in S$ of rank k,*

$$\mathbf{E}[R_x] = 1 - \frac{1}{k}$$

and

$$\mathbf{E}[L_x] = 1 - \frac{1}{n-k+1}.$$

PROOF: We prove only the first result. The second result follows by symmetry since the rank of x becomes $n-k+1$ if we invert the total order underlying the key values. We will demonstrate that the distribution of R_x is the same as that of the value of Game D with the choice of characters $\mathcal{P} = S^- \backslash \{x\}$, $\mathcal{T} = \{x\}$, and $\mathcal{B} = S^+ \backslash \{x\}$, where $S^- = \{y \in S \mid y \leq x\}$ and $S^+ = \{y \in S \mid y \geq x\}$ as before. Since we now have $p = k-1$, $t = 1$, and $b = n-k$, Lemma 8.4 implies that

$$\mathbf{E}[R_x] = D^1_{k-1} = H_{k-1} + H_1 - H_k = 1 - \frac{1}{k}.$$

To relate the length of the right spine of the left sub-tree of x to Game D, we make the following claim: an element $z < x$ lies on the right spine of the left sub-tree of x if and only if z is inserted after x, and all elements whose values lie between z and x are inserted after z. The proof relies on the memoryless property of treaps.

We first prove the backward implication in the claim. Consider the path followed by the insertion procedure in locating the leaf at which z is inserted. This path must go through the node containing x, since the only way to distinguish between z and x is via a comparison with some element that lies between them, and all such elements are inserted after z. Since z is smaller than x and inserted after x, it must lie in the left sub-tree of x. Moreover, since all the elements in the left sub-tree of x are smaller than x, and z is the largest of these at the time of its insertion, z must lie on the right spine of this sub-tree.

The forward implication in the claim is proved similarly. Since z lies in the left sub-tree of x, it must have been inserted after x and be of value smaller than x. Moreover, all elements with value between those of z and x must be in the left sub-tree of x, and since z lies on the right spine these elements must have been inserted after z. □

The following theorem summarizes the performance bounds for random treaps. The proof is an easy consequence of the preceding lemmas and is left as an exercise. Note that the search time for a key $x \notin S$ is essentially the search time for the elements of S that would have been its predecessor or successor had it belonged to S.

Theorem 8.8: *Let T be a random treap for a set S of size n.*

1. *The expected time for a* FIND, INSERT, *or* DELETE *operation on T is* $O(\log n)$.

2. *The expected number of rotations required during an* INSERT *or* DELETE *operation is at most* 2.

3. *The expected time for a* JOIN, PASTE, *or* SPLIT *operation involving sets S_1 and S_2 of sizes n and m, respectively, is* $O(\log n + \log m)$.

8.3. Skip Lists

We now turn to another elegant randomized data structure called *skip lists*. Consider a set $S = \{x_1 < x_2 < \cdots < x_n\}$ drawn from a totally ordered universe.

► **Definition 8.1:** A *leveling with r levels* of an ordered set S is a sequence of nested subsets (called *levels*)

$$L_r \subseteq L_{r-1} \subseteq \cdots \subseteq L_2 \subseteq L_1$$

such that $L_r = \emptyset$ and $L_1 = S$.

► **Definition 8.2:** Given an ordered set S and a leveling for it, the level of any element $x \in S$ is defined as

$$l(x) = \max\{i \mid x \in L_i\}.$$

Given any leveling of the set S, we can define an ordered list data structure as follows. For convenience, we will assume that two special elements $-\infty$ and $+\infty$ belong to each of the levels, where $-\infty$ is smaller than all elements in S and $+\infty$ is larger than all elements in S. Observe that both $-\infty$ and $+\infty$ are of level r. The level L_1 is stored in a sorted linked list, and each node x in this linked list has a pile of $l(x) - 1$ nodes sitting above it. There are horizontal and vertical pointers between nodes as illustrated in Figure 8.4. This data structure is the skip list corresponding to a specific leveling of S.

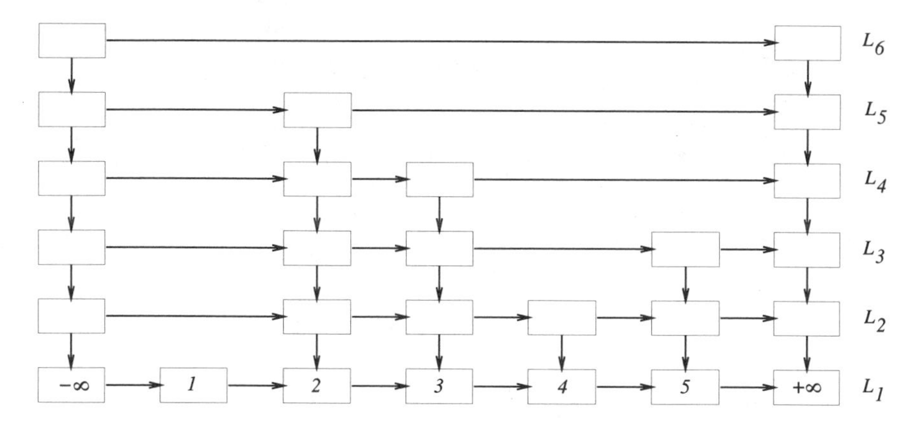

Figure 8.4: A skip list.

In Figure 8.4, the skip list represents the set $S = \{1, 2, 3, 4, 5\}$, and the leveling that determines this skip list consists of the following 6 levels: $L_6 = \emptyset$, $L_5 = \{2\}$, $L_4 = \{2, 3\}$, $L_3 = \{2, 3, 5\}$, $L_2 = \{2, 3, 4, 5\}$, and $L_1 = \{1, 2, 3, 4, 5\}$. A pile of $l(x)$ nodes is created for an element x of S. Further, starting at the ith node from the bottom in the left-most column of nodes and following the horizontal pointers will yield a set of nodes corresponding to the elements of the level L_i.

► **Definition 8.3:** An *interval* at level i is the set of elements of S spanned by a specific horizontal pointer at level i.

The sequence of levels L_i can be viewed as successively coarser partitions of S into a collection of intervals. In the example shown in Figure 8.4, we can view the levels as determining the following successive partitions:

$$\begin{aligned}
L_1 &= [-\infty, 1] \cup [1, 2] \cup [2, 3] \cup [3, 4] \cup [4, 5] \cup [5, +\infty] \\
L_2 &= [-\infty, 2] \cup [2, 3] \cup [3, 4] \cup [4, 5] \cup [5, +\infty] \\
L_3 &= [-\infty, 2] \cup [2, 3] \cup [3, 5] \cup [5, +\infty] \\
L_4 &= [-\infty, 2] \cup [2, 3] \cup [3, +\infty] \\
L_5 &= [-\infty, 2] \cup [2, +\infty] \\
L_6 &= [-\infty, +\infty]
\end{aligned}$$

The interval partition structure is more conveniently viewed as a tree (see Figure 8.5) where each node corresponds to an interval, and all intervals at the same level are represented by nodes at the same level in the tree. If an interval J at level $i+1$ contains as a subset an interval I at the level i, then node J is the parent of node I in the tree. For an interval I at level $i+1$, $c(I)$ denotes the number of children it has at level i. Since $c(I)$ can be arbitrarily large, the tree is not binary in general. The skip list representation can be viewed as a threaded version of this tree, where each thread is a series of pointers forming an ordered linked list of the nodes in a level. In Figure 8.5, the horizontal pointers correspond to the threads.

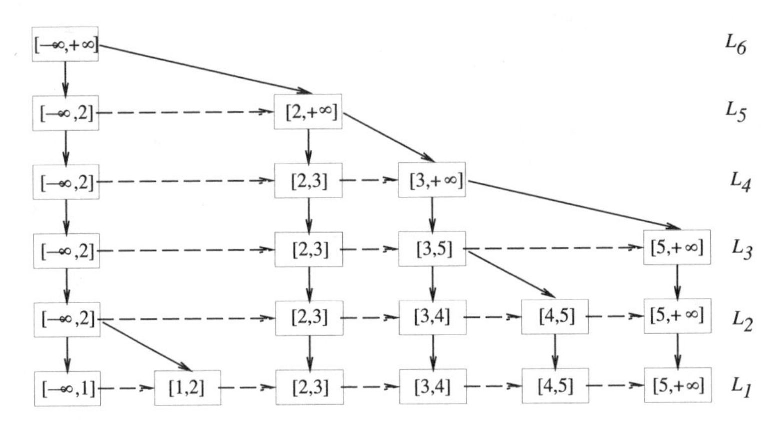

Figure 8.5: Tree representation of a skip list.

Consider an element y, which is not necessarily a member of S. Define $I_j(y)$ as the interval at level j that contains y. If y lies on the boundary between two intervals, we assign it to the left-most one. We can now view the nested sequence of intervals $I_r(y) \subseteq I_{r-1}(y) \subseteq \cdots \subseteq I_1(y)$ containing y as a root–leaf path in the tree representation of the skip list. To complete the description of a skip list, we have to specify the choice of the leveling that underlies it. The basic idea is to choose a random leveling, thereby defining a random skip list. The analysis will show that there is a high probability that the search tree corresponding to a random skip list is balanced.

8.3.1. Analyzing Random Skip Lists

A *random leveling* of the set S is defined as follows: given the choice of the level L_i, the level L_{i+1} is defined by independently choosing to retain each element $x \in L_i$ with probability $1/2$. This process starts with $L_1 = S$, and it terminates when, for the first time, a newly constructed level is empty. An alternate view of this construction is as follows: let the levels $l(x)$ for $x \in S$ be independent random variables, each with the geometric distribution with parameter $p = 1/2$. Let r be one more than the maximum of these random variables. Place x in each of the levels $L_1, \ldots, L_{l(x)}$. As was the case with the random priorities in treaps, a random level is chosen for every element of S upon its insertion, and this remains fixed until the element is deleted.

Exercise 8.8: Show that the expected space requirement of a random skip list for a set S of size n is $O(n)$.

Lemma 8.9: *The number of levels r in a random leveling of a set S of size n has expected value $\mathbf{E}[r] = O(\log n)$. Moreover, $r = O(\log n)$ with high probability.*

PROOF: We prove only the high probability result; the bound on the expected value is left as an exercise. The number of levels $r = 1 + \max_{x \in S} l(x)$, and the levels $l(x)$ are i.i.d. random variables distributed geometrically with parameter $p = 1/2$. We may thus view the levels of the members of S as independent geometrically distributed random variables $X_1, \ldots, X_n$. It is easy to verify that $\mathbf{Pr}[X_i > t] \leq (1-p)^t$ and, therefore,

$$\mathbf{Pr}[\max_i X_i > t] \leq n(1-p)^t = \frac{n}{2^t},$$

since $p = 1/2$ in this case. Using $t = \alpha \log n$ and $r = 1 + \max_i X_i$, we obtain the desired result that

$$\mathbf{Pr}[r > \alpha \log n] \leq \frac{1}{n^{\alpha-1}},$$

for any $\alpha > 1$. □

Exercise 8.9:

1. Use the ideas in the proof of Lemma 8.9 to show that $\mathbf{E}[r] = O(\log n)$.
2. Use Theorem 1.3 to show that $\mathbf{E}[r] = O(\log n)$.

This result implies that the tree representing the skip list has height $O(\log n)$ with high probability. Unfortunately, since the tree need not be binary, it does not immediately follow that the search time is similarly bounded. To understand this, we first specify an efficient implementation of the FIND operation.

We will describe the implementations of all operations in terms of the tree representation of skip lists and then translate this description back into the skip list representation. The implementation of FIND(x, S) corresponds to walking down the path $I_r(y) \subseteq I_{r-1}(y) \subseteq \cdots \subseteq I_1(y)$, as follows: at level j, starting at the node $I_j(y)$, use a vertical pointer to descend to the leftmost child of the current interval; then, using the horizontal pointers, move rightward till the node $I_j(y)$ is reached. It is easy to determine whether y belongs to a given interval, or to an interval to its right. Also, in the original skip list representation, the vertical pointers allow access to only the left-most child of an interval, and hence it is essential to use the horizontal pointers to traverse the list of its children.

The cost of the FIND(y, S) operation is proportional to the number of levels as well as the number of intervals (or nodes) visited at each level. The number of nodes visited at level j does not exceed the number of children of the interval $I_{j+1}(y)$. It is now clear that the cost of a FIND operation depends not only on the number of levels, but is proportional to the total number of children of the nodes on the search path. This cost can be bounded by

$$O\left(\sum_{j=1}^{r}(1 + c(I_j(y)))\right).$$

Fortunately, as shown in the following lemma, this quantity has expectation bounded by $O(\log n)$ as well.

Lemma 8.10: *Let y be any element and consider the search path $I_r(y), \ldots, I_1(y)$ followed by* FIND(y, S) *in a random skip list for the set S of size n. Then,*

$$\mathbf{E}[\sum_{j=1}^{r}(1 + c(I_j(y)))] = O(\log n).$$

PROOF: We will show that for any specific interval I in a random skip list, $\mathbf{E}[c(I)] = O(1)$. Since Lemma 8.9 guarantees that $r = O(\log n)$ with high probability, this will yield the desired result. Note that we do need the high probability bound on r – it is not correct to multiply the expectation of r with that of $1 + c(I)$ since the two random variables are not independent. On the other hand, since we know that $r > \alpha \log n$ with probability at most $1/n^{\alpha-1}$, and since $\sum_j(1 + c(I_j(y))) = O(n)$, we can argue that the case $r > 2\log n$ does not contribute significantly to the expectation of $\sum_j c(I_j(y))$.

Let J be any interval at level i of the skip list. We will prove that the expected number of siblings of J (children of its parent) is bounded by a constant, and this will imply that the expected number of children of an interval is bounded by a constant. In fact, it will suffice to prove that the number of siblings of J to its right is bounded by a constant.

Let the intervals to the right of J be $J_1 = [x_1, x_2]$, $J_2 = [x_2, x_3]$, ..., $J_k = [x_k, +\infty]$. These intervals exist at level i if and only if each of the elements x_1, ..., x_k belong to L_i. If J has s siblings to its right, then it must be the case that $x_1, \ldots, x_s \notin L_{i+1}$, and $x_{s+1} \in L_{i+1}$. Since each element of L_i is independently chosen to be in L_{i+1} with probability 1/2, the number of right siblings of J is stochastically dominated by a random variable that is geometrically distributed with parameter 1/2. It follows that the expected number of right siblings of J is at most 2. □

In Problem 8.14 we suggest a different approach, which leads to a precise determination of the expected cost of the FIND operation.

We now describe the implementation of the update operations on a skip list. Consider the operation INSERT(y, S), and assume that a random level $l(y)$ is chosen for y as described earlier. If the value of $l(y)$ exceeds r, then start by creating new levels from $r + 1$ to $l(y)$ in the original skip list. This can be done in time $O(r)$ since the new levels are all empty prior to the insertion of y. Then, perform the operation FIND(y, S) and determine the search path $I_r(y)$, ..., $I_1(y)$, where r is updated to its new value if necessary. Given the search path, the actual insertion process can be accomplished in time $O(l(y))$ since all it requires is the splitting around y of the intervals $I_1(y), \ldots, I_{l(y)}(y)$, and of course updating the pointers as appropriate. The DELETE operation is the converse of the INSERT operation, and it involves performing FIND(y, S) followed by the collapsing of the intervals that have y as an end-point. In addition to the cost of a FIND operation, both operations require additional work proportional to $l(y)$. Combining this with Lemmas 8.9 and 8.10, we obtain the following theorem.

Theorem 8.11: *In a random skip list for a set S of size n, the operations* FIND, INSERT, *and* DELETE *can be performed in expected time* $O(\log n)$.

These results extend to the other operations described in treaps.

Exercise 8.10: Describe an implementation of operations JOIN, PASTE, and SPLIT for random skip lists. Analyze the running time of your implementation, and compare the result with the same operations in the case of treaps.

8.4. Hash Tables

In the rest of this chapter, we restrict ourselves to the following special cases of the data-structuring problem considered in the previous sections:

1. In the *static dictionary* problem we are given a set of keys S and must organize it into a data structure that supports the efficient processing of FIND queries.

2. In the *dynamic dictionary* problem the set S is not provided in advance. Instead it is constructed by a series of INSERT and DELETE operations that are intermingled with the FIND queries.

These problems can be solved using data structures discussed earlier, i.e., balanced search trees, random treaps, and random skip lists. For a set S of size s, these data structures require $\Omega(\log s)$ time (worst-case or expected) to process any search or update operation. The time bounds achieved are optimal in the sense that for data structures based on pointers and search trees, we are faced with a logarithmic lower bound on the cost of a search. These lower bounds are based on the assumption that the only computation we can perform over the keys is to compare them and thereby determine their relationship in the underlying total order.

We now present an entirely different approach that allows us to circumvent this lower bound and achieve O(1) search time. We mention briefly the reasons why the logarithmic lower bounds will not apply to the dictionary problem we will consider. We will assume that the keys in S are chosen from a totally ordered universe M of size m; without loss of generality, we define $M = \{0, \ldots, m-1\}$. We will also assume that the keys are represented as integers in a manner that permits us to perform arithmetic operations over them. Finally, we will choose to work in the RAM model of computation in its full generality.

To better understand the difference in the models, we describe a scheme that enables us to obtain search and update times that are bounded independently of the size of S. In this scheme, we create a table T of size m; a table is simply an array supporting random access. For each $k \in M$, we set $T[k] = 1$ if and only if $k \in S$. We can perform search or update operations for a key in unit time by accessing the corresponding entry in the table. The problem with this approach is that the key space is typically many orders of magnitude larger than the set S. For example, in a 32-bit machine we have $m = 2^{32}$, so such a table of size m will consume the entire memory of the machine. In fact, the preprocessing cost of initializing the table is equally large in this solution. Even though this approach is impractical, it serves to illustrate the point that the new model permits us to get around the comparison-based lower bounds on searching in a totally ordered set. This is because we are now making use of the full power of the RAM model of computation including random access and indirect indexing (which permits an m-way branch in a single step), not to mention the dual use of key values as table indices.

In this section we focus on the dynamic dictionary problem, and our goal is to obtain a more practical version of the table-based scheme. The main issue is that of reducing the size of the table to a value close to $|S|$, while maintaining the property that a search or update operation can be performed in O(1) time. To this end, we introduce *hashing*, a data structuring technique in which we use a fingerprint function (see Chapter 7) to determine where a key should be located in the table.

A *hash table* is a data structure for the dictionary problem that consists of the following components: a table T consisting of n *cells* indexed by $N = \{0, 1, \ldots, n-1\}$, and a *hash function* h, which is a mapping from M into N. We assume that n is smaller than m, since otherwise the dictionary problem is trivial. Each cell is a memory word that can hold exactly as many bits as required to encode an element of M, i.e., the word size is $\log m$. The hash function is a fingerprint function for the keys in M, and it specifies a location in the table for each element of M. Ideally, we would want the hash function to map distinct keys in S to distinct locations in the table. A *collision* is said to occur between two distinct keys x and y if $h(x) = h(y)$ and they are said to collide at the corresponding location in T.

► **Definition 8.4:** A hash function $h : M \rightarrow N$ is said to be *perfect* for a set $S \subseteq M$ if h does not cause any collisions among the keys of the set S.

Exercise 8.11: Show that a perfect hash function can be constructed for any set S of size at most n.

Given a perfect hash function for a set S, we can use the hash table to process a sequence of FIND operations in O(1) time each: store each element $k \in S$ at the location $T[h(k)]$; to search for a key q, just check whether $T[h(q)] = q$. A problem arises when we try to use this hash function to process updates. The problem is that no hash function can be perfect for all possible sets $S \subseteq M$; this follows from the observation that for $n < m$, any function h must map some two elements of M to the same location, and so it cannot be perfect for any set containing those two elements. Thus, perfect hash functions are useless for the dynamic dictionary problem. It is still possible that they can be used to obtain a good solution to the static dictionary problem, and we will return to this issue in Section 8.5.

A natural approach to solving the dynamic dictionary problem is to relax the definition of perfect hash functions to that of "near-perfect" hash functions, which are allowed to cause a small number of collisions at each location in the table. There has been great deal of research into the design of such near-perfect hash functions, but typically this is under the assumption that the sequence of operations to be performed is drawn from some well-behaved probability distribution. Under this assumption, it is possible to come up with simple hash functions that cause only O(1) collisions on the average at any table location, provided the number of items present in the hash table is bounded by some linear function in the table size n. The keys colliding at any given location are usually organized into a secondary data structure accessible from that location, or they can be rehashed into a secondary hash table using a new hash function. To process any operation, the hash function is used to determine the appropriate location in the table, and the operation is then performed on the secondary data structure associated with that location. Since the expected size of the secondary data structure is O(1), it follows that each operation has expected cost O(1) in addition to the cost of evaluating the hash function. Hash functions are chosen so that they can be evaluated in O(1) time.

We will present a randomized hashing scheme for the *dynamic* dictionary problem that processes search and update operations in *expected* time O(1), without making any probabilistic assumptions about the operation sequence. The expectation is with respect to the random choices internal to the hash table.

8.4.1. Universal Hash Families

Our solution requires the construction of a class of hash functions that have found a surprisingly large number of applications in areas far removed from the original problem, such as routing in networks and complexity theory. The idea is to choose a *family* of hash functions $H = \{h : M \rightarrow N\}$, where each $h \in H$ is easily represented and evaluated. While any one function $h \in H$ may not be perfect for very many choices of the set S, we can ensure that for every set S of small cardinality, a large fraction of the hash functions in H are near-perfect for S in the sense that the number of collisions is small. Thus, for any particular set S, a random choice of $h \in H$ will give the desired performance. The hash functions described here can also be used to solve some of the problems discussed in earlier sections.

► **Definition 8.5:** Let $M = \{0, 1, \ldots, m-1\}$ and $N = \{0, 1, \ldots, n-1\}$, with $m \geq n$. A family H of functions from M into N is said to be *2-universal* if, for all x, $y \in M$ such that $x \neq y$, and for h chosen uniformly at random from H,

$$\mathbf{Pr}[h(x) = h(y)] \leq \frac{1}{n}.$$

A totally random mapping from M to N has a collision probability of exactly $1/n$; thus, a random choice from a 2-universal family, of hash functions gives a seemingly random function. The collection of all possible functions from M to N is a 2-universal family, but it has several disadvantages. Picking a random function requires $\Omega(m \log n)$ random bits. This is also the number of bits of storage required to represent the chosen function. Our goal is to obtain smaller 2-universal families of functions that require a small amount of space and are easy to evaluate; in particular, we would like to construct 2-universal families containing only a small subset of all possible functions. The reason this is possible is that a randomly chosen function $h \in H$ is required to behave like a random function only with respect to pairs of elements. In fact, as x ranges over M, the values $h(x)$ behave somewhat like pairwise independent random variables, which is precisely the reason for the name "2-universal." On the other hand, for a purely random function f, the values $f(x)$ have complete independence. In Section 8.4.4 we will discuss "strong" 2-universal hash families, which have an exact correspondence with pairwise independent random variables.

From here on fix the sets M, N, and H as in Definition 8.5. For any $x, y \in M$ and $h \in H$, define the following indicator function for a collision between the keys x and y under the hash function h:

$$\delta(x,y,h) = \begin{cases} 1 & \text{for } h(x) = h(y) \text{ and } x \neq y \\ 0 & \text{otherwise.} \end{cases}$$

For all $X, Y \subseteq M$, define the following extensions of the indicator function δ:

$$\begin{aligned} \delta(x,y,H) &= \sum_{h \in H} \delta(x,y,h), \\ \delta(x,Y,h) &= \sum_{y \in Y} \delta(x,y,h), \\ \delta(X,Y,h) &= \sum_{x \in X} \delta(x,Y,h), \\ \delta(x,Y,H) &= \sum_{y \in Y} \delta(x,y,H), \\ \delta(X,Y,H) &= \sum_{h \in H} \delta(X,Y,h). \end{aligned}$$

For a 2-universal family H and any $x \neq y$, we have $\delta(x,y,H) \leq |H|/n$.

The following theorem shows that our definition of 2-universality is essentially the best possible, since a significantly smaller collision probability cannot be obtained for $m \gg n$.

Theorem 8.12: *For any family H of functions from M to N, there exist $x, y \in M$ such that*

$$\delta(x,y,H) > \frac{|H|}{n} - \frac{|H|}{m}.$$

PROOF: Fix some function $h \in H$, and for each $z \in N$ define the set of elements of M mapped to z as

$$A_z = \{x \in M \mid h(x) = z\}.$$

The sets A_z, for $z \in N$, form a partition of M. It is easy to verify that

$$\delta(A_w, A_z, h) = \begin{cases} 0 & w \neq z \\ |A_z|(|A_z| - 1) & w = z. \end{cases}$$

This is because any two elements that collide must belong to the same set A_z, and the number of collisions within the elements of A_z is exactly $|A_z|(|A_z| - 1)$. The total number of collisions between all possible pairs of elements is minimized when these sets A_z are all of the same size. We obtain

$$\begin{aligned} \delta(M,M,h) &= \sum_{z \in N} |A_z|(|A_z| - 1) \\ &\geq n\left(\frac{m}{n}\left(\frac{m}{n} - 1\right)\right) = m^2\left(\frac{1}{n} - \frac{1}{m}\right). \end{aligned}$$

This calculation was for any fixed choice of $h \in H$, and so $\delta(M, M, H) = \sum_{h \in H} \delta(M, M, h) \geq |H| m^2(1/n - 1/m)$. By the pigeonhole principle there must exist a pair of elements x, $y \in M$ such that

$$\begin{aligned}
\delta(x, y, H) &\geq \frac{\delta(M, M, H)}{m^2} \\
&= \frac{|H| \delta(M, M, h)}{m^2} \\
&\geq \frac{|H| m^2 \left(1/n - 1/m\right)}{m^2} \\
&= |H| \left(\frac{1}{n} - \frac{1}{m}\right).
\end{aligned}$$

□

8.4.2. Application to Dynamic Dictionaries

Before we provide a construction for a 2-universal hash family, let us see why it gives a good solution to the dynamic dictionary problem. The following lemma will prove useful in the analysis of a dynamic dictionary scheme based on a 2-universal family H.

Lemma 8.13: *For all $x \in M$, $S \subseteq M$, and random $h \in H$,*

$$\mathbf{E}[\delta(x, S, h)] \leq \frac{|S|}{n}.$$

PROOF: The following simple calculation constitutes the proof.

$$\begin{aligned}
\mathbf{E}[\delta(x, S, h)] &= \sum_{h \in H} \frac{\delta(x, S, h)}{|H|} \\
&= \frac{1}{|H|} \sum_{h \in H} \sum_{y \in S} \delta(x, y, h) \\
&= \frac{1}{|H|} \sum_{y \in S} \sum_{h \in H} \delta(x, y, h) \\
&= \frac{1}{|H|} \sum_{y \in S} \delta(x, y, H) \\
&\leq \frac{1}{|H|} \sum_{y \in S} \frac{|H|}{n} \\
&= \frac{|S|}{n}.
\end{aligned}$$

□

Our dynamic dictionary scheme first chooses a hash function $h \in H$ uniformly at random, and then processes the entire sequence of updates and queries using h. Note that the hash function remains fixed during any given invocation of the hash table. An inserted key x is stored at the location $h(x)$, and due to collisions there could be other keys also stored at that location. The keys colliding at a given location are organized into a linked list and a pointer to the head of the list is maintained in that cell. The time to perform an INSERT, DELETE, or FIND operation involving a key x is essentially determined by the time required to perform that operation on the linked list stored at the location $h(x)$, and the latter is at most the length of the list itself. Assuming that the set of keys currently stored in the table is $S \subseteq M$, the length of the linked list is $\delta(x, S, h)$, which has expectation $|S|/n$. Of course, we could use a balanced binary search tree instead of a linked list to reduce the cost of each operation to $O(\log \delta(x, S, h))$, but this does not seem worthwhile given that we expect that the number of collisions at each location will be fairly small.

Consider a request sequence $R = R_1 R_2 \ldots R_r$ of update and search operations starting with an empty hash table. Suppose that this sequence contains s INSERT operations; then, the table will never contain more than s keys. Let $\rho(h, R)$ denote the total cost of processing these requests using the hash function $h \in H$ and the linked list scheme for collision resolution. The following theorem is easy to prove.

Theorem 8.14: *For any sequence R of length r with s INSERTs, and h chosen uniformly at random from a 2-universal family H,*

$$\mathbf{E}[\rho(h, R)] \leq r\left(1 + \frac{s}{n}\right).$$

If we pick the table size n to be larger than the maximum number of elements ever present in the table, we conclude that the expected time per operation is at most 2. By the Markov inequality, the probability that the total cost of the request sequence will exceed $2rt$ is at most $1/t$. We emphasize that this analysis does not assume anything about the request sequence R except a bound on the table occupancy.

8.4.3. Constructing Universal Hash Families

We now turn to the task of devising explicit constructions of 2-universal hash families. Our construction of a 2-universal family is algebraic. Fix m and n, and choose a prime $p \geq m$. We will work over the field $\mathbb{Z}_p = \{0, 1, \ldots, p-1\}$. Let $g : \mathbb{Z}_p \to N$ be the function given by $g(x) = x \bmod n$. For all a, $b \in \mathbb{Z}_p$ define the linear function $f_{a,b} : \mathbb{Z}_p \to \mathbb{Z}_p$ and the hash function $h_{a,b} : \mathbb{Z}_p \to N$ as follows.

$$\begin{aligned} f_{a,b}(x) &= ax + b \bmod p, \\ h_{a,b}(x) &= g\left(f_{a,b}(x)\right). \end{aligned}$$

We define the family of hash functions $H = \{h_{a,b} \mid a, b \in \mathbb{Z}_p \text{ with } a \neq 0\}$ and claim that it is 2-universal. Although H uses $\mathbb{Z}_p$ as its domain, the claim applies to the restriction of H to any subset of $\mathbb{Z}_p$, in particular to the domain M.

Lemma 8.15: *For all $x, y \in \mathbb{Z}_p$ such that $x \neq y$,*

$$\delta(x, y, H) = \delta(\mathbb{Z}_p, \mathbb{Z}_p, g).$$

PROOF: We show that the number of hash functions in H that cause x and y to collide is determined by the size of the residue classes of $\mathbb{Z}_p$ modulo n. Suppose that x and y collide under a specific function $h_{a,b}$. Let $f_{a,b}(x) = r$ and $f_{a,b}(y) = s$, and observe that $r \neq s$ since $a \neq 0$ and $x \neq y$. A collision takes place if and only if $g(r) = g(s)$, or equivalently, $r \equiv s \pmod{n}$. Now, having fixed x and y, for each such choice of $r \neq s$ the values of a and b are uniquely determined as the solution to the following system of linear equations over the field $\mathbb{Z}_p$.

$$\begin{aligned} ax + b &\equiv r \pmod{p} \\ ay + b &\equiv s \pmod{p} \end{aligned}$$

Thus, the number of hash functions $h_{a,b}$ that cause x and y to collide is exactly the number of choices of $r \neq s$ such that $r \equiv s \pmod{n}$. The latter is given by $\delta(\mathbb{Z}_p, \mathbb{Z}_p, g)$. □

Given the similarity of the definition of 2-universality to pairwise independence, it is not surprising that the constructions and their proofs are also very similar (see Section 3.4).

Theorem 8.16: *The family $H = \{h_{a,b} \mid a, b \in \mathbb{Z}_p \text{ with } a \neq 0\}$ is a 2-universal family.*

PROOF: For each $z \in N$, let $A_z = \{x \in \mathbb{Z}_p \mid g(x) = z\}$; it is clear that $|A_z| \leq \lceil p/n \rceil$. In other words, for every $r \in \mathbb{Z}_p$ there are at most $\lceil p/n \rceil$ different choices of $s \in \mathbb{Z}_p$ such that $g(r) = g(s)$. Since there are p different choices of $r \in \mathbb{Z}_p$ to start with,

$$\delta(\mathbb{Z}_p, \mathbb{Z}_p, g) \leq p\left(\left\lceil \frac{p}{n} \right\rceil - 1\right) \leq \frac{p(p-1)}{n}.$$

Lemma 8.15 now implies that for any distinct x and y in $\mathbb{Z}_p$, $\delta(x, y, H) \leq p(p-1)/n$. Since the size of $|H|$ is exactly $p(p-1)$, this gives the desired result that $\delta(x, y, H) \leq |H|/n$. □

A well-known result in number theory called Bertrand's Postulate states that for any number m, there exists a prime between m and $2m$. Thus we can choose $p = O(m)$, and the number of random bits needed to sample a hash function from H is no more than $2 \log p = O(\log m)$. Choosing, storing, and evaluating hash functions from H is remarkably simple and efficient. Pick a and b independently and uniformly at random from $\mathbb{Z}_p$. These are stored using very little memory, and computing $h_{a,b}$ is a trivial task. Contrast this with the use of a totally random function as a hash function.

8.4.4. Strongly Universal Hash Families

The definition of 2-universality merely constrains the probability that two distinct keys get mapped to the same location. This does not fully capture the pairwise independence property (Section 3.4) inherent in the construction of 2-universal hash functions presented in Section 8.4.3. In fact, essentially the same construction gives the stronger guarantee required by the following definition.

► **Definition 8.6:** Let $M = \{0, 1, \ldots, m-1\}$ and $N = \{0, 1, \ldots, n-1\}$, with $m \geq n$. A family H of functions from M into N is said to be *strongly 2-universal* if for all $x_1 \neq x_2 \in M$, any $y_1, y_2 \in N$, and h chosen uniformly at random from H,

$$\mathbf{Pr}[h(x_1) = y_1 \text{ and } h(x_2) = y_2] = \frac{1}{n^2}.$$

Note the similarity to pairwise independence and use this to solve the following exercise.

Exercise 8.12: Assume that $n = m = p$ is a prime number. Show that the hash function family $H = \{h_{a,b} \mid a, b \in \mathbb{Z}_p\}$ is strongly 2-universal.

Most known constructions of 2-universal hash families actually yield a strongly 2-universal hash family. For this reason, the two definitions are generally not distinguished from one another. This definition generalizes to strongly k-universal hash families in the obvious way: for any set S containing k distinct elements from M, and any set T containing k elements from N, the probability that a random hash function $h \in H$ maps the ith element of S to the ith element of T is $1/n^k$. This is closely related to k-wise independent random variables (see Section 3.4).

8.5. Hashing with O(1) Search Time

While the hashing scheme described in Section 8.4 achieves a bounded expected search time for $|S| = O(n)$, it has the disadvantage of requiring unbounded time in the worst case. In this section, we describe a hashing scheme that processes the FIND operation using O(1) time in the *worst case.* The catch is that our solution applies only to the static dictionary problem, i.e., we assume that a set S of size s is fixed in advance and that we only need to support the FIND operation.

Recall that if we do not restrict the table size, there is a trivial solution that takes unit time per query, although it does have the disadvantage of requiring $\Omega(m)$ time for the preprocessing. Our goal is to devise a hashing strategy that uses *linear* space, while guaranteeing bounded search cost and a polynomially bounded preprocessing cost. Therefore, in the ensuing discussion we will focus exclusively on tables of size $n = O(s)$.

8.5.1. Nearly Perfect Hash Families

One way of achieving our goal is to use a hash function h that is perfect for S. Since a hash function cannot be perfect for every possible set S, we will actually need a family of perfect hash functions.

► **Definition 8.7:** A family of hash functions $H = \{h : M \to N\}$ is said to be a *perfect hash family* if for each set $S \subseteq M$ of size $s < n$ there exists a hash function $h \in H$ that is perfect for S.

For notational convenience, we do not explicitly specify the parameters m, n, and s that go into the definition of perfect hash functions and some of the related definitions that follow. The reader should keep in mind that the notion of perfection is defined only with reference to these values.

It is clear that perfect hash families exist: for example, the family of all possible functions from M to T is a perfect hash family. Given a perfect hash family H, we solve the static dictionary by finding $h \in H$ perfect for S, storing each key $x \in S$ at the location $T[h(x)]$, and then responding to a search query for a key q by examining the contents of $T[h(q)]$. The preprocessing cost depends on the cost of identifying a perfect hash function for a specific choice of S, while the search cost depends on the time required to evaluate the hash function. Moreover, since the choice of the hash function will depend on the set S, its description must also be stored in the table. We assume that some auxiliary cells are added to T for just this purpose. Suppose that the size of the perfect hash family H is r. Then, storing the description of a hash function from H will require $\Omega(\log r)$ bits. Since we cannot afford to spend more than $O(1)$ time per search, it is essential that the description of the hash function should fit into $O(1)$ locations in the table T. A cell in the table is only large enough to accommodate a key from M, and so it can be used to encode at most $\log m$ bits of information; therefore, we will only be interested in constructing hash families whose size r is bounded by a polynomial in m. It is also essential that given an encoding of a hash function into $O(\log m)$ bits, we should be able to evaluate this hash function efficiently on arbitrary keys.

Consider the universal hash function family H defined in Section 8.4.3: each hash function $h_{a,b}$ is determined by the elements $a, b \in \mathbb{Z}_p$. Given a choice of p reasonably close to m, the functions $h_{a,b}$ can be stored in $O(1)$ cells in the table; given a and b, the hash function $h_{a,b}$ can be evaluated in $O(1)$ time. The only problem is that the universal hash family is not a perfect hash family. Let us try to determine the conditions under which a perfect hash family can be shown to exist, ignoring for now the issue of efficient storage and evaluation.

Exercise 8.13: Assume for simplicity that $n = s$. Show that for $m = 2^{\Omega(s)}$, there exist perfect hash families of size polynomial in m. (**Hint:** Use the probabilistic method.)

The existence of a perfect hash family is guaranteed only for values of m that are extremely large relative to n. This stems from the requirement that the hash family should have size polynomial in m. The following exercise shows that this restriction is unavoidable and that the bound in the Exercise 8.13 is close to the best possible.

Exercise 8.14: Assuming that $n = s$, show that any perfect hash family must have size $2^{\Omega(s)}$.

Thus, we need to have $m = 2^{\Omega(s)}$, or $s = O(\log m)$, to guarantee even the existence of a perfect hash family of size polynomial in m. Unfortunately, in practice the case $s = O(\log m)$ is not very interesting for typical values of m, e.g., for $m = 2^{32}$.

To circumvent this inherent problem in the use of perfect hash functions, we will employ the strategy of double hashing. The idea is to relax the property of perfection and allow for a few collisions; the keys that are hashed to a particular location of the primary table are handled by using a new hash function to map them into a secondary hash table associated with that location. The set of keys colliding at a specific location of the primary hash table is called a bin. In fact, we can view the application of a hash function $h : M \to N$ to the data set S as a partition of S into n bins (some of which may be empty).

► **Definition 8.8:** Let $S \subset M$ and $h : M \to N$. For each table location $0 \leq i \leq n-1$, we define the bin

$$B_i(h, S) = \{x \in S \mid h(x) = i\}.$$

The size of a bin is denoted by $b_i(h, S) = |B_i(h, S)|$.

A perfect hash function ensures that all bins are of size at most 1. Consider the following generalization of perfect hash functions.

► **Definition 8.9:** A hash function h is *b-perfect* for S if $b_i(h, S) \leq b$, for each i. A family of hash functions $H = \{h : M \to N\}$ is said to be a *b-perfect hash family* if for each $S \subseteq M$ of size s there exists a hash function $h \in H$ that is b-perfect for S.

Exercise 8.15: Show that there exists a b-perfect hash family H such that $b = O(\log n)$ and $|H| \leq m$, for any $m \geq n$. (**Hint:** Use the probabilistic method.)

Using the preceding exercise, we can now outline a scheme for double hashing. At the first level we use a $(\log m)$-perfect hash function h to map S into the primary table T. The description of h can be stored in one auxiliary cell.

Consider the bin B_i consisting of all keys from S mapped into a particular cell $T[i]$. In this cell we store the description of a secondary hash function h_i, which is used to map the elements of the bin B_i into the secondary table T_i associated with that location. Since the size of B_i is bounded by b, we know from the earlier discussion that we can find a hash function h_i that is perfect for B_i provided 2^b is polynomially bounded in m. For $b = \mathrm{O}(\log m)$ this condition holds, and so the double hashing scheme can be implemented with O(1) query time, for any $m \geq n$.

One problem with this approach is that it uses $\Omega(s \log m)$ space, since there must be a secondary table of size $\mathrm{O}(\log m)$ for each of the $n = \mathrm{O}(s)$ locations in the primary table. While the space bound could possibly be reduced using clever memory allocation schemes, a more serious concern is the issue of efficient construction and evaluation of the hash functions being used. Both the primary and secondary hash families are shown to exist via the probabilistic method, and we do not know of any efficient construction. But we can infer the following crucial insight from this scheme: the goal of the primary hash functions should be to create bins small enough that some perfect hash functions can be used as the secondary hash functions. The following exercise describes how we may ensure the existence of suitable secondary hash functions.

Exercise 8.16: Consider a table of size r indexed by $R = \{0, \ldots, r-1\}$, Show that there exists a perfect hash family $H = \{h : M \rightarrow R\}$ with $|H| \leq m$ provided that $r = \Omega(s^2)$, for all $m \geq s$.

We are now ready to describe our final solution. We will use a primary table of size $n = s$, choosing a primary hash function that ensures that the bin sizes are small; the perfect hash functions from Exercise 8.16 are then used to resolve the collisions by using secondary hash tables of size quadratic in the *bin sizes*, thereby guaranteeing perfect hashing at the secondary level. It follows that total space required by the double hashing scheme is

$$s + \mathrm{O}\left(\sum_{i=0}^{s-1} b_i^2\right).$$

This is linear space provided the sum of the squares of the bin sizes is linearly bounded in s. Also, the time required for a search operation is clearly O(1).

8.5.2. Achieving Bounded Query Time

Our goal now is to find primary hash functions which ensure that the sum of the squares of the bin sizes is linear, and perfect hash functions for the secondary tables, which use at most quadratic space. It turns out that the nearly-2-universal hash functions discussed in Problem 8.22 are the appropriate choice for both primary and secondary hashing.

The following notation will be used for these hash functions. For the sake of simplicity, we assume that $p = m + 1$ is a prime number.

► **Definition 8.10:** Consider any $V \subseteq M$ with $|V| = v$, and let $R = \{0, \ldots, r-1\}$ with $r \geq v$. For $1 \leq k \leq p-1$, define the function $h_k : M \rightarrow R$ as follows,

$$h_k(x) = (kx \bmod p) \bmod r.$$

For each $i \in R$, the bins corresponding to the keys colliding at i are denoted as

$$B_i(k, r, V) = \{x \in V \mid h_k(x) = i\}$$

and their sizes are denoted by $b_i(k, r, V) = |B_i(k, r, V)|$.

We include r as a parameter in the bin sizes since we do not assume that r is linearly related to v, unlike in Definition 8.8 where we had $n = \mathrm{O}(s)$. The hash functions h_k have a rather simple description since they are completely determined by the value of k. Since $k \in \{1, \ldots, m\}$, this description can be encoded into a key value in $M = \{0, \ldots, m-1\}$ and stored in a single cell in the table. (The function h_0 is identically 0, and this is why we choose k from the set $\{1, \ldots, m\}$ instead of from M.) The following lemma summarizes the critical property of these hash functions that motivates their use in this application. For $b_i < 2$, we define $\binom{b_i}{2}$ to be 0.

Lemma 8.17: *For all $V \subseteq M$ of size v, and all $r \geq v$,*

$$\sum_{k=1}^{p-1} \sum_{i=0}^{r-1} \binom{b_i(k, r, V)}{2} < \frac{(p-1)v^2}{r} = \frac{mv^2}{r}. \tag{8.2}$$

PROOF: The left-hand side of (8.2) counts the number of tuples $(k, \{x, y\})$ such that h_k causes x and y to collide. Equivalently, it is the number of tuples that satisfy the following two conditions:

1. $x, y \in V$ with $x \neq y$, and
2. $((kx \bmod p) \bmod r) = ((ky \bmod p) \bmod r)$.

Fix any (unordered) pair $\{x, y\} \subseteq V$ with $x \neq y$. The total contribution of this pair to the summation is the number of choices of k satisfying the second condition. In other words, this pair's contribution is the number of choices of k such that

$$k(x - y) \bmod p \in \{\pm r, \pm 2r, \pm 3r, \ldots, \pm \lfloor (p-1)/r \rfloor r\}.$$

Since p is a prime and $\mathbb{Z}_p$ is a field, for any fixed value of $x - y$ there is a unique solution for k satisfying the equation

$$k(x - y) \bmod p = jr$$

for any value of j. This immediately implies that the number of values of k that cause a collision between x and y is at most $2(p-1)/r$.

Finally, noting that the number of choices of the pair $\{x, y\}$ is $\binom{v}{2}$, we obtain

$$\sum_{k=1}^{p-1}\sum_{i=0}^{r-1}\binom{b_i(k,r,V)}{2} \leq \binom{v}{2}\frac{2(p-1)}{r} < \frac{(p-1)v^2}{r}.$$

□

The pigeonhole principle immediately yields the following corollary.

Corollary 8.18: *For all $V \subseteq M$ of size v, and all $r \geq v$, there exists $k \in \{1, \ldots, m\}$ such that*

$$\sum_{i=0}^{r-1}\binom{b_i(k,r,V)}{2} < \frac{v^2}{r}.$$

The primary hash function h_k maps a set $S \subseteq M$ of size s into a hash table T of size $n = s$. The keys in $B_i(k,r,V)$ (the elements of S that are mapped to $T[i]$) are then hashed into a secondary table of size $b_i(k,r,V)^2 = |B_i(k,r,V)|^2$ using the secondary hash function h_{k_i}, which is guaranteed to be perfect. The processing of a search query works in the obvious way. The performance of this scheme is summarized in the following theorem, which guarantees the existence of k, k_1, …, $k_s \in \{1, \ldots, m\}$ with the desired properties.

Theorem 8.19: *For any $S \subseteq M$ with $|S| = s$ and $m \geq s$, there exists a hash table representation of S that uses space* O(s) *and permits the processing of a* FIND *operation in* O(1) *time.*

PROOF: The double hashing scheme is as described above, and all that remains to be shown is that there are choices of the primary hash function h_k and the secondary hash functions h_{k_1}, …, h_{k_s} that ensure the promised performance bounds.

Consider first the primary hash function h_k. The only property desired of this function is that the sum of squares of the colliding sets (the bins) be linear in n to ensure that the space used by the secondary hash tables is O(s). Applying Corollary 8.18 to the case where $V = S$ and $R = T$, implying that $v = r = s$, we obtain that there exists a $k \in \{1, \ldots, m\}$ such that

$$\sum_{i=0}^{s-1}\binom{b_i(k,s,S)}{2} < s$$

or that

$$\sum_{i=0}^{s-1} b_i(k,s,S)[b_i(k,s,S) - 1] < 2s.$$

Since $\cup_{i=0}^{s-1} B_i(k,s,S) = S$ and $\sum_{i=0}^{s-1} b_i(k,s,S) = s$,

$$\sum_{i=0}^{s-1} b_i(k,s,S)^2 < 2s + \sum_{i=0}^{s-1} b_i(k,s,S) = 3s.$$

Consider now the secondary hash function h_{k_i} for the set $S_i = B_i(k, s, S)$ of size s_i. Applying Corollary 8.18 to the case where $V = S_i$ (or $v = s_i$) and using a secondary hash table of size $r = s_i^2$, it follows that there exists a $k_i \in \{1, \ldots, m\}$ such that

$$\sum_{j=0}^{s_i^2-1} \binom{b_j(k_i, s_i^2, S_i)}{2} < 1,$$

where $b_j(k_i, s_i^2, S_i)$ is the number of collisions at the jth location of the secondary hash table for $T[i]$. This can be the case only when each term of the summation is zero, implying that $b_j(k_i, s_i^2, S_i) \leq 1$ for all j. Thus, it follows that there exists a perfect secondary hash function h_{k_i}.

This scheme requires a total of $6s + 1$ cells: $s + 1$ cells for the primary hash table and the description of the primary hash function, $3s$ cells for the secondary hash tables, and $2s$ cells to store the size of the secondary tables and the description of their hash functions. The processing of a query consists of examining 5 cells: the value of k and one cell in the primary hash table, the cells storing the size and hash function for the secondary hash table, as well as the actual location in that table. A bounded number of arithmetic operations suffices for computing the two hash functions. Finally, the entire data structure can be stored in an array of size $6s + 1$, provided $m > 6s + 1$ to ensure that it is possible to encode pointers to secondary tables as keys in the primary table. □

► **Example 8.1:** We illustrate the hashing scheme for the following setting: $m = 30$, $p = 31$, $s = 6$, and $S = \{2, 4, 5, 15, 18, 30\}$. The key for the primary hash function is $k = 2$, and the keys for the various secondary hash functions are shown in Figure 8.6. Notice that the entire data structure is stored in one array of size 25. The pointer entries are merely an index to the location in the array where the appropriate secondary table begins.

Consider the query for the key $q = 30$. We compute the location in the primary hash table as follows: $h_2(30) = (2 \times 30 \bmod 31) \bmod 6 = 5$. Following the pointer at the location $T[5]$, we reach the appropriate secondary table. Noting that $k_5 = 3$ and that the square of the secondary table size is 4, we compute that location in the secondary hash table as follows: $h_3(30) = (3 \times 30 \bmod 31) \bmod 4 = 0$. Examining cell 0 in this table shows that $30 \in S$.

Consider now the query for the key $q = 8$. We compute the location in the primary hash table as follows: $h_2(8) = (2 \times 8 \bmod 31) \bmod 6 = 4$. Following the pointer at the location $T[4]$, we reach the appropriate secondary table. Noting that $k_4 = 1$ and that the square of the secondary table size is 4, we compute that location in the secondary hash table as follows: $h_1(8) = (1 \times 8 \bmod 31) \bmod 4 = 0$. Examining cell 0 in this table shows that $8 \notin S$.

All aspects of this scheme are realistic and efficient, barring one minor detail. The previous theorem guarantees only the existence of good primary and

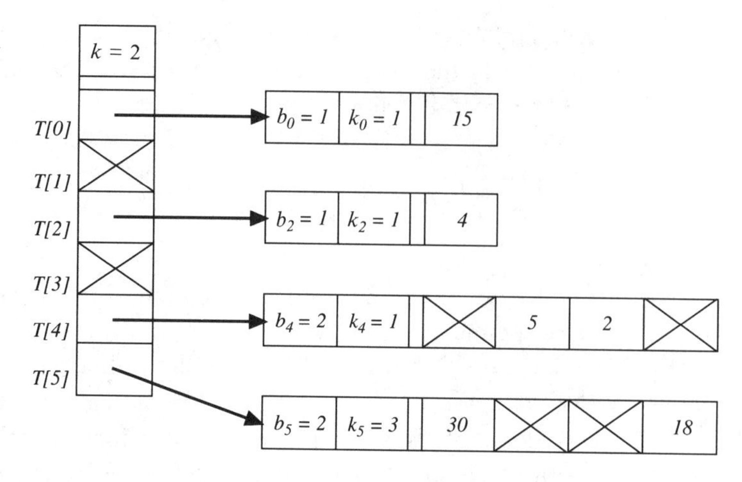

Figure 8.6: An example of double hashing.

secondary hash functions, but gives no clue as to how these may be identified. Of course, since we know the set S a priori, we could exhaustively try all possible keys in $\{1,\ldots,m\}$ as potential choices for k by computing the sizes of the collision bins, and repeating the procedure for the secondary keys. However, for the primary key alone, this will require work at least linear in m. But the value of m could be super-polynomial in s, and having such a large preprocessing cost is impractical. Fortunately, a simple trick using randomization can reduce the total preprocessing cost to a polynomial in s at the expense of increasing the space requirement by a small constant factor. This trick is based on the following modification of Corollary 8.18. The proof is left as Problem 8.25.

Corollary 8.20: *For all $V \subseteq M$ of size v, and all $r \geq v$,*

$$\sum_{i=0}^{r-1} \binom{b_i(k,r,V)}{2} < 2\frac{v^2}{r}$$

for at least one-half of the choices of $k \in \{1,\ldots,m\}$.

A value k satisfying the inequality in the corollary can be found in expected time $O(v)$ by random sampling from $\{1,\ldots,m\}$, since the validity of the inequality for a specific value of k is easily verified in $O(v)$ time by applying h_k to all elements of V and keeping track of the bucket sizes. Problem 8.26 requires you to show that the weaker inequality in this corollary does not affect the validity of Theorem 8.19, except that it increases the space bound by a small constant factor.

Notes

Comprehensive descriptions of balanced search trees may be found in most textbooks on data structures. Self-adjusting binary search trees (or splay trees) are due to Sleator and Tarjan [380]. Tarjan [391] gives an excellent description of splay trees, balanced search trees, and other related data structures. The material on random treaps is drawn from the work of Aragon and Seidel [30], and the games used in the analysis are based on the techniques of Mulmuley [315]. Skip lists are due to Pugh [339].

Knuth's book [260] gives information on early work on hashing, especially under the assumption of a distribution on the input elements. The issue of using hashing to exploit the power of the RAM model, and thereby circumventing the logarithmic lower bound on searching, was first raised by Yao [420]. Perfect hash functions were defined by Sprugnoli [385]. Some efficient constructions of perfect hash families and bounds on were provided by Yao [420], Tarjan and Yao [392], Graham (cited in [420]), and Fredman and Komlós [155]. The paper of Tarjan and Yao also gives a solution to the hashing problems for small key space size, i.e., when the value of m is polynomially bounded in n.

Universal hash functions were defined by Carter and Wegman [88], with the stronger definition given in the paper by Wegman and Carter [414]. Universal hashing has found application in a wide variety of areas; for example, see Nisan [320] for an application to pseudo-random generation and complexity theory. Section 8.5 is based on the work of Fredman, Komlós, and Szemerédi [156]. A version of the hash table for dynamic dictionaries has been provided by Dietzfelbinger, Karlin, Mehlhorn, Meyer auf der Heide, Rohnert, and Tarjan [124]. Their data structure guarantees constant search time, and the update time is bounded by a constant only in the amortized and expected sense. They also prove lower bounds showing that the worst-case amortized time for an update must be at least logarithmic, unless one is willing to increase the search time.

Problems

8.1 Prove Lemma 8.4.

8.2 Prove Lemma 8.5.

8.3 (Due to K. Mulmuley [315].) Consider the following version of the Mulmuley games. The pool consists of the sets $\mathcal{P}$, $\mathcal{B}$, $\mathcal{T}$, and $\mathcal{S}$, where $\mathcal{P}$ is a set of p players, $\mathcal{B}$ a set of b bystanders, $\mathcal{T}$ a set of t triggers, and $\mathcal{S}$ a set of s stoppers. Assume that the players are totally ordered and that all sets are non-empty and pairwise disjoint. The game consists of picking random elements of the pool, without replacement, until the pool is empty. The value of the game, $G_p^{t,s}$, is defined as the expected value of the following quantity: after *all* triggers have been chosen, and before any stopper has been chosen, the number of players who, when chosen, are larger than all previously chosen players. This is the same as Game E except for the requirement that we start counting only after all triggers have been picked.

Determine the expected value of $G_p^{t,s}$.

8.4 Given a set of keys $S = \{k_1, k_2, \ldots, k_n\}$, consider constructing a random treap for S where we do not introduce the dummy leaves needed for the endogenous property. Is every element of S equally likely to be a leaf in this treap? Discuss the implications of your result for the performance of a treap.

8.5 We have shown that for any element in a set S of size n, the expected depth of a random treap for S is $O(\log n)$. Show that the depth is $O(\log n)$ with high probability. Conclude a similar high probability bound on the height of a random treap. (**Hint:** One of way achieving this bound is to derive a Chernoff-type bound on the tail of the distribution of the value of Game A.)

8.6 Let T be a random treap for a set S of size n. Determine the expected size of the sub-tree rooted at an element $x \in S$ whose rank is k.

8.7 (Due to C.R. Aragon and R.G. Seidel [30].) Let T be a random treap for the set S, and let $x, y \in S$ be two elements whose ranks differ by r. Prove that the expected length of the (unique) path from x to y in T is $O(\log r)$.

8.8 While the Mulmuley games are useful for explaining the analysis of random treaps, they are easily dispensed with. To see this, attempt to provide a direct proof of Lemmas 8.6 and 8.7.

8.9 A *finger* search tree is a binary search tree with a special pointer (the finger) associated with it. The finger always points to the last item accessed in the tree. Describe how you would implement the FIND operation starting from the finger, rather than the root. Finger search trees perform especially well on a sequence of FINDs that has some locality of reference. Analyze the performance of a random treap in terms of the ranks of the keys accessed during a sequence of FIND operations. (The result in Problem 8.7 may be useful for this purpose.)

8.10 (Due to C.R. Aragon and R.G. Seidel [30].) Another important property of random treaps is that they adapt well to scenarios where the elements have specific access frequencies. Suppose that each key in S will be accessed a prespecified number of times, but the exact order of the accesses is unknown. Equivalently, consider accesses that involve an element of S chosen at random according to a specific distribution that is not necessarily uniform. In either case, the following notion of a weighted treap provides an optimal solution to the resulting data-structuring problem.

(a) Consider a random treap T for a set S. Associate a positive integer weight f_x with each $x \in S$, and define $F = \sum_{x \in S} f_x$. Define a random *weighted* treap as a treap obtained by choosing priorities for each $x \in S$ as follows: p_x is the maximum of f_x independent samples from a continuous distribution $\mathcal{D}$. Describe how you will maintain a random weighted treap under the full set of operations supported by an unweighted treap.

(b) Prove the following performance bounds for random weighted treaps with an arbitrary choice of the weights f_x.

1. The expected time for a FIND, INSERT, or DELETE operation involving a key x is

$$O\left(1 + \log \frac{F}{\min\{f_x, f_y, f_z\}}\right),$$

where F includes the weight of x, and the keys y and z are the predecessor and successor of x in the set S.

2. The expected number of rotations needed for an INSERT or DELETE operation involving a key x is

$$O\left(1 + \log \frac{f_y + f_x}{f_y} + \log \frac{f_z + f_x}{f_z}\right),$$

where the keys y and z are the predecessor and successor of x in the set S.

3. The expected time to perform a JOIN, PASTE, or SPLIT operation involving sets S_1 and S_2 of total weight F_1 and F_2, respectively, is

$$O\left(1 + \log \frac{F_1}{f_x} + \log \frac{F_2}{f_y}\right),$$

where x is the largest key in S_1 and y is the smallest key in S_2.

8.11 In Problem 8.10, it was assumed that the access frequency or probability is known in advance, and this knowledge was important in the choice of an appropriate distribution for the elements' priorities. Explain how weighted treaps can be made to adapt to the *observed* frequency of access of the elements in the treaps. There is a solution that does not explicitly keep track of the observed frequency and will use no more random bits than in the case where the frequencies are known in advance.

8.12 Let us now analyze the number of random bits needed to implement the operations of a treap. Suppose we pick each priority p_i uniformly at random from the unit interval $[0, 1]$. Then, the binary representation of each p_i can be generated as a (potentially infinite) sequence of bits that are the outcome of unbiased coin flips. The idea is to generate only as many bits in this sequence as is necessary for resolving comparisons between different priorities. Suppose we have only generated some prefixes of the binary representations of the priorities of the elements in the treap T. Now, while inserting an item y, we compare its priority p_y to others' priorities to determine how y should be rotated. While comparing p_y to some p_i, if their current partial binary representation can resolve the comparison, then we are done. Otherwise, they have the same partial binary representation and we keep generating more bits for each till they first differ.

Compute a tight upper bound on the expected number of coin flips or random bits needed for each update operation. (See also Problem 1.5.)

8.13 Compute a tight upper bound on the expected number of coin flips or random bits needed for each update operation for random skip lists.

8.14 In Lemma 8.10 we gave an upper bound on the expected cost of a FIND operation in a random skip list. Determine the expectation of this random variable as precisely as you can. (**Hint:** We suggest the following approach. For each element x_i, determine the probability that it lies on the search path for a particular query y, and sum this over i to get the desired expectation. To determine the probability, find a characterization of the level numbers that will lead to x_i being on the search path.)

8.15 We have shown that the expected cost of a FIND operation in a random skip list is $O(\log n)$. Prove that the cost is bounded by $O(\log n)$ with high probability, using a Chernoff-type bound for the sum of geometrically distributed random variables. Can you prove a similar probability bound for the INSERT and DELETE operations?

8.16 Give a high probability bound on the space requirement of a random skip list for a set S of size n.

8.17 (Due to W. Pugh [339].) In defining a random leveling for a skip list, we sampled the elements from L_i with probability 1/2 to determine the next level L_{i+1}. Consider instead the skip list obtained by performing the sampling with probability p (at each level), where $0 < p < 1$.

(a) Determine the expectation of the number of levels r, and prove a high probability bound on the value of r.

(b) Determine as precisely as you can the expected cost of each operation in this skip list.

(c) Discuss the relation between the choice of the value p and the performance of the skip list in practice.

8.18 Formulate and prove results similar to those in Problems 8.7 and 8.9 for random skip lists.

8.19 Consider the scenario described in Problem 8.10 for random treaps. Adapt the random skip list structure to prove similar results, and compare the bounds obtained in the two cases.

8.20 (Due to M.N. Wegman and J.L. Carter [414]; see also M. Blum and S. Kannan [66].) Consider the problem of deciding whether two integer *multisets* S_1 and S_2 are identical in the sense that each integer occurs the same number of times in both sets. This problem can be solved by sorting the two sets in $O(n \log n)$ time, where n is the cardinality of the multisets. In Problem 7.4, we considered applying the randomized techniques for verifying polynomial identities to the solution of the multiset identity problem. Suggest a randomized algorithm for solving this problem using universal hash functions. Compare your solution with the randomized algorithm suggested in Problem 7.4.

8.21 (Due to J.L. Carter and M.N. Wegman [88].) Suppose that $M = \{0,1\}^m$ and $N = \{0,1\}^n$. Let $\mathcal{M} = \{0,1\}^{(m+1)\times n}$ denote the space of Boolean matrices with $m+1$ rows and n columns. For any $x \in M$, denote by $x^{(1)}$ the $(m+1)$-bit vector obtained by appending a 1 to the end of x. For $A \in \mathcal{M}$, define $h_A(x) = x^{(1)}A \bmod 2$. Show that $H = \{h_A \mid A \in \mathcal{M}\}$ is a 2-universal hash family. Is it also strongly 2-universal? Why did we augment the vector x to $x^{(1)}$? Compare the complexity and the use of randomness in this construction with that of the construction described in Section 8.4.

8.22 (Due to J.L. Carter and M.N. Wegman [88].) In this problem we consider a weakening of the notion of 2-universal families of hash functions. Let $g(x) = x \bmod n$ be as before. For each $a \in \mathbb{Z}_p$, define the function $f_a(x) = ax \bmod p$, and $h_a(x) = g(f_a(x))$, and let $H = \{h_a \mid a \in \mathbb{Z}_p,\ a \neq 0\}$. Show that H is nearly-2-universal in that, for all $x \neq y$,

$$\delta(x, y, H) \leq \frac{2|H|}{n}.$$

Also, show that the bound on the collision probability is close to the best possible for this family of hash functions.

8.23 (Due to M.N. Wegman and J.L. Carter [414].) Define a super-strong universal hash family to be a family of hash functions from M to N that is strongly k-universal for all values of k (simultaneously). Provide a complete characterization of function families that satisfy this definition.

8.24 (Due to N. Nisan [320].) An interesting property of a strongly 2-universal hash function is the following. For any $A \subseteq M$ define $\rho(A) = |A|/|M|$; similarly, for any $B \subseteq N$, define $\rho(B) = |B|/|N|$. For any $\epsilon > 0$, $A \subset M$, and $B \subset N$, a hash function $h : M \to N$ is said to be ϵ-good for A and B if for x chosen uniformly at random from M

$$|\mathbf{Pr}[x \in A \text{ and } h(x) \in B] - \rho(A)\rho(B)| \leq \epsilon.$$

Let h be chosen uniformly at random from a strongly 2-universal hash family H. Show that for any $\epsilon > 0$, $A \subset M$, and $B \subset N$, the probability that h is not ϵ-good for A and B is at most

$$\frac{\rho(A)\rho(B)(1 - \rho(B))}{\epsilon^2 |M|}.$$

8.25 Prove Corollary 8.20.

8.26 (Due to M.L. Fredman, J. Komlós, and E. Szemerédi [156].) Show that the hash table representation analyzed in Theorem 8.19 can be constructed with expected $O(s^2)$ preprocessing time, using $13s + 1$ cells and the same search time.

8.27 (Due to M.L. Fredman, J. Komlós, and E. Szemerédi [156].) Show that the hash table representation described in Theorem 8.19 can be constructed with worst-case $O(s^3 \log s)$ preprocessing time, using $13s + 1$ cells and the same search time.

8.28 (Due to M.L. Fredman, J. Komlós, and E. Szemerédi [156].) Show that the hashing scheme of Section 8.5 can be modified to use space $s + o(s)$ while still requiring only polynomial preprocessing time and constant query time. (**Hint:** Increase the size of the primary hash table and observe that most of the bins will be empty. Find an efficient scheme for packing together the non-empty bins, while creating secondary hash tables only for the bins of size greater than 1.)

CHAPTER 9
Geometric Algorithms and Linear Programming

IN this chapter we consider algorithms that manipulate geometric objects such as points, lines, and planes. In Chapter 1 we encountered one such algorithm: the **RandAuto** algorithm for line segments in the plane. We will use the RAM of Section 1.5.1, with the following additional observations. We will deal with points whose coordinates are real numbers; we assume that we can compare these coordinates and perform arithmetic operations (including the square-root operation) in constant time. Similarly, we can check in constant time whether or not two line segments intersect. Unless otherwise specified, we use the Euclidean metric, by which the distance between points (x_1, y_1) and (x_2, y_2) is $\sqrt{(x_1 - x_2)^2 + (y_1 - y_2)^2}$. Our use of randomness will as usual be "discrete" rather than "continuous": we will use random numbers to select objects at random from a finite population (say the points or lines that constitute an instance of a geometric problem), but not to choose, say, a random point from the interior of a triangle.

9.1. Randomized Incremental Construction

In many computational problems, the use of randomization yields algorithms that are substantially faster than their known deterministic counterparts. In computational geometry, however, randomized algorithms often only match the running times of known deterministic algorithms, but are usually much simpler to understand and implement.

One strikingly simple approach to designing randomized geometric algorithms is that of *randomized incremental construction.* Here the n objects comprising the input to the problem are considered one at a time, in a random order, and the effect of each added object on the solution is computed. For many geometric problems, this paradigm bears a strong resemblance to algorithms favored (and used) by programmers, except that programmers process the objects in the order present in the input rather than in a random order.

Before proceeding to geometric problems, we give a simple non-geometric algorithm that motivates randomized incremental construction. Consider *randomized incremental sorting:* given n numbers to be sorted, we use the following scheme to sort them. After the ith of n steps $(1 \leq i \leq n)$, we will make sure that we have i of the input numbers in a sorted list. Clearly these i sorted numbers will partition the ranks of the remaining $n - i$ (yet unsorted) numbers into $i + 1$ intervals. The $(i + 1)$th step consists of choosing one of the $n - i$ yet unsorted numbers uniformly at random, and inserting it into the sorted list. After n such insertion steps, we are left with a list of all the input numbers, in sorted order.

There are many ways of performing this insertion step, and we will study one that is simple to understand and analyze. Throughout the algorithm, we maintain a pointer for each number yet to be inserted into the sorted list. After the ith step, the pointer for each uninserted number specifies which of the $i + 1$ intervals in the sorted list it would be inserted into, if it were the next to be inserted (assume for the moment that all the numbers in the input are distinct). The pointers are bidirectional, so that given an interval we can determine the numbers whose pointers point to it. What is the work required to maintain these pointers? Suppose we insert a number x whose pointer points to interval I. On inserting x, we have three tasks: (i) find all numbers whose pointers point to I; (ii) update the pointers of all numbers whose pointers point to I; (iii) delete the pointer from x to I. The important task is (ii). The update task consists of changing each of the pointers to point to one of the two new sub-intervals of I created by the insertion of x. Clearly, the work done in this update step is proportional to the number of pointers pointing to I.

Consider the work done in the ith step when the objects in the input are considered in a random order. While we could directly analyze this random variable, we use this occasion to introduce *backwards analysis*, a tool that will often prove useful. In this view of things, we imagine that the algorithm is run backwards starting from the sorted list we have at the end. Thus, in analyzing the ith step, we imagine that we are deleting one of the i numbers in the sorted list and updating the pointers. A moment's thought shows that the work done in updating the pointers in this case is the same as if we had run the algorithm forward as usual. There is a second crucial component to backwards analysis: since the numbers were added in random order in the original algorithm, in the backwards analysis we may assume that each of the i numbers in the sorted list is equally likely to be deleted at this step. What is the expected number of pointers to be updated at this step? Since there are i intervals and $n - i + 1$ pointers remaining after the deletion, the expected number of pointers that were altered at the ith step is $O((n - i)/i)$, which is $O(n/i)$. Now, we use linearity of expectation to sum the work done over all the steps, to obtain a bound of $O(\sum_i n/i) = O(n \log n)$ on the expectation of the total work.

Viewed as yet another variant of quicksort, the above may not be especially interesting. However, it paves the way for our study of randomized incremental algorithms for a number of geometric problems.

9.2. Convex Hulls in the Plane

Given a set S of n points, their *convex hull* is the smallest convex set that contains all of the n points (see Figure 9.1). In the plane, intuitively, if we were to surround the points of S by a large, stretched rubber band, the convex hull is the (convex) polygonal shape that would be enclosed by the band when released. Similarly, for points in three dimensions the analogy would be one of "gift-wrapping" the points in S to form their convex hull. We will be interested in algorithms for computing the convex hull of S given S. We denote by $conv(S)$ the convex hull of S. We begin with the case when the points in S are in the plane.

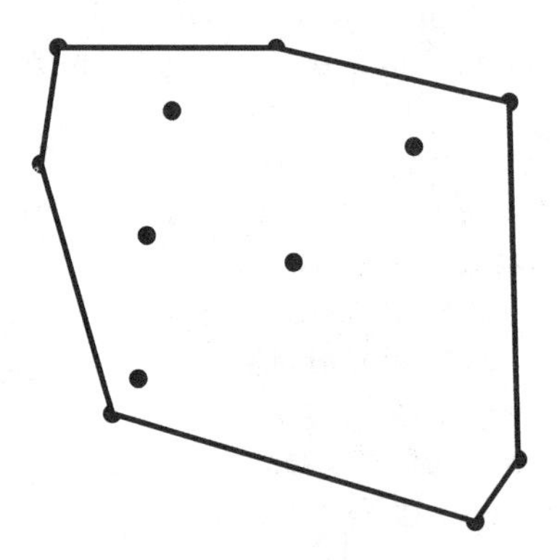

Figure 9.1: The convex hull of 12 points in the plane.

The boundary of $conv(S)$ forms a convex polygon whose vertices are a subset of S; whenever there is no risk of confusion, we will refer to the polygon as $conv(S)$. The problem of computing a convex hull in the plane is then the following: given S, we are to compute the polygon (bounding) $conv(S)$. The output is to be given as a list containing the points of S that appear as vertices of $conv(S)$, in counterclockwise order as they appear on the polygon; the starting point for the list may be arbitrary. For definiteness, we prescribe that the first point in this ordering is the point in S with the smallest x-coordinate. Assume that no three points in S lie on a straight line. This assumption can be dispensed with in an implementation by exercising due care. We now show that the randomized incremental paradigm described above in the context of sorting can be applied to this problem.

Before we describe the algorithm, we note some basic facts about computing convex hulls in the plane.

Exercise 9.1: By making use of the fact that sorting n numbers requires $\Omega(n \log n)$ steps in our model of computation, prove that finding the convex hull of n points requires $\Omega(n \log n)$ steps. Indeed, the lower bound for sorting (and as a consequence of this exercise, finding the convex hull) holds even for randomized algorithms.

Exercise 9.2: Let S be a set of n points in the plane each represented by a pair of coordinates. Given another point $p = (x, y)$, how many steps suffice to determine whether p lies in the convex hull of S?

The algorithm first randomly permutes the points in the input set S; let p_i be the ith point in this random ordering, for $1 \leq i \leq n$. Let S_i denote the set $\{p_1, \ldots, p_i\}$. Next, the algorithm proceeds through n stages. After the ith step, the algorithm will have computed $conv(S_i)$. During the ith step, it adds p_i to $conv(S_{i-1})$, forming $conv(S_i)$ in the process. We now specify the details of this update step.

We maintain at all times a point in the interior of $conv(S)$; in particular, we could utilize the centroid of $conv(S_3)$ (which can be computed in constant time) for this purpose. Call this point p_0. We also maintain after the ith step a (circular) linked list containing the vertices of $conv(S_i)$. In addition, for simplicity of description, we imagine that this linked list also contains the edges joining successive vertices in this list (this can easily be avoided in an implementation, with minor additional work). Let $S \backslash S_i$ denote the set of points yet to be added after the ith step, for $3 \leq i \leq n-1$. For each such point $p \in S \backslash S_i$, we maintain a (bidirectional) pointer from p to the edge of $conv(S_i)$ cut by the ray emanating from p_0, and passing through p. We say that p *cuts* this edge of $conv(S_i)$. Thus, given any edge of $conv(S_i)$, we can enumerate all points p that cut the edge in time linear in the number of such points.

Having specified the data structures, we describe the actions required to update these structures at each step. The point p_i inserted at the ith step is either inside or outside $conv(S_{i-1})$. Using the line segment $\overline{p_i p_0}$ and the associated pointer, we can in constant time detect which of these two cases holds (our assumption that no three points are collinear precludes the possibility that p_i lies on the boundary of $conv(S_{i-1})$). If p_i is inside $conv(S_{i-1})$, we delete the pointer from p_i and proceed to step $i+1$. On the other hand, if p_i is outside $conv(S_{i-1})$, we must update the linked list representing the polygon bounding the hull. The vertices of $conv(S_{i-1})$ are partitioned into three sets by the addition of p_i:

1. Vertices of $conv(S_{i-1})$ that have to be deleted because they are not vertices of $conv(S_i)$.
2. Two vertices of $conv(S_{i-1})$ that become the neighbors of p_i on $conv(S_i)$. Let us denote these vertices v_1 and v_2.
3. Vertices of $conv(S_{i-1})$ that remain in $conv(S_i)$ with their incident edges unchanged.

Clearly the end-points of the edge η intersected by the line-segment $\overline{p_i p_0}$ are of type (1) or (2). By marching away from η (on both sides) along the linked list representing $conv(S_{i-1})$, we can detect the vertices of types (1) and (2). We do so in time linear in the number of such vertices. As we do so, we detect the points in $S \backslash S_i$ that cut the edges being deleted, and update their pointers to either the edge $\overline{p_i v_1}$ or $\overline{p_i v_2}$. This takes constant time (since we have to check only two edges $\overline{p_i v_1}$ and $\overline{p_i v_2}$) for each point of $S \backslash S_i$ whose pointer needs to be updated (see Figure 9.2).

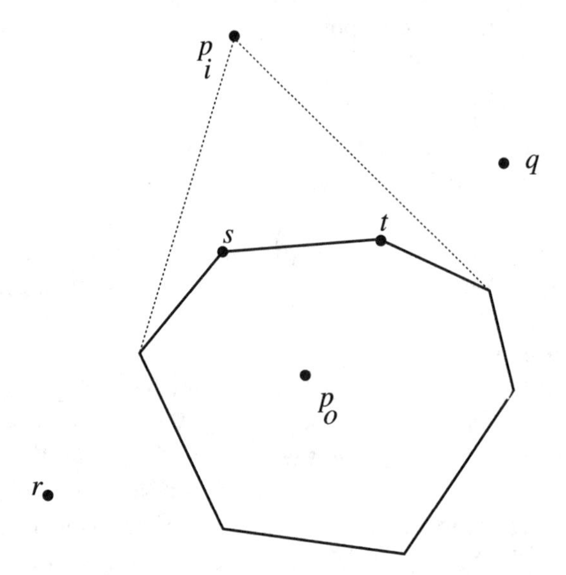

Figure 9.2: The addition of p_i results in the deletion of vertices s and t, and the pointer for q requires updating while that for r does not.

What is the total work done at the ith step? The cost of deleting an edge of $conv(S_{i-1})$ can be charged against the cost of creating it, since an edge can be deleted only once after being created. Since only two edges are created at each step, the total number of these edge creations/deletions (over all steps) is at most $2n$. What about the cost of updating the pointers at the ith step? This is the number of points p in $S \backslash S_i$ such that $\overline{pp_0}$ cuts an edge that is deleted during the step. To bound the expectation of this random variable, we resort to backwards analysis. Imagine running the algorithm backward, and deleting a point of $conv(S_i \backslash S_3)$ to form $conv(S_{i-1})$. Then, the number of pointers updated in the ith step of the original algorithm is the same as the number deleted in the corresponding step of the backward algorithm. We show that the expected number of pointers updated is $O(n/i)$, conditioned on any fixed set of points $S_i \backslash S_3$ from which we delete a random point in the backward step. Since this upper bound holds for any set of i points, the conditioning on a particular set $S_i \backslash S_3$ can be removed.

For a point $p \in S \backslash S_i$, let e_p be the edge of $conv(S_i)$ cut by $\overline{pp_0}$. The probability that p's pointer is updated is precisely the probability that e_p is deleted as a result of the deletion step. Now, e_p is deleted if one of its two end-points is deleted in the backward step. Since the point being deleted from S_i is chosen uniformly from the $i-3$ points in $S_i \backslash S_3$, this probability is $O(1/i)$. The expected number of pointers updated is $O((n-i)/i)$, so that the total work done at this step is $O(n/i)$. A crucial point is that in the deletion step of the backward algorithm, we delete a random point in S_i, not a random vertex of $conv(S_i)$. We now invoke linearity of expectation to bound the expected running time of the algorithm by $O(n \log n)$.

Theorem 9.1: *The expected running time of the above randomized incremental algorithm for computing the convex hull of n points in the plane is* $O(n \log n)$.

We should stress again that the chief advantage of the above algorithm is its extreme simplicity of implementation. An incremental approach such as this is natural to program. The (expected) running time is asymptotically the same as that of many known deterministic convex hull algorithms and matches the lower bound. More importantly, the same simple approach lends itself to computing convex hulls of points in higher dimensions, where deterministic algorithms are rather complicated. Before we proceed to the three-dimensional case, we introduce the notion of geometric duality.

9.3. Duality

The notion of *geometric duality* is fundamental to computational geometry and plays a key role in designing algorithms. The dual of the point $p = (a, b)$ in the plane is the straight line whose equation is $ax + by + 1 = 0$; conversely, the dual of the straight line defined by $ax + by + 1 = 0$ is the point (a, b). Thus duality in the plane maps points to lines, and lines to points. The mapping is involutary: the dual of the dual of a point is the point itself, and a similar statement holds for a line. A simple calculation shows that if a point p is at distance d from the origin, its dual (a line ℓ) is perpendicular to the line joining p to the origin. Further, the distance between the origin and the closest point on ℓ is $1/d$, and ℓ does not pass through the quadrant containing p. Figure 9.3 illustrates this. In this definition, we disallow lines through the origin and points at infinity. We also disallow the point (0,0).

Exercise 9.3: Let p_1 and p_2 be two points, and l_1 and l_2 be their respective dual lines. Show that the line ℓ passing through p_1 and p_2 is the dual of the point of intersection of l_1 and l_2.

We will apply the duality relationship to map the convex hull problem into another geometric problem in the plane. The *half-plane intersection problem* is the following: the input is a set H of half-planes $\{h_1, h_2, \ldots, h_n\}$; we are to determine the intersection of these half-planes. This will be a convex polygon

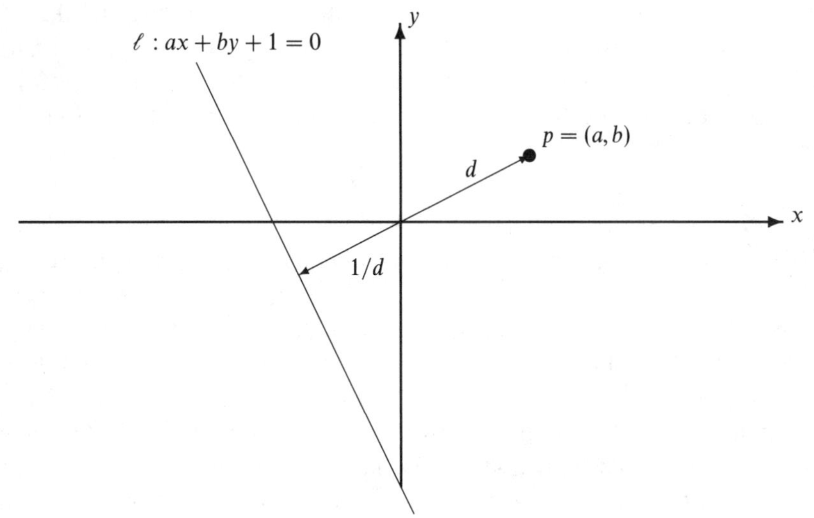

Figure 9.3: Duality between a point and a line.

if the intersection is non-empty, and we ask for the algorithm to output it as a linked list of vertices much as we did in the convex hull problem.

We will show that, in a sense, the half-plane intersection problem is the dual of the convex hull problem. Assume for the moment that the convex hull of the given set S contains the origin of the coordinate system (see Exercise 9.4 below) and that the origin is not one of the input points. Given a line l in the plane that does not pass through the origin, we let l^+ denote the half-plane bounded by l containing the origin. Throughout this chapter, all half-planes/half-spaces will be open half-planes/half-spaces. Let l_i be the dual of $p_i \in S$, and $h_i = l_i^+$. The proof of the following theorem is elementary, and is a consequence of the result in Exercise 9.3.

Theorem 9.2: *Let the convex hull of S contain the origin, and let the origin not be one of the points in S. Let p_{i_1}, p_{i_2}, and p_{i_3}, be three vertices of the convex hull of S, occurring in that order in the output. Then h_{i_1}, h_{i_2}, and h_{i_3} bound the intersection of the half-spaces h_i, appearing on the boundary of the intersection in that order.*

Exercise 9.4: Give a linear-time transformation that shifts the points of S to ensure that the origin lies inside their convex hull. Once we perform this operation, it is easy to satisfy the condition that the origin not be in S: since the origin is inside the convex hull of S, it need no longer be considered for computing the convex hull and can therefore be deleted from S even if it occurs in S.

Each h_i can be determined from p_i in constant time. Given the intersection of the half-spaces, we can identify in linear time the line segments (and hence the lines) that actually appear on the boundary of the intersection. Each line bounding the intersection now corresponds to a point on the convex hull of S, and we can read these off in order in linear time. In other words, an algorithm that computes the intersection of half-planes yields an algorithm that computes the convex hull of points in the plane.

Given an algorithm, data structure, or analysis that works in the "primal" space (in this case, points whose convex hull we wish to compute), there is a corresponding algorithm, data structure, or analysis that works in the dual space (in this case, half-planes whose intersections we wish to compute). Indeed, in Problem 9.2 we derive a randomized incremental algorithm for computing the intersection of n given half-planes.

In the next section we will exploit the notion of duality in higher dimensions. The following exercise will pave the way for computing convex hulls in three dimensions, by reducing the problem to computing half-space intersections in three dimensions.

Exercise 9.5: Extend the notion of duality to three dimensions, working through the statements of Exercises 9.3 and 9.4, and of Theorem 9.2. In fact, the correspondence can be made in $d > 3$ dimensions as well.

9.4. Half-space Intersections

The goal of this section is to develop a randomized incremental algorithm for computing the intersection of n half-spaces in three dimensions. The algorithm will be shown to have an expected running time of $O(n \log n)$; by applying the results of Exercise 9.5, we will then have an algorithm for computing the convex hull of n points in three dimensions with an expected running time of $O(n \log n)$.

Given a set S of n half-spaces in three dimensions, their intersection $inter(S)$ is a (possibly empty) convex polyhedral set in space. Note that the intersection need not be bounded. Every facet of this polyhedron is contained in a plane bounding one of the half-spaces. We assume that each half-space is given to us as a linear inequality whose variables are the coordinates; the corresponding equality gives the equation defining the plane bounding the half-space. Since $inter(S)$ is a polyhedron (when non-empty), we can represent it as a graph each of whose vertices corresponds to a vertex of this polyhedron, with vertices of the graph being adjacent if the corresponding vertices on the polyhedron are joined by a line formed on its surface by the intersection of two half-spaces in S. When $inter(S)$ is unbounded, we assume for convenience that there is a point at "infinity" that is the common end-point of all semi-infinite edges of the polyhedron. Given S, our goal is to compute the graph representing the facets of the polyhedron $inter(S)$; we represent this graph by giving the positions (in space) of all its vertices, together with the adjacencies between vertices.

Since every facet of this polyhedron is contained in a plane bounding one of the half-spaces and no plane contains more than one facet, the number of facets is at most n. Further, the graph representation of *inter*(S) is a planar graph, in which the number of vertices and the number of edges are both $O(n)$. We assume that no four such bounding planes pass through a common point, so that every vertex of the polyhedron/graph (except possibly the "infinity" vertex, when necessary) has degree three. Just as we speak of the edges adjacent to a vertex, we may also speak of the facets of the polyhedron (corresponding to the faces of the graph) adjacent to a vertex; thus there are three facets adjacent to each (finite) vertex of *inter*(S). Likewise, we may speak of the edges bounding a facet, and of the two facets on either side of an edge.

The randomized algorithm for computing *inter*(S) is very similar to the one we have described for computing the convex hull of points in the plane, in Section 9.2. The algorithm first randomly permutes the half-spaces in the input set S; let h_i be the ith half-space in this random ordering, for $1 \leq i \leq n$. Let S_i denote the set $\{h_1, \ldots, h_i\}$. Next, the algorithm proceeds through n stages. After the ith step, the algorithm will have computed *inter*(S_i). During the ith step, it adds h_i to *inter*(S_{i-1}), forming *inter*(S_i) in the process. Geometrically, this can be viewed as cutting away the portion of *inter*(S_{i-1}) not contained in h_i. In the process, some vertices of the polyhedron *inter*(S_{i-1}) are deleted, and some new vertices are added. We describe the details of this addition process now, and then give the analysis. We assume first for simplicity that the intersection of $\{h_1, h_2, h_3, h_4\}$ is bounded; thus *inter*(S_i) will be a bounded polyhedron throughout the execution of the algorithm. This assumption can easily be removed and is the subject of Exercise 9.8.

Let $S \backslash S_i$ denote the set of half-spaces yet to be added after the ith step. In the following description, we concern ourselves only with half-spaces in $S \backslash S_i$ whose bounding plane intersects *inter*(S_{i-1}); it will be clear that the remaining half-spaces are easily dealt with. For any half-space h, let $\overline{h}$ denote the complement of h. For a half-space h, we say that a vertex of *inter*(S_{i-1}) *conflicts* with h if that vertex is in $\overline{h}$.

Assume for the moment that for each half-space $h \in S \backslash S_i$, we have a (bidirectional) pointer to some vertex of *inter*(S_{i-1}) that conflicts with h. (The precise choice of this vertex will become apparent from the discussion following Exercise 9.7.) Under this assumption, the details of the algorithm are fairly straightforward. The process of adding h_i to form *inter*(S_i) begins at the vertex of *inter*(S_{i-1}) that conflicts with h_i. Starting at this vertex, we search the graph representing *inter*(S_{i-1}), ensuring throughout that we do not "enter" *inter*$(S_{i-1}) \cap h_i$. In the course of this search, we determine the vertices and the edges of *inter*(S_{i-1}) that are destroyed by the addition of h_i, and the newly created vertices of *inter*(S_i) (all of which lie on the plane bounding h_i).

Exercise 9.6: Show that the vertices destroyed by the addition of h_i form a connected component of the graph representing *inter*(S_{i-1}).

Clearly, the cost of this search is proportional to the sum of the number of vertices destroyed and the number of vertices created. As in our analysis of the convex hull algorithm in two dimensions, we may ignore the cost of the deletions, since a vertex is deleted at most once and thus it suffices to count vertices when they are created. To analyze the expected number of vertices created by the addition of h_i, we resort to backwards analysis again. Thus, we imagine that we have *inter*(S_i), from which we delete a randomly chosen half-space. Using the fact that the number of vertices and edges in a planar graph with k faces is $O(k)$, the following exercise requires an analysis very similar to that in Section 9.2. The approach once more is to first derive the result conditioned on S_i being a fixed set of half-spaces one of which (chosen at random) is deleted, and then removing the conditioning by noting that the result is independent of the set S_i we start with.

Exercise 9.7: The expected number of vertices created at any step of the randomized incremental half-space intersection algorithm is a constant.

It remains to substantiate the assumption that for each half-space $h \in S \backslash S_i$, we have a (bidirectional) pointer to a vertex of *inter*(S_{i-1}) that conflicts with h. We now describe how this information can be maintained, and then analyze the cost of doing so. In particular, we must specify how the pointers for the half-spaces in $S \backslash S_i$ are updated following the addition of h_i.

When we destroy a vertex v of *inter*(S_{i-1}) during the addition of h_i, we check whether there are any pointers from v to half-spaces in $S \backslash S_i$ (recall that our pointers are bidirectional). For each such pointer (pointing to a half-space $h \in S \backslash S_i$), we must shift it to a new vertex $w \in h \cap$ *inter*(S_i). How do we find such a vertex w? The process is similar to that used in updating *inter*(S_{i-1}) to form *inter*(S_i). Note that the vertex v is in $\overline{h} \cap \overline{h_i}$. We perform a walk on the graph representing *inter*(S_{i-1}) starting at v, taking care never to enter h, until we first arrive at a vertex of *inter*(S_i). On arriving at such a vertex of *inter*(S_i), we have found the new vertex w we seek, since it is in $\overline{h}$ and thus conflicts with h. We move the bidirectional conflict pointer for h to point to w.

It remains to analyze the cost of this search. As in the analysis yielding the statement of Exercise 9.7, we use the fact that every vertex of the graph has degree 3. Therefore, the cost of this search is proportional to the number of vertices in $\overline{h} \cap \overline{h_i} \cap$ *inter*(S_{i-1}). Equivalently, this is the number of destroyed vertices of *inter*(S_{i-1}) in conflict with h, plus the number of newly created vertices of *inter*(S_i) in conflict with h. In considering the asymptotic total cost for maintaining the pointer for h, it suffices to count only the newly created vertices, since any vertex that is destroyed has been counted once when created.

We now wish to bound the expected number of such newly created vertices in conflict with h, summed over all $h \in S \backslash S_i$. This is exactly

$$\sum_v |\{h \in S \backslash S_i : h \text{ conflicts with } v\}|, \tag{9.1}$$

the summation being taken over the set of the vertices of $inter(S_i)$ newly created by the addition of h_i. We bound the expectation of (9.1).

For a set of half-spaces H, let $c(H,h)$ denote the number of vertices of $inter(H)$ in conflict with h. Resorting again to backwards analysis, we consider first a fixed set S_i from which a random half-space is deleted to give $inter(S_{i-1})$. Noting that each vertex of $inter(S_i)$ has degree 3, the expectation of (9.1) is thus bounded by

$$\frac{3}{i}\sum_{h\in S\setminus S_i} c(S_i,h). \tag{9.2}$$

Since h_{i+1} is chosen uniformly at random from $S\setminus S_i$,

$$\mathbf{E}[c(S_i,h_{i+1})] = \frac{1}{n-i}\sum_{h\in S\setminus S_i} c(S_i,h). \tag{9.3}$$

Combining (9.2) and (9.3), the expectation of (9.1) is bounded above by

$$\frac{3(n-i)}{i}\mathbf{E}[c(S_i,h_{i+1})].$$

The random variable $c(S_i,h_{i+1})$ also counts the expected number of vertices destroyed by the addition of h_{i+1}, the half-space added at step $i+1$. Thus, the expectation of the sum over all i of (9.1) (which measures the total work in updating pointers over the course of the entire algorithm) is bounded above by

$$\sum_{i=1}^{n}\frac{3(n-i)}{i}\mathbf{E}[\text{ Number of vertices destroyed at time } i+1]. \tag{9.4}$$

For a vertex v created in the course of the algorithm, let $t_c(v)$ denote the time (step number) at which it is created, and $t_d(v)$ the time at which it is destroyed. Then, (9.4) can be rewritten as

$$\sum_{v}\frac{3(n-t_d(v)-1)}{t_d(v)-1}, \tag{9.5}$$

where v ranges over all vertices ever created during and execution of the algorithm. Since $t_c(v)\le t_d(v)-1$, we can bound (9.5) from above by

$$\sum_{v}\frac{3(n-t_c(v))}{t_c(v)} \le \sum_{i=1}^{n}\frac{3(n-1)}{i}\mathbf{E}[|\{v \mid t_c(v)=i\}|].$$

But we have already seen in Exercise 9.7 that $\mathbf{E}[|\{v \mid t_c(v)=i\}|]$ is a constant. We thus have:

Theorem 9.3: *The expected running time of the randomized incremental algorithm for computing the intersection of n half-spaces in three dimensions is* $O(n\log n)$.

Exercise 9.8: In the above description, we assumed that the intersection $inter(S_i)$ was bounded for all $i \geq 4$. How can this assumption be removed?

9.5. Delaunay Triangulations

Let $P = \{p_1, \ldots, p_n\}$ be a set of n points in the plane. For a point $p_i \in P$, let $cell(p_i)$ denote the set of points in the plane that are closer to p_i than to any $p_j \in P$, for $j \neq i$.

Exercise 9.9: Show that $cell(p_i)$ is a (possibly unbounded) convex polygonal region for each i, and that the regions $cell(p_i)$ form a decomposition of the plane into n open convex polygonal regions.

The partition of the plane described in Exercise 9.9 is known as the *Voronoi diagram* of P, and we will denote it by $vor(P)$. The convex polygonal region $cell(p_i)$ corresponding to p_i is known as the *Voronoi cell* of p_i. The notion of Voronoi cells and diagrams can in fact be readily formulated for points in higher dimensional space, but we will focus on points in the plane here.

The Voronoi diagram of a set of points is a fundamental structure in computational geometry, and has many applications. We will be interested in algorithms for constructing $vor(P)$ and related structures, given P. We assume henceforth that no four points of P lie on any circle, and that no three lie on any straight line. These assumptions greatly simplify the descriptions of the algorithms discussed below and may be removed with some care. The Voronoi diagram of a set of points in the plane has a number of properties that are easy to verify:

Exercise 9.10:

1. Show that the boundary between any two cells (known as a Voronoi edge) is the locus of points equidistant from two points of P.
2. Viewing $vor(P)$ as a planar graph, show that every vertex of the graph has degree 3.
3. Show that if $cell(p_i)$, $cell(p_j)$, and $cell(p_k)$ share a vertex in the Voronoi diagram, then the circle passing through p_i, p_j, and p_k contains no other points of P.
4. Show that if p_i is a point of P on the convex hull of P, then $cell(p_i)$ is unbounded. Is the converse also true?

Let us view $vor(P)$ as a planar graph, each of whose faces corresponds to a point $p_i \in P$. Consider the planar dual of this graph, with a vertex at each point $p_i \in P$ (representing the face $cell(p_i)$), and an edge between two vertices if the corresponding cells share an edge in $vor(P)$. This dual graph is known as the *Delaunay triangulation* of P, which we denote by $del(P)$ (see Figure 9.4). From property 2 of Exercise 9.10, it follows that $del(P)$ is indeed a triangulation (i.e., each of its facets except for the outermost one is a triangle). Clearly, given P and $vor(P)$, we can construct $del(P)$ in time $O(n)$.

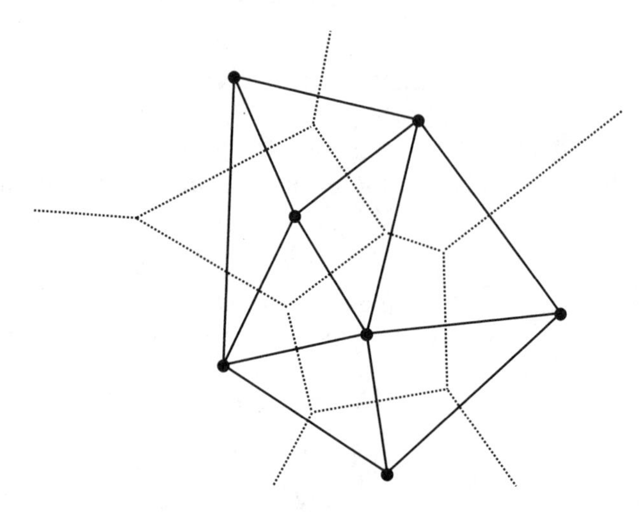

Figure 9.4: A Voronoi diagram (dashed lines) and the corresponding Delaunay triangulation (solid lines), for a set of seven points in the plane.

Exercise 9.11: Show that $vor(P)$ can be constructed from $del(P)$ in time $O(n)$.

In the remainder of this section, we concentrate on algorithms for constructing $del(P)$; by Exercise 9.11 above, this will readily imply an algorithm for computing $vor(P)$. We first describe a parabolic transformation that reduces the problem of computing $del(P)$ to one of computing the intersection of n half-spaces in three dimensions. Given the points of P in the xy-plane, consider the paraboloid $z = x^2 + y^2$. Denote by q_i the point $(x_i, y_i, x_i^2 + y_i^2)$ on the surface of the paraboloid that is directly "above" $p_i = (x_i, y_i, 0)$. Let h_i denote the half-space that is above the plane tangent to the paraboloid at q_i (see Figure 9.5). Consider the polyhedron formed by the intersection of the h_i.

Exercise 9.12: Let p be a point in the xy-plane at distance d_i from p_i, and let q be the point on the paraboloid directly above p. Show that vertical distance between q and the tangent plane bounding h_i is d_i^2.

Exercise 9.12 has the following consequence, which is easy to prove; a detailed proof may be found in any of the texts on computational geometry listed in the Notes section.

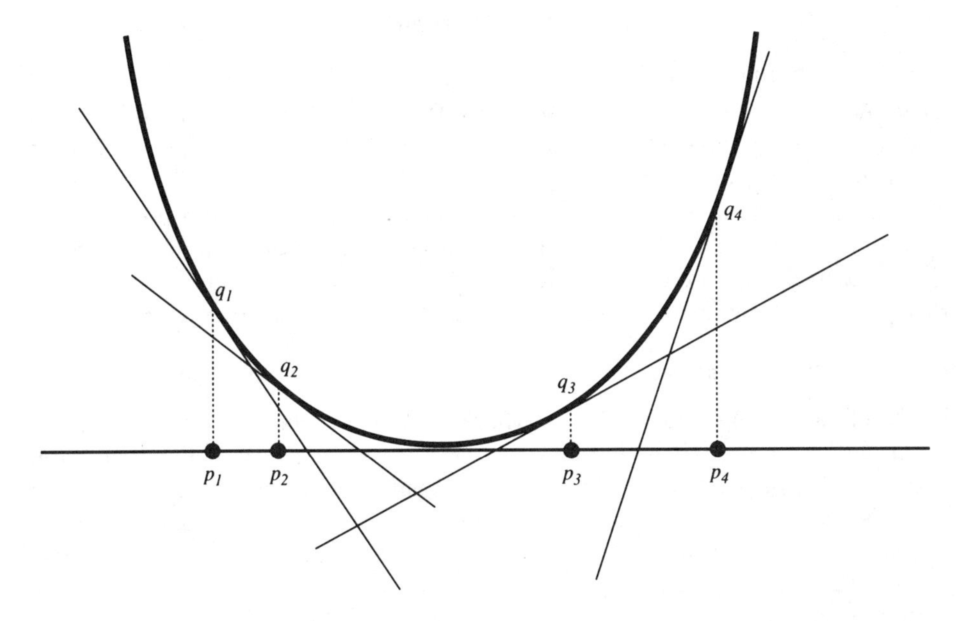

Figure 9.5: The parabolic transformation.

Theorem 9.4: *Given $P = \{p_1, \ldots, p_n\}$, let $H = \{h_1, \ldots, h_n\}$ as described above. Let inter(H) denote the intersection of the half-spaces above the tangents at the points in H. The Voronoi diagram of P results from projecting the edges of inter(H) vertically down to the xy-plane.*

Corollary 9.5: *Given inter(H), we can compute del(P) in time* $O(n)$.

By Corollary 9.5, we thus have a randomized incremental algorithm for computing *del*(P) that runs in expected time $O(n \log n)$: we transform P to H using the parabolic transformation and invoke the algorithm of Section 9.4 to compute *inter*(H).

We now focus on a special case of the problem of computing *del*(P), in which the points of P are the vertices of a convex polygon. We will show in Section 9.5.1 below that a simple randomized algorithm runs in expected time $O(n)$ for this case. Before we do so, we will require the following easy consequence of Exercise 9.6.

Exercise 9.13: Let *del*(P) denote the Delaunay triangulation of a set P of points in the plane. Consider the addition of a new point q; the Delaunay triangulation of $P \cup \{q\}$ can be formed by deleting some triangles of *del*(P), and retriangulating the affected region. Show that the set of triangles destroyed forms a connected component of the graph *del*(P).

9.5.1. Chew's Algorithm

We now show that a simple randomized algorithm computes *del*(P) in expected time $O(n)$ when the points of P are the vertices of a convex polygon. We will require the points of P to be given to us in the order in which they appear on this convex polygon.

The algorithm is recursive, and is as follows. We pick a random point $p \in P$; let q and r denote its neighbors on the boundary of the given convex polygon. We recursively compute *del*$(P \backslash \{p\})$, while $|P \backslash \{p\}| > 3$. Having computed *del*$(P \backslash \{p\})$, we augment it to form *del*(P) by the following three steps:

1. Add the triangle pqr to *del*$(P \backslash \{p\})$. Let D denote the resulting graph.
2. Identify all triangles of *del*$(P \backslash \{p\})$ whose circumcircle contains p (such triangles can no longer be Delaunay triangles), by a depth-first search of the dual graph of *del*$(P \backslash \{p\})$, much as in the search for conflicting vertices in the half-space intersection algorithm of Section 9.4. By Exercise 9.13, these triangles form a connected component of *del*$(P \backslash \{p\})$. Let S denote the set consisting of these "bad" triangles together with the triangle pqr.
3. Remove from D all edges that have triangles of S on both sides and retriangulate the resulting face by introducing all diagonals that have p as an end-point.

The second step above can be performed in time linear in the number of triangles in S. This number, in turn, is one more than the number of edges introduced in the third (retriangulation) step above. Thus the expected cost of the update is proportional to the expected degree of the vertex p in *del*(P). Since *del*(P) is a planar graph, this expected degree is a constant (since p was chosen uniformly at random from the n points in P, and *del*(P) has $O(n)$ edges). Summing this expected cost over the $n-3$ recursive steps, we have:

Theorem 9.6: *The above algorithm computes del*(P) *in expected time* $O(n)$, *provided the points of* P *are vertices of a convex polygon given in the order in which they appear on the boundary of the polygon.*

Exercise 9.14: Why does the above running time guarantee fail if the vertices of P are not vertices of a convex polygon?

9.6. Trapezoidal Decompositions

Our next example of a randomized incremental algorithm (sometimes also known as the *vertical decomposition*) comes from the construction of a *trapezoidal decomposition* for a set of line segments in the plane. The trapezoidal decomposition is a basic structure for representing and manipulating an arrangement of line segments. Let S denote a set of n (possibly intersecting) line segments in the plane; we assume that the x-coordinates of the segments are all distinct. Let k denote the number of points at which two or more segments intersect. Imagine passing a vertical line through each end-point of each segment of S, as well as through each of the k intersection points. These vertical lines continue until they hit one of the other segments, where they stop. Some of these lines will continue to infinity, because they do not hit any other line segments.

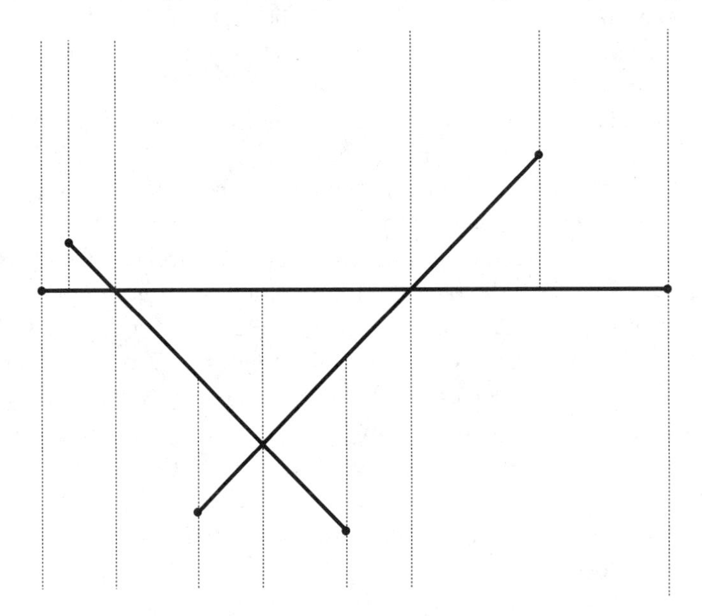

Figure 9.6: A trapezoidal decomposition of three segments.

The resulting decomposition of the plane is known as a *trapezoidal decomposition* (see Figure 9.6); each of the regions into which the plane is partitioned is in general a trapezium with two parallel vertical sides. Some regions are infinite, of course. By imagining that the region containing the segments in S is enclosed in a large rectangular "bounding box," we can view the trapezoidal decomposition of S as a planar graph each of whose vertices is either (i) an end-point of a segment in S, or (ii) a point at which two or more segments of S intersect, or (iii) a point at which the vertical line through a vertex of type (i) or (ii) hits a line segment or the bounding box. It is important to note that a face of this planar graph may have an arbitrary number of vertices, even though it is geometrically a trapezium.

Exercise 9.15: Consider computing the trapezoidal decomposition of S, and representing the output as a planar graph. The size of this graph is $\Omega(n+k)$, which is clearly a lower bound on the number of steps in the computation. Show that the computation also requires $\Omega(n \log n)$ comparisons.

Let *trap*(S) denote the trapezoidal decomposition of S, represented as a planar graph. We now give a simple randomized incremental algorithm for computing *trap*(S), with expected running time $O(n \log n + k)$. By the result of Exercise 9.15, this is the best possible.

Assume without loss of generality that no line in the input is vertical. The algorithm first randomly permutes the line segments in S; let s_i be the ith segment in this random ordering. Let S_i denote $\{s_1, \ldots, s_i\}$. The algorithm proceeds through n stages, after the ith of which it will have computed *trap*(S_i). During the ith step, $i > 1$, it adds s_i to *trap*(S_{i-1}), forming *trap*(S_i) in the process. We first specify the details of this update step, and then proceed to analyze the running time.

For $i > 1$, let $S \backslash S_i$ denote the set of points of S to be added in the incremental construction after the ith step (i.e., the set $\{s_{i+1}, s_{i+2}, \ldots s_n\}$). For each segment in $S \backslash S_i$, we maintain a bidirectional pointer to the face of *trap*(S_i) containing its left end-point. Thus, given a face of *trap*(S_i), we can read off the segments in $S \backslash S_i$ contained in that face in time linear in the number of such points.

Next, we describe how *trap*(S_{i-1}) is updated to *trap*(S_i) by the addition of s_i. We begin by identifying the face of *trap*(S_{i-1}) containing the left end-point of s_i. We then march along s_i to its other end-point, updating the data structures as we go along. Let us consider the different update actions that may be necessary. We first pass a vertical line through the left end-point of s_i, determining the upper and lower end-points of this vertical line (the points above and below where it first hits a segment in S_{i-1}, or a horizontal edge of the bounding box); let us refer to the resulting vertical line segment(s) as the vertical attachment(s) for the left end-point of s_i.

As we proceed along s_i, we have to split each face of S_{i-1} that it cuts into two faces. In particular, whenever a segment in S_{i-1} is cut by s_i, vertical attachments are computed for the point of intersection (Figure 9.7). On arriving at the right end-point of s_i, vertical attachment(s) are again computed for this point.

Having computed the new vertical attachments resulting from the addition of s_i, we make a second pass through the resulting planar graph (call it G_i). Whenever s_i cuts a vertical edge of *trap*(S_{i-1}), one portion of that vertical edge is deleted, and consequently two faces of G_i are merged (Figure 9.8).

The final update step involves updating the bidirectional pointers of the segments in $S \backslash S_i$. We need only update the pointers of segments whose left end-points were contained in faces of *trap*(S_{i-1}) intersected by s_i.

For a face f of *trap*(S_{i-1}), let $n(f)$ denote the number of vertices of *trap*(S_{i-1}) bounding f, and let $\ell(f)$ denote the number of segments of $S \backslash S_i$ whose left end-points lie in f. We use backwards analysis to analyze the expected cost of updating *trap*(S_{i-1}) to obtain *trap*(S_i). Imagine that at step i line segment in *trap*(S_i) chosen uniformly at random is deleted. As before, this is valid since any of the i segments in S_i is equally likely to have been labeled s_i in the initial random permutation. The following is an easy consequence of the preceding discussion, and we invite the reader to verify it:

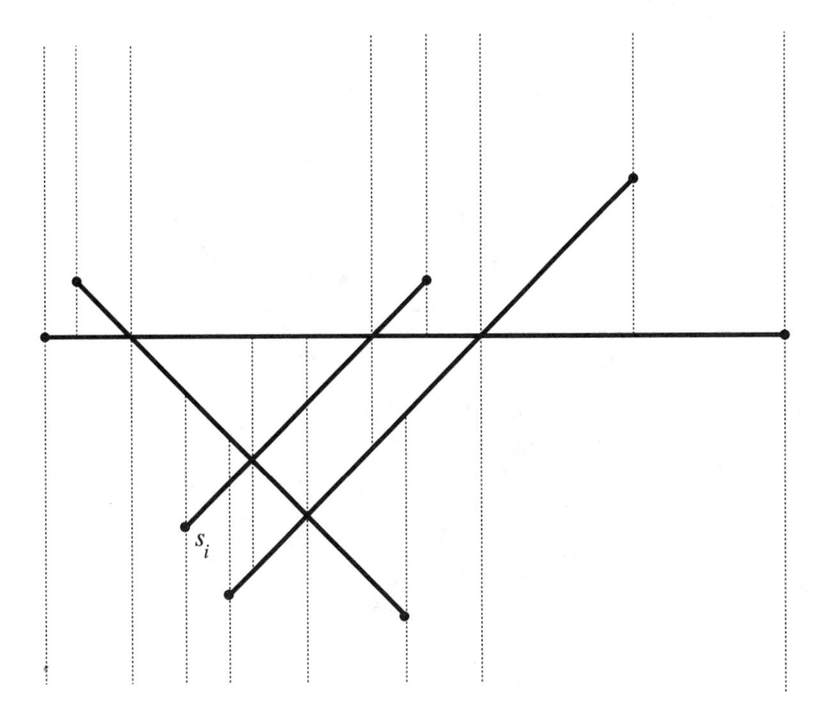

Figure 9.7: The planar graph G_i resulting from the addition of s_i.

Exercise 9.16: The expected update cost on adding s_i is proportional to

$$\frac{1}{i}\sum_{s\in S_i}\sum_{f\in F(s)}[n(f)+\ell(f)], \tag{9.6}$$

where $F(s)$ is the set of those faces of $trap(S_i)$ whose boundary contains at least one point of the segment s.

It remains to bound the expression in (9.6) in terms of n and k. Clearly, the term $\sum_{s\in S_i}\sum_{f\in F(s)}\ell(f)$ is proportional to the total number of pointers for segments in $S\backslash S_i$, which is $n-i$ (no two segments have end-points with the same x-coordinate, so that a face f occurs in $F(s)$ for at most four segments s). We next observe that $\sum_{s\in S_i}\sum_{f\in F(s)} n(f)$ is proportional to $i+k_i$, where k_i is the number of points at which two or more segments of S_i intersect. Thus the expected update cost when adding s_i is proportional to $(n+\mathbf{E}[k_i])/i$. It remains to compute the expectation of k_i, given that S_i is a random subset of i segments from S. Let x be one of the k points at which two segments (say r and s) of S intersect. Now, x occurs in $trap(S_i)$ if and only if both the segments r and s are in S_i. The probability of this is proportional to i^2/n^2. By linearity of expectation over the k possible choices of x, it follows that $\mathbf{E}[k_i]$ is $O(ki^2/n^2)$. Here we

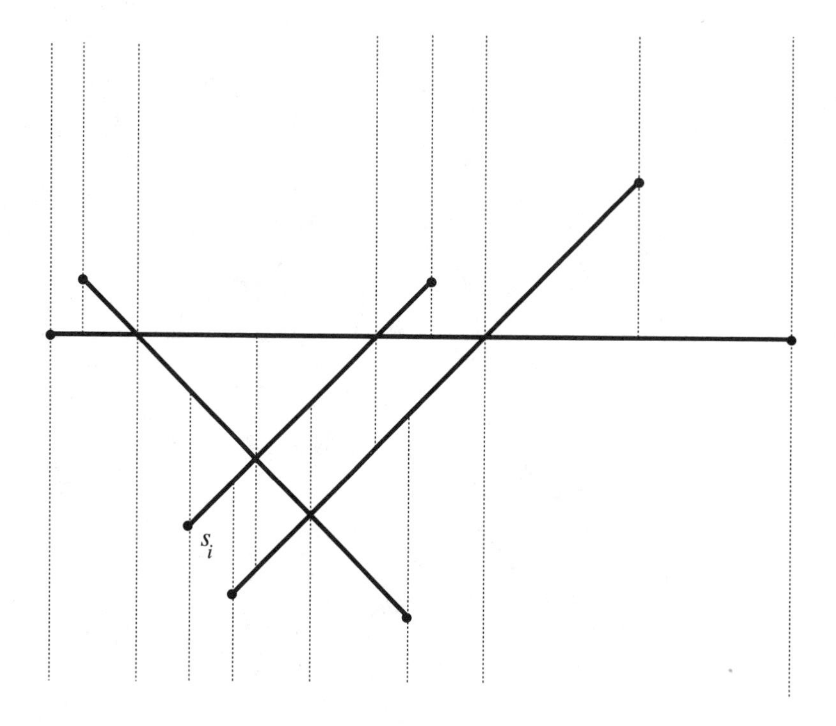

Figure 9.8: Truncating vertical attachments to convert G_i to $trap(S_i)$.

explicitly make use of the fact that S_i is a random subset of S, a fact we did not explicitly use in our previous backwards analyses (where we used only the fact that given a set S_i, a random element is deleted from it for the backwards analysis). Summing the update costs over all the steps, we then have:

Theorem 9.7: *The expected cost of building the trapezoidal decomposition of n line segments in the plane is $O(n \log n + k)$, where k is the number of points at which two or more of the segments intersect.*

9.7. Binary Space Partitions

In this section we study the binary space partition problem (Section 1.3) in three dimensions. We begin with a different analysis of the **RandAuto** algorithm of Section 1.3, making use of a notion known as *free cuts*. Although this will afford no asymptotic performance improvement in the planar case, it will be of crucial importance in the three-dimensional case that we will consider next.

Recall that in the binary planar partition problem, we are given a set $S = \{s_1, s_2, \ldots, s_n\}$ of non-intersecting line segments in the plane. We wish to find a binary planar partition such that every region in the partition contains one line segment, or a portion of one line segment. The **RandAuto** algorithm considers the lines one at a time, in a random order. When a line segment is chosen, it is extended until it partitions the region containing it into two regions.

Suppose that at some stage of the **RandAuto** algorithm we have a region R and a segment s that passes right through R. Clearly it is advantageous to partition R along s immediately (Figure 9.9), since this prevents $s \cap R$ from ever being cut at some later stage. Further, we can make this cut at no additional increase in the number of segments that are cut, since s partitions R. Such a cut is called a *free cut*.

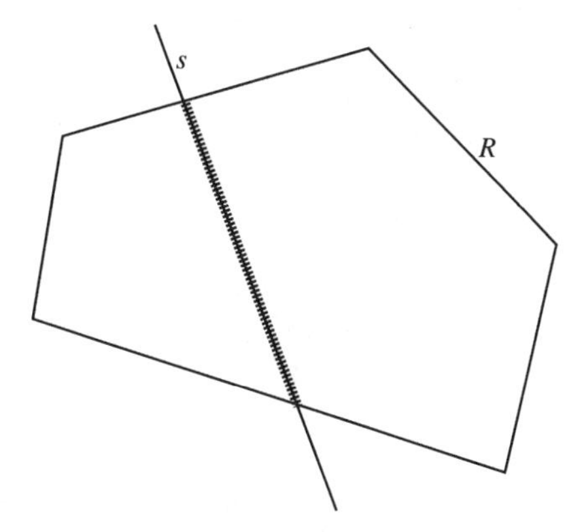

Figure 9.9: An example of a free cut.

The observation that we make no extra cuts (and thus do not increase the size of our binary autopartition tree) by making use of free cuts implies (by Theorem 1.2) that **RandAuto** augmented by the use of free cuts produces an autopartition whose expected size is $O(n \log n)$. However, it is instructive to prove this directly in preparation for the three-dimensional case.

Theorem 9.8: *The expected size of the autopartition produced by* **RandAuto** *with free cuts is* $O(n \log n)$.

PROOF: As in Section 1.3, we denote by P_π the autopartition induced by the permutation π. For an input segment s, consider those segments u such that $l(u)$ intersects s, and label them $u_1, u_2, \ldots, u_k$ based on the left-to-right order of the intersections of the lines $l(u_i)$ with s. We study how many of these are likely to cut s in P_π.

Consider Figure 9.10. Suppose that the ordering induced by the randomly chosen permutation π is u_1, u_3, u_4, u_2, v. Then v is cut by u_1, u_3, and u_4 but not by u_2. When v has been cut by u_1 and u_3, the part of v between these cuts partitions a region and therefore makes a free cut of that region. It is helpful to think of an input segment in the problem (such as v) as being rigidly moored at its end-points – when two cuts are made on v, the portion in between the cuts "falls off" and drops out for the remainder of the problem; it will never be cut again. Two pieces of v remain, each moored at one end-point; in the course

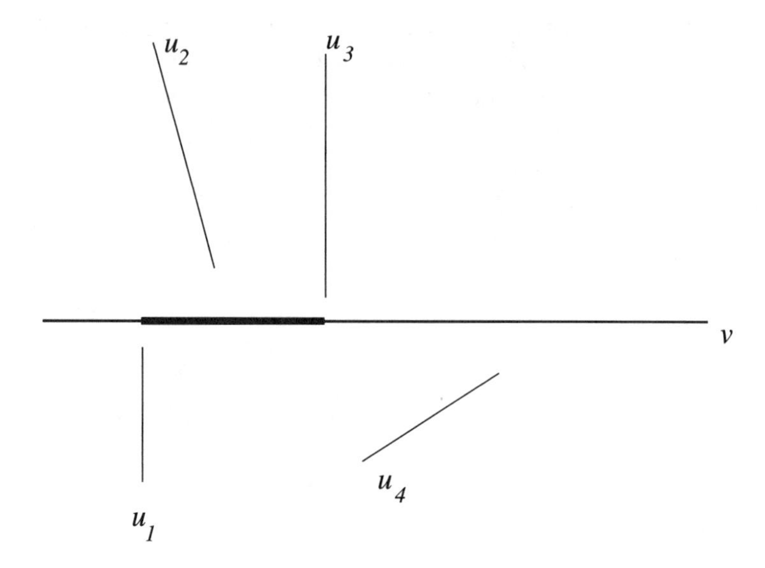

Figure 9.10: The effect of free cuts.

of further processing, each piece may lose more pieces because of cuts causing the unmoored end to "fall off." This continues until v occurs in π, at which point $l(v)$ becomes a partitioning line for the region(s) containing the surviving piece(s) of v, and v is not cut again.

Thus, $l(u_i)$ cuts v only if u_i precedes all of $v, u_1, u_2, \ldots, u_{i-1}$ or if u_i precedes all of $v, u_{i+1}, \ldots, u_k$ in π. The probability of the former event is $1/(i+1)$, and that of the latter is $1/(k-i+2)$. Both events include the event that u_i is the first of $v, u_1, u_2, \ldots, u_k$ in the order induced by π, which has probability $1/(k+1)$. As in Section 1.3, we use the notation $u \dashv v$ to mean that during the execution, an extension of segment u cuts the segment v. Therefore,

$$\mathbf{Pr}[u_i \dashv v] \leq \frac{1}{i+1} + \frac{1}{k-i+2} - \frac{1}{k+1}.$$

Summing this over all v and all u_i yields $O(n \log n)$ for the expected number of cuts, as in Theorem 1.2. □

We now consider the three-dimensional version of the binary partition problem. The input is a set S of n non-intersecting triangles $\{f_1, f_2, \ldots, f_n\}$. We assume that more complex polyhedral scenes are first decomposed into such triangles, just as we assumed that a planar scene had been broken up into line segments. For a triangle f, we define $h(f)$ to be the (infinite) two-dimensional plane containing f.

One interesting aspect of the three-dimensional problem is the following. In three dimensions, unlike the two-dimensional case, a total ordering of the triangles may not exist with respect to the occlusion relation; cyclic dependencies may exist. We will nevertheless be able to build a binary partition in three-dimensional space by methods very similar to those used in the two-dimensional case.

In an analogous way to a binary partition of the plane, we can speak of a binary partition in three-dimensional space. The partition consists of a binary tree together with the following additional information. Associated with a node v of the tree is a convex polyhedral region $r(v)$. Associated with each internal node v of the tree is a plane $h(v)$ that intersects $r(v)$. The region corresponding to the root is all of three-dimensional space. The region $r(v)$ is partitioned by $h(v)$ into two regions $r_1(v)$ and $r_2(v)$, which are associated with the two children of v. We use a random permutation π of $\{1,2,\ldots,n\}$ and free cuts to obtain a partition of expected size $O(n^2)$ of the planes $\{h(f_1),h(f_2),\ldots,h(f_n)\}$. Thus the algorithm for three dimensions is the obvious extension of the **RandAuto** algorithm with free cuts.

Theorem 9.9: *The expected size of the autopartition generated by a random permutation π with free cuts is $O(n^2)$.*

PROOF: In three dimensions, when a plane $h(u)$ intersects a triangle v, it can cut a number of sub-facets of v that lie in different regions of the partition created so far. Let Y_k be the total number of additional cuts created by $u_{\pi(k)}$, and let Y_{ku} be the number of these on input triangle $u \in \{u_1,u_2,\ldots,u_n\}\backslash\{u_{\pi(k)}\}$. Thus the total "fragmentation" – the number of cuts – is $\sum_k Y_k = \sum_k \sum_u Y_{ku}$. The goal is to show that $\mathbf{E}[Y_{ku}]$ is $O(1)$, and the result then follows from linearity of expectation.

To calculate Y_{ku}, we consider the sub-facets of u that are cut by $h(u_{\pi(k)})$. Consider the arrangement $L_{\pi,k}$ of line segments $\{l_{\pi(1)},l_{\pi(2)},\ldots,l_{\pi(k)}\}$ on the triangle u, where the line segment $l_{\pi(i)}$ is the intersection of $h(u_{\pi(i)})$ with triangle u, for $1 \le i \le k$ (see Figure 9.11). Without free cuts, the sub-facets would be exactly those regions of $L_{\pi,k-1}$ intersected by $l_{\pi(k)}$. However, because of free cuts by u, any of the *internal* sub-facets of $L_{\pi,k-1}$ would have already "dropped out." Thus Y_{ku} is the number of *external* regions intersected by $l_{\pi(k)}$.

For an arrangement L of k lines $l_1,l_2,\ldots,l_k$ on triangle u and for $1 \le i \le k$, let $x(L,i)$ denote the number of external regions in the arrangement $L-\{l_i\}$ that are cut by l_i. Observe that $\sum_{i=1}^k x(L,i)$ equals the total number of edges bounding the external sub-facets of L. In Figure 9.11, for instance, $\sum_{i=1}^4 x(L,i) = 12$. We now invoke a standard result in combinatorial geometry (see the Notes section for a reference): $\sum_{i=1}^k x(L,i) = O(k)$ for any arrangement L on a triangle u.

Since π is a random permutation, $l_{\pi(k)}$ is equiprobably any of the lines in the arrangement L. Thus

$$\mathbf{E}[Y_{ku}] = \frac{1}{k}\sum_{i=1}^{k} x(L,i) = O(1). \tag{9.7}$$

□

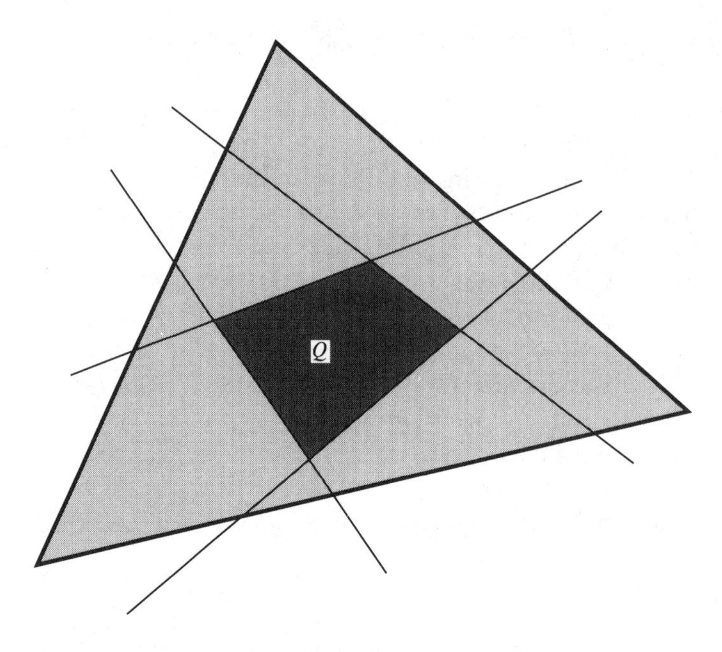

Figure 9.11: An arrangement of four lines on a triangle; only region Q is internal.

Exercise 9.17: Consider using a random permutation π to obtain an autopartition of a set of n triangles in three dimensions, but without using free cuts. Show that the expected size of the autopartition is $O(n^3)$. How does the proof depend on the fact that π is randomly chosen?

9.8. The Diameter of a Point Set

Given a set S of n points in three dimensions, the *diameter* of S, denoted $D(S)$, is the distance between the points in S that are furthest apart. (The definition could be made for points in any number of dimensions, and with any distance metric defined between pairs of points.) In this section, we will study a fast randomized algorithm for computing the diameter of a set of points in three dimensions. Thus, unlike the algorithms of the previous sections, which built a geometric structure on the input points, here we seek to determine a single number. However, we will build a geometric structure in order to compute this quantity. In particular, we will show that constructing the intersection of a set of suitably defined spheres provides a key tool in the computation of the diameter.

For a positive real number ρ, let $I_\rho(S)$ denote the convex body formed by the intersection of the n closed spheres centered at the n points of S, each of radius ρ. For a point $p \in S$, let $F(p)$ denote the distance between p and the point in S

that is farthest from p. Consider the spherical intersection $I_\rho(S)$ when $\rho = F(p)$ for some $p \in S$. For any point $q \in S$, if q is in $I_\rho(S)$, we have $F(q) \leq F(p) \leq D(S)$. On the other hand, if q lies outside $I_\rho(S)$, we $F(p) < F(q) \leq D(S)$.

The following randomized algorithm now suggests itself:

1. Pick a point $p \in S$ at random. In time $O(n)$ we compute $F(p)$ and we set $\rho = F(p)$.
2. Compute $I_\rho(S)$.
3. Find the points of S outside $I_\rho(S)$; denote this subset by S_O.
4. If S_O is empty, we know that the diameter is ρ and can stop. If not, we recur on S_O.

Clearly the running time of a single pass through Steps 1–4 is dominated by Steps 2 and 3. In addition, we must consider the effect of the randomized recursive call in Step 4. In particular, we must determine the expectation of $|S_O|$.

Consider an ordering of the points of S in non-increasing order of the values $F(p)$. Since Step 1 chooses the point p uniformly at random, the rank of $F(p)$ is uniformly distributed on $[1, n]$ (ties are broken arbitrarily); thus $|S_O|$ is uniformly distributed on $[0, n-1]$. Let $T(n)$ denote the expected running time of the algorithm when $|S| = n$, and $T_{23}(n)$ denote the corresponding cost of Steps 2 and 3. Then, we have

$$T(n) \leq cn + T_{23}(n) + \frac{\sum_{i=1}^{n-1} T(i)}{n}. \tag{9.8}$$

What can we say about $T_{23}(n)$? In Problem 9.5 we will show that these steps can be performed in expected time $O(n \log n)$ for the Euclidean metric in three dimensions, by adapting the half-space intersection algorithm of Section 9.4. Here we will consider the simpler case of the L_1 metric in three dimensions. A sphere in the L_1 metric in three dimensions is a polyhedron with eight facets and six vertices; the polyhedron can be thought of as the intersection of eight half-spaces.

Exercise 9.18: Show that the half-space intersection algorithm of Section 9.4 can be adapted to find the intersection $I_\rho(S)$ of L_1 spheres and also to determine the set S_O in expected time $O(n \log n)$, for $|S| = n$.

Using this result in (9.8), it follows that the expected running time of the randomized L_1 diameter algorithm is $O(n \log n)$. In fact, for the L_1 metric, it is not necessary to resort to the half-space intersection algorithm of Section 9.4 in order to perform Steps 2 and 3 of our diameter algorithm. A simpler algorithm running in time $O(n)$ will be considered in Problem 9.6. In this case, the recurrence (9.8) solves to $T(n) = O(n)$. From these observations, we have:

Theorem 9.10: *The above scheme for computing the diameter of n points runs in expected time $O(n \log n)$ for the L_2 metric, and in expected time $O(n)$ for the L_1 metric.*

9.9. Random Sampling

There are situations for which randomized incremental construction might not be appropriate. For instance, randomized incremental construction is inherently sequential, and may thus be unsuitable for designing parallel randomized geometric algorithms. In addition, one often builds a geometric structure (such as the Voronoi diagram) not as an end in itself, but as a means for solving search problems. For instance, the Voronoi diagram serves as the basis for *nearest-neighbor queries:* each query is a point in the space containing the n input points, and we are required to report the input point that is closest to the query point. In such cases, we wish to build, not only the geometric structure itself, but some additional structures that will enable rapid query processing. Here, again, randomized incremental construction by itself often does not suffice. We now turn to a different paradigm for designing randomized geometric algorithms, known variously as *random sampling* or as *randomized divide-and-conquer.* We first give a high-level outline of the technique, and then illustrate it using a point-location problem.

We begin with a familiar non-geometric problem. Suppose that we are given a set S of n numbers, and wish to answer *membership queries:* a query is a number, and we are to report whether or not the query number is a member of S. Consider the following approach, which is a simple generalization of the standard binary search tree. We pick a random sample R of r numbers from S, where r is a constant whose choice will become apparent from the following analysis. We sort the elements of R (in constant time), and then partition $S \backslash R$ (in time $O(n)$) into $r+1$ subsets; the ith subset contains those elements of $S \backslash R$ that are larger than exactly i elements of R. Let us call the sample R *good* if every one of the $r+1$ resulting subsets of $S \backslash R$ has size at most $(an \log r)/r$, for a fixed, suitable (as will be clear from the analysis below) constant a.

Exercise 9.19: Show that R is good with probability at least 1/2, for a suitably large constant a.

The solution to Exercise 9.19 may also be obtained by adapting the proof of Lemma 9.11 below. Given a sample R, we can check whether it is good in time $O(n)$. Thus, by Exercise 9.19, in expected time $O(n)$ we can find a good sample (by repeating the sample process whenever the sample chosen is not good).

For each subset containing more than b elements, for a suitable constant value of $b > r$, we recur by again choosing a random subset of r elements from it, and so on. This process induces a search-tree in a natural fashion, and the search process for a query is clear. Given a query q, we identify (in constant time) one of the $r+1$ subsets of S in which to continue the search for q. We search recursively in the sub-tree associated with this subset.

Exercise 9.20: Show that the expected number of steps to construct the entire search structure is $O(n \log n)$.

Given the above search structure, what is the cost of a search? Letting $Q(n)$ be the cost of a search on a set containing n elements, we have the recurrence

$$Q(n) \leq c + Q\left(\frac{an \log r}{r}\right), \tag{9.9}$$

where a is small compared with $r/\log r$ and c is a constant representing the cost of descending one level of the tree. This is easily seen to solve to $Q(n) = O(\log n)$. Notice that this bound on the search cost is a fixed constant and not a random variable. This is because in the process of constructing the search tree, we ensured that the random sample at every level was good.

Although the above example does not have a geometric flavor, it captures the essence of random sampling methods in the construction of geometric search structures. We now give a geometric example that uses random sampling and illustrates the major principles of the technique.

9.9.1. Point Location in Arrangements

Let L be a set of n lines in the plane. The lines in L partition the plane into $O(n^2)$ convex polygonal regions (some of which may be unbounded). The resulting structure is known as an *arrangement of lines.* Our description will be simplified by assuming that we are only interested in the portion of this arrangement that lies within a fixed triangle τ that contains in its interior all points of intersection between lines in L. This can be viewed as a planar graph as follows. There is a vertex of the graph for each point at which two lines meet (for simplicity, we assume for the remainder of the section that no three lines of L meet at a point). In addition, there is also a vertex for each point at which a line of L intersects the boundary of τ. An edge between two vertices corresponds in the natural sense to the line segment between two vertices that are adjacent in the arrangement. Each face of this planar graph is one of the polygonal regions into which τ is partitioned by the lines in L. We study the following query problem: given a query point q in the plane, what facet of this graph contains the query point? This is known as the point location problem in an arrangement of lines.

For convenience, we will triangulate each facet of the planar graph. We will refer to this as a triangular arrangement of the lines in L, and denote it by $\mathcal{T}(L)$. We note that this notation is slightly ambiguous, since the precise geometric structure $\mathcal{T}(L)$ depends on the large triangle within which we enclose the intersection points of the lines in L. However, we tolerate this imprecision for the following reasons: (1) in the point location problem, the identity of the facet within which a point lies is unaffected by the choice of the bounding triangle, even though the exact shape of the facet may vary; (2) for the most part (in the description of the algorithm below), the enclosing triangle will be implicit and unique.

Exercise 9.21: Show that given L, a triangular arrangement of the lines in L can be computed in time $O(n^2)$.

We now turn to the problem of point location in the triangular arrangement of lines $\mathcal{T}(L)$. The algorithm and data structure are as follows:

1. Pick a random sample R of r lines from L, where r is a suitably large constant that can be determined from the analysis below. Construct the arrangement $\mathcal{T}(R)$. The number of facets in $\mathcal{T}(R)$ is $O(r^2)$, and is thus a constant.
2. For each (triangular) facet f in $\mathcal{T}(R)$, determine the set of lines of $L \setminus R$ intersecting f; denote by L_f this set of lines. This can be done in time $O(nr^2)$. We say a facet f is *good* if it is intersected by no more than $(an \log r)/r$ lines of L for a suitable constant a. We say the random sample R is good if every facet of $\mathcal{T}(R)$ is good. If the chosen sample R is not good, we repeatedly pick samples R until we get a good sample R.
3. For each facet f of $\mathcal{T}(R)$ for which $|L_f| > b$ for a constant b, we recur on this process. Note that in the recursive steps, the enclosing triangle is just the triangle bounding the facet f. We maintain a pointer from each facet f to the triangular arrangement of the recursive random sample of lines intersecting f. These pointers will facilitate the search process.

Exercise 9.22: Show that step **2.** can in fact be implemented in time $O(nr)$.

Before we analyze the expected runing time of the above construction procedure, we explain the search process. Given the query point q, we determine (in time $O(1)$) the facet f of $\mathcal{T}(R)$ that contains q. We then recursively continue the search within $\mathcal{T}(L_f)$. Since we know that $|L_f| \leq (an \log r)/r$, we immediately know that the search time $Q(n)$ satisfies the recursion (9.9), so that $Q(n) = O(\log n)$. We stress again that this upper bound on the query time is an absolute guarantee, and not an expectation.

We turn now to the cost of constructing and storing the recursive search structure. We first establish the analog of Exercise 9.19 for the present problem.

Lemma 9.11: *The probability that any facet of $\mathcal{T}(R)$ is intersected by more than $(an \log r)/r$ lines of L is less than $1/2$, for a suitably large constant a.*

PROOF: Let S denote the set of all points at which either two lines of L intersect, or a line of L intersects the perimeter of the bounding triangle. Let Δ denote the set of all triplets of points from S. What is the probability that the triangle defined by a triplet from Δ occurs in $\mathcal{T}(R)$, and is intersected by more than

$(an \log r)/r$ lines of L? Given a triplet $\delta \in \Delta$, let $I(\delta)$ denote the set of lines of L that intersect the triangle induced by δ. Let $G(\delta)$ denote the lines of L that form the points in δ (clearly $|G(\delta)| \leq 6$). To bound the probability that the triplet δ defines a facet of $\mathcal{T}(R)$, we write it as the product of two probabilities as follows. Let $\mathcal{E}_1(\delta)$ denote the event that all lines of $G(\delta)$ are in R, and $\mathcal{E}_2(\delta)$ denote the event that none of the lines in $I(\delta)$ are in R. Clearly both $\mathcal{E}_1(\delta)$ and $\mathcal{E}_2(\delta)$ must occur in order for δ to define a facet of R (although these events are not sufficient – why?). Then,

$$\mathbf{Pr}[\delta \text{ appears as a facet of } \mathcal{T}(R)] \leq \mathbf{Pr}[\mathcal{E}_1(\delta)]\mathbf{Pr}[\mathcal{E}_2(\delta)|\mathcal{E}_1(\delta)].$$

We now bound $\mathbf{Pr}[\mathcal{E}_2(\delta)|\mathcal{E}_1(\delta)]$: having picked the lines in $G(\delta)$, we consider what happens on the remaining $r - |G(\delta)|$ drawings of R. In particular, consider the probability that none of the $r - |G(\delta)|$ remaining drawings picks any line in $I(\delta)$. This is bounded by

$$\prod_{i=0}^{r-|G(\delta)|-1} \left(1 - \frac{|I(\delta)|}{n - |G(\delta)| - i}\right) \leq \left(1 - \frac{|I(\delta)|}{n}\right)^{r-|G(\delta)|} \leq e^{-rI(\delta)/2n}$$

for any value of $r > 12$ (since $|G(\delta)| \leq 6$). We are only interested in δ such that $I(\delta) > (an \log r)/r$; call these *large* triplets. Thus, for large triplets we have $\mathbf{Pr}[\mathcal{E}_2(\delta)] < r^{-a/2}$. Then,

$$\begin{aligned} &\mathbf{Pr}[\text{A large triplet appears as a facet of } \mathcal{T}(R)] \\ &\qquad \leq r^{-a/2} \sum_{\text{large triplets } \delta} \mathbf{Pr}[\mathcal{E}_1(\delta)]. \end{aligned} \tag{9.10}$$

Now, the summation in (9.10) is exactly the expected number of large triplets in R. Since R is an arrangement of r lines, and each point of a triplet is formed by at most 2 lines, it follows that this summation is never more than r^6. Then, for $a > 12$ the lemma follows. □

Corollary 9.12: *The expected number of trials before we obtain a good sample R is at most 2.*

We now complete the analysis of the construction of the data structure. By the preceding discussion, the construction time satisfies the recurrence

$$T(n) \leq n^2 + cr^2 T\left(\frac{an \log r}{r}\right),$$

where c is a constant and $T(k)$ denotes the upper bound on the expected cost of constructing the data structure for an arrangement of k lines. This solves to $T(n) = O(n^{2+\epsilon(r)})$, where $\epsilon(r)$ is a positive constant that becomes smaller as r gets larger.

Theorem 9.13: *The above algorithm constructs a data structure in expected time $O(n^{2+\epsilon})$ for a set of n lines in the plane for any fixed $\epsilon > 0$, and this data structure can support point location queries in time $O(\log n)$.*

Exercise 9.23: What are the effects of increasing r on the construction time for the search structure, and on the query time?

9.10. Linear Programming

We continue the study of random sampling by considering the linear programming problem. The linear programming problem is a particularly notable example of the two main benefits of randomization – simplicity and speed. In Section 9.10.1 we will study randomized incremental algorithms for this problem.

The linear programming problem is to find the extremum of a linear objective function of several real variables, subject to constraints that are linear functions of these variables. Hereafter, we will let d denote the number of variables, and n the number of constraints. Each of the n constraints may be thought of as delineating a half-space in d-dimensional space, stipulating that our extremization is restricted to points in this half-space. The intersection of these half-spaces is a polyhedron in d-dimensional space (which may be empty, or possibly unbounded), which we will refer to as the *feasible region.* Throughout, we will measure the amount of computation we perform by the number of arithmetic operations, treating the operands as real numbers on which an arithmetic operation can be performed in constant time. This is consistent with our view throughout this chapter, but the reader is cautioned that much of the work in the linear programming literature deals with operands of finite precision. For such finite precision operands, there has been considerable work on the number of bit operations performed by various algorithms. We will not concern ourselves with such bit operations, but will treat all numbers as atomic operands.

Let $x_1, \ldots, x_d$ denote the d variables in the linear program. Let $c_1, \ldots, c_d$ denote the coefficients of these variables in the objective function, and let A_{ij}, $1 \leq i \leq n$ and $1 \leq j \leq d$ denote the coefficient of x_j in the ith constraint. Letting $\boldsymbol{A}$ denote the matrix (A_{ij}), $\boldsymbol{c}$ the vector $(c_1, \ldots, c_d)$, and $\boldsymbol{x}$ the vector $(x_1, \ldots, x_d)$, the linear programming problem may be expressed as

$$\text{minimize } \boldsymbol{c}^T \boldsymbol{x} \tag{9.11}$$

subject to

$$\boldsymbol{A}\boldsymbol{x} \leq \boldsymbol{b}, \tag{9.12}$$

where $\boldsymbol{b}$ is a column vector of constants.

We denote by $\mathcal{F}(\boldsymbol{A}, \boldsymbol{b})$ the feasible region defined by $\boldsymbol{A}$ and $\boldsymbol{b}$. The vector $\boldsymbol{c}$ specifies a direction in d-space. Geometrically, we seek the furthest point in $\mathcal{F}(\boldsymbol{A}, \boldsymbol{b})$ in the direction opposite to $\boldsymbol{c}$ (since we are minimizing), if such a finite point exists. The linear programming problem has a long history, a partial summary of which is given in the Notes section. The starting point in our treatment will be the following set of assumptions, which is known (see the Notes section and the references therein) to capture the general linear programming problem; these assumptions do not specialize or simplify the problem from the standpoint of designing algorithms. All of these assumptions can be removed by standard techniques; this will be explored further in Problem 9.8.

1. The polyhedron $\mathcal{F}(\boldsymbol{A}, \boldsymbol{b})$ is non-empty and bounded. Note that we are not assuming that we can *test* an arbitrary polyhedron for non-emptiness or boundedness; this is known to be equivalent to solving a linear program. We only make this assumption about $\mathcal{F}(\boldsymbol{A}, \boldsymbol{b})$.
2. The objective function we are minimizing is x_1; in other words, $\boldsymbol{c} = (1, 0, \ldots, 0)$. Thus we seek a point of $\mathcal{F}(\boldsymbol{A}, \boldsymbol{b})$ with the minimum value of x_1.
3. The minimum we seek occurs at a unique point which is a vertex of $\mathcal{F}(\boldsymbol{A}, \boldsymbol{b})$.
4. Each vertex of $\mathcal{F}(\boldsymbol{A}, \boldsymbol{b})$ is defined by exactly d constraints.

Let H denote the set of constraints defined by $\boldsymbol{A}$ and $\boldsymbol{b}$. Let $S \subseteq H$ be a subset of constraints from H. We will frequently consider the linear program defined by such a subset S, together with $\boldsymbol{c}$. When such a linear program attains a finite minimum, we will assume that versions of assumptions 3–4 above still hold: (i) the minimum occurs at a unique point; (ii) each vertex of the feasible region is defined by d constraints. We denote by $\mathcal{O}(S)$ the value of the objective function for the linear programming problem defined by $\boldsymbol{c}$ and S (it is possible that $\mathcal{O}(S) = -\infty$). A *basis* is a set of constraints, B, such that $\mathcal{O}(B) > -\infty$ and $\mathcal{O}(B') < \mathcal{O}(B)$ for any $B' \subset B$. The *basis of* H, denoted $\mathcal{B}(H)$, is a minimal subset $B \subseteq H$ with $\mathcal{O}(B) = \mathcal{O}(H)$. Our goal is to find $\mathcal{B}(H)$. Since $\mathcal{B}(H)$ defines the optimal vertex of our linear program, we will sometimes refer to $\mathcal{B}(H)$ or to $\mathcal{O}(\mathcal{B}(H))$ as the optimum of the linear program.

One approach to solving the linear programming problem would be to use a half-space intersection algorithm to compute $\mathcal{F}(\boldsymbol{A}, \boldsymbol{b})$ and to then evaluate the objective function at each vertex of the polyhedron $\mathcal{F}(\boldsymbol{A}, \boldsymbol{b})$. Such an exhaustive evaluation process could in general be very slow, since the number of vertices of $\mathcal{F}(\boldsymbol{A}, \boldsymbol{b})$ may be $\Omega(n^{\lceil d/2 \rceil})$. We therefore seek algorithms that do not enumerate the vertices of $\mathcal{F}(\boldsymbol{A}, \boldsymbol{b})$.

Before proceeding to our study of randomized algorithms for linear programming, we will recall the elements of the classic simplex algorithm. This is a deterministic algorithm that starts from a vertex of $\mathcal{F}(\boldsymbol{A}, \boldsymbol{b})$ and, at each subsequent iteration, proceeds to a neighboring vertex at which the objective function has a lower value. If no such vertex exists, we have reached the minimum we seek. While this is the essential idea of the simplex algorithm, a number of complications arise when adjacent vertices have the same objective function value, and from problems with no finite minimum. We will avoid a detailed discussion of the simplex algorithm; in our discussion it will suffice to assume the existence of a function **Simplex** that will solve linear programs by visiting the vertices of $\mathcal{F}(\boldsymbol{A}, \boldsymbol{b})$ in turn until the optimum is found, if one exists.

We call a constraint $h \in H$ *extreme* if $\mathcal{O}(H \setminus \{h\}) < \mathcal{O}(H)$; thus these are the constraints in $\mathcal{B}(H)$. Intuitively, the constraints of H that are not extreme are redundant constraints whose absence would not alter the optimum. Our first algorithm **SampLP** uses random sampling to throw away redundant constraints quickly. Starting from the empty set, **SampLP** builds up a set S of constraints over a series of phases. In each phase, a set $V \subset H \backslash S$ is added to S. The set V will have two important properties: (i) it will be small, and (ii) it will contain at least one extreme constraint from $\mathcal{B}(H)$ that is not in S. Since $|\mathcal{B}(H)| = d$, we terminate after at most d phases.

We will describe **SampLP** in pseudocode below, and then proceed to the more sophisticated algorithm **IterSampLP**. We will finish by analyzing **IterSampLP**.

```
Algorithm SampLP:

Input: A set of constraints H.
Output: The optimum B(H).

1. S ← φ;
2. if n < 9d^2
     return Simplex (H)
   else

     2.1. V ← H; S ← φ;
     2.2. while |V| > 0
            Choose R ⊂ H\S at random, with |R| = r = min{d√n, |H\S|};
            x ← SampLP(R ∪ S);
            V ← {h ∈ H | vertex defined by x violates h};
            if |V| ≤ 2√n
            then S ← S ∪ V;
     2.3. return x;
```

Thus, for $n > 9d^2$ **SampLP** chooses a random subset R of r constraints. The value of r is normally $d\sqrt{n}$, unless $H \backslash S$ contains fewer than $d\sqrt{n}$ constraints. It recursively solves the linear program defined by $R \cup S$, and determines the set $V \subset H$ of constraints that are violated by this optimum; note that these violated constraints will in fact be from $H \backslash S$. If V has no more than $2\sqrt{n}$ elements (we will argue that this is likely), we add V to S. When V becomes empty (meaning that $\mathcal{B}(H)$ is contained in S), we return x.

Exercise 9.24: Construct a simple example to show that after one pass through the **while** loop of **SampLP**, V may not contain all of $\mathcal{B}(H)$. Hence, we may only infer that V contains at least one constraint of $\mathcal{B}(H)$ that is not already in S.

The routine **Simplex** is invoked only with $9d^2$ or fewer constraints. For such "small" linear programming problems, we may bound the cost of invoking

Simplex as follows. The total number of vertices in the polyhedron for such a problem is no more than $\binom{9d^2}{\lceil d/2 \rceil}$, which is at most $(49d)^{\lceil d/2 \rceil}$. There is a constant a such that the simplex algorithm spends at most time d^a at each vertex, so that we have:

Lemma 9.14: *The total cost in an invocation of* **Simplex** *with* $9d^2$ *or fewer constraints is* $O(d^{d/2+a})$.

Next, we wish to argue that V, the set of constraints that violate x, is small.

Lemma 9.15: *Let $S \subseteq H$, and let $R \subseteq H \backslash S$ be a random subset of size r. Let m denote $|H \backslash S|$. The expected number of constraints of H violated by $\mathcal{O}(R \cup S)$ is no more than $d(m-r+1)/(r-d)$.*

PROOF: We define two sets of optima for linear programs formed by subsets of the constraints. Let $\mathcal{C}_H$ denote the set of optima $\{\mathcal{O}(T \cup S) \mid T \subseteq H \backslash S\}$. Thus, the call the **SampLP**$(R \cup S)$ returns an element of this set. Similarly, we define $\mathcal{C}_R$ to be the set of optima $\{\mathcal{O}(T \cup S) \mid T \subseteq R\}$ for a particular subset R. Now, $\mathcal{O}(R \cup S)$ is the unique element in $\mathcal{C}_R$ that satisfies every constraint in R. For each element $x \in \mathcal{C}_H$, let v_x denote the number of constraints of H violated by x. Let the indicator i_x be 1 whenever x is $\mathcal{O}(R \cup S)$, and 0 otherwise.

We may now write

$$\mathbf{E}[|V|] = \mathbf{E}[\sum_{x \in \mathcal{C}_H} v_x i_x] = \sum_{x \in \mathcal{C}_H} v_x \mathbf{E}[i_x]. \tag{9.13}$$

Now, $\mathbf{E}[i_x]$ is simply the probability that x is the optimum $\mathcal{O}(R \cup S)$. For this event to occur, d given constraints must be in R, and the remaining $r-d$ constraints of R must be from among the $m - v_x - d$ constraints of $H \backslash S$ that neither define nor are violated by x. Thus

$$\mathbf{E}[i_x] = \frac{\binom{m-v_x-d}{r-d}}{\binom{m}{r}}. \tag{9.14}$$

Exercise 9.25: By combining (9.13) and (9.14) and simplifying, show that

$$\mathbf{E}[|V|] \leq \frac{m-r+1}{r-d} \sum_{x \in \mathcal{C}_H} v_x \frac{\binom{m-v_x-d}{r-d-1}}{\binom{m}{r}}. \tag{9.15}$$

We will complete the proof by showing that the summation on the right-hand side of (9.15) is no more than d. The factor $\binom{m-v_x-d}{r-d-1}/\binom{m}{r}$ is the probability that x is an element of $\mathcal{C}_R$ that violates exactly one constraint of R. Weighting this by v_x and summing yields the expected number of elements of $\mathcal{C}_R$ that violate exactly one constraint of R. However, the number of such elements is at most d, since each such element is the optimum of the set $R \cup S \backslash \{h\}$ for a constraint h

that defines the optimum $\mathcal{O}(R \cup S)$. There are d constraints defining the optimum $\mathcal{O}(R \cup S)$. □

With this bound on the expected number of violated constraints, the Markov inequality now implies that following any random sample in **SampLP**, $\mathbf{Pr}[|V| > 2\sqrt{n}] \leq 1/2$. It follows that the expected number of iterations of Step 2.2 between augmentations to S is at most 2. Let $T(n)$ denote the maximum expected running time of **SampLP**. The set S is initially empty, and in each of d phases adds at most $2\sqrt{n}$ constraints. Thus, $|R \cup S|$ never exceeds $3d\sqrt{n}$. For each of d phases, we perform at most n constraint violation tests at a cost of $O(d)$ for each test; thus the total work in constraint checking is $O(d^2 n)$. When in a recursive call the number of constraints drops to $9d^2$ or less, we resort to the time bound on the call to **Simplex** (Lemma 9.14). Putting these observations together, we have

$$T(n) \leq 2dT(3d\sqrt{n}) + O(d^2 n), \quad \text{for } n > 9d^2. \tag{9.16}$$

Exercise 9.26: Derive the best possible upper bound on $T(n)$ in (9.16), in conjunction with Lemma 9.14.

We now describe the algorithm **IterSampLP**. Rather than try to discover $\mathcal{B}(H)$ little by little, it uses a technique known as *iterative reweighting* to increase the probability of including a useful constraint in the sample. We choose a random subset of constraints R and determine the subset $V \subset H$ of constraints violated by the optimum of the linear program defined by R. Instead of adding V to a set S as in **SampLP**, we put the constraints of V back in H after first increasing the probability that they are chosen in future rounds. Intuitively, the constraints of $\mathcal{B}(H)$ will repeatedly find themselves in V, and hence their probabilities of being included in R increase rapidly. After relatively few such iterations (as we will show), all the constraints of $\mathcal{B}(H)$ are likely to be in R, and we terminate. A detailed description of **IterSampLP** follows. We will associate a positive integral *weight* w_h with each constraint $h \in H$; the constraint h will be put in R with probability proportional to the current value of w_h.

In Step 2.2, the probability that a constraint h is chosen is proportional to w_h. We turn to the analysis of **IterSampLP**.

Call an execution of the **while** loop *successful* if

$$\sum_{h \in V} w_h \leq (2 \sum_{h \in H} w_h)/(9d - 1)$$

(thus, we double w_h for each $h \in V$).

```
Algorithm IterSampLP:

Input: A set of constraints H.
Output: The optimum B(H).

1. ∀h ∈ H, set w_h ← 1;
2. if n < 9d^2
   return Simplex (H)
   else

     2.1. V ← H;
     2.2. while |V| > 0
            Choose R ⊂ H at random, with |R| = r = 9d^2;
            x ← Simplex(R);
            V ← {h ∈ H|x violates h};
            if Σ_{h∈V} w_h ≤ (2 Σ_{h∈H} w_h)/(9d − 1)
            then ∀h ∈ V set w_h ← 2w_h;
     2.3. return x;
```

Lemma 9.16: *The expected number of iterations of the* **while** *loop between successful iterations is at most 2.*

Note that we cannot directly invoke the result of Lemma 9.15 for the analysis of **IterSampLP**, since the constraints in the random subset R are not chosen equiprobably. The proof of Lemma 9.16 is an extension of the analysis leading to Lemma 9.15; the reader may follow the hint in Problem 9.9 to complete the proof.

Theorem 9.17: *There exist constants c_1, c_2, and c_3 such that the expected running time of* **IterSampLP** *is at most*

$$c_1 d^2 n \log n + (c_2 d \log n) d^{d/2+c_3}.$$

PROOF: We will argue that the expected number of executions of the **while** loop is $O(d \log n)$. The idea is that $\sum_{h \in \mathcal{B}(H)} w_h$ grows much faster than $\sum_{h \in H} w_h$, so that after $d \log n$ iterations $V = \phi$ unless $\sum_{h \in \mathcal{B}(H)} w_h > \sum_{h \in H} w_h$, which would be a contradiction.

After each successful execution of the loop, the weight w_h is doubled for at least one constraint $h \in \mathcal{B}(H)$ (since V must contain at least one constraint $h \in \mathcal{B}(H)$). Following kd successful executions of the loop, we have $\sum_{h \in \mathcal{B}(H)} w_h = \sum_{h \in \mathcal{B}(H)} 2^{n_h}$, where n_h is the number of times h entered V. Clearly $\sum_{h \in \mathcal{B}(H)} n_h \geq kd$.

These facts together imply that

$$\sum_{h \in \mathcal{B}(H)} w_h \geq d2^k. \tag{9.17}$$

On the other hand, after each successful execution of the **while** loop, the net increase in $\sum_{h \in H} w_h$ is no more than $(2\sum_{h \in H} w_h)/(9d-1)$. Initially $\sum_{h \in H} w_h = n$. Following kd successful iterations it is no more than

$$n[1 + 2/(9d-1)]^{kd} \leq n \exp[2kd/(9d-1)]. \tag{9.18}$$

Comparing (9.17) and (9.18), it follows that after $O(d \log n)$ iterations we drop out of the loop.

How much time do we spend between successful iterations of the **while** loop? By Lemma 9.16, the expected number of iterations between successful iterations is 2. During each iteration, we incur the cost of a **Simplex** call (whose running time we have bounded in Lemma 9.14 above), and determine V in time $O(nd)$. Putting these facts together yields the theorem. □

9.10.1. Incremental Linear Programming

We have so far studied linear programming algorithms based on random sampling. We now explore randomized incremental algorithms for linear programming. The following algorithm suggests itself immediately: add the n constraints in random order, one at a time. After adding each constraint, determine the optimum of the constraints added so far. This algorithm may also be viewed in the following "backward" manner, which will prove useful in the sequel.

Algorithm SeideLP:

Input: A set of constraints H.

Output: The optimum of the LP defined by H.

0. **if** $|H| = d$, output $\mathcal{B}(H) = H$.

1. Pick a random constraint $h \in H$;
Recursively find $\mathcal{B}(H \backslash \{h\})$;

2.1. **if** $\mathcal{B}(H \backslash \{h\})$ does not violate h, output $\mathcal{B}(H \backslash \{h\})$ to be the optimum $\mathcal{B}(H)$;

2.2. **else** project all the constraints of $H \backslash \{h\}$ onto h and recursively solve this new linear programming problem;

The idea of the algorithm is simple. Either h (the constraint chosen randomly in Step 1) is redundant (in which case we execute Step 2.1), or it is not. In the latter case, we know that the vertex formed by $\mathcal{B}(H)$ must lie on the hyperplane bounding h. In this case, we project all the constraints of $H \backslash \{h\}$ onto h and solve this new linear programming problem (which has dimension $d-1$). When the number of constraints is down to d, **SeideLP** stops recurring.

Since there are at most d extreme constraints in H, the probability that the randomly chosen constraint h is one of the extreme constraints we seek is at most d/n. Let $T(n,d)$ denote an upper bound on the expected running time of the algorithm for any problem with n constraints in d dimensions. Then, we may write

$$T(n,d) \leq T(n-1,d) + \mathrm{O}(d) + \frac{d}{n}[\mathrm{O}(dn) + T(n-1,d-1)]. \qquad (9.19)$$

In (9.19), the first term on the right denotes the cost of recursively solving the linear program defined by the constraints in $H \backslash \{h\}$. The second accounts for the cost of checking whether h violates $\mathcal{B}(H \backslash \{h\})$. With probability d/n it does, and this is captured by the bracketed expression, whose first term counts the cost of projecting all the constraints onto h. The second counts the cost of (recursively) solving the projected problem, which has one fewer constraint and dimension. The following theorem may be verified by substitution, and proved by induction.

Theorem 9.18: *There is a constant b such that the recurrence (9.19) satisfies the solution $T(n,d) \leq bnd!$.*

The above incremental algorithm is thus likely to be slow unless d is rather small. The reader may wonder why, when solving the problem of dimension $d-1$ in Step 2.2, we completely discard any information obtained from the solution of the linear program $H \backslash \{h\}$ (Step 1). We now proceed to a more sophisticated algorithm that retains such information carefully. Before doing so, the following exercise is provided to strengthen the reader's intuition.

Exercise 9.27: Consider the algorithm **SeideLP**. Construct an example to show that the optimum of the linear program defined by the constraints in $\mathcal{B}(H \backslash h) \cup \{h\}$ may be different from the optimum of the linear program defined by H. Thus, if the test in Step 2.1 fails and we proceed to Step 2.2, it does not suffice to consider the constraints in $\mathcal{B}(H \backslash h) \cup \{h\}$ alone.

By the above exercise, it follows that we must once again consider *all* the constraints in H in Step 2.2 of **SeideLP**. However, it is still reasonable to hope that $\mathcal{B}(H \backslash h)$ will in fact contain many of the constraints in $\mathcal{B}(H)$. Could we somehow use $\mathcal{B}(H \backslash h)$ to "jump-start" the recursive call in Step 2.2 of **SeideLP**? The result of this idea is the algorithm **BasisLP**, which is invoked with two arguments, a set $G \subseteq H$ of constraints, and a basis $T \subseteq G$ (not in general the basis of G). **BasisLP** returns the basis of G.

Algorithm BasisLP:

Input: G, T.

Output: A basis B for G.

0. If $G = T$, output T;

1. Pick a random constraint $h \in G \backslash T$;
$T' =$ **BasisLP**$(G \backslash \{h\}, T)$;

2.1. **if** h does not violate T', output T';

2.2. **else** output **BasisLP**$(G,$ **Basis**$(T' \cup \{h\}))$;

The function **Basis** returns a basis for a set of $d + 1$ or fewer constraints, if such a basis exists. In our algorithm, we always invoke **Basis** on a given basis T' with d constraints, together with a new constraint h. By computing the intersection of h with each of the d subsets of T' that have cardinality $d - 1$, and evaluating $\mathcal{O}$ at each of these d points, we may determine **Basis** $(T' \cup \{h\})$.

Exercise 9.28: Show that the above description of **Basis** will terminate in $O(d^4)$ steps. (Note that a system of d linear equations can be solved in $O(d^3)$ steps.)

Exercise 9.29: The routine **BasisLP** requires a basis T as one of the inputs. Suggest a scheme for starting the algorithm initially with a suitable basis, so that when finished we have the optimum $\mathcal{O}(H)$. (**Hint:** Use a bounding box.)

Each invocation of **Basis** is preceded by a violation test (in the **if** statement). In our analysis below we will bound the number of violation tests, and from this infer a bound on the number of invocations of **Basis** and thus the overall running time. What is the probability that we fail a violation test in a given execution of **BasisLP**? Suppose that $|G| = i$. We are reintroducing a constraint $h \in G \backslash T$ that was chosen at random, and wish to bound the probability that h violates the optimum of $G \backslash \{h\}$. Clearly this is at most $d/(i - |T|)$, since at most d constraints of G determine $\mathcal{B}(G)$ and h is equally likely to be any of the $i - |T|$ constraints in $G \backslash T$. We now refine this estimate on the probability. The intuition is that this probability decreases further if T contains some of the constraints of $\mathcal{B}(G)$; indeed, this was our motivation for refining **SeideLP** to obtain **BasisLP**. To this end, we introduce some additional notions.

Given $T \subseteq G \subseteq H$, we call a constraint $h \in G$ *enforcing* in (G, T) if $\mathcal{O}(G \backslash \{h\}) < \mathcal{O}(T)$. This concept is illustrated in Figure 9.12. In this figure, there are four constraints, numbered $1, 2, 3,$ and 4. Each constraint is a line that allows the half-plane above itself as the feasible region. Clearly constraints 1 and 4 are the extreme constraints for the set $\{1, 2, 3, 4\}$. Consider for the moment a view of **BasisLP** played "backward," and a situation in which the constraints are added back in the order $1, 2, 3, 4$. Observe that constraint 1 is not enforcing in G, T for $G = \{1, 2, 3, 4\}$ and $T = \{1, 2\}$.

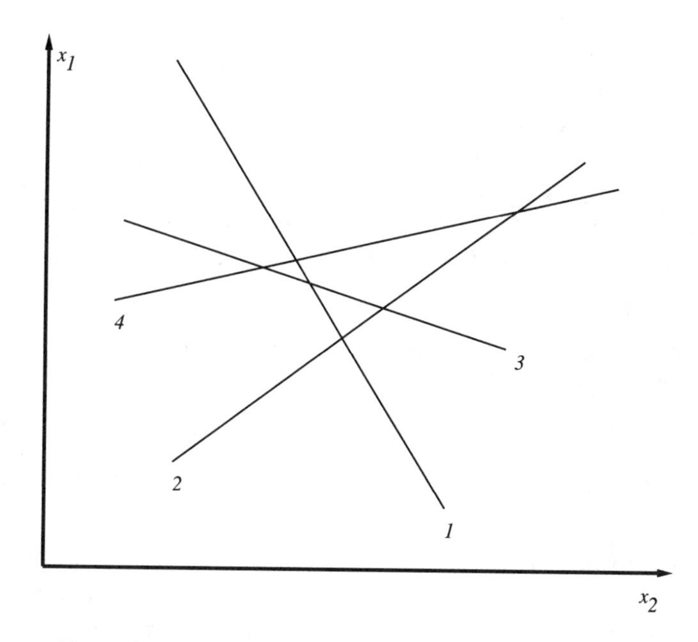

Figure 9.12: Extreme and enforcing constraints.

Exercise 9.30: If the constraints are deleted in the order 4,3,2,1, trace the course of the call to **BasisLP**$(G, \{1, 2\})$, determining the arguments of the various recursive calls. Repeat this if the order of deletion of constraints is 1,4,3,2.

Exercise 9.31: If h is enforcing in (H, T), show that (i) $h \in T$, and (ii) h is extreme in all G such that $T \subseteq G \subseteq H$.

If all d constraints in T are enforcing in (G, T), we have $T = \mathcal{B}(G)$. Given $T \subseteq G \subseteq H$, let $\Delta_{G,T}$ denote d minus the number of constraints that are enforcing in (G, T). We call $\Delta_{G,T}$ the *hidden dimension* of (G, T). The number of constraints of $\mathcal{B}(G)$ that are not already in T. From the above discussion, the probability that a violation occurs in the **if** statement can be bounded by $\Delta_{G,T}/(i - |T|)$. We will first establish that the hidden dimension decreases by at least 1 at each recursive call in Step 2.2; later, we will improve this by arguing that it is likely to decrease much faster.

Exercise 9.32: Let $T \subseteq F \subseteq G \subseteq H$, and let $h \in F \backslash T$ be an extreme constraint in F. Let S be a basis of $\mathcal{B}(F \backslash \{h\}) \cup \{h\}$. Show that

1. any constraint g that is enforcing in (G, T) is also enforcing in (F, S);
2. h is enforcing in (F, S);
3. $\Delta_{F,S} \leq \Delta_{G,T} - 1$.

Thus, as we proceed down the recursion (in a sequence of executions of Step 2.2), the numerator of the probability bound decreases by at least 1 at each execution. We will now show that the decrease in the hidden dimension (and thus the decrease in the probability) is likely to be faster. Given sets F and T such that $T \subset F \subseteq G$, and a random $h \in F \setminus T$, we bound the probability that the addition of h to $F \setminus \{h\}$ causes a recursive call. When it does, we study the probability distribution of the hidden dimension of the arguments of such a call.

Exercise 9.33: Let $g_1, g_2, \ldots g_s$ be the extreme constraints of F that are not in T, numbered so that

$$\mathcal{O}(F \setminus \{g_1\}) \leq \mathcal{O}(F \setminus \{g_2\}) \leq \cdots$$

Show that for all ℓ and for $1 \leq j \leq \ell$, g_j is enforcing in $(F, \mathbf{Basis}(\mathcal{B}(F \setminus \{g_\ell\}) \cup \{g_\ell\}))$.

In other words, when $h = g_\ell$, all of $\{g_1, g_2, \ldots, g_\ell\}$ will be enforcing in $(F, \mathbf{Basis}(\mathcal{B}(F \setminus \{h\}) \cup \{h\}))$. Then, the arguments of the recursive call will have hidden dimension $\Delta_{G,T} - \ell$. The crucial observation is that since any of the g_i is equally likely to be h (by backwards analysis!), ℓ is uniformly distributed on the integers in $[1, s]$. Thus the hidden dimension of the arguments of the recursive call is uniformly distributed on the integers in $[0, s-1]$.

For a call to **BasisLP** with arguments (G, T), where $|G| = m$ and $\Delta_{G,T} = k$, let us denote by $T(m,k)$ the maximum expected number of violation tests (executions of the **if** statement).

Exercise 9.34: Show that $T(m, 0) = m - d$.

For $m \geq d+1$ and $k \geq 1$, the above discussion on the probability distribution of the hidden dimension yields the following recurrence:

$$T(m,k) \leq T(m-1,k) + 1 + \frac{T(m,0) + T(m,1) + \cdots + T(m,k-1)}{m-d}. \tag{9.20}$$

Exercise 9.35: Verify that $T(m,k) \leq 2^k(m-d)$.

By combining the results of Exercises 9.29 and 9.35, we have:

Theorem 9.19: *The expected running time of* **BasisLP** *on a problem with n constraints in d dimensions is* $O(d^4 2^d n)$.

Note the improvement over Theorem 9.18. By a slightly more careful analysis, and a more complicated analysis of the recurrence that results, the time bound of Theorem 9.19 can be improved considerably. This will be discussed briefly in the Notes section.

Notes

The first algorithms for all of the geometric problems we have considered were deterministic; rather than give sources for each of these deterministic algorithms, we refer the reader to textbooks on computational geometry [133, 336]. A comprehensive introduction to the design and analysis of randomized geometric algorithms is the book by Mulmuley [316]. Rabin's [341] description of a randomized algorithm for the problem of finding nearest neighbors in a set of n points is perhaps the earliest use of randomization in a geometric algorithm. The systematic use of randomization in geometric algorithms was pioneered in a series of papers by Clarkson [101, 102, 103, 105], Clarkson and Shor [106, 107], and Mulmuley [315]. Below, we give more detailed pointers to the various problems and algorithms we have studied.

The **RandAuto** algorithm for binary space partitions is due to Paterson and Yao (see [329] and references therein). They also prove that there are inputs for the three-dimensional case for which every autopartition has size $\Omega(n^2)$. The result used in the proof of Theorem 9.9 concerning the number of edges bounding external sub-facets is described in the book by Edelsbrunner [133].

► **Research Problem 9.1:** Paterson and Yao show that in the case where the line segments are all parallel to two (orthogonal) axes, a binary partition of size $O(n)$ can be found. Is it always possible to find a partition of size $O(n)$? Is there a configuration of n segments that forces a lower bound of $\Omega(n \log n)$ on the size of any autopartition for that configuration?

► **Research Problem 9.2:** Since any partition must have size $\Omega(n)$ and we can find one of size $O(n \log n)$ using the **RandAuto**, it is clear that we find a partition whose size is within $O(\log n)$ of the optimal size. Can we prove something stronger, say, find a partition of size is within a constant (or any factor better than $\log n$) of the optimum? It is plausible that this question can be answered independently of Research Problem 9.1.

► **Research Problem 9.3:** Can we give a high confidence estimate for the size of the autopartition produced by the random permutation algorithm (with free cuts) in three dimensions? In other words, we require a statement of the form "with probability $1 - f(n)$, the size of the autopartition does not exceed $g(n)$."

► **Research Problem 9.4:** As in the two-dimensional case, can we say whether our algorithm is *provably good* in that it always finds a partition whose size is within some provable factor of the optimum? Notice that there is more room for leeway here than in the planar case – the optimum could be anywhere from $n - 1$ to $\Omega(n^2)$.

Randomized incremental constructions are simple to implement, and their power was demonstrated in a series of papers by Clarkson, Shor, Mulmuley, and others [107, 315, 368, 369]; the algorithms we have described for convex hulls and for trapezoidal decompositions appear in these papers. Prior to this work, Chazelle and Edelsbrunner [90] gave a deterministic but relatively complicated algorithm for trapezoidal decompositions with running time $O(n \log n + k)$. The key idea of backwards analysis appeared first in a paper by Chew [94]; the algorithm of Section 9.5.1 for finding the Delaunay triangulation of the vertices of a convex polygon is from this paper. However, the generality and widespread applicability of this idea (to geometric as well as non-geometric problems) went unnoticed prior to the work of Seidel [371]. Guibas, Knuth, and Sharir [187] showed that this paradigm can be applied directly to the construction of Voronoi diagrams. The incremental construction paradigm has been applied to a diverse collection of geometric problems; the interested reader should consult Mulmuley's treatise [316] for further pointers. The use of random sampling was pioneered by Clarkson [102], who proved a general version of Lemma 9.11; this paper also describes the data structure of Section 9.9.1 for point location in an arrangement of lines. The application of sampling to geometric problems owes its origins to a paper by Haussler and Welzl [197]. A variant of the random sampling technique has been used by Chazelle and Friedman [92], improving the expected running time from $O(n^{2+\epsilon})$ to $O(n^2)$. Random sampling, too, has been applied to a large number of geometric problems, and the reader may again consult Mulmuley [316] for further pointers. One theoretical benefit of randomized geometric algorithms is that they can be derandomized to yield deterministic algorithms that are faster than known algorithms. Chazelle and Friedman [91] pioneered this study; see also the survey by Matoušek [294].

The linear programming problem has a long and rich history; the reader is referred to treatises by Chvátal [100] and by Schrijver [366] for the history of the problem and the classical Simplex algorithm invoked in Section 9.10. These books (as well as several of the papers we mention below) also discuss how to remove the assumptions we have made at the beginning of Section 9.10. Megiddo [307] gave a deterministic algorithm for linear programming running in time $O\left(n2^{2^d}\right)$. Much subsequent work focused on reducing the 2^{2^d} term in the running time, and indeed all the algorithms we have described have variants whose running time can be bounded as $O(nf(d))$ where $f(d)$ is some (typically exponential) function of d. This also applies to the random sampling algorithms of Section 9.10; these algorithms are due to Clarkson [104]. The iterative reweighting technique of Section 9.10 was first applied to geometric algorithms by Welzl [417]. The **SeideLP** algorithm of Section 9.10.1 is due to Seidel [369].

In the discussion leading to Lemma 9.14, we invoked a bound on the maximum number of vertices that a polyhedron with $9d^2$ constraints can have; this bound is a special case of general bounds on the number of vertices of a polyhedron. Such bounds are given, for instance, in Edelsbrunner's book [133].

The **BasisLP** algorithm and its analysis are due to Sharir and Welzl [374]. Kalai [226] achieved a breakthrough by giving a randomized algorithm whose expected running time is at most

$$\min\{n^{a\sqrt{d/\log d}}, n2^{a\sqrt{n-d}\sqrt{\log d}}\}$$

for an absolute constant a. Following this, Matoušek, Sharir, and Welzl [295] showed that the **BasisLP** algorithm in [374] in fact runs in time

$$O\left(nd\exp(\sqrt{d}\ln(n+1)\right).$$

By augmenting the analysis of [295] with Clarkson's sampling technique, it is possible to obtain the slightly improved time bound of

$$\mathrm{O}\left(d^2 n + b^{\sqrt{d \log d}} \log n\right)$$

for an absolute constant b. Goldwasser [177] gives an eminently readable account of the algorithms and analyses of Kalai [226] and of Matoušek, Sharir, and Welzl [295]. In fact, he points out that the algorithm of Matoušek, Sharir, and Welzl is exactly dual (in the sense of linear programming duality [100]) to one variant of Kalai's.

Sharir and Welzl [374] in fact describe their algorithm as being applicable to a general class of *abstract optimization problems* that includes linear programming as a special case. We explore this theme further in Problem 9.11. Gärtner [163] extended this approach and applied it to obtain sub-exponential algorithms for such problems as finding the minimum distance between two polytopes in d dimensions.

The **Random Simplex** algorithm is the following: starting from any vertex of $\mathcal{F}(\boldsymbol{A}, \boldsymbol{b})$, proceed to a random adjacent vertex of $\mathcal{F}(\boldsymbol{A}, \boldsymbol{b})$ that improves the objective function. Algorithms that only move between adjacent vertices of $\mathcal{F}(\boldsymbol{A}, \boldsymbol{b})$ are generally known as *simplex algorithms*, following Danzig [119, 120].

► **Research Problem 9.5:** Derive a sub-exponential upper bound on the expected running time of the **Random Simplex** algorithm.

Gärtner and Ziegler [164] have established a tight, polynomial upper bound for a restricted class of polytopes known as *Klee-Minty cubes*. Any simplex algorithm is condemned to incur a running time that is at least the diameter of the polytope $\mathcal{F}(\boldsymbol{A}, \boldsymbol{b})$. The best upper bound known on the diameter of polytopes defined by n constraints in d dimensions is $n^{2+\log d}$, due to Kalai and Kleitman [227]. The major open problem left open by these papers is:

► **Research Problem 9.6:** Devise a randomized algorithm for linear programming that runs in expected time polynomial in n and d.

Thus, in order to resolve Research Problem 9.6 one either has to improve the Kalai-Kleitman diameter bound, or devise a non-simplex algorithm.

Problems

9.1 Prove Theorem 8.8 using backwards analysis.

9.2 By "dualizing" the randomized incremental algorithm for convex hulls in the plane (Section 9.2), derive a randomized incremental algorithm for computing the intersection of n given half-planes. Show that its expected running time is $\mathrm{O}(n \log n)$.

9.3 Use the Mulmuley games of Section 8.2.1 to derive Theorem 9.8.

9.4 The object of this problem is to show that the time bound in Theorem 9.1

holds with high probability. For a point $p \in S$, define the indicator variable $X_j(p)$ as follows:

$$X_j(p) = \begin{cases} 1 & \text{if } p\text{'s pointer is updated at the } j\text{th step;} \\ 0 & \text{otherwise} \end{cases}$$

Thus the total work done in updating p's pointer is $\sum_j X_j(p)$. By showing that $\sum_j X_j(p)$ is $O(\log n)$ with probability $1 - n^{-2}$, show that the total work is $O(n \log n)$ with high probability.

9.5 Show that the randomized incremental half-space intersection algorithm of Section 9.4 can be adapted to construct $I_p(S)$, the intersection of n spheres in three dimensions, in expected time $n \log n$.

9.6 Show that the set S_O resulting from Steps 2 and 3 in the randomized diameter algorithm (Section 9.8) can be found in time linear in the size of S, for the L_1 metric.

9.7 Let S be a set of n points in the plane. For any positive integer $k < n$, show that there is a subset S_k consisting of k points in S with the property that no triangle in $del(S_k)$ contains more than $(cn \log k)/k$ points, for a suitably chosen constant c.

9.8 In this problem, we discuss the removal of the simplifying assumptions made at the beginning of our discussion of linear programming algorithms. We focus on the non-degeneracy assumptions 3–4. Consider a set of $d+1$ constraints whose defining hyperplanes intersect at a common point p; without loss of generality, let these be defined by the first $d+1$ rows of $\boldsymbol{A}$ (together with the first $d+1$ components of $\boldsymbol{b}$). Consider adding ϵ^i to the ith component of $\boldsymbol{b}$, for $1 \le i \le d+1$, where ϵ is a small positive real. Show that for every choice of $\boldsymbol{A}$ and $\boldsymbol{b}$, there is a choice ϵ such that (i) the hyperplanes intersecting at p no longer intersect at a single point, and (ii) if p were the optimum of the linear program determined by $\boldsymbol{A}$ and $\boldsymbol{b}$, the new optimum is defined by d of the constraints that originally intersected at p.

9.9 Prove Lemma 9.16. (**Hint:** For every constraint h of weight $w_h > 1$, replace it by w_h "virtual copies" of h each of weight 1, and consider sampling this multiset.)

9.10 The Boolean n-cube is an undirected graph that has $N = 2^n$ nodes connected in the following manner. Let $(i_0, \ldots, i_{n-1})$ be the (ordered) binary representation of vertex i, i.e., $i = \sum_{j=0}^{n-1} i_j 2^j$, $i_j \in \{0, 1\}$. Then there is an edge between vertex i and vertex j if and only if $(i_0, \ldots, i_{n-1})$ and $(j_0, \ldots, j_{n-1})$ differ in exactly one position. Thus every vertex in the n-cube has degree $n = \log_2 N$. An *acyclic orientation* of the cube is an assignment of a direction to each edge, such that the resulting directed graph is acyclic. A *sink* in the digraph is a node with no edges directed out of it. Consider a random walk on an n-cube with an acyclic orientation: at each step, the walk proceeds along an outgoing edge chosen uniformly at random. Show that for every n, there is an acyclic orientation of the n-cube and a starting vertex such that expected number of steps for the walk to reach a sink is $2^{\Omega(n)}$.

This has the following significance. The n-cube can be realized as a polyhedron defined by the intersection of $2n$ half-spaces in n-dimensions. Consider the **Random Simplex** algorithm on this polyhedron. The directions on the edges are meant to model directions of improving objective function. The above lower bound suggests that if we had to give a sub-exponential upper bound on the performance of the **Random Simplex** algorithm, we would have to take into account the geometry of the polytope, using it to preclude the kind of arbitrary acyclic orientation that led to the lower bound.

9.11 In this problem, we consider the extension of the **BasisLP** algorithm to optimization problems more general than linear programming. Consider the following framework for an *abstract optimization problem*. There is a set H of n constraints, and a function $\mathcal{O}$ that maps every subset G of H to the real numbers; we think of $\mathcal{O}$ as the optimum value for G. Let $F \subseteq G \subseteq H$, and $h \in H$. For any such F, G, and h, we further require that

1. $\mathcal{O}(F) \leq \mathcal{O}(G)$, and
2. $\mathcal{O}(F) = \mathcal{O}(G)$ implies that

$$\mathcal{O}(F \cup \{h\}) > \mathcal{O}(F) \Leftrightarrow \mathcal{O}(G \cup \{h\}) > \mathcal{O}(G).$$

Defining the concept of a *basis* as for linear programming, let us call the maximum cardinality of any basis as the *combinatorial dimension* of the instance.

Modify the **BasisLP** algorithm so that it works for such abstract optimization problems, and show that the analysis of **BasisLP** may be applied with d replaced by the combinatorial dimension.

9.12 Consider the *smallest enclosing ball* problem: given n points in d-dimensional space, find the radius of the smallest ball that contains all n points. By showing that this fits the paradigm of an *abstract optimization problem*, show that a suitably modified version of the **BasisLP** algorithm can be used to solve it.

CHAPTER 10

Graph Algorithms

In this chapter we consider several fundamental optimization problems involving graphs: all-pairs shortest paths, minimum cuts, and minimum spanning trees. In each case, deterministic polynomial time algorithms are known, but the use of randomization allows us to obtain significantly faster solutions for these problems. We show that the problem of computing all-pairs shortest paths is reducible, via a randomized reduction, to the problem of multiplying two integer matrices. We present a fast randomized algorithm for the min-cut problem in undirected graphs, thereby providing evidence that this problem may be easier than the max-flow problem. Finally, we present a linear-time randomized algorithm for the problem of finding minimum spanning trees.

Unless stated otherwise, all the graphs we consider are assumed to be undirected and without multiple edges or self-loops. For shortest paths and min-cuts we restrict our attention to unweighted graphs, although in some cases the results generalize to weighted graphs; we give references in the discussion at the end of the chapter.

10.1. All-pairs Shortest Paths

Let $G(V, E)$ be an undirected, connected graph with $V = \{1, \ldots, n\}$ and $|E| = m$. The adjacency matrix $\boldsymbol{A}$ is an $n \times n$ 0-1 matrix with $A_{ij} = A_{ji} = 1$ if and only if the edge (i, j) is present in E. Given $\boldsymbol{A}$, we define the *distance matrix* $\boldsymbol{D}$ as an $n \times n$ matrix with non-negative integer entries such that D_{ij} equals the length of a shortest path from vertex i to vertex j. The diagonal entries in both $\boldsymbol{A}$ and $\boldsymbol{D}$ are zeroes. Since G is connected, all entries in $\boldsymbol{D}$ are finite; this is not a restrictive assumption since a graph can be decomposed easily into connected components in linear time.

The *all-pairs shortest paths* (APSP) problem is to compute a representation of the shortest paths between all pairs of vertices, i.e., the paths that determine the entries in the distance matrix. To make this precise, we will compute an implicit representation of the shortest paths such that for any specific pair of vertices, the shortest path between them can be determined in time proportional to its length. A restricted version of this problem requires us to compute only the distance matrix; we refer to this as the *all-pairs distances* (APD) problem.

The APSP problem can be solved in $O(nm)$ time, as follows: from each vertex $i \in V$, compute the breadth-first search tree T_i rooted at i. Each such tree can be computed in $O(m)$ time, and, in any tree T_i, the (unique) path from i to any vertex j is the shortest path between them. Given the collection of breadth-first search trees, the distance matrix can be computed in $O(n^2)$ time by assigning level numbers to the vertices in each tree.

We consider only unweighted graphs, although the above definitions have obvious generalizations to the case where the edges have real-valued weights (or lengths). The classical algorithms of Dijkstra, Floyd-Warshall, and Johnson solve APSP in $O(n^3)$ time; the first and the last of these can actually be implemented in $O(nm + n^2 \log n)$ time.

While it is clear that the APSP or APD problem would require $\Omega(n^2)$ time in the worst case, there is no reason to believe that the $O(nm)$ time bound (which can be as much as $\Theta(n^3)$) is even close to the best possible. We now show that a substantial improvement can be obtained for the unweighted case with the use of randomization and fast matrix multiplication. While these results do not generalize completely to the weighted case, there is some indication that this should be possible.

What does matrix multiplication have to do with the shortest path problem? Consider first the problem of Boolean matrix multiplication: given $n \times n$ Boolean matrices $\boldsymbol{A}$ and $\boldsymbol{B}$, their product $\boldsymbol{C}$ has entries

$$C_{ij} = \sum_{k=1}^{n} A_{ik} B_{kj}$$

where the product of two Boolean values denotes the Boolean AND operation, and the sum denotes the Boolean OR operation. Suppose that $\boldsymbol{A} = \boldsymbol{B}$ is the adjacency matrix of the graph G. Then the product $\boldsymbol{C} = \boldsymbol{A}^2$ has its (i,j) entry equal to 1 if and only if there is a path of length 2 between the vertices i and j; the matrix $\boldsymbol{A}^\ell$ corresponds to paths of length ℓ. A related concept is that of the *closure* of a Boolean matrix $\boldsymbol{A}$, which is defined as the infinite sum $\boldsymbol{A}^* = \sum_{l=0}^{\infty} \boldsymbol{A}^l$, where $\boldsymbol{A}^0$ is the identity matrix. The closure matrix $\boldsymbol{A}^*$ has its (i,j) entry equal to 1 if and only if there is some path between the vertices i and j.

Computing all powers of $\boldsymbol{A}$ from 1 to n will thus enable us to solve the APD problem. Unfortunately, this takes time $O(n^4)$ using the obvious Boolean matrix multiplication algorithm, which runs in time $O(n^3)$. On the other hand, computing the closure $\boldsymbol{A}^*$ of the Boolean matrix $\boldsymbol{A}$ requires only as much time as a single Boolean matrix multiplication (see Problem 10.1).

Actually, it is possible to embed Boolean matrix multiplication into integer matrix multiplication by treating the Boolean entries as the integers 0 and 1. This corresponds to embedding the closed semiring of Boolean algebra into the ring of integers. Let MM(n) denote the time required to multiply two

$n \times n$ matrices with integer entries. All known integer matrix multiplication algorithms are applicable to an arbitrary ring, rather than the ring of integers alone.

Exercise 10.1: Show that Boolean matrix multiplication for $n \times n$ matrices can be performed via integer matrix multiplication in time $O(\text{MM}(n))$. How large are the integer values that arise during this computation?

Currently, the best integer matrix multiplication algorithm runs in time $O(n^{2.376})$. By the preceding exercise this result carries over to Boolean matrix multiplication. Unfortunately, even the use of this observation gives a super-cubic algorithm for the APD problem in unweighted graphs. There is, however, another trick that permits the solution of the APD problem in time $O(\text{MM}(n))$. The idea is to reduce the problem of computing the distance matrix for a graph to a matrix multiplication over the closed semiring of the reals augmented with ∞, where scalar addition is replaced by the "min" operator, and scalar multiplication is replaced by scalar addition. Let A now be the matrix in which the (i, j) entry is the weight of the edge (i, j) if it exists, and ∞ otherwise. The semiring product of matrices A and B has entries

$$C_{ij} = \min_{1 \le k \le n} (A_{ik} + B_{kj}).$$

It can be verified that the closure matrix A^* is exactly the solution to the APD problem. Some non-trivial ideas are needed to show that the semiring closure can be computed via integer matrix multiplication; we omit the details. This technique applies to weighted graphs too.

There are two serious deficiencies in the solution described in the previous paragraph – the algorithm does not generalize from the APD problem to the APSP problem and, more importantly, the reduction to integer matrix multiplication creates integer matrices whose entries are integers whose *length* is super-linear in n. In any real machine, this implies that each arithmetic operation takes super-linear time, and the usual unit-cost assumption for basic arithmetic operations is invalid. We present a different approach for reducing the APD problem to integer matrix multiplication using integers of only *logarithmic* length. Then, we show that this can be extended, via randomization, to actually solve the APSP problem using a black-box for matrix multiplication. The algorithm is practical to the extent that the fast matrix multiplication algorithm being invoked is practical.

10.1.1. Computing Distances

Our first goal is to present a (deterministic) algorithm to solve the APD problem using a black-box for integer matrix multiplication. In the ensuing discussion, all matrix multiplications are over the ring of integers and the adjacency matrix is treated as an integer matrix.

Let $G'(V, E')$ be the graph obtained from $G(V, E)$ by placing an edge between every pair of vertices $i \neq j \in V$ that are at distance 1 or 2 in G. The graph G is a subgraph of G', and we could view G' as the "square" of the graph G. For G', let $\boldsymbol{A}'$ denote the adjacency matrix and $\boldsymbol{D}'$ denote the distance matrix. The proof of the following lemma is left as an exercise.

Lemma 10.1: *Let $\boldsymbol{Z} = \boldsymbol{A}^2$, where $\boldsymbol{A}$ is the adjacency matrix of the graph G. Then there is a path of length 2 in G between a pair of vertices i and j if and only if $Z_{ij} > 0$. Further, the value of Z_{ij} is the number of distinct length 2 paths between i and j.*

The matrix $\boldsymbol{Z} = \boldsymbol{A}^2$ can be computed in $O(\text{MM}(n))$ time, and if we know $\boldsymbol{A}$ and $\boldsymbol{Z}$ it is easy to determine the matrix $\boldsymbol{A}'$ in $O(n^2)$ time. The diagonal entries in $\boldsymbol{Z} = \boldsymbol{A}^2$ will be non-zero in general (corresponding to cycles of length 2), and care must be taken in constructing $\boldsymbol{A}'$ to ensure that it has a zero diagonal. In particular, we compute $\boldsymbol{A}'$ by setting $A'_{ij} = 1$ if and only if $i \neq j$ and at least one of A_{ij} and Z_{ij} is non-zero.

Observe that G' is complete if (and only if) G has diameter at most 2, where the diameter of a graph is the maximum shortest path length over all pairs of vertices. In this case, the APD matrix $\boldsymbol{D} = 2\boldsymbol{A}' - \boldsymbol{A}$ is easily obtained from $\boldsymbol{A}$ and $\boldsymbol{A}'$ in time $O(n^2)$.

In general, of course, the graph G could have arbitrarily large diameter. The following sequence of observations will allow us to handle the general case. The proof of the next lemma is left as an exercise.

Lemma 10.2: *Consider any pair of vertices $i, j \in V$.*

- *If D_{ij} is even then $D_{ij} = 2D'_{ij}$.*
- *If D_{ij} is odd then $D_{ij} = 2D'_{ij} - 1$.*

An immediate implication of this lemma is that given the APD matrix $\boldsymbol{D}'$ for G', the APD matrix $\boldsymbol{D}$ for G can be computed quickly provided we know the parity of each of the shortest path lengths in $\boldsymbol{D}$. This suggests a recursive algorithm for APD that first computes $\boldsymbol{A}'$ and G', uses recursion to determine $\boldsymbol{D}'$, and then computes $\boldsymbol{D}$ from $\boldsymbol{D}'$ using the observation in Lemma 10.2. The only remaining detail is the method for computing the parities of the shortest path lengths. The proof of the next lemma is an easy exercise.

Lemma 10.3: *Consider any pair of distinct vertices i and j in G.*

- *For any neighbor k of i, $D_{ij} - 1 \leq D_{kj} \leq D_{ij} + 1$.*
- *There exists a neighbor k of i such that $D_{kj} = D_{ij} - 1$.*

We now present a structural property of shortest paths that allows us to compute the parities of their lengths.

Lemma 10.4: *Consider any pair of distinct vertices i and j in G.*

- *If D_{ij} is even, then $D'_{kj} \geq D'_{ij}$ for every neighbor k of i in G.*
- *If D_{ij} is odd, then $D'_{kj} \leq D'_{ij}$ for every neighbor k of i in G. Moreover, there exists a neighbor k of i in G such that $D'_{kj} < D'_{ij}$.*

PROOF: Consider first the case where $D_{ij} = 2\ell$ is even. By Lemma 10.3, for any neighbor k of i we have $D_{kj} \geq 2\ell - 1$. Lemma 10.2 implies that $D'_{ij} = \ell$. Also by Lemma 10.2 we have $D'_{kj} \geq D_{kj}/2 \geq \ell - 1/2$, and since distances are integral we conclude that $D'_{kj} \geq \ell = D'_{ij}$.

A similar argument applies in the case where $D_{ij} = 2\ell - 1$ is odd. By Lemma 10.3 we have $D_{kj} \leq 2\ell$ for any neighbor k of i, and therefore, by Lemma 10.2, $D'_{kj} \leq (D_{kj} + 1)/2 \leq \ell + 1/2$. By integrality it follows that $D'_{kj} \leq \ell$, and by Lemma 10.2 we have $D'_{ij} = \ell$, implying the desired result that $D'_{kj} \leq D'_{ij}$. Further, there exists a neighbor k of i such that $D_{kj} = D_{ij} - 1 = 2\ell - 2$, and therefore Lemma 10.2 yields $D'_{kj} = \ell - 1 < \ell = D'_{ij}$. □

Let $\Gamma(i)$ denote the set of neighbors of i in G, and let $d(i)$ be the degree of i. Note that $Z_{ii} = d(i)$, for all i. Summing the inequalities in Lemma 10.4 over the neighbors of the vertex i, and noting that the two resulting inequalities are mutually exclusive, we obtain the following result.

Lemma 10.5: *Consider any pair of distinct vertices i and j in G.*

- *D_{ij} is even if and only if $\sum_{k \in \Gamma(i)} D'_{kj} \geq D'_{ij} d(i)$.*
- *D_{ij} is odd if and only if $\sum_{k \in \Gamma(i)} D'_{kj} < D'_{ij} d(i)$.*

This gives us an efficient method for determining the parities of the shortest path lengths in G. The resulting recursive algorithm is summarized in Algorithm **APD**.

In Step 5 we are using matrix multiplication to compute

$$\sum_{k \in \Gamma(i)} D'_{kj} = \sum_{k=1}^{n} A_{ik} D'_{kj} = S_{ij}.$$

The correctness of the algorithm follows from the preceding discussion. We summarize the running time analysis in the following theorem. The length of the integers in the matrices will never exceed $O(\log n)$.

Algorithm APD:

Input: Graph $G(V, E)$ in form of an adjacency matrix $\boldsymbol{A}$.

Output: The APD matrix $\boldsymbol{D}$ for G.

1. $\boldsymbol{Z} \leftarrow \boldsymbol{A}^2$.
2. **compute** matrix $\boldsymbol{A}'$ such that $A'_{ij} = 1$ if and only if
$$i \neq j \text{ and } (A_{ij} = 1 \text{ or } Z_{ij} > 0).$$
3. **if** $A'_{ij} = 1$ for all $i \neq j$ **then return** $\boldsymbol{D} = 2\boldsymbol{A}' - \boldsymbol{A}$.
4. Recursively compute the APD matrix $\boldsymbol{D}'$ for the graph G' with adjacency matrix $\boldsymbol{A}'$.
5. $\boldsymbol{S} \leftarrow \boldsymbol{A}\boldsymbol{D}'$.
6. **return** matrix $\boldsymbol{D}$ with $D_{ij} = \begin{cases} 2D'_{ij} & \text{if } S_{ij} \geq D'_{ij} Z_{ii} \\ 2D'_{ij} - 1 & \text{if } S_{ij} < D'_{ij} Z_{ii} \end{cases}$

Theorem 10.6: *The* **APD** *algorithm computes the distance matrix for an n-vertex graph G in time* $O(\text{MM}(n) \log n)$ *using integer matrix multiplication on matrices with entries of value bounded by* $O(n^2)$.

PROOF: Suppose that the graph G has diameter δ. Then the graph G' has diameter $\lceil \delta/2 \rceil$. Let $T(n, \delta)$ denote the running time of the **APD** algorithm on input graphs with n vertices and diameter δ. In the case $\delta = 1$, G is a complete graph, and in the case $\delta = 2$ we have that $T(n, \delta) = \text{MM}(n) + O(n^2)$.

Exercise 10.2: Verify that $T(n, \delta)$ satisfies the following recurrence for $\delta > 2$,

$$T(n, \delta) = 2\text{MM}(n) + T(n, \lceil \delta/2 \rceil) + O(n^2).$$

Noting that $\delta < n$ and $\text{MM}(n) = \Omega(n^2)$, and that the recursion depth is $O(\log n)$, the desired result follows immediately. Finally, since the integers in the distance matrices are bounded by n, it follows that the integers in the $\boldsymbol{S}$ matrices are bounded by n^2. □

10.1.2. Witnessing Boolean Matrix Multiplication

We now extend the above technique to solving the APSP problem; this is where randomization proves useful. The extension is based on solving the problem of finding "witnesses" for Boolean matrix multiplication. Suppose $\boldsymbol{A}$ and $\boldsymbol{B}$ are $n \times n$ Boolean (or, 0-1) matrices and $\boldsymbol{P} = \boldsymbol{AB}$ is their product under *Boolean* matrix multiplication. A witness for P_{ij} is an index $k \in \{1, \ldots, n\}$ such that

$A_{ik} = B_{kj} = 1$. Observe that $P_{ij} = 1$ if and only if it has some witness k. A *Boolean product witness matrix* (BPWM) for $\boldsymbol{P}$ is an integer matrix $\boldsymbol{W}$ such that each entry W_{ij} contains a witness k for P_{ij} if any, and is 0 if there is no such witness. The matrix $\boldsymbol{W}$ has entries drawn from the set $\{0, 1, \ldots, n\}$. The BPWM problem is to find a witness matrix $\boldsymbol{W}$, given the matrices $\boldsymbol{A}$ and $\boldsymbol{B}$ (and, if necessary, also the matrix $\boldsymbol{P}$).

There could be as many as n witnesses for each entry in $\boldsymbol{P}$. In fact, the *integer* matrix multiplication of $\boldsymbol{A}$ and $\boldsymbol{B}$, treating their entries as the integers 0 and 1, yields a matrix $\boldsymbol{C}$ whose entry C_{ij} corresponds exactly to the number of witnesses for the Boolean matrix entry P_{ij}.

Recall that if $\boldsymbol{A} = \boldsymbol{B}$ is the adjacency matrix of a graph G, then $P_{ij} = 1$ if and only if there exists a path of length 2 from i to j, and C_{ij} is the number of such paths. A witness k for P_{ij} is the intermediate vertex on a length-2 path from i to j. It thus appears that finding witnesses for Boolean matrix multiplication is closely related to the issue of extending the **APD** algorithm to finding the shortest paths. The problem is that the obvious brute-force approach of trying each $k \in \{1, \ldots, n\}$ as a potential witness for P_{ij} requires $\Omega(n)$ time and gives only an $O(n^3)$ time algorithm for the BPWM problem.

Consider first the issue of finding a witness matrix when there is a *unique* witness for each entry in $\boldsymbol{P}$. There is a simple reduction of the BPWM problem to integer matrix multiplication in this case, as suggested in the following exercise. In the rest of this section, except in the computation of $\boldsymbol{P}$, all matrix products involve integer matrix multiplication.

Exercise 10.3: Consider the matrix $\widehat{A}$ obtained by setting $\widehat{A}_{ik} = kA_{ik}$. Show that the integer matrix multiplication of $\widehat{A}$ and $\boldsymbol{B}$ yields a matrix that contains the witness for all entries in the matrix $\boldsymbol{P}$ that have a unique witness. In particular, if each entry of $\boldsymbol{P}$ has a unique witness, then $\boldsymbol{W} = \widehat{A}\boldsymbol{B}$ is a solution to the BPWM problem.

Of course, there is no a priori guarantee that there is a unique witness for any particular entry in $\boldsymbol{P}$. However, we can use randomization to achieve the effect of such a guarantee for a sufficiently large number of entries in $\boldsymbol{P}$. This approach bears some resemblance to the use of the *isolating lemma* used in devising a parallel algorithm for maximum matching, described in Section 12.4.

Let us focus our attention on a specific entry P_{ij}. Assume that the number of witnesses for this entry has been determined to be w. We may find the number of witnesses w by using integer matrix multiplication to compute $\boldsymbol{C} = \boldsymbol{AB}$, and then looking at the entry C_{ij}. We assume that $w \geq 2$, since it is easy to find the witness (if any), otherwise. Let r be an integer such that $n/2 \leq wr \leq n$. We claim that a random set of indices $R \subseteq \{1, \ldots, n\}$ of cardinality r is very likely to contain a unique witness for P_{ij}. To verify this claim, consider an urn containing n balls, one for each of the n indices; the balls corresponding to witnesses are colored white, and the rest are colored black. The following lemma then shows that the probability that R contains a unique witness is reasonably large.

Lemma 10.7: *Suppose an urn contains n balls of which w are white, and $n - w$ are black. Consider choosing r balls at random (without replacement), where $n/2 \leq wr \leq n$. Then*

$$\mathbf{Pr}[\textit{exactly one white ball is chosen}] \geq \frac{1}{2e}.$$

PROOF: By elementary counting, the desired probability can be bounded as follows.

$$\begin{aligned}
\frac{\binom{w}{1}\binom{n-w}{r-1}}{\binom{n}{r}} &= w\frac{r!}{(r-1)!}\frac{(n-w)!}{n!}\frac{(n-r)!}{(n-w-r+1)!} \\
&= wr\left(\prod_{i=0}^{w-1}\frac{1}{n-i}\right)\left(\prod_{j=0}^{w-2}(n-r-j)\right) \\
&= \frac{wr}{n}\left(\prod_{j=0}^{w-2}\frac{n-r-j}{n-1-j}\right) \\
&\geq \frac{wr}{n}\left(\prod_{j=0}^{w-2}\frac{n-r-j-(w-j-1)}{n-1-j-(w-j-1)}\right) \\
&= \frac{wr}{n}\left(\prod_{j=0}^{w-2}\frac{n-w-(r-1)}{n-w}\right) \\
&= \frac{wr}{n}\left(1-\frac{r-1}{n-w}\right)^{w-1} \\
&\geq \frac{1}{2}\left(1-\frac{1}{w}\right)^{w-1}
\end{aligned}$$

The last inequality follows from the observations that $wr/n \geq 1/2$ and $(r-1)/(n-w) \leq 1/w$, which in turn follow from the assumption that $n/2 \leq wr \leq n$. Finally, applying Proposition B.3, the last expression is bounded by $1/2e$. □

Assuming that the set R contains a unique witness for P_{ij}, it is easy to modify the technique described in Exercise 10.3 to identify this witness. Suppose that R is represented as an incidence vector that has $R_k = 1$ for $k \in R$ and $R_k = 0$ for $k \notin R$. Let $\boldsymbol{A}^R$ be the matrix obtained from $\boldsymbol{A}$ by setting $A_{ik}^R = kR_kA_{ik}$; further, let $\boldsymbol{B}^R$ be the matrix obtained from $\boldsymbol{B}$ by setting $B_{kj}^R = R_kB_{kj}$. The only difference between $\boldsymbol{A}^R$ and $\boldsymbol{B}^R$ and the two matrices used in Exercise 10.3 is that each column of $\boldsymbol{A}^R$ and each row of $\boldsymbol{B}^R$ corresponding to the indices not chosen in R is turned into an all-zero vector. The reason behind this construction is explicated in the next exercise.

Exercise 10.4: Suppose that the entry P_{ij} has a unique witness in the set R. Show that the corresponding entry in the integer matrix multiplication of $\boldsymbol{A}^R$ and $\boldsymbol{B}^R$ is the index of this unique witness.

A key point is that the product of A^R and B^R yields witnesses for all entries in P that have a unique witness in R. By Lemma 10.7, there is a constant probability that a random set R of size r has a unique witness for an entry in P with w witnesses, where $n/2r \leq w \leq n/r$. Repeating this for $O(\log n)$ independent choices of R makes it extremely unlikely that witnesses are not identified for such entries in P, and these missing witnesses can be found by brute-force enumeration. Of course, we will need to use several different values of r to take care of the range of values possible for w, but it suffices to try only those values of r that are powers of 2 between 1 and n. The resulting algorithm is presented below.

Algorithm BPWM:

Input: Two $n \times n$ 0-1 matrices A and B.

Output: Witness matrix W for the Boolean matrix $P = AB$.

1. $W \leftarrow -AB$.
2. **for** $t = 0, \ldots, \lfloor \log n \rfloor$ **do**

    2.1. $r \leftarrow 2^t$.

    2.2. **repeat** $\lceil 3.77 \log n \rceil$ **times**

        2.2.1. **choose** random $R \subseteq \{1, \ldots, n\}$ with $|R| = r$.

        2.2.2. **compute** A^R and B^R.

        2.2.3. $Z \leftarrow A^R B^R$.

        2.2.4. **for all** (i, j) **do**
        if $W_{ij} < 0$ and Z_{ij} is witness **then** $W_{ij} \leftarrow Z_{ij}$.

3. **for all** (i, j) **do**
    if $W_{ij} < 0$ **then** find witness W_{ij} by brute force.

The initial setting of W ensures that the only negative entries are those where the value of P_{ij} is non-zero and there is a need to find a witness. Thereafter, the negative entries mark the locations in P for which witnesses have not yet been found. The brute-force search in the last step for the witnesses not identified by the randomized strategy ensures that the algorithm is Las Vegas. We now turn to the task of analyzing the expected running time.

Theorem 10.8: *The* **BPWM** *algorithm is a Las Vegas algorithm for the BPWM problem with expected running time* $O\left(\text{MM}(n) \log^2 n\right)$.

PROOF: Step 1 takes time $\text{MM}(n)$. There are $O\left(\log^2 n\right)$ iterations of the innermost loop body in Step 2, and the most expensive operation performed there is an integer matrix multiplication of matrices of dimension at most $n \times n$. This would yield the desired time bound, provided that the brute-force computations in Step 3 are not too expensive. We claim that for any non-zero P_{ij}, a witness is found in Step 2 with probability at least $1-1/n$. This implies that the expected number of witnesses remaining to be found at the start of Step 3 is n, and since each of these is then found by brute force in $O(n)$ time, it follows that the expected cost of Step 3 is $O(n^2)$.

To verify the claim, consider any specific non-zero P_{ij} and assume that it has w witnesses. There will be at least one iteration of the outer loop with a value r such that $n/2 \leq wr \leq n$. During that iteration, the probability that a random choice of R does not have a unique witness for P_{ij} is at most $1 - 1/2e$, by Lemma 10.7. Since the inner loop is repeated $3.77 \log n$ times, it follows that the probability that no witness is found for this entry before the end of Step 2 is at most $(1 - 1/2e)^{3.77 \log n} \leq 1/n$. □

10.1.3. Determining Shortest Paths

Finally, we show how the Algorithms **APD** and **BPWM** can be used to solve the APSP problem. The first problem we face is that there exist graphs with many pairs of vertices for which the shortest path length is linear in n, and so any explicit representation of all-pairs shortest paths will require $\Omega(n^3)$ time to compute.

Exercise 10.5: Construct an n-vertex graph with $\Omega(n^2)$ pairs of vertices at distance $\Omega(n)$.

To circumvent this problem, we will compute an implicit representation of the shortest paths such that for any specific pair of vertices their shortest path can be extracted in time proportional to its length.

► **Definition 10.1:** A *successor* matrix S for an n-vertex graph G is an $n \times n$ matrix such that S_{ij} is the index k of a neighbor of vertex i that lies on a shortest path from i to j.

Exercise 10.6: Given a successor matrix S and a pair of vertices i, j, explain how you would obtain an explicit representation of the shortest path from i to j in time proportional to the length of the path.

Suppose we are provided with the adjacency matrix A and the distance matrix D for a graph G. Consider a pair of vertices i and j that are at distance d from each other. The entry S_{ij} can be k if and only if $D_{kj} = d - 1$ and $D_{ik} = 1$ (or $A_{ik} = 1$). Let B^d denote the $n \times n$ 0-1 matrix in which $B^d_{kj} = 1$ if and only if

$D_{kj} = d - 1$. Observe that $\boldsymbol{B}^d$ can be computed from $\boldsymbol{D}$ in $O(n^2)$ time. As the following exercise indicates, finding the successor entry for any pair i and j at distance d is easy given the matrix $\boldsymbol{B}^d$.

Exercise 10.7: Applying the **BPWM** algorithm to compute the witness matrix for the Boolean matrices $\boldsymbol{A}$ and $\boldsymbol{B}^d$, show that the successor matrix entries for all pairs of vertices at distance d can be simultaneously determined in expected time $O(\text{MM}(n) \log^2 n)$.

The only problem with this approach is that the entire process must be repeated for the n different values of d, leading to a super-cubic algorithm for APSP. However, a simple observation leads to a reduction of the number of witness matrix computations from n down to 3.

Recall from Lemma 10.3 that for any pair of vertices i and j, and any neighbor k of i, it must be the case that $D_{ij} - 1 \leq D_{kj} \leq D_{ij} + 1$. Furthermore, any neighbor k with $D_{kj} = D_{ij} - 1$ is a valid candidate for the successor matrix entry S_{ij}. It follows that any k such that $A_{ik} = 1$ and $D_{kj} \equiv D_{ij} - 1 \pmod 3$ is a valid candidate for S_{ij}.

For $s \in \{0, 1, 2\}$, define the $n \times n$ 0-1 matrix $D^{(s)}$ to be such that $D^{(s)}_{kj} = 1$ if and only if $D_{kj} + 1 \equiv s \pmod 3$. The successor matrix can be computed by finding the witnesses of the Boolean matrix multiplication of $\boldsymbol{A}$ with each of $D^{(0)}$, $D^{(1)}$, and $D^{(2)}$, as described in Algorithm **APSP**.

Algorithm APSP:

Input: An $n \times n$ adjacency matrix $\boldsymbol{A}$ for a graph G.

Output: The successor matrix S for G.

1. **compute** the distance matrix $\boldsymbol{D} = \mathbf{APD}(\boldsymbol{A})$.
2. **for** $s = \{0, 1, 2\}$ **do**
   - 2.1. **compute** 0-1 matrix $\boldsymbol{D}^{(s)}$ such that $D^{(s)}_{kj} = 1$ if and only if $D_{kj} + 1 \equiv s \pmod 3$.
   - 2.2. **compute** the witness matrix $\boldsymbol{W}^{(s)} = \mathbf{BPWM}(\boldsymbol{A}, \boldsymbol{D}^{(s)})$.
3. **compute** successor matrix S such that $S_{ij} = W^{(D_{ij} \bmod 3)}_{ij}$.

Given the performance bounds on the algorithms **APD** and **BPWM**, the following theorem is easily verified.

Theorem 10.9: *Algorithm* **APSP** *computes the successor matrix for an n-vertex graph G in expected time* $O(\text{MM}(n) \log^2 n)$.

10.2. The Min-Cut Problem

We now return to the *min-cut* problem considered in Section 1.1. Let $G(V,E)$ be an undirected multigraph with n vertices and m edges. A multigraph is permitted more than one edge between any given pair of vertices. A *cut* in G is a partition of the vertices $V = (C, \overline{C})$ into two non-empty sets; we refer to this as the cut C with the understanding that $\overline{C}$ is $V \setminus C$.

The *value* or size of a cut C is the number of edges crossing the cut, i.e., edges with one end-point in each of the two sets C and $\overline{C}$. A multiple edge will contribute its multiplicity to the value of the cut. A min-cut is a cut of minimum value; the min-cut problem is that of finding a min-cut in an input graph G. The value of a min-cut is sometimes referred to as the edge connectivity of the graph, as it is the minimum number of edges that must be removed from the graph to render it disconnected.

We assume that the input graph G is connected, since otherwise the problem is trivially solved by determining the connected components of G in time $O(m)$. The above definitions generalize to weighted graphs, where the value of a cut is defined to be the sum of the weights of the edges crossing the cut. We restrict ourselves to non-negative edge weights. Permitting negative edge weights would make the problem ***NP***-complete since it would then include as a special case the max-cut problem, a classical ***NP***-complete problem.

The min-cut problem should be contrasted with the *s-t min-cut* problem. In the latter, two distinguished vertices s and t are specified in the input, and the solutions are restricted to the cuts C with the property that $s \in C$ and $t \notin C$.

Exercise 10.8: Show that the min-cut problem for a graph G can be solved via a polynomial number of invocations of an s-t min-cut algorithm applied to the same graph.

The classical duality result in network flows states that the value of a maximum s-t flow in a network equals the value of a s-t min-cut. In fact, computing a maximum s-t flow yields an s-t min-cut as a side-effect. It follows that the min-cut problem can be solved via a polynomial number of invocations of a maximum flow algorithm. Actually, it can be shown that $n-1$ flow computations suffice for this purpose. Since the best deterministic maximum flow algorithm runs in time $O(mn \log(n^2/m))$, this approach to the min-cut problem would require $\Omega(mn^2)$ time. Fortunately, the $n-1$ maximum flow computations needed for the min-cut problem can be implemented in time proportional to the cost of a single maximum flow computation, and so we can compute a min-cut in time $O(mn \log(n^2/m))$.

A very interesting question is whether the s-t min-cut problem can be solved faster than the s-t max-flow problem. Note that whereas a flow computation immediately yields the cut, the converse does not seem to be true. In this section we show that at least for the min-cut problem (without the s-t requirement), there is an efficient randomized algorithm running in $O\left(n^2 \log^{O(1)} n\right)$ time. For dense graphs this is significantly better than the running time of the best-known max-flow algorithm.

10.2.1. The Contraction Algorithm Revisited

We start by reviewing the the contraction algorithm described in Section 1.1. Actually, we present only an abstract version of this algorithm and leave the implementation details as an exercise.

Given an edge (x, y) in a multigraph $G(V, E)$, a *contraction* of the edge (x, y) corresponds to replacing the vertices x and y by a new vertex z, and for each $v \notin \{x, y\}$ replacing any edge (x, v) or (y, v) by the edge (z, v); the rest of the graph remains unchanged. Any multiple edges created are to be retained. The graph obtained by this contraction is denoted by $G/(x, y)$.

Given a collection of edges $F \subseteq E$, the effect of contracting the edges in F is independent of the order of contraction, and the resulting graph is denoted by G/F. The vertex set and edge set of a graph G/F are denoted by V/F and E/F. The "meta-vertices" in V/F correspond to a (connected) set of vertices in V, and the edges in E/F are exactly those edges in E whose end-points do not get collapsed into the same meta-vertex in V/F. In Problem 10.9, the reader is asked to show that it is possible to maintain the graph G/F under an online sequence of edge contractions at a cost of $O(n)$ time per contraction, keeping track of the correspondence between the elements of V/F and V, and E/F and E.

The basic idea behind the contraction algorithm is summarized below. We assume that the Algorithm **Contract** uses the data structure developed in Problem 10.9 to implement the edge contractions.

Algorithm Contract:

Input: A multigraph $G(V, E)$.

Output: A cut C.

1. $H \leftarrow G$.

2. **while** H has more than 2 vertices **do**

  2.1. **choose** an edge (x, y) uniformly at random from the edges in H.

  2.2. $F \leftarrow F \cup \{(x, y)\}$.

  2.3. $H \leftarrow H/(x, y)$.

3. $(C, \overline{C}) \leftarrow$ the sets of vertices corresponding to the two meta-vertices in $H = G/F$.

The only implementation issue remaining in this algorithm is the selection of the edge (x, y) uniformly at random from the set of all edges in the graph H. In Problem 10.10, the reader is asked to show that this can be done in $O(n)$ time per random selection. The results from Problems 10.9–10.11 yield the following theorem.

Theorem 10.10: *Algorithm* **Contract** *can be implemented to run in* $O(n^2)$ *time on any n-vertex multigraph G.*

The running time of this algorithm is independent of the number of (multi) edges in the graph G. This may seem surprising at first since the number of such edges is not bounded by $\binom{n}{2}$. However, as suggested in Problem 10.9, the multiplicity of an edge can be represented by an integer weight on the edge and hence the number of edges can effectively be bounded by $\binom{n}{2}$.

Of course, this just shows that the **Contract** algorithm terminates in $O(n^2)$ time with a cut C. There is no guarantee that the cut will indeed be a min-cut. We now briefly review the argument from Section 1.1 that established that this algorithm finds a min-cut with a non-negligible probability.

Lemma 10.11: *A cut C is produced as output by Algorithm* **Contract** *if and only if none of the edges crossing this cut is contracted by the algorithm.*

Fix any one min-cut K in the graph G. Let k denote the value of a min-cut in G; in particular, k is the value of the cut K. We would like to compute the probability that K is produced as the output of Algorithm **Contract**. By Lemma 10.11, this will happen if and only if none of the k edges crossing the cut is contracted during the course of the algorithm's execution. To determine the probability of this event, we make use of the following obvious facts.

Lemma 10.12: *In an n-vertex multigraph G with min-cut value k, no vertex has degree smaller than k. Further, the total number of edges in the graph satisfies* $m \geq nk/2$.

Lemma 10.13: *Given an edge* (x, y) *in a graph G, the min-cut value in* $G/(x, y)$ *is at least as large as the min-cut value in G.*

The number of vertices in the graph H decreases by exactly one during each iteration of Algorithm **Contract**. After $n-2$ iterations the number of vertices is reduced from n to 2. At the ith iteration, there are $n_i = n - i + 1$ vertices in H. Suppose that none of the edges in K is contracted during the first $i-1$ iterations. Since K is also a cut in H, Lemma 10.13 implies that H has min-cut value k, and then Lemma 10.12 implies that the number of edges in H is at least $n_i k/2$. Thus, the probability that any edge of K is contracted during this iteration is at most $2/n_i$. It follows that the probability that no edge of K is ever contracted can be bounded as follows.

$$\begin{aligned}
\mathbf{Pr}[K \text{ is output by Algorithm } \mathbf{Contract}] &\geq \prod_{i=1}^{n-2}\left(1-\frac{2}{n_i}\right)\\
&= \prod_{i=1}^{n-2}\left(1-\frac{2}{n-i+1}\right)\\
&= \prod_{j=n}^{3}\left(\frac{j-2}{j}\right)\\
&= 1\Big/\binom{n}{2} = \Omega(n^{-2}).
\end{aligned}$$

We have established the following theorem.

Theorem 10.14: *Any specific min-cut K is output by Algorithm* **Contract** *with probability $\Omega(n^{-2})$.*

Since the graph must have at least one min-cut, it follows that the probability of success of this algorithm is $\Omega(n^{-2})$. Repeating the algorithm $O(n^2 \log n)$ times gives a reasonable probability that some invocation of the algorithm produces a min-cut; then, the smallest cut produced by these invocations is very likely to be the min-cut. This gives a Monte Carlo algorithm running in $O(n^4 \log n)$ time. Before trying to improve this result, we note the following variant of Theorem 10.14.

Lemma 10.15: *Suppose that the Algorithm* **Contract** *is terminated when the number of vertices remaining in the contracted graph is exactly t. Then any specific min-cut K survives in the resulting contracted graph with probability at least*

$$\binom{t}{2}\Big/\binom{n}{2} = \Omega\left(\left(\frac{t}{n}\right)^2\right).$$

10.2.2. A Faster Min-Cut Algorithm

We now modify the implementation of the contraction algorithm to reduce its running time to $O\left(n^2 \log^{O(1)} n\right)$. The basic problem with Algorithm **Contract** is that it succeeds in finding a min-cut only with probability $\Omega(n^{-2})$. This entails running the algorithm at least $\Omega(n^2)$ times to ensure a reasonable probability of success. Thus, the obvious approach to improving the running time is to increase the probability that a min-cut is produced by Algorithm **Contract**.

Suppose we focus our attention on a specific min-cut K and wish to have the algorithm produce this as its output. The initial contractions are quite unlikely to involve the edges crossing the cut K; in particular, the very first iteration will contract an edge of K with probability at most $2/n$. The key insight is that it is only toward the end of the contraction process that there is any non-negligible probability that an edge of K gets contracted; in particular, this probability could be as large as $2/3$ in the very last iteration.

This suggests that we contract the edges until the number of vertices decreases, but not by too much, and then use some slower algorithm that guarantees a higher probability of success. The first stage guarantees that the slower algorithm will not require too much time to find a min-cut, but at the same time, since the contractions are performed on graphs with a large number of vertices, the probability that one of K's edges gets contracted is reasonably small. Unfortunately, the best deterministic algorithm known requires $O(n^3)$ time, and the following exercise shows that the above approach will fail to achieve a running time close to $O(n^2)$.

Exercise 10.9: Consider running the contraction algorithm until the number of vertices is reduced to t and then using a cubic-time algorithm to find the min-cut in the contracted graph. Show that repeating this process as many times as necessary to ensure a probability of success at least $1/2$ leads to an algorithm with running time $\Omega(n^{8/3})$.

The crucial insight is that instead of using a slower deterministic algorithm, it is better to use *two* independent invocations of the Algorithm **Contract** itself on the contracted graph with t vertices. This is because the two repetitions boost the probability of success on the smaller instance, while the cost of the repetition on this instance is not as much as the cost of repeating the entire algorithm; in fact, this effect multiplies with each successive stage of the recursion. We now specify the algorithm more precisely: first use contractions to reduce the number of vertices to roughly $n/\sqrt{2}$, and then recursively compute the min-cut in the resulting graph; perform this twice and choose the smaller of the two min-cuts obtained as the final output. The resulting recursive algorithm is summarized below, and the reasons behind this precise choice of the parameters will become clear shortly.

Algorithm FastCut:

Input: A multigraph $G(V, E)$.

Output: A cut C.

1. $n \leftarrow |V|$.

2. if $n \leq 6$ **then compute** min-cut of G by brute-force enumeration **else**

> **2.1.** $t \leftarrow \lceil 1 + n/\sqrt{2} \rceil$.
>
> **2.2.** Using Algorithm **Contract**, perform two independent contraction sequences to obtain graphs H_1 and H_2 each with t vertices.
>
> **2.3.** Recursively compute min-cuts in each of H_1 and H_2.
>
> **2.4. return** the smaller of the two min-cuts.

The recursion is stopped when $n \leq 6$ since at that point t will not be smaller than n. An intuitive way of viewing this algorithm is in terms of a binary computation tree. The root corresponds to the graph G. For any node of this tree with an associated graph H, we associate with the two children the graphs H_1 and H_2 obtained by performing independent sequences of contractions that reduce the number of vertices in H by a factor of $\sqrt{2}$. The depth of the tree is roughly $2 \log n$, and the number of leaves is $O(n^2)$. In contrast, the $O(n^2)$ independent iterations of Algorithm **Contract** can be viewed as a tree of depth 1 with one root and $O(n^2)$ leaves that are direct descendants of the root. Thus, the speed-up in this algorithm does not come from generating a smaller set of potential min-cuts, but instead it is due to the sharing of work between the various contraction sequences required to generate these potential min-cuts.

Algorithm **FastCut** is guaranteed to return some cut in G. We first bound the time and space requirements of this algorithm.

Theorem 10.16: *Algorithm* **FastCut** *runs in* $O(n^2 \log n)$ *time and uses* $O(n^2)$ *space.*

PROOF: The depth of recursion is $O(\log n)$ since the size of the graph is reduced by a constant factor at each level of recursion. Algorithm **Contract** uses $O(n^2)$ time to reduce an n-vertex graph to a 2-vertex graph, and so it can certainly perform a partial reduction to both H_1 and H_2 in $O(n^2)$ time. We obtain the following recurrence for the running time $T(n)$ of Algorithm **FastCut** when given an n-vertex graph as input:

$$T(n) = 2T\left(\left\lceil 1 + \frac{n}{\sqrt{2}} \right\rceil\right) + O(n^2).$$

The solution to this recurrence is given by $T(n) = O(n^2 \log n)$.

Turning to the space requirement, observe that at any time only one graph needs to be stored at each level of recursion. Since the graphs at depth d of recursion have $O(n/2^{d/2})$ vertices, it follows that the total space needed is bounded by

$$O\left(\sum_{d=0}^{\infty} \frac{n^2}{2^d}\right) = O(n^2).$$

We also have to keep track of the best min-cut found at each level of the recursion, but this can certainly be done with space $O(n^2)$. This completes the proof. □

It remains to show that this algorithm has reasonably high probability of returning a min-cut.

Theorem 10.17: *Algorithm* **FastCut** *succeeds in finding a min-cut with probability* $\Omega(1/\log n)$.

PROOF: Suppose that the input graph G has min-cut value k. Assume that a cut of value k has survived up to some point in the recursion where the size of the residual graph H is t. This can be viewed as a node labeled by the graph H in the recursion tree discussed earlier. Let H_1 and H_2 be the graphs associated with the children of the node associated with H; these are the two contracted versions of H on which the algorithm will recur further.

The invocation of the recursive algorithm on graph H will return a min-cut for G provided the following two conditions are met: a cut value of k survives one of the two contraction sequences leading to H_1 and H_2; and, the **FastCut** algorithm succeeds in finding the min-cut in that same graph H_i.

By Lemma 10.15, the probability that any specific min-cut in H (which must also be a min-cut in G) survives a contraction sequence that reduces the number of vertices from t to $\lceil 1 + t/\sqrt{2} \rceil$ is at least

$$\frac{\lceil 1 + t/\sqrt{2} \rceil (\lceil 1 + t/\sqrt{2} \rceil - 1)}{t(t-1)} \geq \frac{1}{2}.$$

Let $P(t)$ denote the probability that Algorithm **FastCut** succeeds in finding a min-cut in a graph with t vertices. It follows that

$$P(t) \geq 1 - \left(1 - \frac{1}{2} P\left(\lceil 1 + t/\sqrt{2} \rceil\right)\right)^2.$$

To solve this recurrence, it will be convenient to perform a change of variables and turn it into an equality. Let $k = \Theta(\log t)$ denote the depth of recursion, and $p(k)$ be a lower bound on the success probability. Then, we have $p(0) = 1$ and the recurrence:

$$p(k+1) = p(k) - \frac{p(k)^2}{4}.$$

A further change of variables with $q(k) = 4/p(k) - 1$, or $p(k) = 4/(q(k) + 1)$, yields the following upon simplification:

$$q(k+1) = q(k) + 1 + \frac{1}{q(k)}.$$

A simple inductive argument now establishes that

$$k < q(k) < k + H_{k-1} + 4,$$

where H_i is the ith Harmonic number and is $\Theta(\log i)$. It follows that $q(k) = k + \Theta(\log k)$, implying that $p(k) = \Theta(1/k)$, and this in turn implies that $P(t) = \Theta(1/\log t)$. Using n instead of t in the last expression gives the desired result. □

A reader familiar with the theory of branching processes may see that this proof is essentially bounding the probability of extinction of the graphs having min-cut value exactly that of the original graph G. Finally, we leave it as an exercise to verify that this algorithm can be implemented in the promised time bounds as was done for Algorithm **Contract** in Problems 10.9–10.11.

10.3. Minimum Spanning Trees

Let $G(V,E)$ be a connected graph with real-valued edge weights $w : E \to \mathbb{R}$, having n vertices and m edges. A spanning tree in G is an acyclic subgraph of G that includes every vertex of G and is connected; every spanning tree has exactly $n-1$ edges. The weight of a tree is defined to be the sum of the weights of its edges. A *minimum spanning tree* (MST) is a spanning tree of minimum weight. The *minimum spanning tree problem* (MSTP) is: given G, find an MST of G.

The algorithm we present here will recurse on subgraphs that are not necessarily connected. When the input graph G is not connected, a spanning tree does not exist and we generalize the notion of a minimum spanning tree to that of a *minimum spanning forest* (MSF). A forest F is an acyclic subgraph of G that consists of a collection of disjoint trees in G; we treat isolated vertices in F as trees of size 1. A spanning forest is a forest whose trees are spanning trees for the connected components of the graph G. A spanning forest is a spanning tree if and only if the graph is connected. The weight of a forest is the sum of the weights of its edges, and a minimum spanning forest is a spanning forest of minimum weight. By considering each connected component of G separately, it is easy to modify any algorithm for the MSTP to compute the MSF.

We will assume that all edge weights in G are distinct. This is not a restrictive assumption since we can use any canonical numbering of the edges to resolve ties when edge weights are being compared. Given the distinctness of the edge weights, it follows that the minimum spanning tree must be unique.

The exact weight of the edges will be irrelevant to the following discussion since the algorithms will work in the unit-cost RAM model and only perform comparisons between the edge weights; in particular, these algorithms only depend upon the total ordering of the edge weights and are otherwise insensitive to the values of the weights.

The MSTP is one of the best-studied problems in combinatorial optimization. A variety of algorithms have been developed for this problem, most of which are based on a greedy strategy and run in near-linear $O(m \log n)$ time, e.g., Borůvka's algorithm, Kruskal's algorithm, and Prim's algorithm. Currently, the best deterministic algorithm runs in time $O(m \log \beta(m,n))$, where $\beta(m,n) = \min\{i \mid \log^{(i)} n \leq m/n\}$ and $\log^{(i)} n$ denotes the ith iterated logarithm of n. While this is a linear time algorithm for all practical purposes, the data structures are complicated enough that the simpler algorithms running in time $O(m \log n)$ are preferable to use. In any case, there is still the theoretical issue of devising a linear time algorithm for this problem. In this section, we present a randomized algorithm for the MSTP and show that its expected running time is $O(m)$. In fact, the running time of this algorithm is $O(m)$ with high probability, but we omit this high-probability analysis in our discussion (see the Notes section).

The randomized algorithm we present requires a black-box access to an *MST verification* algorithm. A verification algorithm takes as input a graph G and a spanning tree T, and determines whether T is an MST for the graph G. Clearly, the verification problem for MST should be no harder than the MSTP. Indeed, several deterministic linear-time verification algorithms are known. We omit the details of these algorithms and use them as black boxes (see the Notes section). An important property of some of these linear-time verification algorithms is that when T is not an MST, they produce a list of edges in G any of which can be used to improve T. We will make this more precise later.

10.3.1. Borůvka's Algorithm

We start by describing a particular greedy strategy for MST called Borůvka's algorithm, which runs in time $O(m \log n)$. Later we will show that using randomization in conjunction with this algorithm leads to a linear-time algorithm. Borůvka's algorithm is based on the following simple observation.

Exercise 10.10: Let $v \in V$ be any vertex in G. Show that the MST for G must contain the edge (v, w) that is the minimum-weight edge incident on v.

The basic idea in Borůvka's algorithm is to contract simultaneously the minimum weight edges incident on each of the vertices in G. Recall from Section 10.2 that contracting an edge (v, w) involves collapsing the two endpoints into a single vertex that has all the incident edges of both vertices, except that self-loops are eliminated. In fact, a contraction can create multiple edges between some pairs of vertices but only the minimum weight edge needs to be retained out of any set of multiple edges. This process of contracting the minimum-weight incident edge for each vertex in the graph is called a *Borůvka phase.* A good implementation of a Borůvka phase is the following: mark the edges to be contracted; determine the connected components formed by the marked edges; replace each connected component by a single vertex; and, finally, eliminate the self-loops and multiple edges created by these contractions.

Exercise 10.11: Given a graph G with n vertices and m edges, show that a Borůvka phase can be implemented in time $O(n + m)$.

Exercise 10.12: Show that the set of edges marked for contraction during a Borůvka phase induces a forest in G.

We claim that the graph G' obtained from the Borůvka phase has at most $n/2$ vertices. This is because each contracted edge can be the minimum incident edge on at most two vertices. The number of marked edges is thus at least $n/2$. Since each vertex chooses exactly one edge to mark, it is easy to verify that each marked edge must eliminate a distinct vertex. The number of edges in G' is no more than m since no new edges are created during this process.

Let us now examine the benefit of performing a Borůvka phase. By Exercise 10.10, each of the contracted edges must belong to the MST of G. In fact, the forest induced by the edges marked for contraction is a subgraph of the MST.

Exercise 10.13: Let G' be the graph obtained from G after a Borůvka phase. Show that the MST of G is the union of the edges marked for contraction during this phase with the edges in the MST of G'.

Borůvka's algorithm thus reduces the MST problem in an n-vertex graph with m edges to the MST problem in an $(n/2)$-vertex graph with at most m edges. The time required for the reduction is only $O(m+n)$. It follows that the worst-case running time of this algorithm is $O(m \log n)$.

10.3.2. Heavy Edges and MST Verification

Before describing how randomization can be used to speed up Borůvka's algorithm, we develop a technical lemma on random sampling of edges from the graph G.

Fix a forest F in G and consider any pair of vertices u, $v \in V$. If they lie in the same connected component (i.e., tree) of F, there exists a unique path $P(u,v)$ between them in the graph F. Let $w_F(u,v)$ denote the maximum weight of an edge on the path $P(u,v)$ if it exists, and set $w_F(u,v) = \infty$ when u and v are disconnected in F. The value $w_F(u,v)$ should not be confused with the weight $w(u,v)$ of the edge (u,v) in G, if indeed such an edge exists.

► **Definition 10.2:** An edge $(u,v) \in E$ is said to be *F-heavy* if $w(u,v) > w_F(u,v)$. The edge (u,v) is said to be *F-light* if $w(u,v) \le w_F(u,v)$.

Note that all edges in F must be F-light. An edge (u,v) is F-heavy if the forest F contains a path from u to v using only edges of weight smaller than that of (u,v) itself. The following exercise illustrates the importance of this notion. The crucial point is that the choice of the forest F is irrelevant to the result in this exercise.

Exercise 10.14: Let F be *any* forest in the graph G. Show that if an edge (u,v) is F-heavy, then it does not lie in the MST for G. Verify that the converse is not true.

An edge "improves" a forest if adding it to the forest either reduces the number of trees in that forest, or removing the edge of largest weight in the unique cycle created by its addition leads to a forest of weight no larger than F. An F-light edge can be used to improve the forest F, while an F-heavy edge cannot. It is possible to design a greedy algorithm (essentially, Kruskal's algorithm) that starts with an empty forest F and, considering the edges of G

in order of increasing weight, checks whether each successive edge is F-light, in which case the edge is used to improve the current forest.

A verification algorithm for the MST can be viewed as taking as input a tree T in a graph G, and checking that the only T-light edges are the edges in T itself. It should be clear that this is equivalent to verifying that T is an MST. Such verification algorithms are easily adapted to verifying minimum spanning forests. In fact, there exist linear-time verification algorithms that can be adapted to go a step further and identify all F-heavy and F-light edges with respect to any forest F. We omit the details of these algorithms and instead only summarize their performance in the following theorem.

Theorem 10.18: *Given a graph G and a forest F, all F-heavy edges in G can be identified in time* $O(n+m)$.

10.3.3. Random Sampling for MSTs

The only use of randomization in the MST algorithm to be presented shortly is in the use of random sampling to identify and eliminate edges that are guaranteed not to belong to the MST. Consider a (random) graph $G(p)$ obtained by independently including each edge of G in $G(p)$ with probability p. The graph $G(p)$ has n vertices and expected number of edges mp. There is no guarantee that $G(p)$ will be connected.

Let F be the minimum spanning forest for $G(p)$. For reasonably large values of p, the forest F should be a good approximation to the MST for G. More precisely, we expect very few edges in G to be F-light. This intuition is made concrete in the lemma presented below.

We first review some elementary probability theory. Recall that a random variable X has the *negative binomial distribution* with parameters n and p if it corresponds to the number of independent trials required for n successes when each trial has a probability of success p (see Appendix C); further, the expectation of X is given by n/p. A random variable X *stochastically dominates* another random variable Y if, for all $z \in \mathbb{R}$, $\mathbf{Pr}[X > z] \geq \mathbf{Pr}[Y > z]$. Proposition C.7 states that if X stochastically dominates Y, then $\mathbf{E}[X] \geq \mathbf{E}[Y]$.

Exercise 10.15: Let X have the negative binomial distribution with parameters n_1 and p, and Y have the negative binomial distribution with parameters n_2 and p. For $n_1 \geq n_2$, show that X stochastically dominates Y.

Lemma 10.19: *Let F be the minimum spanning forest in the random graph $G(p)$ obtained by independently including each edge of G with probability p. Then the number of F-light edges in G is stochastically dominated by a random variable X that has the negative binomial distribution with parameters n and p. In particular, the expected number of F-light edges in G is at most n/p.*

PROOF: Let $e_1, \ldots, e_m$ be the edges of G arranged in order of increasing weight. Suppose that we construct $G(p)$ by traversing the list of edges in this order, flipping a coin with probability of HEADS equal to p for each edge in turn, and including an edge e_i in $G(p)$ if the ith coin flip turns up HEADS. (This is an application of the Principle of Deferred Decisions from Section 3.5.)

The minimum spanning forest F for $G(p)$ can be constructed online during this process. Initially F is empty. At step i, after we flip the coin for the edge $e_i = (u, v)$, if e_i is chosen for $G(p)$, we consider e_i for inclusion in F. The edge is added to F if and only if the two end-points u and v belong to different connected components of F. Recall that $e_i = (u, v)$ is F-light if and only if F does not contain a path from u to v consisting entirely of edges of smaller weight than e_i; given the order of examination of the edges, an edge is F-light when examined if and only if its end-points lie in different connected components.

The crucial observations are:

- the F-lightness of e_i depends only on the outcome of the coin flips for the edges preceding it in the ordering;
- edges are never removed from F during this process;
- and the edge e_i is F-light at the end if and only if it is F-light at the start of step i.

Define phase k as starting after the forest F has $k - 1$ edges and continuing until it has k edges. Every edge that is F-light during this phase has probability p of being included in $G(p)$, and hence of being added to F. The phase ends exactly when an F-light edge is added to $G(p)$ for the first time during the phase. It follows that the number of F-light edges considered during this phase has the geometric distribution with parameter p (see Appendix C). The F-heavy edges processed during this phase are entirely irrelevant.

Suppose the forest F grows in size from 0 to s. It follows that the total number of F-light edges processed till the end of phase s is distributed as the sum of s independent geometrically distributed random variables, each with parameter p. To account for the F-light edges processed after that but not chosen for $G(p)$, we continue flipping coins (for dummy edges) until a total of n HEADS have appeared. The total number of coin flips is a random variable which has the negative binomial distribution with parameters n and p (see Appendix C). Since s is at most $n - 1$, it follows that the total number of F-light edges is stochastically dominated by the random variable which represents the total number of coin flips. The expected number of F-light edges is bounded from above by the expectation of this random variable, which is n/p. □

10.3.4. The Linear-Time MST Algorithm

The randomized linear time MST algorithm interleaves Borůvka phases that reduce the number of vertices with random sampling phases that reduce the number of edges. After a random sampling phase, the minimum spanning forest F of the sampled edges is computed using recursion, and the verification algorithm is used to eliminate all but the F-light edges. Then, the MST with respect to the residual (F-light) edges is computed using another recursive invocation of the algorithm. This is summarized in Algorithm **MST**.

Although we refer to this algorithm as **MST**, it actually computes a minimum spanning forest and does not require that the input graph be connected.

Algorithm MST:

Input: Weighted, undirected graph G with n vertices and m edges.

Output: Minimum spanning forest F for G.

1. Using three applications of Borůvka phases interleaved with simplification of the contracted graphs, compute a graph G_1 with at most $n/8$ vertices and let C be the set of edges contracted during the three phases. **If** G is empty **then** exit and **return** $F = C$.
2. Let $G_2 = G_1(p)$ be a randomly sampled subgraph of G_1, where $p = 1/2$.
3. Recursively applying Algorithm **MST**, compute the minimum spanning forest F_2 of the graph G_2.
4. Using a linear-time verification algorithm, identify the F_2-heavy edges in G_1 and delete them to obtain a graph G_3.
5. Recursively applying Algorithm **MST**, compute the minimum spanning forest F_3 for the graph G_3.
6. **return** forest $F = C \cup F_3$.

We now prove that this algorithm has linear expected running time. In Problem 10.21 the reader is asked to show that it has the same worst-case running time as Borůvka's algorithm.

Theorem 10.20: *The expected running time of Algorithm* **MST** *is* $O(n + m)$.

PROOF: Let $T(n, m)$ be the expected running time of Algorithm **MST** on graphs with n vertices and m edges. Consider the cost of the various steps in this algorithm for such input graphs.

Step 1 uses three applications of Borůvka's algorithm, which runs in $O(n + m)$ time, and produces a graph G_1 with at most $n/8$ vertices and m edges. Step 2 performs a random sampling to produce the graph $G_2 = G_1(1/2)$ with $n/8$ vertices and an expected number of edges equal to $m/2$, and this also runs in $O(n + m)$ time. Finding the minimum spanning forest of G_2 has expected cost $T(n/8, m/2)$, by induction and linearity of expectation. The verification in Step 4 runs in time $O(n + m)$ and produces a graph G_3 with at most $n/8$ vertices and an expected number of edges at most $n/4$, by Lemma 10.19. Finding the minimum spanning forest of G_3 in Step 5 has expected cost $T(n/8, n/4)$. Finally, $O(n)$ time suffices for Step 6.

Putting all this together, we obtain that

$$T(n,m) \leq T(n/8, m/2) + T(n/8, n/4) + c(n+m),$$

for some constant c. A solution to this recurrence is given by $2c(n+m)$, implying that the expected running time of the **MST** algorithm is $O(n+m)$. □

Notes

The various algorithms for all-pairs shortest paths mentioned above (Dijkstra [125], Floyd-Warshall [150, 413], and Johnson [215]) are discussed in detail in the books by Aho, Hopcroft, and Ullman [5], Cormen, Leiserson, and Rivest [114], and Tarjan [391]. The issue of matrix multiplication over closed semirings or rings, and the applications to shortest path problems, is discussed in the book by Aho, Hopcroft, and Ullman [5] (see also Pan [322]). The best known algorithm for (unweighted) all-pairs shortest paths that does not resort to matrix multiplication is due to Feder and Motwani [140] and this runs in time $O\left(n^3/\log n\right)$; it runs in $O(nm)$ time for sparse graphs. The matrix multiplication algorithm running in time $O\left(n^{2.376}\right)$ is due to Coppersmith and Winograd [113]. The idea of using integer matrix multiplication for solving the all-pairs distances problem, using integer entries of super-logarithmic length, has been explored by Romani [359] and Yuval [421].

The results on the all-pairs shortest paths problem described here originated in the work of Alon, Galil, and Margalit [21]. They show how to solve the APD problem in $O(\text{MM}(n)\log n)$ time for undirected graphs, and in $O\left(\sqrt{\text{MM}(n)n^3}\log^3 n\right)$ time for directed graphs. These results generalize to integer edge weights of absolute value bounded by L while increasing the number of vertices by a factor of L with a concomitant increase in the running time. The randomized algorithm described here is an adaptation of an algorithm due to Seidel [370]; similar algorithms have been designed by Alon, Galil, Margalit, and Naor [22], and Karger (see [370]). Alon, Galil, Margalit, and Naor [22] have also derandomized the **BPWM** algorithm at the cost of an increase by polylogarithmic factors in the running time.

► **Research Problem 10.1:** Devise an algorithm for the all-pairs shortest paths problem that does not use matrix multiplication and runs in time $O\left(n^{3-\epsilon}\right)$ for a positive constant ϵ.

► **Research Problem 10.2:** Devise an algorithm for computing the diameter of an unweighted graph that does not use matrix multiplication and runs in time $O\left(n^{3-\epsilon}\right)$ for a positive constant ϵ.

The early algorithms for finding min-cuts (or s-t min-cuts) relied on the duality to maximum flows in networks. The flow-cut duality was first observed by Elias, Feinstein, and Shannon [136], and Ford and Fulkerson [152, 223]. The observation that min-cuts could be computed by performing $n-1$ maximum flow computations is due to Gomory and Hu [180]. It was shown that in the unweighted case the cost of the flow computations could be reduced to just $O(nm)$ by Podderyugin [334], Karzanov and Timofeev [252], and Matula [299]. Later, Hao and Orlin [192] obtained essentially the same bounds for the weighted case by showing that a min-cut could be computed in roughly the same time as a max-flow. Currently, the faster maximum flow algorithms all derive from the push-relabel algorithm of Goldberg and Tarjan [171]; their time bound of $O(nm \log(n^2/m))$ has been improved slightly by King, Rao, and Tarjan [256], and by Phillips and Westbrook [332].

The contraction algorithm is based on a deterministic algorithm for min-cuts with running time $O(mn + n^2 \log n)$ due to Nagamochi and Ibaraki [318]. Algorithm **Contract** is due to Karger [231], and Algorithm **FastCut** is due to Karger and Stein [234]. The last two papers also gave fast parallel implementations of the randomized contraction-based algorithm, and Karger and Motwani [233] derandomized a variant of these algorithms to obtain a fast deterministic parallel algorithm for min-cuts (see also the Notes section of Chapter 12).

► **Research Problem 10.3:** Devise a Las Vegas or a deterministic algorithm for min-cuts with running time close to $O(n^2)$.

► **Research Problem 10.4:** Is there a randomized algorithm for min-cuts with expected running time close to $O(m)$?

An excellent treatment of network optimization problems, including minimum spanning trees, can be found in the books by Ahuja, Magnanti and Orlin [7] and by Tarjan [391]. The reader may refer to the survey article by Graham and Hell [181] for a history of developments concerning the minimum spanning tree problem up to 1985. Borůvka's algorithm [80] is perhaps the earliest complete description of an MST algorithm. The other classical algorithms are due to Kruskal [270] and Prim [337] (see also Dijkstra [125]). The current best deterministic algorithm, requiring $O(m \log \beta(m,n))$ time, is due to Gabow, Galil, and Spencer [160, 159]. Deterministic linear-time algorithms are known for more powerful models of computation that permit bit-manipulation of the representation of the edge weights (see Fredman and Willard [154]).

Tarjan [390] gave an efficient algorithm for MST verification that has running time $O(m\alpha(m,n))$, where $\alpha(m,n)$ is the inverse Ackerman function. The first linear-time verification algorithm is due to Komlós [268] – this performs only $O(m)$ edge weight comparisons, but requires super-linear time to choose the comparisons. The first completely linear-time verification algorithm is due to Dixon, Rauch, and Tarjan [127], but this algorithm is complex and combines ideas from the previous verification algorithm with a table look-up strategy. A substantially simpler linear-time algorithm, based on the work of Komlós [268], has been devised by King [255]. The latter two algorithms have the desired features of being able to identify all F-heavy edges, as discussed above.

The randomized linear-time **MST** algorithm is based on an approach due to Karger [229]; Karger originally proved only a super-linear running time bound for this algorithm, and the linear-time analysis is based on the work of Klein and Tarjan [257]. A complete description of this algorithm and its analysis can be found in the article by Karger, Klein, and Tarjan [232].

► **Research Problem 10.5:** Devise a simple randomized MST verification algorithm with expected running time $O(n + m)$.

► **Research Problem 10.6:** Is there a deterministic MST algorithm with running time $O(n+m)$?

Problems

10.1 Suppose that the time required for Boolean matrix multiplication is $\text{BM}(n)$. Show that the closure of a Boolean matrix can be computed in time $O(\text{BM}(n))$.

10.2 Prove Lemma 10.1.

10.3 Prove Lemma 10.2.

10.4 Prove Lemma 10.3

10.5 Prove Lemma 10.5.

10.6 Modify the **BPWM** algorithm so as to obtain a high probability bound on its running time.

10.7 Show that the product of A^R and B^R can be computed in time $O\left((n/r)^2\text{MM}(r)\right)$ by omitting the columns of A^R and the rows of B^R corresponding to the indices not present in R, and then multiplying these $n \times r$ and $r \times n$ matrices in blocks of $r \times r$ matrices.

10.8 Suppose that $\text{MM}(n) = \Omega(n^{2+\epsilon})$ for some $\epsilon > 0$. Show that it is possible to implement Algorithm **BPWM** such that its expected running time becomes $O(\text{MM}(n)\log n)$. Why does this not work for $\text{MM}(n) = O\left(n^2\right)$? (**Hint:** Use the idea suggested in Problem 10.7.)

10.9 Let $G(V,E)$ be a multigraph. Devise a data structure that processes any arbitrary sequence of edge contractions in G, such that at any given point where the set of edges contracted is F, the graph G/F is available in the adjacency matrix format. Furthermore, it should possible to efficiently determine for any edge in E/F the corresponding edge in E. Your data structure should require $O(n)$ time per contraction and use a polynomial amount of space. Can you modify this to provide the adjacency list format for G/F using only $O(m)$ space?

Remark: Note that the time bound is independent of the number of edges. For this, the multigraph needs to be represented as a graph with integer edge weights that represent the multiplicities of the edges. You may assume that the number of edges in the multigraph is polynomial in n, although this is not strictly necessary.

10.10 Given a multigraph $G(V,E)$, show that an edge can be selected uniformly at random from E in time $O(n)$, given access to a source of random bits. (See the remark in Problem 10.9.)

10.11 Combining the solutions to Problems 10.9 and 10.10, prove Theorem 10.10. What is the space requirement for this implementation?

10.12 Prove Lemma 10.15.

10.13 (Due to D.R. Karger [231].) For any $\alpha \geq 1$, define an α-approximate cut in a multigraph G as any cut whose cardinality is within a multiplicative factor α of the cardinality of a min-cut in G. Determine the probability that a single iteration of the randomized algorithm for min-cuts will produce as output some α-approximate cut in G.

10.14 (Due to D.R. Karger [231].)
(a) Using the analysis of the randomized min-cut algorithm, show that the number of distinct min-cuts in a multigraph G cannot exceed $n(n-1)/2$, where n is the number of vertices in G.

(b) Formulate and prove a similar result for the number of α-approximate cuts in a multigraph G (see Problem 10.16).

10.15 Consider the min-cut problem in weighted graphs. Describe how you would generalize Algorithm **Contract** to this case. What is the running time and space requirement for your implementation?

10.16 Suppose that the edges of a graph are presented in an arbitrary order, and the number of edges m is not known in advance. Using the idea for a greedy algorithm described in Section 10.3.2, devise an online MST algorithm that runs in time $O(m \log n)$.

10.17 Show that Borůvka's algorithm can be implemented to run in time $O(\min\{m \log n, n^2\})$.

10.18 Show that the Algorithm MST has the same worst-case running time as Borůvka 's algorithm, i.e., $O(\min\{m \log n, n^2\})$.

CHAPTER 11
Approximate Counting

In this chapter we apply randomization to hard counting problems. After defining the class $\#P$, we present several $\#P$-complete problems. We present a (randomized) polynomial time approximation scheme for the problem of counting the number of satisfying truth assignments for a DNF formula. The problem of approximate counting of perfect matchings in a bipartite graph is shown to be reducible to that of the uniform generation of perfect matchings. We describe a solution to the latter problem using the rapid mixing property of a suitably defined random walk, provided the input graph is sufficiently dense. We conclude with an overview of the estimation of the volume of a convex body.

We say that a decision problem Π is in NP if for any YES-instance I of Π, there exists a proof that I is a YES-instance that can be verified in polynomial time. Equivalently, we can cast the decision problem as a language recognition problem, where the language consists of suitable encodings of all YES-instances of Π. A proof now certifies the membership in the language of an encoded instance of the problem. Usually the proof of membership corresponds to a "solution" to the search version of the decision problem Π: for instance, if Π were the problem of deciding whether a given graph is Hamiltonian, a possible proof of this for a Hamiltonian graph (YES-instance) would be a Hamiltonian cycle in the graph. In the counting version of this problem, we wish to compute the *number* of proofs that an instance I is a YES-instance. Thus we would be interested in how many Hamiltonian cycles, if any, the input graph contains. In Section 7.7.2 we encountered a counting version of the 3-SAT problem.

An algorithm for a counting problem takes as input an instance I of the decision problem Π, and produces as output a non-negative integer that is the number of solutions (or proofs) for the instance I. If Π is in NP, then the maximum possible number of solutions is $O(\exp(p(n)))$, where n is the size of the input and $p(n)$ is a polynomial. Thus the output of the counting algorithm is of length polynomial in the input size. A closely related class of problems is that of listing the solutions rather than merely counting them. Our focus will be on the counting problems associated with ***NP*** decision problems.

While counting problems are of interest for various purely theoretical reasons, they also arise naturally in a range of applications. One application of such counting problems stems from the study of *network reliability* problems: we are given an undirected graph, together with a probability of failure p_e for each edge e. We are interested in questions such as the following: what is the probability that the graph remains connected if each edge e fails independently with probability p_e? This provides the motivation behind the first problem we will study – the problem of counting the number of satisfying truth assignments for a Boolean formula in the disjunctive normal form (DNF) formula. A second application comes from statistical physics, and this motivates the second problem we study – counting the number of perfect matchings in a bipartite graph.

Clearly, a counting problem is at least as hard as the corresponding decision problem. Thus the counting problem associated with an ***NP***-complete decision problem is ***NP***-hard. What about the counting problem associated with decision problems in ***P***? Consider for example the decision problem of verifying the connectivity of an input graph. This problem can be solved in polynomial time. A proof of connectivity corresponds to a spanning tree in the input graph. The associated counting problem can also be solved in polynomial time: by a classical result, the number of spanning trees in a graph equals the determinant of a matrix derived from the adjacency matrix of the graph. On the other hand, while the problem of deciding whether a graph has a perfect matching is in ***P***, the associated counting problem is not believed to be in ***P***. Interestingly, the number of perfect matchings in a bipartite graph equals the permanent of the matrix of adjacencies between the vertices on the two sides of the graph. While the determinant is easy to compute, computing the closely related permanent function is extremely difficult. There are other decision problems in ***P*** whose associated counting problems are not known to have polynomial time algorithms.

The class of counting problems associated with ***NP*** decision problems is denoted by #***P***. Intuitively, the class #***P*** consists of all counting problems associated with the decision problems in ***NP***. Formally, a problem Π belongs to #***P*** if there is a non-deterministic polynomial time Turing machine that, for any instance I, has a number of accepting computations that is exactly equal to the number of distinct solutions to instance I. We say that Π is *#**P**-complete* if for any problem Π' in #***P***, Π' can be reduced to Π by a polynomial time Turing machine.

While there are "easy" problems in #***P*** such as counting spanning trees (where polynomial time algorithms are known), a large number of such counting problems appear to be intractable. Quite clearly, a #***P***-complete problem can be solved in polynomial time only if ***P*** = ***NP***, implying that it is quite unlikely that we can efficiently solve such problems. In the face of this apparent intractability, it is natural to ask whether instead we can compute approximate solutions to such counting problems. Unfortunately, we do not know of a good deterministic approximation algorithms for any #***P***-complete problem. However, the situation changes appreciably if we permit ourselves the use of randomization in the approximation algorithm. The rest of this chapter is devoted to presenting such algorithms.

11.1. Randomized Approximation Schemes

We start by introducing the notion of an approximation scheme. Consider a problem Π, and let $\#(I)$ denote the number of distinct solutions for an instance I of Π. For example, when Π is the problem of testing for Hamiltonian cycles, for an input graph I we denote by $\#(I)$ the number of such cycles in the graph. An approximation algorithm $\mathcal{A}$ takes as input I and outputs an integer $A(I)$, which is purported to be close to $\#(I)$.

► **Definition 11.1:** A *polynomial approximation scheme* (PAS) for a counting problem Π is a deterministic algorithm $\mathcal{A}$ that takes an input instance I and a real number $\epsilon > 0$, and in time polynomial in $n = |I|$ produces an output $A(I)$ such that

$$(1-\epsilon)\#(I) \leq A(I) \leq (1+\epsilon)\#(I).$$

A *fully polynomial approximation scheme* (FPAS) is a polynomial approximation scheme whose running time is polynomially bounded in both n and $1/\epsilon$.

The output $A(I)$ is called an ϵ-approximation to $\#(I)$. Suppose that $\epsilon < 1$. The length of the description of ϵ only adds a factor of $\Theta(\log 1/\epsilon)$ to the size of the input, yet we allow the approximation algorithm $\mathcal{A}$ to run in time polynomial in $1/\epsilon$.

Exercise 11.1: Show that if we were to modify the definition of an approximation scheme to read "polynomial in n and $\log 1/\epsilon$," the existence of such an approximation scheme for a $\#P$-complete problem would imply that $P = \#P$.

Since only a multiplicative error is permitted in an ϵ-approximation, it can be used to distinguish between the case $\#(I) = 0$ and the case $\#(I) > 0$, thereby implying a polynomial time algorithm for the decision version of the problem. Thus, such schemes can only be devised for counting problems whose decision versions are in P. Unless $P = NP$, it would be necessary to relax this definition (possibly by permitting some *additive* error also) to enable its applicability to counting versions of NP-complete problems.

No deterministic approximation schemes are known for $\#P$-complete problems. However, randomized versions of such approximation schemes are known, and so we make the following definition.

► **Definition 11.2:** A *polynomial randomized approximation scheme* (PRAS) for a counting problem Π is a randomized algorithm $\mathcal{A}$ that takes an input instance I and a real number $\epsilon > 0$, and in time polynomial in $n = |I|$ produces an output $A(I)$ such that

$$\mathbf{Pr}\left[(1-\epsilon)\#(I) \leq A(I) \leq (1+\epsilon)\#(I)\right] \geq \frac{3}{4}.$$

A *fully polynomial randomized approximation scheme* (FPRAS) is a polynomial randomized approximation scheme whose running time is polynomially bounded in both n and $1/\epsilon$.

The probability is taken over the random choices of the algorithm. Notice that when $\#(I)$ is not in the range $[A(I)(1-\epsilon), A(I)(1+\epsilon)]$, an event that occurs with probability at most 1/4, we assume nothing about how far $A(I)$ is from $\#(I)$. By an argument similar to that required in Exercise 11.1, modifying the running time requirement to "polynomial in n and $\log 1/\epsilon$" would preclude a randomized approximation scheme for a $\#\boldsymbol{P}$-complete problem unless $\boldsymbol{BPP} = \#\boldsymbol{P}$.

Exercise 11.2: The quantity 3/4 for the success probability in the definition of a randomized approximation scheme is somewhat arbitrary; in fact, we could replace it by practically any value that exceeds 1/2 by a constant. Devise a "bootstrapping scheme" which, given any $\delta \in (0, 1]$, invokes a randomized approximation scheme N times and outputs an integer $B(I)$ such that $\#(I) \in [B(I)(1-\epsilon), B(I)(1+\epsilon)]$ with probability at least $1-\delta$, where N is polynomial in $\log 1/\delta$. (**Hint:** Consider the median of the results of independent repetitions.)

A randomized approximation scheme can be used to distinguish between the case $\#(I) = 0$ and the case $\#(I) > 0$, thereby implying a randomized polynomial time algorithm for the decision version of the problem. Thus, such schemes can only be devised for counting problems whose decision versions are in $\boldsymbol{BPP}$. Since it is unlikely that $\boldsymbol{NP}$ is contained in $\boldsymbol{BPP}$, we do not expect to find such schemes for counting versions of $\boldsymbol{NP}$-complete problems.

► **Definition 11.3:** An (ϵ, δ)-FPRASfor a counting problem Π is a fully polynomial randomized approximation scheme that takes an input instance I and computes an ϵ-approximation to $\#(I)$ with probability at least $1-\delta$ in time polynomial in n, $1/\epsilon$, and $\log 1/\delta$.

Approximate counting is an area in which randomization makes a dramatic difference in our ability to (approximately) solve problems. Indeed, there are problems (such as the volume estimation problem in Section 11.4) for which randomization results in efficient algorithms where no efficient deterministic algorithm is possible. In the sequel, we describe such schemes for some counting problems that are $\#\boldsymbol{P}$-complete. Observe that such approximation schemes are

Monte Carlo. (Why is it difficult to convert this into a Las Vegas approximation scheme?)

11.2. The DNF Counting Problem

Let $F(X_1, \ldots, X_n)$ be a Boolean formula in disjunctive normal form (DNF) over the n Boolean variables $X_1, \ldots, X_n$. In other words, F is a disjunction $C_1 \vee \cdots \vee C_m$ of clauses C_i, where each clause C_i is a conjunction $L_1 \wedge \cdots \wedge L_{r_i}$ of r_i literals. Each literal L_j is either a variable X_k or its negation $\overline{X_k}$. We may assume that each variable occurs at most once in any given clause.

The variables are to be assigned values in $\{0, 1\}$, where 0 corresponds to FALSE and 1 corresponds to TRUE. A truth assignment $\boldsymbol{a} = (a_1, \ldots, a_n)$ is an assignment of value a_i to the variable X_i for each i. A truth assignment $\boldsymbol{a}$ is said to satisfy F if $F(a_1, \ldots, a_n)$ evaluates to 1 or TRUE. We denote by $\#F$ the number of distinct satisfying assignments of a given formula F. Clearly, $0 < \#F \leq 2^n$.

The DNF counting problem is to compute the value of $\#F$. This problem is known to be $\#\boldsymbol{P}$-complete and hence it is unlikely to have an exact polynomial time algorithm. We describe an (ϵ, δ)-FPRAS for this problem. The input size is at most nm. We desire that the approximation scheme have a running time that is polynomial in n, m, $1/\epsilon$, and $\log 1/\delta$.

11.2.1. An Unsuccessful Attempt

To understand the difficulty of finding an (ϵ, δ)-FPRAS for the DNF counting problem, we formulate a more abstract problem.

Let U be a finite set of known size, and let $f : U \rightarrow \{0, 1\}$ be a Boolean function over U. We define the set $G = \{u \in U \mid f(u) = 1\}$ as the pre-image of 1. Assume that given a particular $u \in U$, $f(u)$ can be computed quickly. Assume also that it is possible to sample uniformly at random from U. In our abstraction, both of these operations can be assumed to take unit time. The problem is to estimate the size of G.

This formulation includes the DNF counting problem as a special case. Let $U = \{0, 1\}^n$ be the set of all 2^n truth assignments, and define $f(\boldsymbol{a}) = F(\boldsymbol{a})$ for each $\boldsymbol{a} \in U$. Now, the set G consists of all satisfying truth assignments for F. It is easy to verify that we can compute f and sample from U in polynomial time.

An obvious randomized approach to estimating $|G|$ is to use the classical Monte Carlo method. This involves choosing N independent samples from U, say $u_1, \ldots, u_N$, and using the value of f on these samples to estimate the probability that a random choice will lie in G. More formally, define the random variables $Y_1, \ldots, Y_N$ as follows:

$$Y_i = \begin{cases} 1 & \text{if } f(u_i) = 1 \\ 0 & \text{otherwise.} \end{cases}$$

By this definition, $Y_i = 1$ if and only if $u_i \in G$. Finally, define the estimator random variable

$$Z = |U| \sum_{i=1}^{N} \frac{Y_i}{N}.$$

It is easy to verify that $\mathbf{E}[Z] = |G|$ and we might hope that with high probability the value of Z is an ϵ-approximation to $|G|$. Of course, the probability that the approximation is good depends upon the choice of N. The following theorem relates the value of N to ϵ and δ.

Theorem 11.1 (Estimator Theorem): *Let $\rho = |G|/|U|$. Then the Monte Carlo method yields an ϵ-approximation to $|G|$ with probability at least $1 - \delta$ provided*

$$N \geq \frac{4}{\epsilon^2 \rho} \ln \frac{2}{\delta}.$$

PROOF: Fix some $\epsilon \in (0, 1]$ and $\delta \in (0, 1]$. Notice that the random variables Y_i have the Bernoulli distribution with parameter ρ. Define $Y = \sum_{i=1}^{N} Y_i$, and observe that this has the binomial distribution with parameters N and ρ. Moreover, the estimator $Z = |U|Y/N$. By a straightforward application of the Chernoff bound (see Theorems 4.2 and 4.3), we obtain that

$$\begin{aligned} & \mathbf{Pr}\left[(1-\epsilon)|G| \leq Z \leq (1+\epsilon)|G|\right] \\ = \; & \mathbf{Pr}\left[(1-\epsilon)N\rho \leq Y \leq (1+\epsilon)N\rho\right] \\ \geq \; & 1 - F^{+}(N\rho, \epsilon) - F^{-}(N\rho, \epsilon) \geq 1 - 2e^{-N\rho\epsilon^2/4}. \end{aligned}$$

It is easy to see that for the given lower bound on N, the latter expression is bounded by $1 - \delta$. □

At this point it may appear that we have the desired approximation scheme. But there is a flaw in this approach – it has a running time of at least N, where $N \geq 1/\rho$. First of all, we do not know the value of ρ; in fact, the problem is exactly that of estimating ρ. However, this problem could be circumvented by using a successively refined lower bound on ρ to determine the number of samples to be chosen. A more disturbing problem is that the running time is inversely proportional to ρ, and at least for the DNF counting problem this could be exponentially large. (Consider for example the case where F only has a polynomial number of satisfying truth assignments.) The following exercise shows that if we were to relax the requirement of obtaining an ϵ-approximation *relative* to the size of G, and instead required only that the approximation have a small error with respect to $|U|$, then the sampling technique is indeed efficient.

Exercise 11.3: Devise a randomized approximation scheme for the DNF counting problem that computes an estimator Z such that

$$\mathbf{Pr}\left[\left|\frac{|G|}{|U|} - \frac{Z}{|U|}\right| \geq \epsilon\right] \leq \delta.$$

The running time should be polynomial in n, m, $1/\epsilon$, and $\log 1/\delta$.

This problem is fundamental to this approach and not an artifact of the analysis, since the Chernoff bound gives a fairly tight estimate of the tail probability of a binomial distribution. Fortunately, there is a standard statistical technique called *importance sampling* for dealing with the following problem: if we sample uniformly from a large population to estimate the size of a small subset of the population, it is necessary that the number of samples be extremely large to ensure that the estimator is a good relative approximation. The idea is to modify the process from a uniform sampling of the population to a skewed sampling that concentrates the probability on the sub-population of interest (the area of "importance"). We now apply this idea to our problem.

11.2.2. The Coverage Algorithm

We want to reduce the size of the sample space so as to ensure that the ratio ρ is relatively large, while ensuring that the set G is still completely represented. We start by formulating a slightly different abstract problem – the *union of sets* problem. This formulation captures the essential structure of the DNF counting problem, and has applications to several other problems in reliability.

Let V be a finite universe. We are given m subsets $H_1, \ldots, H_m \subseteq V$ such that the following assumptions are valid.

1. For all i, $|H_i|$ is computable in polynomial time.
2. It is possible to sample uniformly at random from any H_i.
3. For all $v \in V$, it can be determined in polynomial time whether $v \in H_i$.

The goal is to estimate the size of the union $H = H_1 \cup \cdots \cup H_m$. The brute-force approach to computing $|H|$ is inefficient when the universe V and the sets H_i are of large cardinality. The inclusion–exclusion formula (Proposition C.1) is also extremely inefficient for large m, since it requires computing roughly 2^m terms. However, the assumptions 1–3 turn out to be sufficient to enable the design of a Monte Carlo sampling algorithm that does not suffer from the drawbacks of the algorithm in Section 11.2.1.

The DNF counting problem can be cast as a special case of this union of sets problem, as follows. Consider a DNF formula $F(X_1, \ldots, X_n)$ and let the ith clause C_i be a conjunction of r_i literals. The universe V corresponds to the space of all 2^n truth assignments, and a set H_i contains all the truth assignments that satisfy the clause C_i. Since the truth assignments in H_i all assign the same values to variables appearing in C_i and are otherwise unconstrained, it is easy to see that $|H_i| = 2^{n-r_i}$. The same observation implies that it is easy to sample from H_i by assigning the appropriate values to variables appearing in C_i, and choosing the rest at random. Further, verifying that some $v \in V$ is a member of H_i is equivalent to testing whether a truth assignment satisfies a specific clause, and linear time suffices for this operation. Finally, $H = H_1 \cup \cdots \cup H_m$ is the set of all truth assignments that satisfy at least one clause of F, and hence F itself.

We now present a solution to the union of sets problem, and by the preceding argument this will solve the DNF counting problem as a special case.

We define a multiset $U = H_1 \uplus \cdots \uplus H_m$ as the *multiset union* of the sets H_1, $\ldots$, H_m. Recall that the multiset union contains as many copies of $v \in V$ as the number of H_i's that contain that v. We adopt the convention that the elements of U are ordered pairs of the form (v, i), corresponding to $v \in H_i$; in other words,

$$U = \{(v, i) \mid v \in H_i\}.$$

Observe that $|U| = \sum_{j=1}^{m} |H_i| \geq |H|$.

For all $v \in H$, the *coverage set* of v is defined by

$$\text{cov}(v) = \{(v, i) \mid (v, i) \in U\}.$$

The size of the coverage set is exactly the number of H_i's containing v, or the multiplicity of v in the multiset version of U. (In the DNF problem, for a truth assignment $\boldsymbol{a}$, the set $\text{cov}(\boldsymbol{a})$ is the set of clauses satisfied by $\boldsymbol{a}$.) The following observations are immediate.

1. The number of coverage sets is exactly $|H|$, and these coverage sets are easy to compute.
2. The coverage sets *partition* U, i.e., $U = \cup_{v \in H} \text{cov}(v)$.
3. $|U|$ is easily computed as $|U| = \sum_{v \in H} |\text{cov}(v)|$.
4. For all $v \in H$, $|\text{cov}(v)| \leq m$.

The following definition isolates a canonical element in each coverage set.

► **Definition 11.4:** The function $f : U \to \{0, 1\}$ is defined as follows.

$$f\left((v, i)\right) = \begin{cases} 1 & \text{if } i = \min\{j \mid v \in H_j\} \\ 0 & \text{otherwise.} \end{cases}$$

Also, the set G is defined as the inverse image of 1 under f.

$$G = \{(v, j) \in U \mid f\left((v, i)\right) = 1\}.$$

Define the canonical element for a coverage set of v as the element that corresponds to the occurrence of v in the lowest-numbered H_j containing v. The function f evaluates to 1 only on the canonical element of each coverage set. The set G is merely the set of the canonical elements.

The crucial observation is that $|G| = |H|$. This is because the number of coverage sets is $|H|$, and each coverage set contributes exactly one canonical element to G. Our goal then is to estimate the size of $G \subseteq U$ such that $G = f^{-1}(1)$. This is exactly the setting in which we applied Theorem 11.1 based on the naive Monte Carlo sampling technique. We claim that the naive Monte Carlo sampling algorithm gives an (ϵ, δ)-FPRAS for estimating the size of G. The claim follows from the following lemma.

Lemma 11.2: *In the union of sets problem,*

$$\rho = \frac{|G|}{|U|} \geq \frac{1}{m}.$$

PROOF: The proof relies on the observations made above.

$$\begin{aligned} |U| &= \sum_{v \in H} |\text{cov}(v)| \\ &\leq \sum_{v \in H} m \\ &\leq m|H| = m|G| \end{aligned}$$

The lemma follows. □

The following theorem shows that the Monte Carlo sampling technique gives as (ϵ, δ)-FPRAS for $|G|$, and hence also for $|H|$.

Theorem 11.3: *The Monte Carlo method yields an ϵ-approximation to $|G|$ with probability at least $1 - \delta$ provided*

$$N \geq \frac{4m}{\epsilon^2} \ln \frac{2}{\delta}.$$

The running time is polynomial in N.

PROOF: The sampling procedure and the analysis are exactly as in the Theorem 11.1. We merely have to show that f can be computed in polynomial time and that it is possible to sample uniformly from U.

To compute $f((v, i))$ we check whether the truth assignment v satisfies C_i but none of the clauses C_j for $j < i$. Sampling an element (v, i) uniformly from U is performed in two stages. First, choose i such that $1 \leq i \leq m$ and

$$\mathbf{Pr}[i] = \frac{|H_i|}{|U|} = \frac{|H_i|}{\sum_{i=1}^{m} |H_i|}.$$

Then an element $v \in H_i$ is chosen uniformly at random. It is easy to verify that the resulting pair (v, i) is uniform over U. □

Notice that the lemma implies a polynomial bound on the running time.

Why did this new sampling process give the desired result? Our original problem was to estimate $|H|$, the set of all satisfying truth assignments. Sampling uniformly from V, the space of all truth assignments, failed because V's size could be super-polynomially larger than the size of H. In the redesigned sampling process, we chose a random satisfying truth assignment for a randomly chosen clause. Each truth assignment could then be chosen in a number of ways proportional to the number of clauses it satisfies. In effect, this is a non-uniform sample from the set of all satisfying truth assignments. Since each truth assignment can be selected by at most m different clauses, and only one of these corresponds to a "success," we obtain the desired estimation.

Exercise 11.4: Specialize the new sampling procedure to the DNF counting problem and determine the running time of the (ϵ, δ)-FPRAS in terms of n, m, ϵ, and δ.

11.3. Approximating the Permanent

We turn to the problem of counting the number of perfect matchings in a bipartite graph. The input to this problem is a bipartite graph $G(U, V, E)$ with independent sets of vertices $U = \{u_1, \ldots, u_n\}$ and $V = \{v_1, \ldots, v_n\}$. Recall that a matching is a collection of edges $M \subseteq E$ such that each vertex occurs at most once in M. A perfect matching is a matching of size n. The associated decision problem (determining whether the graph has at least one perfect matching) is in $\boldsymbol{P}$. The problem of counting the number of perfect matchings in a given bipartite graph is $\#\boldsymbol{P}$-complete. The problem is particularly interesting because it is equivalent to computing the *permanent* of a 0-1 matrix. This is a classical $\#\boldsymbol{P}$-complete problem with applications to statistical physics.

In Chapter 7 we noted the connection between perfect matchings and the determinant of a matrix derived from the adjacency matrix of G. This was based on a correspondence between perfect matchings in G and the permutations in $\mathbb{S}_n$: the perfect matching corresponding to a permutation $\pi \in \mathbb{S}_n$ is given by the edges $(u_i, v_{\pi(i)})$, for $1 \leq i \leq n$. We now relate the *number* of perfect matchings in G to the *permanent* of such a matrix.

► **Definition 11.5:** Let $\boldsymbol{Q} = (Q_{ij})$ be an $n \times n$ matrix. The permanent of the matrix is defined as

$$\text{per}(\boldsymbol{Q}) = \sum_{\pi \in \mathbb{S}_n} \prod_{i=1}^{n} Q_{i,\pi(i)}, \tag{11.1}$$

where $\mathbb{S}_n$ is the symmetric group of permutations of size n.

Notice the similarity of this definition to that of the determinant of the matrix – the only difference is that in the determinant, we include the sign of the permutation π with each term of the sum.

Given a bipartite graph G, we define a 0-1 matrix $A(G)$ with one row for each vertex of U, and one column for each vertex of V. Let $A_{ij} = 1$ if there is an edge in the graph joining u_i to v_j, and 0 otherwise. It is well-known that the determinant of A can be computed in polynomial time. In comparison, the best-known method for computing the permanent runs in time $O(n2^n)$. It is not hard to show that per(A) is equal to the number of perfect matchings in G.

Exercise 11.5: Let $\#(G)$ denote the number of perfect matchings in the bipartite graph G. Show that $\#(G) = \text{per}(A(G))$.

Thus, computing the permanent of a 0-1 matrix is #***P***-complete. Given the apparent intractability of computing the number of perfect matchings in a bipartite graph, there has been considerable interest in approximating this quantity. Currently, we know only of randomized approximation algorithms for this problem. The scheme we study gives an (ϵ, δ)-FPRAS, but only if the input graph has a minimum degree at least $n/2$. This is still an interesting problem; it can be shown that computing the number of perfect matchings remains #***P***-complete even in this special case. In the Notes section we mention alternative schemes that work for all possible inputs, but have the disadvantage of requiring exponential time.

We will show that estimating the number of perfect matchings in a bipartite graph can be reduced to sampling uniformly at random from all the perfect matchings in the graph. It is not the case that the problem of random generation is substantially easier than the original counting problem. However, it suffices to generate a perfect matching *almost uniformly* from all the perfect matchings in the graph (we will make this notion precise in a moment), and almost uniform generation can in turn be achieved by simulating a certain random walk on a Markov chain derived from the input graph G (this is *not* the same as a random walk on G).

11.3.1. Reduction to Uniform Generation

We show that the approximation of a 0-1 permanent can be reduced to the problem of sampling uniformly at random from all the perfect matchings in a bipartite graph. Let M_k denote the set of distinct matchings of size k in G, and define $m_k = |M_k|$; thus we seek to estimate $m_n = |M_n|$. A *uniform generator* for M_k is a randomized polynomial time algorithm $\mathcal{U}_k$ that takes G as input and returns a matching $m \in M_k$ such that m is uniformly distributed over M_k.

We claim that a uniform generator $\mathcal{U}_k$ can be used to get an (ϵ, δ)-FPRAS for m_k. The idea is to use randomized self-reducibility – this is a randomized reduction of a problem of size i to the same problem with size $i - 1$. Given the graph G, for any edge $e = (u, v)$ define the following quantities: m_e is the number of matchings in M_k that contain the edge e; and m_{ne} is the number of matchings in M_k that do not contain e. Clearly, $m_k = m_e + m_{ne}$.

Assume for the moment that the ratio $r = m_{ne}/m_k$ is not minuscule, say at least $1/n$. We can then use the basic Monte Carlo sampling idea of Section 11.2.1 to obtain an estimator $\widehat{r}$ as follows: use $\mathcal{U}_k$ to choose a suitably large (but polynomially bounded) number of random matchings from M_k, and let the estimator $\widehat{r}$ be the fraction of these matchings that do not contain e. Theorem 11.1 can now be used to show that this is an (ϵ, δ)-FPRAS.

The next step is to obtain an ϵ-approximation $\widehat{m}_{ne}$ to m_{ne}. Consider the graph H obtained by removing the edge e from G. The number of edges in H is one smaller than in G, and the number of matchings of size k in H is exactly m_{ne}. Thus we can recursively estimate the number of k-matchings in H to obtain an

ϵ-approximation $\widehat{m}_{ne}$. Then the ratio $\widehat{m}_k = \widehat{m}_{ne}/\widehat{r}$ is a good approximator for m_k. The missing details of the analysis are left as an exercise.

Exercise 11.6: A problem with the recursive estimation scheme is that both the error in the approximation and the probability of failure add up over the various stages of recursion. This problem can be handled by requiring an $(\epsilon/N, \delta/N)$-FPRAS at each stage of the recursion, where N is an upper bound on the number of such stages. More importantly, we assumed that $r \geq 1/n$, and this is not true in general. However, it is not hard to show by the pigeonhole principle that there exists a choice of the edge e for which this assumption is valid. (This requires the assumption that the number of edges exceeds k, since otherwise a graph containing only the edges of a matching of size k is a counterexample. Therefore this problem can be handled by repeating the overall algorithm for all choices of e and using the various outputs to determine the correct choice of e.

Using these hints, obtain a complete description of the sampling algorithm and prove that it is an (ϵ, δ)-FPRAS.

Theorem 11.4: *Given a uniform generator $\mathcal{U}_k$, there exists an (ϵ, δ)-FPRAS for $|M_k|$.*

As we remarked earlier, it does not appear that the problem of uniform generation is any easier than the original counting problem. However, it is intuitively clear that even a near-uniform generation of matchings would suffice, although it may contribute to the error in the approximation. We now give a formal definition of a near-uniform generator.

► **Definition 11.6:** Given a sample space Ω, a generator $\overline{\mathcal{U}}$ is said to be a near-uniform generator for Ω with error ρ if, for all $\omega \in \Omega$,

$$\frac{|\mathbf{Pr}[U = \omega] - 1/|\Omega||}{1/|\Omega|} \leq \rho.$$

A uniform generator has $\rho = 0$.

Unfortunately, even a near-uniform generator for M_k is hard to construct. However, as we will show in the next section, for some classes of graphs it is possible to obtain a near-uniform generator for $M_k \cup M_{k-1}$. We now modify the preceding reduction to show how approximate counting can be achieved using the new type of near-uniform generator. From here on, $\overline{\mathcal{U}}_k$ will denote a near-uniform generator for $M_k \cup M_{k-1}$.

Our goal will be to estimate the ratios $r_k = m_k/m_{k-1}$, for $1 \leq k \leq n$. (We define $r_1 = m_1$, and this is just the number of edges in the input graph.) Clearly, the product of these ratios gives an estimator for m_n. If we had a *uniform generator* for $M_k \cup M_{k-1}$, then we could be use it to estimate the ratio r_k by taking a large number of random samples from $M_k \cup M_{k-1}$ and using as the estimator the ratio of the number of elements of M_k to the number of elements of M_{k-1} observed in the samples. The following exercise gives us a sense of the number of samples needed to get a good approximation when we actually have a *uniform generator* for $M_k \cup M_{k-1}$.

Exercise 11.7: Let $\alpha \geq 1$ be a real number such that $1/\alpha \leq r_k \leq \alpha$. Take $N = n^7\alpha$ samples uniformly at random from $M_k \cup M_{k-1}$. Let $\widehat{r}_k$ be the ratio observed in the sample of the number of elements of M_k to the number of elements of M_{k-1}. Using an argument similar to that used in the Theorem 11.1 of Section 11.2.1, show that $(1 - 1/n^3)r_k \leq \widehat{r}_k \leq (1 + 1/n^3)r_k$ with probability at least $1 - c^{-n}$ for a constant $c > 1$.

The number of samples needed grows polynomially with α. We must show that α is relatively small, but let us defer this issue for the moment. The next exercise shows the effect on the error when we multiply the estimators of r_k to obtain an estimator for m_n.

Exercise 11.8: Use the results of Exercise 11.7 to show that if we could sample $M_k \cup M_{k-1}$ uniformly at random for all k, we have a procedure that, with high probability, gives an estimate for m_n that lies in the interval $[m_n(1 - 1/n^2), m_n(1 + 1/n^2)]$.

Argue that the same idea leads to an (ϵ, δ)-FPRAS for m_n provided α is bounded above by a polynomial in n.

It is not very hard to show that we do not need to sample $M_k \cup M_{k-1}$ uniformly at random; it suffices to sample almost uniformly at random.

Exercise 11.9: Suppose $\alpha \leq n^2$, and further assume that we have a near-uniform generator $\overline{U}_k$ for $M_k \cup M_{k-1}$ with error $\rho \leq 1/n^4$. Show that, by extending the ideas of Exercises 11.7 and 11.8, we have an (ϵ, δ)-FPRAS for M_n.

We deal with the issue of devising an appropriate near-uniform generator in Section 11.3.2. We conclude this section by showing how to obtain a guarantee that $\alpha \leq n^2$; this is exactly the reason we need to assume that the graph has minimum degree at least $n/2$.

Theorem 11.5: *Let G be a bipartite graph with minimum degree at least $n/2$. Then, for all k, $1/n^2 \leq r_k \leq n^2$.*

PROOF: We first prove the upper bound. Let each matching of size k choose one of its subsets of size $k - 1$ as a canonical subset. At most $(n - k + 1)^2 \leq n^2$ matchings in M_k can choose any matching in M_{k-1} as a canonical matching. This implies that $m_k \leq n^2 m_{k-1}$, or that $r_k \leq n^2$.

Let $m \subseteq E$ be any matching in the graph G of size at most $n - 1$. An augmenting path $p \subseteq E$ is a path in G between two unmatched vertices such that the edges along the path are alternately in m and $E \setminus m$. It is easy to see that the symmetric difference of p and m gives a matching of cardinality $|m|+1$.

We claim that in graphs with minimum degree at least $n/2$, every matching in M_{k-1} has an augmenting path of length at most 3. Fix any matching $m \in M_{k-1}$, and consider any pair of unmatched vertices $u \in U$ and $v \in V$. The neighborhood sets of these vertices, $\Gamma(u) \subseteq V$ and $\Gamma(v) \subseteq U$, are each of size at least $n/2$. If any vertex in $\Gamma(u)$ is unmatched in m, then we have an augmenting path of length 1 from u to that vertex. Thus, we can assume that each vertex in $\Gamma(u)$, and similarly in $\Gamma(v)$, is matched under m. But then, since $|\Gamma(u)|$ and $|\Gamma(v)| \geq n/2$ and $|m| \leq n-1$, it must be the case that some vertex $a \in \Gamma(u)$ and some vertex $b \in \Gamma(v)$ are matched to each other. It follows that (u, a), (a, b), and (b, v) form an augmenting path of length 3.

Fix any matching $m \in M_k$. We claim that there are at most n^2 matchings $m' \in M_{k-1}$ that can be augmented to m via augmenting paths of length at most 3. The matchings m' that can be augmented into m by length 1 paths are subsets of m, and there are at most k such subsets for any m. Moreover, any length 3 augmenting path for m' will determine a unique pair of edges in m, namely the edges that comprise $m \setminus m'$. The number of such pairs of edges is $k(k-1)/2$ and each pair can participate in at most 2 augmenting paths of length 3. Since each $m' \in M_{k-1}$ has at least one augmenting path of length no more than 3, we obtain that no $m \in M_k$ can be the result of more than $k + k(k-1) \leq n^2$ such augmentations. It follows that $m_{k-1} \leq n^2 m_k$, or that $r_k \geq 1/n^2$. □

11.3.2. Near-Uniform Generation of Matchings

From here on, we fix a bipartite graph G with minimum degree at least $n/2$. We are now down to finding a near-uniform generator for $M_k \cup M_{k-1}$ that has error $\rho \leq 1/n^4$. For this, we will devise a Markov chain $\mathcal{C}_k$, each of whose states is an element of $M_k \cup M_{k-1}$, in such a way that the stationary probability of each state is equal to $1/|M_k \cup M_{k-1}|$. Consider now a simulation of this Markov chain for τ steps, starting at an arbitrary state obtained by constructing a matching of size k in G using any polynomial time algorithm for matching. Our goal is to show that for a value of τ that is not too large, the Markov chain will approach its stationary distribution, thereby yielding a near-uniform sample from $M_k \cup M_{k-1}$.

This simulation of $\mathcal{C}_k$ can be thought of as executing a "random walk" on a graph in which each vertex corresponds to an element of $M_k \cup M_{k-1}$ (we have yet to describe the edges of this graph). If we were to begin this random walk in the stationary distribution, we would remain in the stationary distribution, so that at the end of τ steps we would be at an element of $M_k \cup M_{k-1}$ chosen uniformly at random. However, we start at a certain fixed element of $M_k \cup M_{k-1}$, so that the probability distribution of our position at the end of this random walk of τ steps is not guaranteed to be uniform on the elements of $M_k \cup M_{k-1}$. Instead, we will show that it is almost uniform on the elements of $M_k \cup M_{k-1}$, regardless of which element of $M_k \cup M_{k-1}$ we start at. To do this, we will demonstrate that the underlying graph of the Markov chain C_k resembles an expander, and so by the rapid mixing property of random walks on expanders we obtain the desired rate of convergence to the stationary distribution. Since the number of states in C_k may be exponential in n, it is essential that τ be logarithmic in the number of states. This is reminiscent of the random walk on an expander we have seen in Chapter 6.

We first describe the structure of C_k, the edges in the underlying graph, and the corresponding transition probabilities. It is important to keep in mind the distinction between the input graph G and the graph underlying C_k on which we perform the random walk. In particular, C_k could have size exponential in n. However, in the course of executing the algorithm, we will not store the entire Markov chain C_k explicitly. We only generate representations of those states (matchings of size k or $k-1$) that are visited during an execution of the algorithm, and this will remain polynomial in n. Below, we will describe C_n, the Markov chain used for estimating $|M_n|/|M_{n-1}|$; the modifications for $k < n$ are obvious.

Let E denote the set of edges in G. Let A denote a subset of E, and e be an edge in E. Let $A+e$ and $A-e$ denote the sets $A \cup \{e\}$ and $A \setminus e$, respectively. Armed with this notation, we are now ready to describe the transitions and transition probabilities in C_n.

In any state m of C_n, the transitions and transition probabilities are defined as follows. With probability $1/2$, we remain at the state and do nothing; recall from Chapter 6 that this ensures the aperiodicity of C_n. Otherwise (with probability $1/2$), we choose an edge $e = (u, v)$ of E uniformly at random and then select the appropriate case from the following.

Reduce: if $m \in M_n$ and $e \in m$, move to state (matching) $m' = m - e$;
Augment: if $m \in M_{n-1}$, with u and v unmatched in m, move to $m' = m + e$;
Rotate: if $m \in M_{n-1}$, with u matched to w and v unmatched in m, move to $m' = (m + e) - f$, where f is the edge (u, w) (there is a symmetrical case in which v is matched to w, and we make the corresponding move);
Idle: otherwise, stay at current state.

Figure 11.1 gives an example of C_2 when the input graph is the complete bipartite graph $K_{2,2}$ with $|U| = |V| = 2$. Each state is represented by a large circle, with the corresponding matching (of size 2 or 1) drawn inside the circle. The possible transitions between the states are also drawn, with all edges shown having transition probability $1/8$. In addition (not shown), there is a "self-loop" from each state to itself. It is instructive to go through Figure 11.1 identifying the edge of $K_{2,2}$ (the input graph) corresponding to each transition of the Markov chain in the figure.

In general, each transition of this Markov chain has an associated probability of $1/(2|E|)$, and the remaining probability mass is placed on the self-loops. Because of the **Idle** move, the self-loop at any state may have some additional probability over $1/2$ (as in the example of Figure 11.1). Notice also that if a transition exists from a state m to m', then the reverse transition also exists and

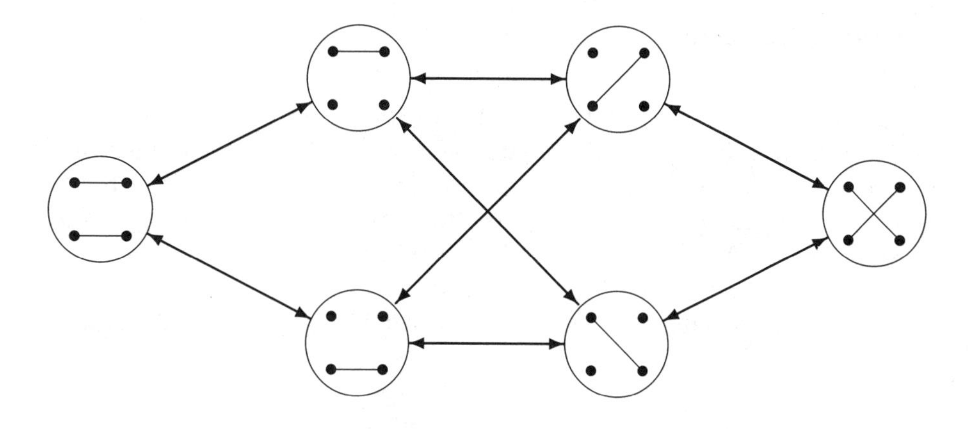

Figure 11.1: An example of C_2 for $G = K_{2,2}$.

has the same probability. Thus, the underlying graph can be viewed as being undirected, a property that we have already observed to be essential for the convergence of the random walk to the stationary distribution.

We now introduce some terminology for showing that the simulation of C_n leads to a state that is almost random in τ steps. Let $\boldsymbol{\pi}$ denote the stationary distribution for C_n. Let X_t denote the state of C_n after t steps of simulation, with X_0 being the state we start in. Let $\boldsymbol{P} = [p_{ij}]$ denote the transition probability matrix of C_n, with p_{ij} being the probability of the transition $i \to j$. Note that if the transition $i \to j$ exists, and $i \neq j$, then $p_{ij} = 1/(2|E|)$; further, each $p_{ii} \geq 1/2$ since there are at most $|E|$ transitions out of any state in C_n. Denote the probability that $X_t = j$ given that $X_0 = i$ by $p_{ij}^{(t)}$, recalling (Chapter 6) that in matrix form, $[p_{ij}^{(t)}] = \boldsymbol{P}^t$.

Theorem 11.6: *The Markov chain C_n is ergodic and its stationary distribution is uniform on $M_n \cup M_{n-1}$.*

PROOF: The irreducibility follows from the observation that we can go from any matching in a graph to any other matching via a suitable sequence of augmentations, reductions, and rotations. Since the self-loop probabilities are all positive, it also follows that the Markov chain is aperiodic. This implies the ergodicity of C_n.

The matrix $\boldsymbol{P}$ is symmetric and therefore doubly stochastic. We have already seen in Chapter 6 that the stationary distribution for a Markov chain with a doubly stochastic transition matrix must be the uniform distribution on its states. □

Exercise 11.10: Let s_1, s_2, s_3, and s_4 be any four states of $\mathcal{C}_n$. Show that under the stationary distribution, the probability of making the transition $s_1 \to s_2$ is equal to the probability of the transition $s_3 \to s_4$.

It remains to be shown that if we were to start from an arbitrary state, our position at the end of a τ-step simulation of $\mathcal{C}_n$ would be distributed almost uniformly over all the states. If we can show this to be the case with τ bounded above by a polynomial in n, then we are done. We will show that, regardless of the starting state $X_0 = i$, the probability distribution of X_τ given by the vector $(p_{ij}^{(\tau)})$ resembles the stationary distribution $\boldsymbol{\pi}$.

To make this precise, we use the notion of relative pointwise distance from Chapter 6:

$$\Delta(t) = \max_{i,j} \frac{|p_{ij}^{(t)} - \pi_j|}{\pi_j}.$$

The idea is to show that $\Delta(t)$ diminishes rapidly, so that in a sense we quickly lose any memory of the state in which the Markov chain was started. In particular, we would like $\Delta(\tau)$ reduced below $1/n^4$, for τ polynomial in n.

Let $\lambda_1 > \lambda_2 \geq \cdots \lambda_N$ be the eigenvalues of $\boldsymbol{P}$, where $N = |M_n \cup M_{n-1}|$ is the number of states in $\mathcal{C}_n$; clearly, $\lambda_1 = 1$ since the matrix $\boldsymbol{P}$ is doubly stochastic (see Section 6.7). The following is a consequence of a refinement of Theorem 6.21 described in Problem 11.7.

Theorem 11.7: $\Delta(t) \leq \dfrac{\lambda_2^t}{\min_j \pi_j} = \lambda_2^t N.$

Theorem 11.7 tells us that the rate at which the relative pointwise distance diminishes depends on how far λ_2 is separated from 1. By Proposition B.3, we have

$$\lambda_2 \leq e^{-(1-\lambda_2)}$$

since λ_2 is bounded away from 1. Choosing τ to be $(4 \ln n)(\ln N)/(1 - \lambda_2)$, we will have $\Delta(\tau) \leq 1/n^4$ as desired; note that $\ln N$ is $O(n^2)$. If we can now show that $1/(1 - \lambda_2)$ is bounded above by a polynomial in n, we will be done.

Obtaining such a bound on $1/(1 - \lambda_2)$ is not an easy matter. To this end, we introduce the concept of the *conductance* of a Markov chain. Let $w_{ij} = \pi_i p_{ij}$ denote the stationary probability of the transition $i \to j$. The reader may verify that since $\boldsymbol{P}$ is doubly stochastic, we have $w_{ij} = w_{ji}$, i.e., the Markov chain is *time reversible* (see Problem 11.7).

► **Definition 11.7:** Let S be a subset of the set of states of $\mathcal{C}_n$ such that $\emptyset \subset S \subset M_n \cup M_{n-1}$, and $\overline{S}$ is its complement.

- The *capacity* of S is defined as $C_S = \sum_{i \in S} \pi_i$.
- The *ergodic flow* out of S is defined as $F_S = \sum_{i \in S, j \in \overline{S}} w_{ij}$.
- The *conductance* of S is defined as $\Phi_S = F_S / C_S$.

The following facts are obvious: $0 < F_S \leq C_S < 1$, $F_S = F_{\overline{S}}$, and $\Phi_S < 1$.

Intuitively, the capacity of S is the probability that the Markov chain is in a state of S, the ergodic flow out of S is the probability that it will leave S, and the conductance of S is the probability that it will leave S conditional upon being inside S. Thus, a high conductance would suggest that the Markov chain will not get stuck inside S. If all sets S have high conductances, then the Markov chain will be rapidly mixing.

► **Definition 11.8:** The *conductance* of a Markov chain with state space Q is defined as

$$\Phi = \min_{\emptyset \subset S \subset Q,\, C_S \leq 1/2} \Phi_S.$$

By the preceding discussion, there should be a relation between the conductance and the second eigenvalue of a Markov chain, since both are closely related to the rapid mixing property. The following lemma provides this relationship; Theorems 6.16, 6.17, and 6.19 provide some intuition on why such a result should hold.

Lemma 11.8: $\lambda_2 \leq 1 - \dfrac{\Phi^2}{2}.$

Going back through our chain of reasoning, it now suffices to prove that $1/\Phi$ (and therefore $2/\Phi^2$) is bounded above by a polynomial in n. The proof is based on the so-called *canonical path argument*, which is described in detail in Section 11.3.3.

11.3.3. The Canonical Path Argument

This section is devoted to the proof of the following theorem.

Theorem 11.9: *For the Markov chain C_n, $\Phi \geq 1/12n^6$.*

The proof proceeds along the following lines. Let H be the graph underlying C_n. By Exercise 11.10, the transition probabilities along all the oriented edges of H are all exactly $1/(2|E|)$, where E is the set of edges in G. We bound the conductance of C_n from below by showing that for any subset S of the vertices of H with $C_S \leq 1/2$, the number of edges between S and $\overline{S}$ is large. To this end, we first specify a *canonical path* between every pair of vertices of H, such that no oriented edge of H occurs in more than bN of these paths. For a subset S of the vertices of H, the number of such canonical paths crossing the cut from S to $\overline{S}$ is

$$|S|(N - |S|) \geq |S|N/2,$$

since we assume that $|S| \leq N/2$. Since at most bN paths pass through each of the edges between S and $\overline{S}$, the number of such edges must be at least $|S|/2b$, so that the conductance of $\mathcal{C}_n$ is at least $1/(4b|E|) \geq 1/(4bn^2)$. In the rest of this section we define a collection of canonical paths for which the value of b is $3n^4$, implying the desired lower bound of $1/12n^6$ on the conductance.

We start by specifying canonical paths for all possible pairs of nodes in the graph H. Recall that although H is a directed graph, we can view it as an undirected graph since for every oriented edge there is an edge in the reverse direction. Further, H is strongly connected.

We associate a unique node (called the *partner*) $\bar{s} \in M_n$ with every node $s \in M_n \cup M_{n-1}$ and choose a canonical path between s and $\bar{s}$. If s is in M_n, then we set $\bar{s} = s$ (and the path between s and $\bar{s}$ is empty). Then, we specify canonical paths between all pairs of nodes in M_n. In general, the canonical path between nodes $s, t \in M_n \cup M_{n-1}$ consists of three consecutive segments: the path between s and $\bar{s}$, the path between $\bar{s}$ and $\bar{t}$, and the path between $\bar{t}$ and t. We now have to specify two different types of paths: *type A* paths between a node $s \in M_{n-1} \cup M_n$ and its partner $\bar{s} \in M_n$; and *type B* paths between pairs of nodes in M_n.

Specifying type A paths is relatively easy, and is handled in three cases. Consider any node $s \in M_n \cup M_{n-1}$. The first case is when s is in M_n, and here we use the empty path since $\bar{s} = s$. The second case is when s is in M_{n-1} and there exists an augmenting path of length 1 for s. In other words, the input graph G has an edge e such that $s + \{e\}$ is a perfect matching. In this case we set $\bar{s} = s + \{e\}$, and it is easy to verify that there is a path of length 1 between s and $\bar{s}$ in H (using an **Augment** transition). Finally, the third case is when s is in M_{n-1} but it has no direct augmentation into a perfect matching. But we have already seen in the proof of Theorem 11.5 that in G every near-perfect matching has an augmenting path of length at most 3. Thus, we now have a path of length 2 from s to some (possibly more than one) perfect matching in H, where this path first uses a **Rotate** transition and then an **Augment** transition (see Figure 11.2). Pick any such perfect matching $\bar{s}$; the path between s and $\bar{s}$ is then uniquely specified.

The type A paths are now completely specified. We now state a useful property of these paths. Let m be any matching in M_n, and define the set $\kappa(m)$ to be the set of all nodes $s \in M_n \cup M_{n-1}$ such that $\bar{s} = m$ and $s \neq m$.

Lemma 11.10: *For any $m \in M_n$, $|\kappa(m)| \leq n^2$.*

PROOF: The only perfect matching that chooses m as its partner is m itself. We further claim that at most $n + n(n-1)$ near-perfect matchings can use m as their partner. To see this, consider any $s \in M_{n-1}$ such that $\bar{s} = m$. Clearly, s must be within distance 2 of m in the graph H. Any near-perfect matching adjacent to m must be connected to m by a **Reduce** transition, and there are n such transitions incident on m in H. The number of near-perfect matchings at distance exactly 2 from m is at most $n(n-1)$, since these matchings must contain exactly $n-2$ edges of m and one other edge not in m. Thus, there are at most $n + n(n-1)$ different near-perfect matchings within distance two of m, and this yields the desired bound on $\kappa(m)$. □

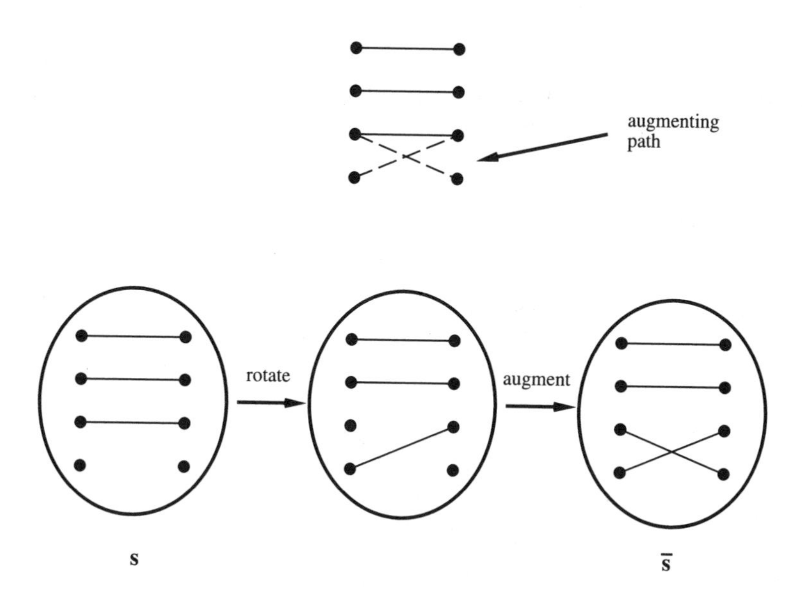

Figure 11.2: Type A path determined by augmenting paths of length 3.

We now specify the type B paths. Fix any two perfect matchings $s, t \in M_n$. Let $d = s \oplus t$ denote the symmetric difference of the edges in these two perfect matchings. It is easy to verify that the edges in d decompose into a collection of *disjoint, even-length, alternating* cycles, each of length at least 4, such that the edges in any such cycle are alternately from s and t.

Assume that the set of even cycles in the graph G is totally ordered, and that a specific vertex in each of these cycles is designated as the *start vertex.* One way to do this is to designate the lowest-numbered vertex in each cycle as its start vertex, and to order the cycles based on the lexicographic ordering on the sequence of vertices visited in the cycles starting with the designated start vertex and moving in the direction of its lowest-numbered neighbor. The reader should keep in mind that the entire notion of canonical path is an artifact of the analysis, and none of this has to be computed by the algorithm under consideration.

Our goal now is to specify a canonical path from s to t. Let $C_1, \ldots, C_r$ be the ordered list of cycles in the symmetric difference d. We first show that it is possible to transform s into t by performing local changes referred to as the *unwinding* of the cycles in d, one by one in the specified order of the cycles. These local changes can then be seen to correspond to transitions along edges of H, thereby yielding a path in H from s to t.

The unwinding of a cycle C_k corresponds to traversing the cycle, starting at the designated start vertex, successively removing the edges of C_k that belong to s and adding the edges that belong to t (see Figure 11.3). The unwinding of each cycle contains precisely one **Reduce** transition (at the start) and one **Augment** transition (at the end). Clearly, if we start with the perfect matching s and unwind all the cycles in d, the result is the perfect matching t (see

Figure 11.4). We leave it as an easy exercise to verify that each step of this sequential unwinding process corresponds precisely to a transition along an edge in the graph H, thereby giving us a unique specification of the type B paths.

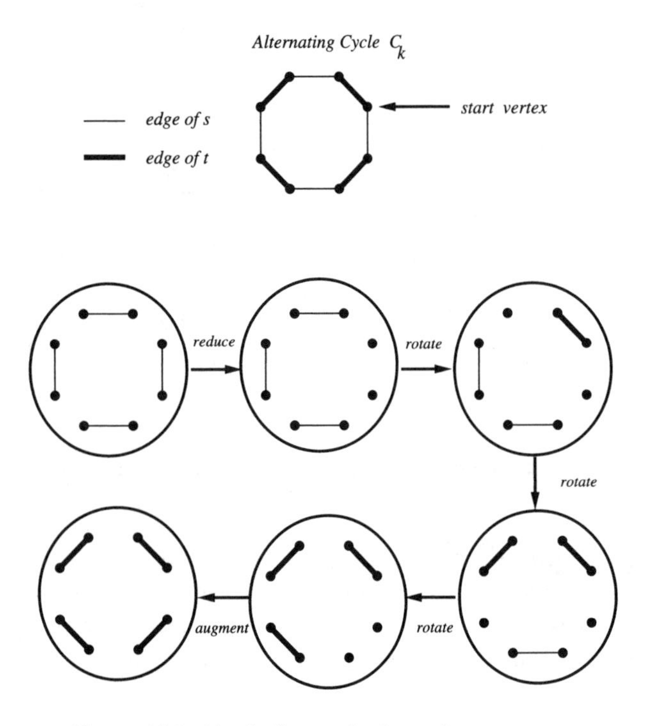

Figure 11.3: Unwinding a single cycle.

Putting all this together gives the desired set of unique canonical paths for all pairs of vertices in H. The following lemma provides the promised bound on the number of canonical paths that contain a specific transition.

Lemma 11.11: *Any transition T in H lies on at most $3n^4N$ distinct canonical paths.*

PROOF: Fix any transition $T = (u,v)$ in the graph H. Now, T can lie on a canonical path from s to t either in the two type A segments, or in the type B segment in the middle. Consider first the case where T lies on the type A segment from s to $\overline{s}$. This path consists of at most two transitions. Verify that if T lies on this path then $v \subseteq \overline{s}$. But for any given $v \in M_n \cup M_{n-1}$, there is at most one perfect matching that contains v as a subset, and in fact this perfect matching is $\overline{v}$. If T lies on the first type A segment of the canonical path from s to t, then $\overline{s} = \overline{v}$. It follows that s must be in the set $\kappa(\overline{v})$. Similarly, if T lies on the second type A segment of the canonical path from $\overline{t}$ to t, then $t \in \kappa(\overline{u})$. Thus, the total number of pairs s and t such that the transition T lies on the type A segments of their canonical path is bounded by $(|\kappa(\overline{v})| + |\kappa(\overline{u})|)N \leq 2n^2N$ (by Lemma 11.10).

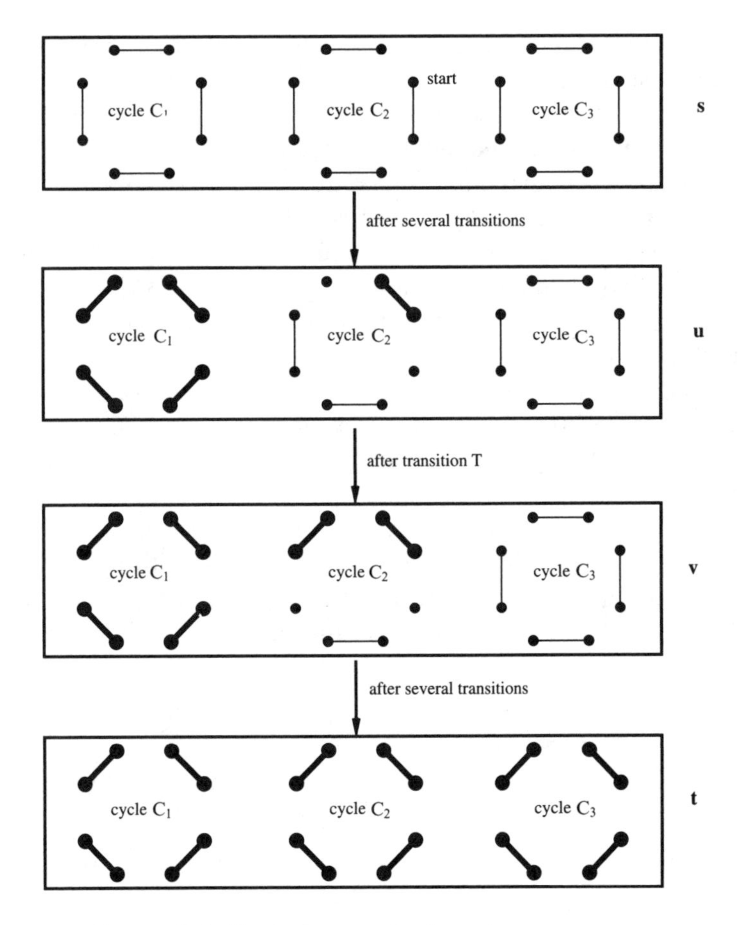

Figure 11.4: Unwinding a collection of cycles.

We now bound the number of canonical paths that contain T in their middle segment of type B. Define the set $CP(T)$ as the set of all pairs of perfect matchings whose canonical path contains the transition T. If T lies on the type B segment of a canonical path from s to t, then it lies on the canonical path from $\bar{s}$ to $\bar{t}$. Since $\kappa(\bar{s})$ and $\kappa(\bar{t})$ contain at most n^2 elements each, we find that the number of canonical paths whose type B segments contain T is at most $n^2 \times |CP(T)| \times n^2 = n^4|CP(T)|$. We now show that $|CP(T)|$ is at most N, implying the desired result that the transition T can lie on at most $2n^2N + n^4N \leq 3n^4N$ canonical paths.

It remains to be shown that $|CP(T)| \leq N$. We will make use of the following encoding mechanism. For each pair of perfect matchings s and t that lie in $CP(T)$, we define an encoding of s and t with respect to T as a matching $\sigma_T(s,t) \in M_n \cup M_{n-1}$. The idea is to ensure that the encoding is unique for each such pair in $CP(T)$, thereby implying that $|CP(T)| \leq |M_n \cup M_{n-1}| = N$.

Let the symmetric difference $d = s \oplus t$ consist of the ordered sequence of cycles $C_1, \ldots, C_r$. Consider a transition T between matchings $u, v \in M_n \cup M_{n-1}$ that occurs during the unwinding of a particular cycle C_k. The encoded matching $\sigma_T(s,t)$ is designed so that it agrees with s on the cycles $C_1, \ldots, C_{k-1}$, as well as the portion of C_k that has already been unwound, and it agrees with t elsewhere. Assume that T is not a **Rotate** transition; then we can set $\sigma_T(s,t) = s \oplus t \oplus (u \cup v)$. It is easy to verify that in this case the encoding is itself a matching from $M_n \cup M_{n-1}$, that it is uniquely defined, and finally that s and t can be completely recovered from the given encoding and the knowledge of u and v. The last property follows from the observation that $s \oplus t = \sigma_T(s,t) \oplus (u \cup v)$, and the fact that the partially unwound cycle C_k (and hence the remaining cycles) can be deduced from T.

The only problem is in the case where T is a **Rotate** transition, as in Figure 11.4. Then there exists a vertex such that $s \oplus t \oplus (u \cup v)$ has two edges incident on it. However, one of these edges (denoted $e_{s,t}$) always has the start vertex of C_k as its other end-point. Thus, when T is a **Rotate** transition we define $\sigma_T(s,t) = s \oplus t \oplus (u \cup v) \oplus e_{s,t}$. It is now easy to verify that all the desired properties of the encoding also hold for this case, with the minor change that $s \oplus t = \sigma_T(s,t) \oplus (u \cup v) \oplus e_{s,t}$. In Figure 11.5, we illustrate the encoding for this case using matchings s and t, and also the transition T, described in Figure 11.4. □

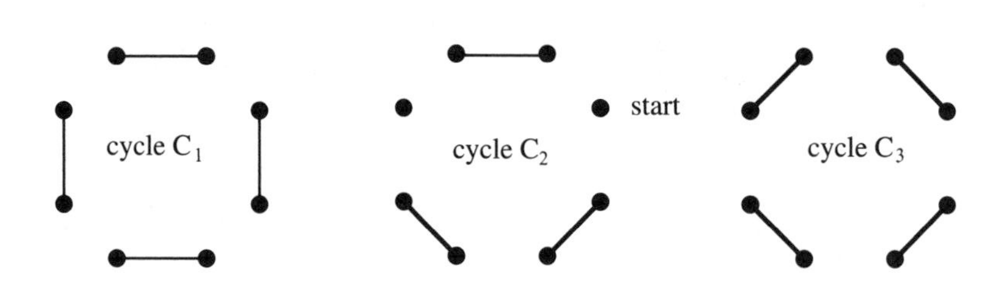

Figure 11.5: The encoding $\sigma_T(s,t)$.

11.3.4. Extension to Arbitrary Size Matchings

Having outlined how to estimate M_n/M_{n-1}, we now describe the estimation of M_k/M_{k-1} for $k < n$. The idea is essentially the same, using the Markov chain $\mathcal{C}_k$ to sample the elements of $M_k \cup M_{k-1}$ almost uniformly. One technical difficulty remains: can the conductance of each $\mathcal{C}_k$ be suitably bounded? There are techniques for showing that $\mathcal{C}_k$ does have high conductance, but we will circumvent this problem by reducing the problem of estimating the ratio r_k to the problem of estimating r_n.

Consider a graph $G(U, V, E)$ with minimum degree $n/2$, and obtain the graph $G_k(U', V', E')$ from G as follows: add $n-k$ new vertices to U to obtain U', and connect each of these new vertices to each vertex in V; similarly, V' is obtained by adding $n-k$ new vertices to V and connecting each new vertex to each vertex in U. It is easy to see that $n' = |U'| = |V'| = 2n-k$ and that the new vertices have degree $n \geq n'/2$ while the old vertices now have degree at least

$3n/2 - k \geq n'/2$. Thus, the new graph has the desired minimum degree property. It can now be shown that knowing the ratio $r_{n'}$ for G_k will enable us to compute the ratio r_k for G; the details are left as an exercise. The ratio $r_{n'}$ for G_k can be estimated as discussed above.

Exercise 11.11: Let R be the ratio of the number of perfect matchings and near-perfect matchings in G_k. Show that

$$R = \frac{m_k}{2(n-k)m_k + m_{k+1} + (n-k+1)^2 m_{k-1}},$$

where the m_i's refer to the original graph G. Using this, show how we can suitably estimate the values of m_k (for all k) with respect to G.

We conclude the following theorem.

Theorem 11.12: *There exists an (ϵ, δ)-FPRAS for the problem of estimating the number of perfect matchings in a bipartite graph of minimum degree at least $n/2$, where n is the number of nodes on each side of the bipartition.*

PROOF: By Exercise 11.9, it suffices to show that we can construct a near-uniform generator that has error bounded by $1/n^4$. In fact, we will now show (by going back through the entire chain of reasoning) that for $\tau = O(n^{15})$ the relative pointwise distance is $\Delta(\tau) = 1/n^4$.

Applying Theorem 11.7, Lemma 11.8, and Theorem 11.9, in succession, we obtain

$$\begin{aligned}
\Delta(\tau) &\leq \lambda_2^\tau N \\
&\leq \left(1 - \frac{\Phi^2}{2}\right)^\tau N \\
&\leq \left(1 - \frac{1}{288n^{12}}\right)^\tau N \\
&\leq e^{-\tau/288n^{12}} N \\
&\leq \frac{1}{n^4}
\end{aligned}$$

where the last inequality follows from the observation that $N \leq 2^{n^2}$ and $\tau \geq n^{15}$. □

The exponent of 15 in the running time can be reduced somewhat by using a more sophisticated algorithm and analysis.

11.4. Volume Estimation

In this section we briefly consider the problem of computing the volume of a given convex body K in n dimensions; we denote this volume by $\Upsilon(K)$. This is a classic problem with numerous applications; for instance, computing the number of linear extensions of a partial order can be reduced to computing the volume of an appropriately derived convex body. We give only an outline of the principal methods here, without any proofs. Pointers to details are given in the Notes section.

Before we examine the complexity of this question, we must pin down the notion of an input to this problem. If we are to allow arbitrary convex bodies as inputs, it is not even clear that there is a description of the body that has finite size. We assume that K is described by means of a *membership oracle*: the algorithm can specify a point p in space, and the oracle responds whether or not p is inside the body K. For technical reasons, we will assume further that K contains a sphere of radius at least r_1 centered at the origin, and in turn is contained in a sphere of radius r_2 centered at the origin. We assume that a call to the oracle takes unit time; other than this, we assume the usual RAM model of computation. We seek an (ϵ, δ)-FPRAS with a running time that is polynomial in $1/\epsilon$, $\log 1/\delta$, and n.

One example of a membership oracle comes from a convex polyhedron that is defined by the intersection of a set of m given half-spaces. Given a point p in the space, we can check in time $O(mn)$ whether p lies in this intersection. We begin with two negative results.

Theorem 11.13: *It is #**P**-hard to compute the volume of a polyhedron defined by the intersection of m half-spaces, each defined by a hyperplane with coefficients in* $\{0,1\}$.

Theorem 11.14: *Suppose that a deterministic polynomial-time algorithm uses the membership oracle for K and generates an upper bound* Υ_u *and a lower bound* Υ_l *on the volume of K. Then, there is a body K and a constant c such that*

$$\frac{\Upsilon_u}{\Upsilon_l} \geq c \left(\frac{n}{\log n} \right)^n .$$

These negative results motivate a FPRAS for this problem; we outline the main ideas below. We begin by explaining the basic idea in the plane. Consider a convex region R in the plane of unit diameter. If we were to enclose R in a rectangle U whose longer side is 1, we could use the following scheme. We pick a sequence of points independently at random from U, and count the fraction f of them that fall within R (using the membership oracle). Since we can easily compute the area $\Upsilon(U)$ of U, we can compute the random variable $f\Upsilon(U)$, which is an estimate of $\Upsilon(R)$ that can be shown to be close to its expectation for a suitably large number of samples. Note (using the Estimator Theorem 11.1) that the number of samples required grows with the ratio $\Upsilon(U)/\Upsilon(R)$.

Exercise 11.12: Show that $\Upsilon(U)/\Upsilon(R) \leq 2$ in the plane.

The main difficulty with extending this approach to higher dimensions is that if we were to enclose the given body K in an n-dimensional cuboid C, the ratio $\Upsilon(C)/\Upsilon(K)$ may be exponential in n. To address this, we make use of the fact that K lies within a sphere of known radius r_2; we denote this sphere by B. Further, it suffices to have a value of r_2 that is small enough that we can define a sequence of convex bodies $B = K_0 \supseteq K_1 \supseteq \cdots \supseteq K_q = K$ such that:

1. There is a constant c such that for $i \geq 1$, $\Upsilon(K_i) \leq c\Upsilon(K_{i-1})$.

2. The length of the sequence q is polynomial in n.

Then, it suffices to devise a FPRAS that will estimate the ratio $\Upsilon(K_i)/\Upsilon(K_{i-1})$ for all i; note the similarity of this approach to that used in estimating the permanent. By appealing to property 1 above, we would like to again pick points at random from K_i and measure the fraction of these that lie in K_{i-1}. Unlike the case of the rectangle U in the plane, picking such a random point in the body K_i is a non-trivial task. The solution is roughly as follows. We impose a suitably fine grid on n-dimensional space. Starting at a fixed point of the grid, we perform a random walk on the nodes of this grid that intersect K_i. The crux of the method is to argue that this random walk is rapidly mixing: in a number of steps that is polynomial in n, its probability distribution is almost uniform on all the grid nodes intersecting K_i. In this manner, we sample points at random from within K_i and measure the fraction that lie within K_{i-1}. We omit almost all the details here – the exact details of the walk, the definition of the sequence K_i, the use of the membership oracle for K for determining membership in K_i, and the proof of the rapid mixing. The reader may find pointers to all of these in the Notes section.

Notes

The class **#P** was first identified by Valiant [401, 402], who also established the **#P**-completeness of the DNF counting problem and the permanent problem. Welsh [416] gives a comprehensive and eminently readable survey of the state of the art in counting algorithms.

The equivalence of counting the number of distinct spanning trees in a graph to a determinant evaluation is due to Kirchhoff [193] and is known as the *matrix-tree theorem.* The approximation of counting problems was initiated in the work of Karp and Luby [247] and Stockmeyer [386]. The randomized approximation scheme for the DNF counting problem, as well as the formal definition of a randomized approximation scheme, is due to Karp and Luby [246]; this work was later extended and improved by Karp, Luby, and Madras [240]. These articles also present applications of the basic approach to network reliability problems (see also [109, 247]).

► **Research Problem 11.1:** Devise a deterministic FPAS for the DNF counting problem.

The best-known deterministic algorithm for computing the permanent of a 0-1 matrix is due to Ryser [361]. The application of the permanent problem to statistical physics is described in the paper by Jerrum and Sinclair [211]. The permanent approximation scheme described in this chapter owes its origins to a paper of Jerrum, Valiant, and Vazirani [212], in which they showed that estimating the number of matchings in a bipartite graph could be reduced to the problem of generating a perfect matching of the graph uniformly at random from all the perfect matchings in the graph. Broder [83] showed that near-uniform generation can be achieved by simulating a random walk on a Markov chain whose states correspond to matchings in the graph. The latter idea follows the lead of Aldous [11]. Broder proved that computing the number of perfect matchings in graphs of minimum degree at least $n/2$ is still $\#\boldsymbol{P}$-complete and showed that, even though the number of states could be exponentially large, the random walk could be efficiently implemented.

The technical portion of Broder's proof contained a subtle error, as was pointed out by Mihail [309]. Jerrum and Sinclair [211] subsequently gave the proof described above establishing that his algorithm was indeed correct. The conductance-based technique for analyzing the rate of convergence of a Markov chain is explored in detail in the paper by Sinclair and Jerrum [376]. Sinclair [377] has shown that it is possible to go directly from the canonical paths argument to a bound on λ_2 without going through the conductance argument. A comprehensive treatment of all these ideas, as well as some further applications, can be found in the monograph by Sinclair [375].

Dagum, Luby, Mihail and Vazirani [118] adopted a slightly different approach to characterizing graphs for which α is bounded above by a polynomial in n, rather than assuming that the graph has minimum degree at least $n/2$. Even with their improved characterization, this approach to estimating M_n works only for a sub-class of bipartite graphs in which every vertex degree is at least a constant fraction of n.

Currently, the most general condition under which it is possible to obtain an approximation to the number of perfect matchings in a bipartite graph is that $|M_{n-1}|/|M_n|$ be polynomially bounded. The following problem remains open.

► **Research Problem 11.2:** Devise a FPRAS for estimating the number of perfect matchings in any bipartite graph (with no restrictions on vertex degrees).

The results of Jerrum and Sinclair [211] and Motwani [312] imply that the Markov chain is rapidly mixing for almost every graph. This involves showing that, with high probability, a random graph [69] (of arbitrary density) has the property that the ratio r_k is polynomially bounded.

Karmarkar, Karp, Lipton, Lovász, and Luby [239] have made some progress in the direction of Research Problem 11.2: they give a randomized approximation algorithm which, with probability $1-\delta$, gives a number guaranteed to be in the interval $[M_n/(1+\epsilon), M_n(1+\epsilon)]$. Their algorithm runs in time $O(\text{poly}(\delta,\epsilon,n)2^{n/2})$, where $\text{poly}(\delta,\epsilon,n)$ is a function that grows polynomially in n. This result was improved considerably by Jerrum and Vazirani [210], who obtained an (ϵ,δ)-approximation with running time $O(\text{poly}(\delta,\epsilon,n)2^{\sqrt{n}\log^2 n})$.

Theorem 11.14 is due to Bárány and Füredi [46]. Dyer, Frieze, and Kannan [132] first gave the method outlined in Section 11.4 for volume estimation. The scheme and running time bounds that result have subsequently been refined in a sequence of papers. The reader is referred to the survey in Welsh's book [416] for pointers to these refinements, and to the host of open research problems remaining in this area.

Problems

11.1 In this problem we will design a Monte Carlo algorithm to estimate the value of π. Consider a circle of diameter 1 enclosed within a square with sides of length 1. We will sample points (uniformly and independently) from the square and set the indicator variable $X_t = 1$ if the tth point is inside the circle, and set $X_t = 0$ otherwise. It is clear that $\mathbf{E}[X] = N\pi/4$, where X is the sum of N of these indicator variables.

Give an upper bound on the value of N for which $4X/N$ gives an estimator of π that is accurate to d digits, with probability at least $1-\delta$.

11.2 (Due to R.M. Karp, M. Luby, and N. Madras [240].) Consider the following variant of the Coverage algorithm for approximating the DNF counting problem. The tth trial of this algorithm first picks a random clause C_t, where the probability of choosing a clause C_j is proportional to the number of satisfying truth assignments for it. Next, it selects a random satisfying truth assignment $\boldsymbol{a}$ for the chosen clause. (So far, this is exactly the same as the sampling procedure described before.) Define the random variable $X_t = 1/|\mathrm{cov}(\boldsymbol{a})|$, where $\mathrm{cov}(\boldsymbol{a})$ denotes the set of clauses that are satisfied by the truth assignment $\boldsymbol{a}$.

The estimator for $\#F$ is the random variable

$$Y = \eta \times \sum_{t=1}^{N} \frac{X_t}{N}$$

where η denotes the sum of the sizes of the coverage sets over all possible truth assignments. Prove that Y is an (ϵ, δ)-approximation for $\#F$ when

$$N = \frac{cm}{\epsilon} \ln \frac{1}{\delta}$$

for some small constant c. (**Hint:** Use the Chernoff-type bound derived in Problem 4.7.)

11.3 Prove the converse of Theorem 11.4. In other words, show that given an algorithm for estimating the number of matchings in a bipartite graph, it is possible to get a near-uniform generator of matchings in the bipartite graph.

11.4 In this problem we will see that the problem of counting the perfect matchings in graphs with large minimum degree is also $\#\boldsymbol{P}$-complete. Suppose there is a polynomial time algorithm A for counting the number of perfect matchings in a graph with minimum degree at least βn, for a constant $0 < \beta < 1$. Show that there must then exist a polynomial time algorithm for counting the number of perfect matchings in an arbitrary bipartite graph.

11.5 Consider the Markov chain induced by a random walk on a connected, undirected graph G on n vertices. How small can the conductance of this Markov chain be, the minimum being taken over connected, undirected graphs on n vertices? How large can it be?

11.6 Let G be a connected, undirected graph on n vertices.

(a) Consider the Markov chain induced by the following random process for moving from one spanning tree of G to another: pick edges e and f

independently and uniformly at random; if the current spanning tree is T and $T' = T + e - f$ is a spanning tree, then move to the new spanning T'; otherwise stay put at T. Show that the conductance of this Markov chain is bounded from below by $1/n^{O(1)}$.

(b) Suggest and analyze an algorithm for approximate counting of the number of spanning trees in a graph G, as an alternative to the matrix-tree theorem.

11.7 (Due to A. Sinclair and M.R. Jerrum [376].) An ergodic Markov chain with transition matrix $\boldsymbol{P}$ is said to be *time reversible* if for all i and j, $P_{ij}\pi_i = P_{ji}\pi_j$. This is equivalent to requiring that the matrix $\boldsymbol{DPD}^{-1}$ is symmetric, where $\boldsymbol{D}$ is a diagonal matrix with $D_{ii} = \sqrt{\pi_i}$. Clearly, the largest eigenvalue of $\boldsymbol{P}$ is $\lambda_1 = 1$; define $\lambda = \max_{i>1} |\lambda_i|$. Show that for any fixed choice of an initial state X_0, the relative pointwise distance of this Markov chain at time t is bounded as follows:

$$\Delta(t) \leq \frac{\lambda^t}{\min_i \pi_i}.$$

What does this imply for the random walk setting considered in Theorem 6.21?

CHAPTER 12

Parallel and Distributed Algorithms

In this chapter we discuss the solution of problems by a number of processors working in concert. In specifying an algorithm for such a setting, we must specify not only the sequence of actions of individual processors, but also the actions they take in response to the actions of other processors. The organization and use of multiple processors has come to be divided into two categories: *parallel processing* and *distributed processing*. In the former, a number of processors are coupled together fairly tightly: they are similar processors running at roughly the same speeds and they frequently exchange information with relatively small delays in the propagation of such information. For such a system, we wish to assert that at the end of a certain time period, all the processors will have terminated and will collectively hold the solution to the problem. In distributed processing, on the other hand, less is assumed about the speeds of the processors or the delays in propagating information between them. Thus, the focus is on establishing that algorithms terminate at all, on guaranteeing the correctness of the results, and on counting the number of messages that are sent between processors in solving a problem. We begin by studying a model for parallel computation. We then describe several parallel algorithms in this model: sorting, finding maximal independent sets in graphs, and finding maximum matchings in graphs. We also describe the randomized solution of two problems in distributed computation: the choice coordination problem and the Byzantine agreement problem.

12.1. The PRAM Model

Our model for parallel computation will be the *synchronous parallel random access machine*, which we will abbreviate by PRAM. The parallel computer will consist of P processors, each of which can be viewed as supporting the RAM model of computation (see Section 1.5.1). There is a *global memory* consisting of M locations; each processor has a (small) constant number of local registers to which it alone has access. Each of the P processors may read from and write into any of the M global memory locations; these global memory locations serve as the only mechanism for communication between the processors. Computation proceeds in a series of synchronous *parallel steps*. In a parallel step, each processor first chooses a global memory location whose contents it reads; next it executes an instruction on the operand fetched, together with any operands in its registers (the allowable instructions are any of those we allow for a conventional single-processor RAM). Finally, the step ends with the processor writing into a memory location of its choice. By our assumption of synchrony, every processor finishes executing step i before any processor begins executing step $i+1$. An instruction for the PRAM is a specification, for each processor, of the actions it is to perform in each of the three phases of a step. A parallel program is a sequence of such instructions.

We now address the important issue of *conflict resolution* in a PRAM: our definition of an instruction permits a number of processors to attempt to read from or write to the same global memory location in a step. Logically, there appears to be no problem in allowing several processors to read the contents of the same global memory location; however, physical limitations make this action difficult to implement in actual hardware. Of greater concern are the difficulties that arise when several processors attempt to write into the same global memory location; which of the (possibly differing) values is actually written into the memory location? A number of solutions have been proposed for this problem of *concurrent writing*. We will adopt the simplest of these: we insist that the parallel program ensure that no execution will ever result in a concurrent write. Thus we deal only with *exclusive write* PRAMs.

As mentioned above, the issue of whether or not to allow concurrent reads is a matter of attention to hardware implementation. These various read/write models for PRAMs are abbreviated as EREW, CREW, and CRCW, where the first two letters denote whether reading is exclusive or concurrent and the last two denote what is permissible for writing. In this chapter, we will only consider EREW and CREW PRAMs.

Of particular theoretical interest is the solution of problems by PRAM algorithms in which the number of processors P is a polynomial function of the input size n, and the number of PRAM steps is bounded by a polylogarithmic function of n. We define the classes ***NC*** and ***RNC*** to capture these notions.

► **Definition 12.1:** The class ***NC*** consists of languages L that have a PRAM algorithm A such that for any $x \in \Sigma^*$

- $x \in L \Rightarrow A(x)$ accepts;
- $x \notin L \Rightarrow A(x)$ rejects;
- the number of processors used by A on x is polynomial in $|x|$;
- the number of steps used by A on x is polylogarithmic in $|x|$.

For randomized PRAM algorithms, we similarly define the class ***RNC***:

► **Definition 12.2:** The class ***RNC*** consists of languages L that have a PRAM algorithm A such that for any $x \in \Sigma^*$

- $x \in L \Rightarrow \mathbf{Pr}[A(x) \text{ accepts}] \geq 1/2$;
- $x \notin L \Rightarrow \mathbf{Pr}[A(x) \text{ accepts}] = 0$;
- the number of processors used by A on x is polynomial in $|x|$;
- the number of steps used by A on x is polylogarithmic in $|x|$.

As in the case of ***RP***, although the definition is in terms of decision or language problems, there is an obvious generalization to function computations. Notice that an ***RNC*** algorithm is Monte Carlo with one-sided error. We can define the two-sided error version analogous to ***BPP***. The Las Vegas version of this class (zero-error and polylogarithmic expected time) is called ***ZNC***, and is defined similar to ***ZPP***.

Exercise 12.1: In the above definitions, we did not distinguish between the various models of concurrent reading and writing. Show that if a problem has a CRCW PRAM algorithm using a number of processors that is polynomial in the input size, and a number of steps that is polylogarithmic, then the problem has an EREW PRAM algorithm using a number of processors that is polynomial in the input size, and a number of steps that is polylogarithmic.

12.2. Sorting on a PRAM

In this section we study algorithms for sorting n numbers on a PRAM with n processors. For convenience, we will assume that the input numbers to be sorted all have distinct values. Our eventual goal will be a randomized (***ZNC***) algorithm that terminates in $O(\log n)$ steps with high probability. Such an algorithm would thus result in a total of $O(n \log n)$ operations among all processors, with high probability.

Consider the implementation of the following variant of randomized quicksort on a CREW PRAM. Initially, each of the n processors contains a distinct input element. We first describe the structure of the algorithm. Following this high-level description, we will break down each stage of this description into a sequence of PRAM steps. Let P_i denote the ith processor.

0. If $n = 1$ stop.

1. We pick a *splitter* uniformly at random from the n input elements.

2. Each processor determines whether its element is bigger or smaller than the splitter.

3. Let j denote the rank of the splitter. If $j \notin [n/4, 3n/4]$, we declare the step a failure and repeat starting at (1) above. If $j \in [n/4, 3n/4]$, the step is a success.

We then move the splitter to P_j. Each element that is smaller than the splitter is moved to a distinct processor P_i for $i < j$. Each element that is larger than the splitter is moved to a distinct processor P_k for $k > j$.

4. We sort the elements recursively in processors P_1 through P_{j-1}, and the elements in processors P_{j+1} through P_n. These recursive sorts are independent of each other.

Let us study the number of CREW PRAM steps taken by each of the above stages. Before we proceed with a detailed analysis, we make a prognosis of what we need in order for the above algorithm to terminate in $O(\log n)$ steps. The best we can hope for is success whenever we split. If we were fortunate enough that this were to happen, every sequence of recursive splits would terminate within $O(\log n)$ stages. Even so, in order for the algorithm to terminate in $O(\log n)$ steps, we would require each split to be implemented in a constant number of steps. Unfortunately we know of no way of doing this.

The second stage in our scheme is trivial and can be implemented in a single step of a CREW PRAM. Let us turn to Stage 3 of the above description. Our goal is to ensure that processor P_i, for $i < j$, contains a distinct input element whose rank is smaller than j, and similarly processor P_k for $k > j$, contains a distinct input element whose rank is larger than j. How many PRAM steps are taken up by this process?

Processor P_i sets a bit b_i in one of its registers to 0 if its element is greater than the splitter, and to 1 otherwise. For all i, let $S_i = \sum_{t \le i} b_t$.

Exercise 12.2: Devise a PRAM algorithm by which, given the b_i, the S_i can be computed (with the result contained in P_i) in $O(\log n)$ steps. Using this, show how Stage 3 of the algorithm can be implemented in $O(\log n)$ steps.

Thus, we see that a single splitting stage can be implemented in $O(\log n)$ steps of a CREW PRAM. In Problem 12.1 we will see that from this, we can infer that the above algorithm terminates in $O(\log^2 n)$ steps with high probability.

The shortcoming of the above scheme is that the splitting work in Stage 3, consuming $O(\log n)$ steps, yielded a relatively small benefit – it cuts the problem size down from n to a constant fraction of n. To improve on this, we consider a more efficient algorithm in which we invest the same amount of work in splitting, but in the process break up the problem into pieces of size $n^{1-\epsilon}$ for a fixed constant ϵ. If we could do this, we could hope for an overall parallel running time of $O(\log n)$ steps: at the next level of recursion, the splitting time would be logarithmic in $n^{1-\epsilon}$, which is a constant fraction of the splitting time at the first level. Thus, the times for proceeding from one level of recursion to the next would form a geometric series summing to $O(\log n)$. The following two exercises pave the way for a concrete scheme for implementing this idea. Exercise 12.3 demonstrates that we can indeed sort in $O(\log n)$ steps if our PRAM were endowed with many more processors than elements to be sorted.

Exercise 12.3: Consider a CREW PRAM having n^2 processors. Suppose that each of the processors P_1 through P_n has an input element to be sorted. Give a deterministic algorithm by which this PRAM can sort these n elements in $O(\log n)$ steps. (**Hint:** We have enough processors to compare all pairs of elements.)

Next, suppose that we have n processors and n elements. Suppose that processors P_1 through P_r contain r of the elements in sorted order, and that processors P_{r+1} through P_n contain the remaining $n-r$ elements. Call the sorted elements in the first r processors the *splitters.* For $1 \leq j \leq r$, let s_j denote the jth largest splitter. Our goal is to "insert" the $n-r$ unsorted elements among the splitters, in the following sense.

1. Each processor should end up with a distinct input element.
2. Let $i(s_j)$ denote the index of the processor containing s_j following the insertion operation. Then, for all $k < i(s_j)$, processor P_k contains an element that is smaller than s_j; similarly, for all $k > i(s_j)$, processor P_k contains an element that is larger than s_j.

In other words, the splitters are contained in processors in increasing order, and the remaining elements are in processors between their "adjacent" splitters.

Exercise 12.4: For n processors, and n elements of which $\sqrt{n}$ are splitters, give a deterministic scheme that completes the above insertion process in $O(\log n)$ steps.

Here are the stages of our parallel sorting algorithm, which we call **BoxSort.** Note that it is a Las Vegas algorithm: it always produces the correct output. Further, it always uses a fixed number of processors; only the number of parallel steps is a random variable. This will be typical of all the parallel algorithms we present. The function **LogSort** is described following Exercise 12.5.

Algorithm BoxSort:

Input: A set of numbers S.

Output: The elements of S sorted in increasing order.

1. Select $\sqrt{n}$ elements at random from the n input elements. Using all n processors, sort them in $O(\log n)$ steps (using the ideas in Exercise 12.3). If two splitters are adjacent in this sorted order, we call them *adjacent splitters.*
2. Using the sorted elements from Stage 1 as splitters, insert the remaining elements among them in $O(\log n)$ steps (using the ideas in Exercise 12.4).
3. Treating the elements that are inserted between adjacent splitters as sub-problems, recur on each sub-problem whose size exceeds $\log n$. For sub-problems of size $\log n$ or less, invoke **LogSort**.

Note that in Step 3 we have available as many processors as elements for each sub-problem on which we recur. The sub-problems that result from the $\sqrt{n}$ splitters have size roughly $\sqrt{n}$, with good probability. This fits with our paradigm for progressing from a problem of size n to one of size $n^{1-\epsilon}$ in $O(\log n)$ steps. As we will see below, with high probability every sub-problem resulting from a splitting operation is small, provided the set being split is itself not too small. We deal with this issue using the following idea. When we have $\log n$ elements to be sorted using $\log n$ processors, we abandon the recursive approach and use brute force:

Exercise 12.5: Show that a CREW PRAM with m processors can sort m elements deterministically in $O(m)$ steps.

Thus, when a sub-problem size is down to $\log n$, we can sort it with the $\log n$ available processors in $O(\log n)$ steps; we call this operation **LogSort**.

We now analyze the use of random sampling for choosing the splitters. Let us call the set of elements that fall between adjacent splitters a *box*. The analysis is similar to the one we used in the analysis of randomized selection in Section 3.3. By invoking the Chernoff bound instead of the Chebyshev bound, the following is an easy consequence:

Exercise 12.6: Consider m splitters chosen uniformly at random from m^2 given distinct elements. Show that the probability that a box has size exceeding bm is at most ma^b, for a constant $a < 1$.

To complete the analysis of the algorithm, we represent an execution of the algorithm by a tree. Each node of the tree is a box that arises during the execution. For this purpose, we will also regard the n input elements as forming a box (of size n), and this is the root of our tree. The children of a node are the boxes that arise when it is partitioned by random splitters. Each leaf is a box of size at most $\log n$.

We are interested in root-leaf paths in this tree. In bounding the running time of algorithm, the quantity of interest is not the length of such root-leaf paths, but rather the number of PRAM steps that elapse as we go down such a path. This is because the time to proceed from a box to one of its children is logarithmic in the size of the box. We will argue that with high probability, the sum of the logarithms of box sizes on any root-leaf path is $O(\log n)$, and this will yield an overall running time of $O(\log n)$.

The idea is to partition the interval $[1, n]$ into sub-intervals $I_0, I_1, \ldots,$ and bound the probability that a box whose size is in I_k has a child whose size is also in I_k. To this end, let γ and d be fixed constants such that $1/2 < \gamma < 1$ and $1 < d < 1/\gamma$. For a positive integer k, define $\tau_k = d^k$, $\rho_k = n^{\gamma^k}$, and the interval $I_k = [\rho_{k+1}, \rho_k]$.

Exercise 12.7: Show that $\rho_k < \log n$ for a value of $k \leq c \log\log n$, for a constant c that depends only on γ.

Thus we confine our attention to $O(\log\log n)$ intervals I_k. For a box B in the tree, we say that $\alpha(B) = k$ if $|B| \in I_k$. In terms of this notation, the time to split B is $O(\log \rho_{\alpha(B)})$. For a root-leaf path $\zeta = (B_1, \ldots, B_t)$, we will study $\sum_{j=1}^{t} \log \rho_{\alpha(B_j)}$, since the overall running time of the algorithm is

$$O\left(\log n + \max_{\zeta} \sum_{j=1}^{t} \log \rho_{\alpha(B_j)}\right).$$

For a path $\zeta = (B_1, \ldots, B_t)$, we say that event $\mathcal{E}_\zeta$ holds if the sequence $\alpha(B_1), \ldots, \alpha(B_t)$ does not contain the value k more than τ_k times, for $1 \leq k \leq c \log\log n$. If $\mathcal{E}_\zeta$ holds, the number of PRAM steps spent on path ζ is at most

$$O\left(\log n + \sum_{k=1}^{\infty} \tau_k \gamma^k \log n\right).$$

Since $\tau_k = d^k$, and $\gamma d < 1$, this sums to $O(\log n)$. Thus it suffices to argue that $\mathcal{E}_\zeta$ holds with high probability for any ζ. This is an easy calculation following the bound from Exercise 12.6.

Lemma 12.1: *There is a constant $\beta > 1$ such that $\mathcal{E}_\zeta$ holds with probability at least $1 - \exp(-\log^\beta n)$.*

The following sequence of three probability calculations establishes Lemma 12.1. These calculations are straightforward, and the reader is asked to perform them in Problem 12.2.

1. Bound the probability that $\alpha(B_{j+1}) = \alpha(B_j)$ using the result of Exercise 12.6.
2. Bound the probability that for any particular k, the value k is contained more than τ_k times in the sequence $\alpha(B_1), \ldots, \alpha(B_t)$.
3. Bound the probability that for $1 \leq k \leq c \log\log n$, the value k is contained more than τ_k times in the sequence $\alpha(B_1), \ldots, \alpha(B_t)$.

Since the number of paths ζ in an execution is at most n, we have:

Theorem 12.2: *There is a constant $b > 0$ such that with probability at least $1 - \exp(-\log^b n)$ the algorithm* **BoxSort** *terminates in $O(\log n)$ steps.*

12.3. Maximal Independent Sets

Let $G(V, E)$ be an undirected graph with n vertices and $m = \Omega(n)$ edges. A subset of vertices $I \subseteq V$ is said to be *independent* in G if no edge in E has both its end-points in I. Equivalently, I is independent if for all $v \in I$, $\Gamma(v) \cap I = \emptyset$. Recall that $\Gamma(v)$ is the set of vertices in V that are adjacent to v and that the degree of v is $d(v) = |\Gamma(v)|$.

An independent set I is *maximal* if it is not properly contained in any other independent set in G. Recall that the problem of finding a *maximum* independent set is ***NP***-hard. In contrast, finding a maximal independent set (MIS) is trivial in the sequential setting. The following greedy algorithm constructs an MIS in $O(m)$ time.

Algorithm Greedy MIS:

Input: Graph $G(V, E)$ with $V = \{1, 2, \ldots, n\}$.

Output: A maximal independent set $I \subseteq V$.

1. $I \leftarrow \emptyset$.
2. **for** $v = 1$ **to** n **do**
   if $I \cap \Gamma(v) = \emptyset$ **then** $I \leftarrow I \cup \{v\}$.

Exercise 12.8: Prove that the **Greedy MIS** algorithm terminates in $O(m)$ time with a maximal independent set, if the input is given in the form of an adjacency list.

A greedy algorithm such as this is inherently sequential. The output of this algorithm is called the *lexicographically first* MIS (LFMIS). It is known that the existence of an ***NC*** (or ***RNC***) algorithm for finding the LFMIS would imply that ***P*** = ***NC*** (respectively, ***P*** = ***RNC***), a consequence that appears almost as unlikely as ***P*** = ***NP***. Thus, we have the somewhat paradoxical situation that the most trivial algorithm finds the LFMIS sequentially, whereas it appears impossible to solve it fast in parallel. However, it turns out that there are simple parallel algorithms for finding an MIS (not necessarily the lexicographically first MIS). We start by describing an ***RNC*** algorithm and later indicate how it can be derandomized to obtain an ***NC*** algorithm. The problem of verifying an MIS is relatively easy to solve in parallel.

Exercise 12.9: Devise a deterministic EREW PRAM algorithm for *verifying* that a set I is an MIS, using $O(m/\log m)$ processors and $O(\log m)$ time.

Consider the variant of the **Greedy MIS** algorithm, which starts with $I = \emptyset$ and repeatedly performs the following step: pick any vertex v, add v to I, and delete v and $\Gamma(v)$ from the graph. The algorithm terminates when all vertices have either been deleted or added to I. Choosing v to be the lowest numbered vertex present in the graph leads to exactly the same outcome as in **Greedy MIS**.

The key idea behind the parallel algorithm is to generalize the basic iterative step in the new algorithm: find an independent set S, add S to I, and delete $S \cup \Gamma(S)$ from the graph. The trick is to ensure that each iteration can be implemented fast in parallel, while also guaranteeing that the total number of iterations is small. One way of ensuring that the number of iterations is small is to choose an independent set S such that $S \cup \Gamma(S)$ is large. This is difficult, but we achieve the same effect by ensuring that the number of edges incident on $S \cup \Gamma(S)$ is a large fraction of the total number of remaining edges; clearly, this will result in an empty graph in a small of number of iterations.

To find such an independent set S, we pick a large random set of vertices $R \subseteq V$. While it is quite unlikely that R will be independent, biasing the sampling in favor of low degree vertices will ensure that there are very few edges with both end-points in R. To obtain the independent set from R we consider each edge of this type and drop the end-point of lower degree. This results in an independent set, and the choice of the end-point retained for S ensures that $\Gamma(S)$ is likely to be large.

This idea is implemented in Algorithm **Parallel MIS**, where the marking of a vertex corresponds to selecting it for the set R. We assume that each vertex (and edge) of G is assigned a dedicated processor that performs the parallel tasks associated with that vertex (or edge). This uses a total of $O(n + m)$ processors.

Algorithm Parallel MIS:

Input: Graph $G(V, E)$.

Output: A maximal independent set $I \subseteq V$.

1. $I \leftarrow \emptyset$.

2. repeat

- **2.1. for all** $v \in V$ **do** (in parallel)
  if $d(v) = 0$ **then** add v to I and delete v from V
  else mark v with probability $1/2d(v)$.
- **2.2. for all** $(u, v) \in E$ **do** (in parallel)
  if both u and v are marked
  then unmark the lower degree vertex.
- **2.3. for all** $v \in V$ **do** (in parallel)
  if v is marked **then** add v to S.
- **2.4.** $I \leftarrow I \cup S$.
- **2.5.** delete $S \cup \Gamma(S)$ from V, and all incident edges from E.

until $V = \emptyset$

Ties are broken arbitrarily in Step 2.2. It is clear that the set S in Step 2.3 is an independent set. The reader should verify that this algorithm is guaranteed to terminate with a maximal independent set in a linear number of iterations. Our goal is to prove that the random choices in Step 2.1 will ensure that the expected number of iterations is in fact $O(\log n)$. We leave the details of implementing each iteration in *NC* as an exercise.

Exercise 12.10: Show that each iteration of the **Parallel MIS** algorithm can be implemented in $O(\log n)$ time using an EREW PRAM with $O(n+m)$ processors.

The analysis is based on showing that the expected fraction of edges removed from E during each iteration is bounded from below by a constant. In fact, we will focus only on a specific class of *good* edges, defined as follows.

► **Definition 12.3:** A vertex $v \in V$ is *good* if it has at least $d(v)/3$ neighbors of degree no more than $d(v)$; otherwise, the vertex is *bad*. An edge is good if at least one of its end-points is a good vertex, and it is bad if both end-points are bad vertices.

In the following discussion, we will analyze only a single iteration of the **Parallel MIS** algorithm. The notion of goodness is with respect to the vertices and edges surviving at the start of that specific iteration. It should be clear that the argument can be applied repeatedly to the successive iterations; together with Theorem 1.3, this implies the result.

We start with an intuitive sketch of the analysis, which is then fleshed out in a sequence of lemmas. A good vertex is quite likely to have one of its lower degree neighbors in S and, thereby be deleted from V. We will show that the number of good edges is large, and since good vertices are likely to be deleted, a large number of edges will be deleted during each iteration.

Lemma 12.3: *Let $v \in V$ be a good vertex with degree $d(v) > 0$. Then, the probability that some vertex $w \in \Gamma(v)$ gets marked is at least $1 - \exp(-1/6)$.*

PROOF: Each vertex $w \in \Gamma(v)$ is marked independently with probability $1/2d(w)$. Since v is good, there exist at least $d(v)/3$ vertices in $\Gamma(v)$ with degree at most $d(v)$. Each of these neighbors gets marked with probability at least $1/2d(v)$. Thus, the probability that none of these neighbors of v gets marked is at most

$$\left(1 - \frac{1}{2d(v)}\right)^{d(v)/3} \leq e^{-1/6}.$$

The remaining neighbors of v can only help in increasing the probability under consideration. □

Lemma 12.4: *During any iteration, if a vertex w is marked then it is selected to be in S with probability at least $1/2$.*

PROOF: The only reason a marked vertex w becomes unmarked, and hence not selected for S, is that one of its neighbors of degree at least $d(w)$ is also marked. Each such neighbor is marked with probability at most $1/2d(w)$, and the number of such neighbors certainly cannot exceed $d(w)$. Thus, the probability that a marked vertex is selected to be in S is at least

$$\begin{aligned} 1 &- \mathbf{Pr}[\exists x \in \Gamma(w) \text{ such that } d(x) \geq d(w) \text{ and } x \text{ is marked}] \\ &\geq 1 - |\{x \in \Gamma(w) \mid d(x) \geq d(w)\}| \times \frac{1}{2d(w)} \\ &\geq 1 - \sum_{x \in \Gamma(w)} \frac{1}{2d(w)} \\ &= 1 - d(w) \times \frac{1}{2d(w)} \\ &= \frac{1}{2}. \end{aligned}$$

□

Let v be a good vertex with $d(v) > 0$, and consider the event $\mathcal{E}$ that some vertex in $\Gamma(v)$ does indeed get marked. Let w be the lowest-numbered marked vertex in $\Gamma(v)$. By Lemma 12.4, the probability that w is selected to be in S is at least $1/2$. Clearly, if $w \in S$, then v must belong to $S \cup \Gamma(S)$. Using the bound on the probability of the event $\mathcal{E}$ from Lemma 12.3, we obtain the following.

Lemma 12.5: *The probability that a good vertex belongs to $S \cup \Gamma(S)$ is at least $(1 - \exp(-1/6))/2$.*

The final step is to bound the number of good edges.

Lemma 12.6: *In a graph $G(V, E)$, the number of good edges is at least $|E|/2$.*

PROOF: Direct the edges in E from the lower degree end-point to the higher degree end-point, breaking ties arbitrarily. Define $d_i(v)$ and $d_o(v)$ as the in-degree and out-degree, respectively, of the vertex v in the resulting digraph. It follows from the definition of goodness that for each bad vertex v,

$$d_o(v) - d_i(v) \geq \frac{d(v)}{3} = \frac{d_o(v) + d_i(v)}{3}.$$

For all S, $T \subseteq V$, define the subset of the (oriented) edges $E(S, T)$ as those edges that are directed from vertices in S to vertices in T; further, define $e(S, T)$ to be $|E(S, T)|$. Let V_G and V_B be the set of good and bad vertices, respectively. The total degree of the bad vertices is given by

$$\begin{aligned} 2e(V_B, V_B) &+ e(V_B, V_G) + e(V_G, V_B) \\ &= \sum_{v \in V_B} (d_o(v) + d_i(v)) \\ &\leq 3 \sum_{v \in V_B} (d_o(v) - d_i(v)) \end{aligned}$$

$$\begin{aligned}
&= 3\sum_{v\in V_G}(d_i(v) - d_o(v)) \\
&= 3[(e(V_B, V_G) + e(V_G, V_G)) - (e(V_G, V_B) + e(V_G, V_G))] \\
&= 3[e(V_B, V_G) - e(V_G, V_B)] \\
&\leq 3[e(V_B, V_G) + e(V_G, V_B)]
\end{aligned}$$

The first and last expressions in this sequence of inequalities imply that $e(V_B, V_B) \leq e(V_B, V_G) + e(V_G, V_B)$. Since every bad edge contributes to the left side and only good edges contribute to the right side, the desired result follows. □

Since a constant fraction of the edges are incident on good vertices, and good vertices get eliminated with a constant probability, it follows that the expected number of edges eliminated during an iteration is a constant fraction of the current set of edges. By Theorem 1.3, this implies that the expected number of iterations of the **Parallel MIS** algorithm is $O(\log n)$.

Theorem 12.7: *The* **Parallel MIS** *algorithm has an EREW PRAM implementation running in expected time* $O\left(\log^2 n\right)$ *using* $O(n+m)$ *processors.*

It is straightforward to obtain a high-probability version of this result.

We briefly describe the construction of an ***NC*** algorithm for MIS obtained by a derandomization of the ***RNC*** algorithm described above. The first step is to show that the preceding analysis works even when the marking of the vertices is not completely independent, but instead is only pairwise independent. Note that the only part of the analysis that uses complete independence is Lemma 12.3. In Problem 12.9 the reader is asked to prove that a marginally weaker version of Lemma 12.5 holds even with pairwise independent marking of vertices. The key advantage of pairwise independence is that only $O(\log n)$ random bits are required to generate the sample points in the corresponding probability space (see the discussion in Section 3.4). In the current application, it is necessary to generate pairwise independent Bernoulli random variables that are not uniform. In Problem 12.10, the reader is asked to modify the earlier construction of pairwise independent probability space to apply to Bernoulli variables that take on the value 1 with non-uniform probabilities, i.e., the marking probabilities of $1/2d(v)$.

The final and most crucial idea is to observe that the total number of choices of the $O(\log n)$ random bits needed for generating pairwise independent marking is polynomially bounded. All such choices can be tried in parallel to see if they yield a good marking, i.e., a marking of vertices that leads to an appropriately large reduction in the number of edges. Note that in each iteration, we are guaranteed that most choices of the random bits will give a good marking; in particular, there exists at least one setting of the $O(\log n)$ random bits that will provide a good marking. Trying all possibilities will (deterministically) identify a good marking. Thus, each iteration can be derandomized and the entire algorithm can be implemented in ***NC***.

12.4. Perfect Matchings

We now turn to the problem of finding an independent set of *edges* (or a *matching*) in a graph. Let $G(V, E)$ be a graph with the vertex set $V = \{1, \ldots, n\}$; without loss of generality, we may assume that n is even. Recall (Chapter 7) that a *matching* in G is a collection of edges $M \subset E$ no two of which are incident on the same vertex. A *maximal matching* is a matching that is not properly contained in any other matching. A *maximum matching* is a matching of maximum cardinality, and a *perfect matching* is one containing an edge incident on every vertex of G.

The matchings in a graph $G(V, E)$ correspond to independent sets in the *line graph* H obtained by creating a vertex for each edge in E, with two such vertices being adjacent if the corresponding edges in E are incident on the same vertex. This implies that the problem of finding matchings is a special case of the independent set problem. A maximal matching can be found sequentially via a greedy algorithm, and on a PRAM, as suggested in Problem 12.6, using the algorithms discussed in Section 12.3. Unlike the case of maximum independent sets, the problem of finding a maximum matching has a polynomial time solution. This raises the possibility of constructing an ***NC*** algorithm for maximum matchings. However, randomization appears to be an essential component of all known fast parallel algorithms for maximum matching, and we devote this section to describing one such ***RNC*** algorithm.

For now we focus on the problem of finding a perfect matching in a graph that is guaranteed to have one, deferring the issue of finding a maximum matching till later. First we show that the decision problem of determining the *existence* of a perfect matching is in ***RNC***. This is based on the algebraic techniques developed in Chapter 7; the reader is advised to review Sections 7.2 and 7.3 from that chapter. We make use of Tutte's Theorem described in Problem 7.8; this is a generalization of Theorem 7.3, which dealt with the case of bipartite matchings.

Theorem 12.8 (Tutte's Theorem): *Let A be the $n \times n$ (skew-symmetric) Tutte matrix of indeterminates obtained from $G(V, E)$ as follows: a distinct indeterminate x_{ij} is associated with the edge (v_i, v_j), where $i < j$, and the corresponding matrix entries are given by $A_{ij} = x_{ij}$ and $A_{ji} = -x_{ij}$, that is,*

$$A_{ij} = \begin{cases} x_{ij} & (v_i, v_j) \in E \text{ and } i < j \\ -x_{ji} & (v_i, v_j) \in E \text{ and } i > j \\ 0 & (u_i, v_j) \notin E \end{cases}$$

Then G has a perfect matching if and only if $\det(A)$ *is not identically zero.*

The ***RNC*** algorithm for deciding the existence of a perfect matching in G first constructs the matrix A with each indeterminate replaced by independently and uniformly chosen random values from a suitably large set of integers, as described in Section 7.2. Then, it evaluates the determinant of the resulting integer matrix. If G has a perfect matching, then with suitably large probability, the determinant will be non-zero. On the other hand, if G does not have any perfect matchings, the determinant will always be zero.

The first stage of this algorithm is easily implemented in ***NC***. Finding the determinant of a matrix in ***NC*** is not trivial, but at least one ***NC*** algorithm is known (see the Notes section). Thus the problem of deciding the existence of a perfect matching is in ***RNC***.

We turn to the task of actually finding a perfect matching in a graph. Once again, the idea is to reduce the search problem to some matrix computations. We summarize known results for parallel matrix computations without attempting to describe the algorithms in any detail.

The (i, j) *minor* of a matrix $\mathbf{U}$, denoted $\mathbf{U}^{ij}$, is the matrix obtained by deleting the ith row and the jth column of $\mathbf{U}$. The *adjoint* $\text{adj}(\mathbf{U})$ of the matrix $\mathbf{U}$ is the matrix $\mathbf{A}$ whose (j, i) entry has absolute value equal to the determinant of the (i, j) minor of $\mathbf{U}$, i.e., $A_{ji} = (-1)^{i+j} \det(\mathbf{U}^{ij})$. It is easy to verify the following relation: $\mathbf{U}\text{adj}(\mathbf{U}) = \det(\mathbf{U})$.

Theorem 12.9: *Let $\mathbf{U}$ be an $n \times n$ matrix whose entries are k-bit integers. Then the determinant, adjoint, and inverse of $\mathbf{U}$ can be computed in* ***NC***. *In particular, let $\text{MM}(n) = O(n^{2.376})$ denote the number of arithmetic operations required to multiply two $n \times n$ matrices. Then the determinant can be computed in $O(\log^2 n)$ time using $O(n^2 \text{MM}(n))$ processors; further, there are* ***RNC*** *algorithms for computing the inverse and the adjoint running in time $O(\log^2 n)$ using $O(n^{3.5} k)$ processors.*

It is instructive to attempt to search for perfect matchings using the decision algorithm described above. It is not very hard to see that this can be done for the special case where the graph has a *unique* perfect matching.

Exercise 12.11: Suppose that G has a unique perfect matching M. Analyze the effect of removing an edge on the determinant of the Tutte matrix, considering both the case where the edge belongs to M and where it does not belong to M. Using this analysis, devise an ***RNC*** algorithm for finding the matching M.

As outlined in Problem 12.15, an ***NC*** algorithm is possible for finding a unique perfect matching. In fact, it is known that there is an ***NC*** algorithm for finding perfect matchings in graphs with a polynomial number of perfect matchings. However, these algorithms break down when the number of perfect matchings in the graph is large.

The problem with having a large number of perfect matchings is that it is necessary to coordinate the processors to search for the same perfect matching. This is the major stumbling block in the parallel solution of the matching problem and is perhaps the main reason why no ***NC*** algorithm is known. If the number of matchings is small, then the processors can easily focus on the same perfect matching. The first ingredient in the ***RNC*** algorithm is to take an arbitrary graph and *isolate* a specific perfect matching. The isolation is achieved by assigning weights to the edges and looking for a minimum weight perfect matching. Of course, there is no reason why there should be a unique minimum weight perfect matching but, as we show in the next section, if the weights are chosen at random there is a good chance that isolation occurs.

12.4.1. The Isolating Lemma

Our goal now is to define a positive integer weight function over the edges of G, say $w : E \to \mathbb{Z}^+$, such that there is a unique minimum weight perfect matching. Observing that the set of all possible perfect matchings can be viewed as a family of subsets of E, we consider a more general setting involving an arbitrary set family.

► **Definition 12.4:** A *set system* $(X, \mathcal{F})$ consists of a finite universe $X = \{x_1, \ldots, x_m\}$ and a family of subsets $\mathcal{F} = \{S_1, \ldots, S_k\}$, where $S_i \subseteq X$ for $1 \leq i \leq k$. The *dimension* of the set system is (the size of the universe) m.

Given a positive integer weight function $w : X \to \mathbb{Z}^+$, we define the weight of a set $S \subseteq X$ as $w(S) = \sum_{x_j \in S} w(x_j)$. The following lemma shows that a random weight function is quite likely to lead to a unique set of $\mathcal{F}$ being of minimum weight.

Lemma 12.10 (Isolating Lemma): *Suppose $(X, \mathcal{F})$ is a set system of dimension m. Let $w : X \to \{1, \ldots, 2m\}$ be a positive integer weight function defined by assigning to each element of X a random weight chosen uniformly and independently from $\{1, \ldots, 2m\}$. Then,*

$$\mathbf{Pr}[\textit{there is a unique minimum weight set in } \mathcal{F}] \geq \frac{1}{2}.$$

Remark: This lemma is truly counterintuitive. First of all, the size of $\mathcal{F}$ is completely irrelevant to the claim. This allows the family $\mathcal{F}$ to be of size as large as 2^m. Since the weights of the sets must lie in the range $\{1, \ldots, 2m^2\}$, one would expect that there could be as many as $2^m/(2m^2)$ sets of any given weight. However, the weights of the sets follow the lattice structure of the family of all subsets of X, thereby ensuring that the weights of the sets are not independent or uniformly distributed.

PROOF: We assume, without loss of generality, that each element of X occurs in at least one of the sets in $\mathcal{F}$. Suppose that we have chosen the (random) weights of all elements of X except one, say x_i. Let W_i be the weight of a minimum weight set containing x_i, computed by ignoring the (undetermined) weight of x_i. Further, let $\overline{W}_i$ be the weight of a minimum weight set not containing the element x_i. Define $\alpha_i = \overline{W}_i - W_i$ and note that α_i could be negative.

Suppose that initially x_i is assigned the weight $-\infty$ (actually, $-2m^2$ will suffice). It is clear that now every set of minimum weight must contain x_i. Consider the effect of increasing the weight of x_i until it reaches $+\infty$ (here, $2m^2$ will suffice). At this point it is clear that no set of minimum weight contains x_i.

We claim that for $w(x_i) < \alpha_i$, every minimum weight set must contain x_i, because there exists a set containing x_i of weight $W_i + w(x_i) < \overline{W}_i$, and all sets not containing x_i must have weight at least $\overline{W}_i$. Similarly, we claim that for $w(x_i) > \alpha_i$, no minimum weight set contains x_i, because any set containing x_i has weight at least $W_i + w(x_i) > \overline{W}_i$, and there exists a set not containing x_i of weight $\overline{W}_i$.

Thus, so long as $w(x_i) \neq \alpha_i$, either every minimum weight set contains x_i or none of them contains x_i. We say that x_i is ambiguous when $w(x_i) = \alpha_i$, since then it cannot be said for certain whether x_i will belong to a minimum weight set chosen arbitrarily. The crucial observation is that since α_i depends only on the weights of the elements other than x_i, and the weights are chosen independently, the random variable α_i is independent of $w(x_i)$. It follows that the probability that x_i is ambiguous is no more than $1/2m$. Note that it is possible that $\alpha_i \notin \{1, \ldots, 2m\}$, in which case the probability is actually zero.

While the ambiguities of the different elements are correlated, it is safe to say that the probability that there exists an ambiguous element in X is at most

$$m \times \frac{1}{2m} = \frac{1}{2}.$$

It follows that with probability at least a half, none of the elements is ambiguous. But if there exist two distinct minimum weight sets, say S_i and S_j, there must be an element that belongs to one of these sets but not the other, i.e., there must be an ambiguous element. Thus, with probability at least a half there is a unique minimum weight set. □

Exercise 12.12: Determine the probability that there is a unique minimum weight set when the weights are chosen from the set $\{1, \ldots, t\}$.

Exercise 12.13: Does a similar result hold for the maximum weight set?

The application of this lemma to the perfect matching problem is obvious. Let X be the set of edges in the graph, and $\mathcal{F}$ the set of perfect matchings. It follows that assigning random weights between 1 and $2m$ to the edges leads to a unique minimum weight perfect matching with probability at least $1/2$. We now turn to the task of identifying this perfect matching.

12.4.2. The Parallel Matching Algorithm

Suppose we have chosen the random weight function w for the edges of G as described above, and let w_{ij} be the weight assigned to the edge (i, j). We will assume that there is a unique minimum weight perfect matching, and that its weight is W. If there is more than one minimum weight perfect matching, the following algorithm will fail (the mode of failure will be evident from the description below). This happens with probability at most $1/2$, and the algorithm can be repeated to reduce the error probability.

Consider the Tutte matrix $\boldsymbol{A}$ corresponding to the graph G. Let $\boldsymbol{B}$ be the matrix obtained from $\boldsymbol{A}$ by setting each indeterminate x_{ij} to the (random) integer value $2^{w_{ij}}$.

Lemma 12.11: *Suppose that there is a unique minimum weight perfect matching and that its weight is W. Then,* $\det(\boldsymbol{B}) \neq 0$ *and, moreover, the highest power of* 2 *that divides* $\det(\boldsymbol{B})$ *is* 2^{2W}.

PROOF: The proof is a generalization of the proof of Tutte's theorem. For each permutation $\sigma \in \mathbb{S}_n$ defined over $V = \{1, \ldots, n\}$, define its *value* with respect to $\boldsymbol{B}$ as $\operatorname{val}(\sigma) = \prod_{i=1}^{n} B_{i\sigma(i)}$. Observe that $\operatorname{val}(\sigma)$ is non-zero if and only if for each $i \in V$, the edge $(i, \sigma(i))$ is present in G. Recall from Section 7.2 that the determinant of the matrix $\boldsymbol{B}$ is given by

$$\det(\boldsymbol{B}) = \sum_{\sigma \in \mathbb{S}_n} \operatorname{sgn}(\sigma) \times \operatorname{val}(\sigma),$$

where $\operatorname{sgn}(\sigma)$ is the sign of a permutation σ. Permutations with sign $+1$ are called even, and those with sign -1 are called odd. The reader should not confuse the sign of a permutation with the sign of its value.

We focus only on the permutations with non-zero value, since the others do not contribute to the determinant. Let us first explicate the structure of the non-zero permutations. The *trail* of a permutation σ of non-zero value is the subgraph of G containing exactly the edges $(i, \sigma(i))$, for $1 \leq i \leq n$. It is convenient to view the edges $(i, \sigma(i))$ as being directed from i to $\sigma(i)$. The n edges corresponding to σ form a multiset where each edge has multiplicity 1 or 2, and the edges of multiplicity 2 occur with both orientations. Each vertex has two edges from the trail incident on it, one incoming and the other outgoing, and these may correspond to the two orientations of the same undirected edge from G. Thus, the trail consists of disjoint cycles and edges, where the isolated edges are those of multiplicity 2. The orientations on the edges are such that the cycles are oriented, and the isolated edges may be viewed as oriented cycles of length 2. Define an odd-cycle permutation as one whose trail contains at least one odd-length cycle, while even-cycle permutations have only even length cycles.

In each odd-cycle permutation σ, fix a canonical odd cycle C as follows; for each cycle, sort the list of vertex indices and use the sorted sequence of indices as label for that cycle; pick the odd cycle whose label is the lexicographically smallest. We can pair off the odd-cycle permutations by associating with such σ the unique odd-cycle permutation $-\sigma$ obtained by reversing the orientation of the edges in the canonical odd cycle C. Given these definitions, both σ and

$-\sigma$ have the same canonical odd cycle and $-(-\sigma) = \sigma$. The skew-symmetric nature of the matrix $\boldsymbol{B}$ implies that $\text{val}(\sigma) = -\text{val}(-\sigma)$, while the identical cycle structure of the two permutations implies that $\text{sgn}(\sigma) = \text{sgn}(-\sigma)$. It follows that their net contribution to $\det(\boldsymbol{B})$ is 0. Thus, the set of odd-cycle permutations has a net contribution of zero toward the value of $\det(\boldsymbol{B})$. This value of the determinant is completely determined by the value of the even-cycle permutations.

Notice that a permutation σ that corresponds to a perfect matching M has a trail consisting exactly of the set of edges in M, and each of these edges has multiplicity 2. Also, for any perfect matching M, the value of the permutation corresponding to it is exactly $(-1)^{n/2}2^{2w(M)}$, where $w(M)$ is the weight of the matching M. If these were the only even-cycle permutations to consider, the result would follow immediately. However, there are even-cycle permutations that do not correspond to any particular perfect matching, although as discussed below they can all be viewed as representing a union of two perfect matchings.

An even-cycle permutation σ consists of a collection of even cycles, and its trail can be partitioned into two perfect matchings, say M_1 and M_2, by considering alternating edges from each cycle.

Exercise 12.14: Verify that $|\text{val}(\sigma)| = 2^{w(M_1)+w(M_2)}$.

When the trail of σ has a cycle of length greater than 2, the two perfect matchings M_1 and M_2 are distinct and, since at most one of these two perfect matchings can be the unique perfect matching of minimum weight, it follows that $|\text{val}(\sigma)| > 2^{2W}$. On the other hand, when the trail has only cycles of length 2, i.e., the permutation corresponds to a perfect matching, we have $M_1 = M_2$ and $|\text{val}(\sigma)| = 2^{2w(M_1)} \geq 2^{2W}$. But note that equality with 2^{2W} is achieved only when $M_1 = M_2$ is the unique minimum weight perfect matching.

Thus, the *absolute* contribution to $\det(\boldsymbol{B})$ from each even-cycle permutation is a power of 2 no smaller than 2^{2W}. Moreover, exactly one of these contributions – the one from the even-cycle permutation corresponding to the unique minimum weight perfect matching – is equal to 2^{2W}. The determinant of $\boldsymbol{B}$ can now be viewed as a sum of powers of 2, possibly negated, such that the exponent of every term but one is strictly greater than $2W$. Since the term of absolute value 2^{2W} cannot cancel out, it follows that $\det(\boldsymbol{B}) \neq 0$ and in fact the largest power of 2 dividing it is 2^{2W}. □

Exercise 12.15: Observe that, after choosing the random weights, both $\boldsymbol{B}$ and $\det(\boldsymbol{B})$ can be computed via *NC* algorithms. Show that the value of W can also be determined in *NC*.

Of course, this only shows how to compute the weight of the minimum weight perfect matching. The following lemma is the basis for actually determining the edges in that matching. Recall that $\boldsymbol{B}^{ij}$ is the minor of $\boldsymbol{B}$ obtained by removing the ith row and the jth column from $\boldsymbol{B}$.

Lemma 12.12: *Let M be the unique minimum weight perfect matching in G, and let its weight be W. An edge (i, j) belongs to M if and only if*

$$\frac{\det(\boldsymbol{B}^{ij})2^{w_{ij}}}{2^{2W}}$$

is odd.

PROOF: Consider the matrix $\boldsymbol{Q}$ obtained from $\boldsymbol{B}$ by zeroing out each entry in the ith row and jth column of $\boldsymbol{B}$, except for B_{ij}. Notice that any permutation of non-zero value with respect to $\boldsymbol{Q}$ must map i to j.

Exercise 12.16: Verify that

$$\det(\boldsymbol{Q}) = (-1)^{i+j} 2^{w_{ij}} \det(\boldsymbol{B}^{ij}) = \sum_{\sigma : \sigma(i) = j} \operatorname{sgn}(\sigma) \times \operatorname{val}(\sigma). \tag{12.1}$$

We can now apply the same argument as in Lemma 12.11 to claim that odd-cycle permutations (mapping i to j) will have a zero net contribution to the sum (12.1). One possible problem with doing so is that the canonical odd cycle in a specific permutation σ may contain the oriented edge going from i to j, in which case its partner $-\sigma$ will invert the orientation on that edge and hence not belong to the set of permutations mapping i to j. This will create problems in the canceling argument. However, note that since n is even, any odd-cycle permutation has at least two odd cycles and so we can choose the canonical cycle to be one not containing the edge from i to j.

If the edge (i, j) belongs to M, then (as before) exactly one even-cycle permutation contributes 2^{2W} to the sum and all others contribute a strictly larger power of 2. This implies that 2^{2W} is the largest power of 2 dividing the sum, and the remainder must be an odd integer. On the other hand, if (i, j) does not belong to M, all even-cycle permutations must contribute powers of 2 strictly larger than 2^{2W}, implying that the sum is divisible by 2^{2W+1} and the remainder of its division by 2^{2W} is an even number. □

It is now easy to determine all the edges in the minimum weight perfect matching M, and the algorithm is summarized below.

Algorithm Parallel Matching:

Input: Graph $G(V, E)$ with at least one perfect matching.

Output: A perfect matching $M \subseteq E$.

1. **for all** edges (i, j), in parallel **do**
   choose random weight w_{ij}.
2. **compute** the Tutte matrix $\boldsymbol{B}$ from w.
3. **compute** $\det(\boldsymbol{B})$.
4. **compute** W such that 2^{2W} is the largest power of 2 dividing $\det(\boldsymbol{B})$.
5. **compute** $\text{adj}(\boldsymbol{B}) = \det(\boldsymbol{B}) \times \boldsymbol{B}^{-1}$ whose (j, i) entry has absolute value $\det(\boldsymbol{B}^{ij})$.
6. **for all** edges (i, j) **do** (in parallel)
   compute $r_{ij} = \det(\boldsymbol{B}^{ij}) 2^{w_{ij}} / 2^{2W}$.
7. **for all** edges (i, j) **do** (in parallel)
   if r_{ij} is odd **then** add (i, j) to M

Exercise 12.17: Verify that each step of this algorithm can be implemented in ***RNC***, implying that it is an ***RNC*** algorithm for finding perfect matchings.

The most expensive computations in this algorithm are those of finding the determinant, inverse, and adjoint of an $n \times n$ matrix whose entries are $O(m)$-bit integers (since the matrix entries have magnitudes that are exponential in the edge weights).

Theorem 12.13: *Given a graph G with at least one perfect matching, the* **Parallel Matching** *algorithm finds a perfect matching with probability at least* $1/2$. *For a graph G with n vertices it requires* $O\left(\log^2 n\right)$ *time and* $O(n^{3.5} m)$ *processors.*

This is a Monte Carlo algorithm with (one-sided) error probability of $1/2$, and this probability can be reduced by repetitions. The only possible error arises when, even though the graph does have a perfect matching, the algorithm determines a set of edges that do not form a perfect matching because the random choice of weights did not yield a unique perfect matching. It is a simple matter to check for this error and convert this into a Las Vegas algorithm. Although we assumed throughout that the number of vertices n is even, it is possible to apply this algorithm to the case of odd n.

Exercise 12.18: In a graph $G(V, E)$ with n vertices, when n is odd we define a perfect matching to be a matching of cardinality $\lfloor n/2 \rfloor$. Explain how the **Parallel Matching** algorithm may be adapted to this case.

Finally, the **Parallel Matching** algorithm can be adapted to obtain a Las Vegas algorithm for finding a maximum matching, as outlined in Problems 12.16–12.18.

12.5. The Choice Coordination Problem

We now move on to distributed computation, in this section and in Section 12.6; we thus no longer use the PRAM model. A problem often arising in parallel and distributed computing is that of destroying the symmetry between a set of possibilities. This may be achieved by the use of randomization as in the case of the *Choice Coordination Problem* (CCP) discussed below. That this is a very "natural" problem is demonstrated by the following situation, which has been studied in biology. A particular class of mites (genus *Myrmoyssus*) reside as parasites on the ear membrane of the moths of family *Phaenidae*. The moths are prey to bats and the only defense they have is that they can hear the sonar used by an approaching bat. Unfortunately, if both ears of the moth are infected by the mites, then their ability to detect the sonar is considerably diminished, thereby severely decreasing the survival chances of both the moth and its colony of mites. The mites would like to ensure the continued survival of their host, and they can do so by infecting only one ear at a time. The mites are therefore faced with a "choice coordination problem": how does any collection of mites infecting a particular ear ensure that every other mite chooses the same ear? The protocol used by these mites involves leaving chemical trails around the ears of the moth.

Our interest in this abstract problem has a more computational motivation. Consider a collection of n identical processors that operate in total asynchrony. They have no global clock and no assumptions can be made about ther relative speeds. The processors have to reach a consensus on a unique choice out of a collection of m identical options. We use the following simple model of communication between the processors. There is a collection of m read-write registers accessible to all n processors. Several processors may simultaneously attempt to access or modify a register. To deal with such conflicts, we assume that the processors use a locking mechanism whereby a unique processor obtains sole access to a register when several processors attempt to access it; moreover, all the remaining processors then wait until the lock is released, and then contend once again for access to the register. The processors are required to run a protocol for picking a unique option out of the m choices. This is achieved by ensuring that at the end of the protocol exactly one of the m registers contains a special symbol $\surd$. The complexity of a choice coordination protocol is measured in terms of the total number of read and write operations performed by the n processors. (Clearly, running time has little meaning in an asynchronous situation.)

It is known that any deterministic protocol for solving this problem will have a complexity of $\Omega(n^{1/3})$ operations. We now illustrate the power of randomization in this context by showing that there is a randomized protocol which, for any $c > 0$, will solve the problem using c operations with a probability of success at least $1 - 2^{-\Omega(c)}$. For simplicity we will consider only the case where $n = m = 2$, although the protocol and the analysis generalize in a rather straightforward manner.

We start by restricting ourselves to the rather simple case where the two processors are synchronous and operate in lock-step according to some global clock. The following protocol is executed by each of the two processors. We index the processors P_i and the possible choices by C_i for $i \in \{0, 1\}$. The processor P_i initially scans the register C_i. Thereafter, the processors exchange registers after every iteration of the protocol. This implies that at no time will the two processors scan the same register. Each processor also maintains a local variable whose value is denoted by B_i.

Algorithm SYNCH-CCP:

Input: Registers C_0 and C_1 initialized to 0.

Output: Exactly one of the two registers has the value $\surd$.

0. P_i is initially scanning the register C_i and has its local variable B_i initialized to 0.

1. Read the current register and obtain a bit R_i.

2. Select one of three cases.

  case: 2.1 $[R_i = \surd]$
    halt;
  case: 2.2 $[R_i = 0,\ B_i = 1]$
    Write $\surd$ into the current register and **halt**;
  case: 2.3 [otherwise]
    Assign an unbiased random bit to B_i and write B_i into the current register;

3. P_i exchanges its current register with P_{1-i} and returns to Step 1.

To verify the correctness of this protocol it suffices to see that at most one register can ever have $\surd$ written into it. Suppose that both registers get the value $\surd$. We claim that both registers must have had $\surd$ written into them during the same iteration; otherwise, Case 2.1 will ensure that the protocol halts before this error takes place. Let us assume that the error takes place during the tth iteration. Denote by $B_i(t)$ and $R_i(t)$ the values used by processor P_i just after Step 1 of the tth iteration of the protocol. By Case 2.3, we know that $R_0(t) = B_1(t)$ and $R_1(t) = B_0(t)$. The only case in which P_i writes $\surd$ during the tth iteration is when $R_i = 0$ and $B_i = 1$; then, $R_{1-i} = 1$ and $B_{1-i} = 0$, and P_{1-i} cannot write $\surd$ during that iteration.

We have shown that the protocol terminates correctly by making a unique choice. But this assumes that the protocol terminates in a finite number of iterations. Why should this happen? Notice that during each iteration, the probability that both the random bits B_0 and B_1 are the same is $1/2$. Moreover, if at any stage these two bits take on distinct values, then the protocol terminates within the next two stages. Thus, the probability that the number of stages exceeds t is $O(1/2^t)$. The computational cost of each iteration is bounded, so that this protocol does $O(t)$ work with probability $1 - O(1/2^t)$.

We now generalize this protocol to the asynchronous case where the two processors may be operating at varying speeds and cannot "exchange" the registers after each iteration. In fact, we no longer assume that the two processors begin by scanning different registers – choosing a unique starting register C_0 or C_1 is in itself an instance of the choice coordination problem. Instead, we assume that each processor chooses its starting register at random. Thus, the two processors could be in a conflict at the very first step and must use the lock mechanism to resolve this conflict. The basic idea is to put time-stamps t_i on the register C_i, and T_i on the local variable B_i. We assume that a read operation on C_i will yield a pair $\langle t_i, R_i \rangle$, where t_i is the time-stamp and R_i is the value of that register. If the processors were to operate synchronously, these time-stamps would be exactly the same as the iteration number t of the previous protocol.

Algorithm ASYNCH-CCP:

Input: Registers C_0 and C_1 initialized to $\langle 0, 0 \rangle$.

Output: Exactly one of the two registers has the value $\surd$.

0. P_i is initially scanning a randomly chosen register. Thereafter, it changes its current register at the end of each iteration. The local variables T_i and B_i are initialized to 0.

1. P_i obtains a lock on the current register and reads $\langle t_i, R_i \rangle$.

2. P_i selects one of five cases.

Case 2.1: $[R_i = \surd]$
    halt;

Case 2.2: $[T_i < t_i]$
    $T_i \leftarrow t_i$ and $B_i \leftarrow R_i$.

Case 2.3: $[T_i > t_i]$
    Write $\surd$ into the current register and **halt**;

Case 2.4: $[T_i = t_i, R_i = 0, B_i = 1]$
    Write $\surd$ into the current register and **halt**;

Case 2.5: [otherwise]
    $T_i \leftarrow T_i + 1$, $t_i \leftarrow t_i + 1$, assign a random (unbiased) bit to B_i and write $\langle t_i, B_i \rangle$ into the current register.

3. P_i releases the lock on its current register, moves to the other register, and returns to Step 1.

Theorem 12.14: *For any $c > 0$, Algorithm* **ASYNCH-CCP** *has total cost exceeding c with probability at most $2^{-\Omega(c)}$.*

PROOF: The only real difference from the previous protocol is in Cases 2.2 and 2.3. A processor in Case 2.2 is playing catch-up with the other processor, and the processor in Case 2.3 realizes that it is ahead of the other processor and is thus free to make the choice. To prove the correctness of this protocol, we consider the two cases where a processor can write $\surd$ into its current cell – these are Cases 2.3 and 2.4. Whenever a processor finishes an iteration, its personal time-stamp T_i equals that of the current register t_i. Further, $\surd$ cannot be written during the very first iteration of either processor.

Suppose P_i has just entered Case 2.3 with time-stamp T_i^* and its current cell is C_i with time-stamp t_i^*, where $t_i^* < T_i^*$. The only possible problem is that P_{1-i} may write (or already have written) $\surd$ into the register C_{1-i}. Suppose this error does indeed occur, and let t_{1-i}^* and T_{1-i}^* be the time-stamps during the iteration of P_{1-i} when it writes $\surd$ into C_{1-i}.

Now P_i comes to C_i with a time-stamp of T_i^*, and so it must have left C_{1-i} with a time-stamp of the same value *before* P_{1-i} could write $\surd$ into it. Since time-stamps cannot decrease, $t_{1-i}^* \geq T_i^*$. Moreover, P_{1-i} cannot have its time-stamp T_{1-i}^* exceeding t_i^*, since it must go to C_{1-i} from C_i and the time-stamp of that register never exceeds t_i. We have established that $T_{1-i}^* \leq t_i^* < T_i^* \leq t_{1-i}^*$. But P_{1-i} must enter Case 2.2 for $T_{1-i}^* < t_{1-i}^*$, contradicting the assumption that it writes $\surd$ into C_{1-i} for these values of the time-stamps.

Case 2.4 can be analyzed similarly, except that we finally obtain that $T_{1-i}^* \leq t_i^* = T_i^* \leq t_{1-i}^*$. This may cause a problem since it allows $T_{1-i}^* = t_{1-i}^*$, and so Case 2.4 can cause P_{1-i} to write $\surd$; however, we can now invoke the analysis of the synchronous case and rule out the possibility of error.

The complexity of this protocol is easy to analyze. The cost is proportional to the largest time-stamp obtained during the execution of this protocol. The time-stamp of a register can go up only in Case 2.5, and this happens only when Case 2.4 fails to apply. Moreover, the processor P_i that raises the time-stamp must have its current B_i value chosen during a visit to the other register. Thus, the analysis of the synchronous case applies. □

12.6. Byzantine Agreement

The subject of this section is the classic *Byzantine agreement problem* in distributed computation. As in Section 12.5, we study a process by which n processors reach an agreement. However, in the scenario we consider here, t of the n processors are faulty processors. We further assume that the faulty processors may collude in order to try and subvert the agreement process. A protocol designed to withstand such strong adversaries should certainly work in the face of weaker faulty behavior arising in practice. The goal is a protocol that achieves agreement while tolerating as large a value of t as possible.

The Byzantine agreement problem is the following. Each of the n processors initially has a *value* that is 0 or 1; let b_i denote the value initially held by the ith processor. There are t faulty processors, and we refer to the remaining $n-t$ identical processors as *good* processors. Following communication according to the rules below, the ith processor ends the protocol with a *decision* $d_i \in \{0,1\}$, which must satisfy the following properties.

1. All the good processors should finish with the same decision.
2. If all the good processors begin with the same value v, then they all finish with their (common) decision equaling v.

The set of faulty processors is determined before the execution of the protocol begins (though of course the good processors do not know the identities of the faulty processors). The agreement protocol proceeds in a sequence of *rounds.* During each round, each processor may send one message to each other processor. Each processor receives a message from each of the remaining processors, before the following round begins. A processor need not send the same message to all the other processors. In the protocol described below, every message will be a single bit. All good processors follow the protocol exactly. A faulty processor may send messages that are totally inconsistent with the protocol, and may send different messages to different processors. In fact, we assume that the t faulty processors work in collusion: at the start of each round, they decide among themselves what messages each of them will send to each good processor, with the goal of inflicting the maximum damage. Agreement is achieved when every good processor has computed its decision consistent with the two properties above. We study the number of rounds it takes to achieve agreement.

It is known (see the Notes section) that any deterministic protocol for agreement in this model requires $t+1$ rounds. We now exhibit a simple randomized algorithm that terminates in a number of steps whose expectation is a constant. The number of rounds is a random variable, but the protocol is always correct in that it results in agreement as defined above; thus we have a Las Vegas protocol. We assume that at each step there is a *global coin toss* that a trusted party performs. The coin toss equiprobably results in a HEADS or a TAILS, and this result (denoted *coin*) is correctly conveyed to all the processors. This assumption can be dispensed with in more complicated protocols, but we do not discuss these here (see the Notes section).

For the remainder of the discussion, the reader may find it convenient to think of $t < n/8$; however, this is not a fundamental barrier, and the protocol in fact works for somewhat larger values of t. (This is the subject of Problem 12.27.) During each round of the protocol, each processor transmits a single bit, called its *vote*, to each other processor. A good processor sends the same vote to all other processors. Faulty processors may send arbitrary, inconsistent votes to good processors. Assume that n is a multiple of 8 for simplicity of exposition;

let L denote $(5n/8) + 1$, H denote $(3n/4) + 1$, and G denote $7n/8$. (In fact, the protocol only requires that $L \geq (n/2) + t + 1$ and $H \geq L + t$ in order to work.) The ith processor executes the following routine, for $1 \leq i \leq n$.

Algorithm ByzGen:

Input: A value b_i.

Output: A decision d_i.

1. *vote* $= b_i$;
2. For each round, **do**
    3. Broadcast *vote*;
    4. Receive votes from all other processors;
    5. *maj* $\leftarrow$ majority value (0 or 1) among votes received including own vote;
    6. *tally* $\leftarrow$ number of occurrences of *maj* among votes received;
    7. **if** *coin* = HEADS **then** *threshold* $\leftarrow L$;
       else *threshold* $\leftarrow H$;
    8. **if** *tally* $\geq$ *threshold* **then** *vote* $\leftarrow$ *maj*;
       else *vote* $\leftarrow 0$;
    9. **if** *tally* $\geq G$ **then** set d_i to *maj* permanently;

We begin the analysis with an easy exercise:

Exercise 12.19: Show that if all the good processors begin a round with the same initial value, they all set their decisions to this value in a constant number of rounds.

The more interesting case for analysis is when the good processors do not all start with the same initial value. In the absence of faulty processors, a solution would be for all processors to broadcast their values, and then set their decisions to the majority of these values. The algorithm **ByzGen** implements this idea in the face of malicious faults.

If two good processors compute different values for *maj* in Step 5, *tally* does not exceed *threshold* regardless of whether L or H was chosen as *threshold*. Then, all good processors set *vote* $= 0$ in Step 8.2. As a result, all good processors set their decisions to 0 in the following round. It thus remains to consider the case when all good processors compute the same value for *maj* in Step 5.

We say that the faulty processors *foil* a threshold $x \in \{L, H\}$ in a round if, by sending different messages to the good processors, they cause *tally* to exceed x for at least one good processor, and to be no more than x for at least one good processor. Since the difference between the two possible thresholds L and H is at least t, the faulty processors can foil at most one threshold in a round. Since the threshold is chosen equiprobably from $\{L, H\}$, it is foiled with probability at most 1/2. Thus, the expected number of rounds before we have an unfoiled threshold is at most 2. If the threshold is not foiled, then all good processors compute the same value v for *vote* in Step 8. In the following round, every good processor receives at least $G > H > L$ votes for v, and sets *maj* to v in Step 5. Then, in Step 9, *tally* exceeds whichever threshold is chosen. When a good processor sets d_i the other good processors must have *tally* $\geq$ *threshold*, since $G \geq H + t$. Therefore they will all vote the same as d_i henceforth.

Theorem 12.15: *The expected number of rounds for* **ByzGen** *to reach agreement is a constant.*

The protocol **ByzGen** above does not include a termination criterion.

Exercise 12.20: Suggest a modification to the protocol **ByzGen** in which all good processors halt upon agreement.

Exercise 12.21: In the protocol **ByzGen**, is it always true that all good processors determine their decisions in the same round?

Notes

Karp and Ramachandran [241] give a comprehensive survey of PRAM algorithms. Some good references for parallel algorithms are the books by JáJá [208] and Leighton [271] and the volume edited by Reif [354]. The **BoxSort** algorithm of Section 12.2 is due to Reischuk [356]. Following Reischuk's work, a number of deterministic sorting algorithms running in $O(\log n)$ steps using n processors have been devised, most notably by Ajtai, Komlós, and Szemerédi [8] with later simplifications and improvements by Paterson [328]; Cole [110] gave a different deterministic parallel algorithm using n processors and $O(\log n)$ steps.

The intractability of the parallel solution of the LFMIS problem was established by Cook [111]. The first ***RNC*** algorithm for MIS is due to Karp and Wigderson [251]; they also provided a derandomized version of their algorithm. This was a complex algorithm requiring a large running time and a high processor count. The **Parallel MIS** algorithm and its derandomization is due to Luby [282]; this paper pioneered the idea of using random variables of limited independence to lead to a deterministic algorithm for a concrete problem (see also the Notes section of Chapter 3). Alon, Babai, and Itai [19] independently gave an ***RNC*** algorithm for the MIS problem and also derandomized it to obtain an *NC* algorithm. A more efficient *NC* algorithm was later provided by Goldberg and Spencer [173]. The paradigm of derandomizing parallel algorithms using limited independence has found a variety of applications. Luby [284] has combined it with the method of conditional probabilities (Section 5.6) to achieve processor efficiency for the maximal independent set problem. Berger and Rompel [55] and Motwani, Naor, and Naor [313] have used a combination of $\log n$-wise independence and the method of conditional probabilities to derive *NC* algorithms for a variety of problems. Karger and

Motwani [233] have used the combination of pairwise independence with the random walk technique for recycling random bits described in Chapter 6 to construct an *NC* algorithm for the min-cut problem. The min-cut problem is closely related to the matching problem – an *NC* algorithm for min-cut in directed graphs would result in an *NC* algorithm for maximum matching in bipartite graphs.

The reader may refer to the survey article by von zur Gathen [412] for a survey of parallel matrix algorithms. The first *NC* algorithm for matrix determinants is due to Csanky [115], but it applies only to fields of characteristic zero. Borodin, von zur Gathen, and Hopcroft [79] gave an *NC* algorithm for the general case (see Berkowitz [56] for a more elegant version). The algorithm due to Chistov [95] is currently the best known solution, and it requires only $O(\log^2 n)$ time. The computation of adjoints and inverses of a matrix can be reduced to the determinant computation at the cost of an increase in time and processor count. The randomized algorithm cited in Theorem 12.9 is due to Pan [323].

The book by Lovász and Plummer [281] is an excellent source for combinatorial and algorithmic results related to matchings, and Vazirani [405] surveys parallel matching algorithms. Section 7.8.3 gives a history of results establishing the connection between matchings and matrix determinants. Israeli and Shiloach [207] give an *NC* algorithm for finding maximal matchings. The *NC* algorithm in the case of a unique perfect matching is due to Rabin and Vazirani [348, 349], and in the case of polynomially small number of perfect matchings is due to Grigoriev and Karpinski [184]. The first ***RNC*** algorithm for matchings was given by Karp, Upfal, and Wigderson [242], and this was subsequently improved by Galil and Pan [162]. This work raised several interesting questions with respect to the parallel complexity of search versus decision problems, and this theme is explored by Karp, Upfal, and Wigderson [250]. The Isolating Lemma and the **Parallel Matching** algorithm are due to Mulmuley, Vazirani, and Vazirani [317]. These Monte Carlo algorithms were converted into Las Vegas algorithms by Karloff [237]. The best known deterministic algorithm using a polynomial number of processors, due to Goldberg, Plotkin, and Vaidya [172], requires $\Omega(n^{2/3})$ time. An interesting special case for which *NC* algorithms are known is that of finding perfect matchings in regular bipartite graphs. Lev, Pippenger, and Valiant [274] derived this result by providing an algorithm for edge coloring (which is a partition into matchings) a bipartite graph of maximum degree Δ with Δ colors. In the non-bipartite case, Karloff and Shmoys gave an ***RNC*** algorithm for approximate edge coloring, and this was derandomized by Berger and Rompel [55] and Motwani, Naor, and Naor [313]. Some interesting open problems are:

► **Research Problem 12.1:** Devise an *NC* algorithm for finding a maximum matching in a given graph.

► **Research Problem 12.2:** Devise an *NC* or an ***RNC*** algorithm for edge coloring a graph of maximum degree Δ using at most $\Delta+1$ colors (see Vizing's Theorem [71]).

► **Research Problem 12.3:** Aggarwal and Anderson [4] have shown that the problem of finding a depth-first search tree in a graph can be solved in ***RNC*** using ***RNC*** algorithms for finding maximum matchings; once again, the issue of an *NC* algorithm is unresolved.

The algorithm for the choice coordination problem in Section 12.5 is due to Rabin [344], and the biological analog is described in a paper by Treat [397]. The Byzantine agreement problem was introduced by Pease, Shostak and Lamport [330]. Fischer and Lynch [148] showed that in our model, any deterministic protocol requires $t+1$ rounds to reach agreement, in the worst case. This lower bound matches an upper bound given in [330]. The **ByzGen** protocol of Section 12.6 is due to Rabin [347]. Our presentation follows Chor and Dwork [96], who give a comprehensive account of the history of the problem, the various models under which it has been studied, and the many variants and improvements of Rabin's scheme. They point out that if the processors do not operate in synchrony, it is impossible to achieve agreement using a deterministic protocol; this result is due to Fischer, Lynch, and Paterson [149]. On the other hand, **ByzGen** and other randomized protocols can be shown to achieve agreement even in an asynchronous setting.

Problems

12.1 Show that the parallel variant of randomized quicksort described in Section 12.2 sorts n elements with n processors on a CREW PRAM, with high probability in $O(\log^2 n)$ steps.

12.2 Prove Lemma 12.1. The following outline is suggested (refer to Section 12.2 for the notation).

1. Bound the probability that $\alpha(B_{j+1}) = \alpha(B_j)$ using the result of Exercise 12.6.
2. Bound the probability that for any particular k, the value k is contained more than τ_k times in the sequence $\alpha(B_1), \ldots, \alpha(B_t)$.
3. Bound the probability that for $1 \leq k \leq c \log\log n$, the value k is contained more than τ_k times in the sequence $\alpha(B_1), \ldots, \alpha(B_t)$.

12.3 Suppose that the random samples in Stage 1 of **BoxSort** are chosen using pairwise independent, rather than completely independent random variables (the choices made by the various boxes are independent of each other, though). Derive the best upper bound you can on the number of parallel steps taken by **BoxSort**.

12.4 Using the ideas of Section 12.2, devise a CREW PRAM algorithm that selects the kth largest of n input numbers in $O(\log n)$ steps using $n/\log n$ processors. Assume that the n input numbers are initially located in global memory locations 1 through n.

12.5 Devise a $\mathcal{ZNC}$ algorithm for generating a random (uniformly distributed) permutation of a set S containing n elements. (**Hint:** Consider assigning random weights to the elements of S. If the weights are drawn from a sufficiently large set, each element will have a distinct weight.)

12.6 A *maximal* matching in a graph is a matching that is not properly contained in any other matching. Use the parallel algorithm for the MIS problem to devise an $\mathcal{RNC}$ algorithm for finding a maximal matching in a graph.

12.7 Consider a graph $G(V, E)$ with maximum degree Δ. Show that a sequential greedy algorithm will color the vertices of the graph using at most $\Delta+1$ colors such that no two adjacent vertices are assigned the same color. Employing the parallel algorithm for MIS, devise an ***RNC*** algorithm for finding a $\Delta + 1$ coloring of a given graph.

12.8 (Due to M. Luby [282].) The *vertex partition* problem is defined as follows: given a graph $G(V, E)$ with edge weights, partition the vertices into sets V_1 and V_2 such that the net weight of the edges crossing the cut (V_1, V_2) is at least a half of the total weight of the edges in the graph. Describe an ***RNC*** algorithm for this problem, and explain how you will convert this into an *NC* algorithm using the idea of pairwise independence.

12.9 (Due to M. Luby [282].) In the **Parallel MIS** algorithm, suppose that the random marking of the vertices is only pairwise independent. Show that the probability that a good vertex belongs to $S \cup \Gamma(S)$ is at least $1/24$.

12.10 (Due to M. Luby [282].) Suppose that you are provided with a collection of n pairwise independent random numbers uniformly distributed over the set $\{0, 1, \ldots, p-1\}$, where $p \geq 2n$. It is desired to construct a collection of n pairwise independent Bernoulli random variables where the ith random variable should take on the value 1 with probability $1/t_i$, for $1 \leq t_i \leq n/8$. Show how you can achieve this goal approximately by constructing a collection of pairwise independent Bernoulli random variables such that the ith variable takes on the value 1 with probability $1/T_i$ where for a constant $c > 1$, T_i satisfies

$$T_i \leq t_i \leq cT_i.$$

12.11 (Due to M. Luby [282].) Combining the results of Problems 12.9 and 12.10, show that the **Parallel MIS** algorithm can be derandomized to yield an *NC* algorithm for the MIS problem. Note that the approach in Problem 12.10 will not work for marking vertices with degree exceeding $n/16$, and these will have to be dealt with separately.

12.12 (Due to M. Luby [282].) In this problem we consider a variant of the **Parallel MIS** algorithm. For each vertex $v \in V$, independently and uniformly choose a random weight $w(v)$ from the set $\{1, \ldots, n^4\}$. Repeatedly strip off an independent set S and its neighbors $\Gamma(S)$ from the graph G, where at each iteration the set S is the set of marked vertices generated by the following process: mark all vertices in V, and then in parallel for each edge in E unmark the end-point of larger weight. Show that this yields an ***RNC*** algorithm for MIS. Can this algorithm be derandomized using pairwise independence?

12.13 (Due to D.R. Karger [231].) Recall the randomized algorithm for min-cuts discussed in Section 1.1 (see also Section 10.2). Describe an ***RNC*** implementation of this algorithm. (**Hint:** While contracting the edges appears to be sequential process, it can be implemented in parallel using the following observation. Consider generating a random permutation on the edges, as described in Problem 12.5 and using this to determine the order in which the edges are contracted. The contraction algorithm will terminate at that point in the permutation where the preceding edges constitute a graph with exactly two connected components. Assume that there is an *NC* algorithm for determining connected components.)

12.14 (Due to M. Luby, J. Naor, and M. Naor [285].) Using the idea of pairwise independence, construct an *RNC* algorithm for the min-cut problem that uses only a polylogarithmic number of random bits (see also Problem 12.13). What implications does this have for placing the min-cut problem in *NC*? (**Hint:** Select a set of edges by choosing each edge pairwise independently with probability $1/c$, where c is the size of the min-cut; see Problem 12.10. In parallel, contract all edges in this set. Repeat this process until the graph is reduced to two vertices.)

12.15 (Due to M.O. Rabin and V.V. Vazirani [349].) Let $G(V, E)$ be a graph with a unique perfect matching. Devise an *NC* algorithm for finding the perfect matching in G. (**Hint:** Consider substituting 1 for each indeterminate in the Tutte matrix. What is the significance of the entries in the adjoint of the Tutte matrix?)

12.16 (Due to K. Mulmuley, U.V. Vazirani, and V.V. Vazirani [317].) Consider the problem of finding a *minimum-weight* perfect matching in a graph $G(V, E)$, given edge-weights $w(e)$ for each edge $e \in E$ in *unary*. Note that it is not possible to apply the Isolating Lemma directly to this case since the random weights chosen there would conflict with the input weights. Explain how you would devise an *RNC* algorithm for this problem. The parallel complexity of the case where the edge-weights are given in *binary* is as yet unresolved – do you see why the *RNC* algorithm does not apply to the case of binary weights? (**Hint:** Start by scaling up the input edge weights by a polynomially large factor. Apply random perturbations to the scaled edge weights and prove a variant of the Isolating Lemma for this situation.)

12.17 (Due to K. Mulmuley, U.V. Vazirani, and V.V. Vazirani [317].) Devise an *RNC* algorithm for the problem of finding a *maximum* matching in a graph. Observe that the **Parallel Matching** algorithm does not work (as stated) when the maximum matching is not a perfect matching.

12.18 (Due to H.J. Karloff [237].) Suppose you are given a Monte Carlo *RNC* algorithm for finding a maximum matching in a bipartite graph. Explain how you would convert this into a Las Vegas algorithm. Can the solution be generalized to the case of non-bipartite graphs? (**Hint:** While this conversion is trivial for perfect matching algorithms, for maximum matching algorithms you will need to devise a parallel algorithm for determining an upper bound on the *size* of a maximum matching in a graph. This requires a non-trivial use of structure theorems for matchings in graphs.)

12.19 This problem explores a different method for converting the Monte Carlo maximum matching into a Las Vegas one. Recall from Problem 7.7 that the rank of the matrix of indeterminates constructed for a bipartite graph is exactly equal to the size of the maximum matching (a similar result holds for the general case). Consider the following approach for determining the size of the maximum matching: replace the indeterminates by random values and compute the rank of the resulting matrix. The rank of an integer matrix can be computed in $\mathcal{NC}$, and one would hope that the random substitution method would preserve the rank with high probability. We would like to use this to verify that the matching algorithm is indeed producing the maximum matching, and thereby obtain a Las Vegas algorithm. Does this method work?

12.20 (Due to R.M. Karp, E. Upfal, and A. Wigderson [242].) In a bipartite graph $G(U, V, E)$, for any set $F \subseteq E$ define the *rank* $r(F)$ as the maximum size of intersection of F with a perfect matching, i.e., $r(F)$ is the largest number of edges in F that appear together in some perfect matching. Devise an $\mathcal{RNC}$ algorithm for computing the rank for any given set F. Can this be generalized to non-bipartite graphs?

12.21 (Due to R.M. Karp, E. Upfal, and A. Wigderson [242].) Assume you are given the algorithm from Problem 12.20. Using this, we will outline the construction of an alternative $\mathcal{RNC}$ algorithm for perfect matchings.

- Assuming that the input graph is sparse in that it has a total of n vertices and fewer than $3n/4$ edges, devise an $\mathcal{NC}$ algorithm for finding a large set S of edges that are guaranteed to belong to every perfect matching in G.
- Suppose now that the input graph has more than $3n/4$ edges. Using the rank algorithm, devise an $\mathcal{RNC}$ algorithm for finding a large set T of edges such that there exists a perfect matching in G none of whose edges belong to T.

Using the above tools, describe an alternative $\mathcal{RNC}$ algorithm for perfect matchings.

12.22 (Due to V.V. Vazirani [405].) Prove that the Isolating Lemma holds even when the weight of a set is defined to be the *product* (instead of *sum*) of the weights of its elements. Can you identify any general family of mappings from the weights of elements to the weights of sets for which the Isolating Lemma is guaranteed to be valid?

12.23 (Due to K. Mulmuley, U.V. Vazirani, and V.V. Vazirani [317].) An intriguing application of the Isolating Lemma is to the class of "uniqueness" problems, i.e., determining whether some problem in $\mathcal{NP}$ has a unique solution. Consider the following two problems, which take as input a graph $G(V, E)$ and a positive integer k:

CLIQUE: Determine whether the graph has a clique of size k.
UNIQUE CLIQUE: Determine whether there is *exactly one* clique of size k.

The complexity of unique solutions has been studied with respect to randomized reductions, which are the natural generalization of polynomial time reductions to allowing randomized polynomial time reductions. Devise a randomized polynomial time reduction from the CLIQUE problem to the UNIQUE CLIQUE.

12.24 (Due to J. Naor.) Let $G(V, E)$ be an unweighted, undirected graph with n vertices and m edges. Under any weight function $w : E \rightarrow \{0, \ldots, W\}$, the length of a path in G is the sum of the weights of the edges in that path. A weight function is said to be *good* if the following two conditions hold for each vertex $x \in V$.

1. For all vertices $y \in V$, the shortest path from x to y is unique.
2. For any pair of vertices $y, z \in V$, the net weight of the shortest path from x to y is different from the net weight of the shortest path from x to z.

What is the smallest value of W (as a function of n and m) for which you can guarantee the *existence* of a good weight assignment?

12.25 (Due to K. Mulmuley, U.V. Vazirani, and V.V. Vazirani [317].) An even more intriguing application of the Isolating Lemma is to the *Exact Matching* problem – given a graph $G(V, E)$ with a subset of edges $R \subseteq E$ colored red, and a positive integer k, determine whether there is a perfect matching using exactly k red edges. This problem is not known to be in ***P***, but can be shown to be in ***RNC*** via a (non-trivial) application of the Isolating Lemma. Devise ***RNC*** algorithms for the decision and search versions of this problem.

12.26 (Due to M.O. Rabin [344].) Show that Algorithm **ASYNCH-CCP** works equally well in the case where the numbers of processors and choices are both greater than 2. How does the complexity depend on the number of processors and choices?

12.27 How large a value of t can the **ByzGen** algorithm tolerate? (Modify the parameters L, H, and G if necessary.)

12.28 Consider what happens if the outcome of the coin toss generated by the trusted party in the **ByzGen** algorithm is corrupted before it reaches some good processors.

(a) Can disagreement occur if different good processors see different outcomes? What happens if, instead of a global coin toss, each processor chooses a random coin independently of other processors, at every round?

(b) Suppose that we were guaranteed that at least H good processors receive the correct outcome of each coin toss. Give a modification for the protocol **ByzGen** that achieves agreement in an expected constant number of rounds, under this assumption.

CHAPTER 13

Online Algorithms

ALL the algorithms we have studied so far receive their entire inputs at one time. We turn our attention to *online algorithms*, which receive and process the input in partial amounts. In a typical setting, an online algorithm receives a sequence of *requests* for service. It must service each request before it receives the next one. In servicing each request, the algorithm has a choice of several alternatives, each with an associated *cost*. The alternative chosen at a step may influence the costs of alternatives on future requests. Examples of such situations arise in data-structuring, resource-allocation in operating systems, finance, and distributed computing.

In an online setting, it is often meaningless to have an absolute performance measure for an algorithm. This is because in most such settings, any algorithm for processing requests can be forced to incur an unbounded cost by appropriately choosing the input sequence (we study examples of this below); thus, it becomes difficult, if not impossible, to perform a comparison of competing strategies. Consequently, we compare the total cost of the online algorithm on a sequence of requests, to the total cost of an *offline* algorithm that services the *same sequence* of requests. We refer to such an analysis of an online algorithm as a *competitive analysis*; we will make these notions formal presently. Intuitively, this form of analysis assumes that there is an inherent cost associated with a request sequence (the cost of the best possible algorithm that knows the entire request sequence in advance and can tailor its responses accordingly), and the performance of an online algorithm on a given sequence is measured in terms of the ratio it achieves with respect to this inherent cost. The worst-case ratio over all possible request sequences is then a natural measure of the quality of the online algorithm. In some practical settings, this approach leads to a meaningful theoretical validation of the difference between competing strategies.

A classical example where this approach has been particularly successful is that of paging in a two-level memory storage system, and we introduce online algorithms through this example. We define three possible scenarios for randomized online algorithms, and then study the relationships between them.

We give optimal algorithms for paging in each of these scenarios. Finally, we present some results for generalizations of the paging problem.

13.1. The Online Paging Problem

We first consider the *paging problem.* Consider a computer memory organized as a two-level store: there is a *cache* or fast memory that can store k memory items, and a slower main memory that can potentially hold an infinite number of items. Each item represents a page of virtual memory (the cache can contain k of these). A paging algorithm decides which k items to retain in the cache at each point in time. We have a sequence of requests, each of which specifies a memory item. If the item requested is currently in the cache, a *hit* is said to occur, and the algorithm incurs no cost on that request. If not, a *miss* occurs and the item must be fetched from the main memory at a unit cost; in addition, one of the k items currently in the cache must be evicted to make room for the incoming item. The cost measure for paging is the number of misses on a sequence of requests. Naturally, the cost incurred depends on the algorithm that decides which k items to retain in the cache at each point in time.

We now examine the actions of an algorithm. When the requested item is fetched from the main memory to the cache and the cache is full, a paging algorithm must invoke an *eviction rule* for deciding which item currently in the cache is evicted to make room for the new item. Intuitively, a paging algorithm will try not to evict items that will be requested again in the near future. An online paging algorithm must make this decision without knowledge of future requests; in contrast, an *offline* algorithm makes each decision with complete knowledge of the future. We first study the basic concepts involved using deterministic algorithms, and then proceed to randomized paging algorithms.

Here are some typical (deterministic) online algorithms that have been used in computer systems.

- *Least Recently Used* (**LRU**): evict the item in the cache whose most recent request occurred furthest in the past.
- *First-in, First-out* (**FIFO**): evict the item that has been in the cache for the longest period.
- *Least Frequently Used* (**LFU**): evict the item in the cache that has been requested least often.

Notice that there is a non-trivial computational cost associated with some of these online algorithms; for instance, **LRU** must maintain a priority queue of time stamps for the k items in the cache.

Let $\boldsymbol{\rho} = (\rho_1, \rho_2, \ldots, \rho_N)$ be a request sequence presented to an online paging algorithm A. Consider the case when A is deterministic. Upon each request, we know exactly how A will respond and, given the sequence $\rho_1, \rho_2, \ldots, \rho_N$, we can deduce the number of times that A misses on this sequence. We can also compute the minimum possible number of misses on this sequence, i.e., the cost of an optimal offline algorithm for this sequence. Let $f_A(\rho_1, \rho_2, \ldots, \rho_N)$ denote the number of times that A misses on the sequence $\rho_1, \rho_2, \ldots, \rho_N$, and let $f_O(\rho_1, \rho_2, \ldots, \rho_N)$ be the minimum number of misses (for an optimal offline algorithm) on the same sequence.

The following (clearly offline) strategy is known to minimize $f_O(\rho_1, \rho_2, \ldots, \rho_N)$ on every request sequence $\rho_1, \rho_2, \ldots, \rho_N$: on a miss, evict that item in the cache whose next request occurs furthest in the future. This offline strategy is known as the **MIN** algorithm. The proof of optimality is non-trivial and a pointer can be found in the Notes section.

Exercise 13.1: In this exercise, we will see that the traditional worst-case performance analysis is meaningless in an online setting such as the paging problem. Consider the rather simple scenario where there are only $k+1$ distinct memory items. Assume whatever is convenient for the initial contents of the cache in each case.

1. Show that for any (deterministic) online paging algorithm A, there exist sequences of arbitrary length such that the algorithm A misses on every request, i.e., $f_A(\rho_1, \rho_2, \ldots, \rho_N) = N$.
2. Show that for the offline paging algorithm **MIN**, the worst-case number of misses on a request sequence of length N is N/k.

Suppose that we wish to study the performance of online algorithms such as **LRU**, **FIFO**, and **LFU**. In Exercise 13.1 we saw that the seemingly natural measure of the worst-case value of $f_A(\rho_1, \rho_2, \ldots, \rho_N)$ is not useful. This motivates the following measure of performance.

► **Definition 13.1:** A deterministic online paging algorithm A is said to be C-*competitive* if there exists a constant b such that on every sequence of requests $\rho_1, \rho_2, \ldots, \rho_N$,

$$f_A(\rho_1, \rho_2, \ldots, \rho_N) - C \times f_O(\rho_1, \rho_2, \ldots, \rho_N) \leq b,$$

where the constant b must be independent of N but may depend on k. The *competitiveness coefficient* of A, denoted C_A, is the infimum of C such that A is C-competitive.

Roughly speaking, competitiveness measures the performance of an online algorithm in terms of the worst-case ratio of its cost to that of the optimal offline algorithm running on the *same* request sequence.

The **LRU** and **FIFO** algorithms mentioned above are known to be k-competitive (see Problems 13.1 and 13.2). In Problem 13.3 we will see that the **LFU** does not achieve a bounded competitiveness coefficient. From Exercise 13.1 we conclude that no deterministic online paging algorithm has competitiveness coefficient smaller than k, thereby obtaining that **LRU** and **FIFO** are optimal deterministic online algorithms. We give an alternate proof of this lower bound on the competitiveness coefficient of deterministic algorithms, so as to develop some tools for the subsequent analysis of randomized algorithms. But first, we define a paging algorithm formally.

A paging algorithm consists of an automaton with a finite set S of states. The response of this automaton to a request is specified by a function F that depends on the current state of the automaton, the k items in the cache, and the newly requested item. It specifies, in general, a new state for the automaton, together with the new set of items in the cache. We impose the following condition on F: the set of items in the cache after the request is serviced must include the item just requested.

Theorem 13.1: *Let A be a deterministic online algorithm for paging. Then $C_A \geq k$.*

PROOF: Imagine that the offline algorithm and A are both managing (separate) caches for the same request sequence. Assume that to start with, both the offline algorithm and A have the same set of k items in their caches.

Consider the following request sequence, which is completely determined by the behavior of A. The first request is to an item not in either cache, and both algorithms incur a miss on this request. Let S be the set of $k+1$ items consisting of the k items initially in the offline algorithm's cache together with the new item. From then on, every request is for the unique item in S not in A's cache. Thus A misses on every request.

We partition the request sequence into *rounds* in a manner described below. We will argue that during each round, A misses at least k times but an optimal offline algorithm has at most one miss. The first round begins with the first request. A round is a maximal sequence of requests in which at most k distinct items are requested; each of these items may be requested any number of times and in any order. A round ends when, after k distinct items have been requested during the round, a new item ρ is requested, and ρ then becomes the first request of the next round. Since the round contains at least k requests and A misses on every one of them, it misses at least k times during the round.

We now argue that there is an offline algorithm that misses only once during a round, in fact on the first request of the round. Since only k distinct items are requested during the round, there is one item that will not be requested until the first request of the following round; denote this item by ρ. When the offline algorithm misses on the first request of the round, it evicts ρ and thereby ensures that there are no further misses in that round (as the **MIN** algorithm would). Because A is deterministic, the offline algorithm can predict the behavior of A during each round. Knowing the initial contents of A's cache (the same as the initial contents of its own cache), it knows the entire request sequence in advance, and in particular the identity of ρ for every round.

At the end of each round, both the online algorithm and the offline algorithm have the same set of items in their caches. Thus this construction can be repeated as many times as desired, proving that there are arbitrarily long sequences on which A has k times as many misses as the offline algorithm. □

We pause to make some observations about the negative result we have just seen. First, the proof uses only the fact that the online algorithm does not know future requests and does not exploit any computational limitation of the online algorithm. Thus the lower bound applies to *any* deterministic online algorithm without any regard for its use of computational resources such as time or space. This is a typical feature of most negative results for online algorithms.

Second, the proof of the lower bound uses only $k+1$ distinct memory items in all. In this lower bound, one can view the offline algorithm as an *adversary* who is not only managing a cache, but is also generating the request sequence. This will be a recurrent theme in the notions of adversaries we will develop for randomized algorithms – that there is an adversary generating requests, in collusion with a reference algorithm that is the yardstick against which the competitiveness of the given online algorithm is being measured. The adversary's goal is to increase the cost to the given online algorithm, while keeping it down for the reference algorithm.

13.2. Adversary Models

Can we overcome the negative result of Theorem 13.1 using randomization? To make this question precise, we must first make precise the notion of the competitiveness of a randomized algorithm. Consider a randomized online paging algorithm R; on a miss, it makes a (possibly random) choice of which of the k items in the cache it will evict. Given a sequence of requests $\rho_1, \rho_2, \ldots, \rho_N$, the number of times that R misses on the sequence is now a random variable, which we will denote by $f_R(\rho_1, \rho_2, \ldots, \rho_N)$. Following the convention in our study of deterministic online paging algorithms, we study the behavior of R when the sequence of requests is generated by an adversary. However, there is no longer a unique notion of an "adversary" for a randomized online algorithm. This section introduces three different possibilities for the notion of an adversary for a randomized online algorithm. The relationships between them will be explored further in Section 13.4.

The central issue is the following question: what does the adversary know in generating each request of the sequence? The weakest adversary we may envision knows the algorithm R in advance, but has no knowledge of the random choices made by R while processing a request sequence. Such an adversary may as well write down the entire request sequence in advance, since it is not influenced in any way by the actual execution of R. Having written down such a "worst case" request sequence for R, the adversary services this sequence optimally using **MIN** and incurs the concomitant cost. This cost of an optimal service strategy is not a random variable, since the sequence is fixed, and so we denote it by

$f_O(\rho_1, \rho_2, \ldots, \rho_N)$. We call such an adversary an *oblivious adversary*, reflecting the fact that the adversary is oblivious to the random choices made by R.

We say that R is C-competitive against the oblivious adversary if for every sequence of requests $\rho_1, \rho_2, \ldots, \rho_N$,

$$\mathbf{E}[f_R(\rho_1, \rho_2, \ldots, \rho_N)] - C \times f_O(\rho_1, \rho_2, \ldots, \rho_N) \leq b,$$

for a constant b independent of N. The *oblivious competitiveness coefficient* of R, denoted C_R^{obl}, is the infimum of C such that R is C-competitive.

What if the adversary were able to choose each request after having observed the previous choices (and thus the current state) of the online algorithm? Whether or not the adversary is allowed to adapt the request sequence to these "run-time" random choices could affect the value of the competitiveness that is achievable. This is not an issue when A is a deterministic online algorithm, since the behavior of A on $\rho_1, \rho_2, \ldots, \rho_i$ is completely predictable and so we could as well assume that the adversary knows of A's responses to these requests when choosing ρ_{i+1}. The response of a randomized algorithm, on the other hand, depends on random choices it makes during its execution.

To study this, we introduce the *adaptive adversary* who chooses ρ_{i+1} after having observed the responses of the randomized online algorithm to $\rho_1, \rho_2, \ldots, \rho_i$. Thus the adaptive adversary is denied information only about the future random choices of the randomized online algorithm R. The cost incurred by R is still a random variable. However, in order to facilitate the definition of the competitiveness of R against an adaptive adversary, we have to specify what we mean by the cost of an optimal algorithm. In the discussion below, it may help the reader to think of the adaptive adversary and the optimal algorithm as working in collusion.

Here there are two possible scenarios. In the first, the adversary generates the sequence adaptively as described above; when the entire sequence has been generated in this fashion, the adversary exhibits its optimal strategy for servicing the sequence (using **MIN**). We refer to this as the *adaptive offline adversary*. Since the request sequence depends on the behavior of the algorithm R, it is a random sequence. Thus both $f_R(\rho_1, \rho_2, \ldots, \rho_N)$ and $f_O(\rho_1, \rho_2, \ldots, \rho_N)$ are random variables. Before defining the competitiveness of R against an adaptive offline adversary, let us look at the second possible scenario involving an adaptive adversary.

Suppose the adversary were to generate the sequence adaptively as before, but in addition was required to concurrently manage a cache *online*. In other words, the adversary generates ρ_{i+1} based on the responses of R to $\rho_1, \rho_2, \ldots, \rho_i$, and immediately exhibits its own response to ρ_{i+1} (but does not reveal it to R, of course). Then, following R's response to ρ_{i+1}, it generates ρ_{i+2}, responds to ρ_{i+2}, and so on. Again both $f_R(\rho_1, \rho_2, \ldots, \rho_N)$ and $f_O(\rho_1, \rho_2, \ldots, \rho_N)$ are random variables. We refer to such an adversary as an *adaptive online adversary*.

Let $\rho_1, \rho_2, \ldots, \rho_N$ be a sequence of requests generated by an adaptive offline adversary. We say that R is C-competitive against the adaptive offline adversary if

$$\mathbf{E}[f_R(\rho_1, \rho_2, \ldots, \rho_N)] - \mathcal{C} \times \mathbf{E}[f_O(\rho_1, \rho_2, \ldots, \rho_N)] \leq b$$

for a constant b independent of N. The *adaptive offline competitiveness coefficient* of R, denoted $\mathcal{C}_R^{aof}$, is the infimum of $\mathcal{C}$ such that R is $\mathcal{C}$-competitive. Likewise, we define the *adaptive online competitiveness coefficient* of R, denoted $\mathcal{C}_R^{aon}$.

Clearly, the adaptive offline adversary is at least as powerful as the adaptive online adversary, which in turn is at least as powerful as the oblivious adversary. It follows that for any algorithm R,

$$\mathcal{C}_R^{obl} \leq \mathcal{C}_R^{aon} \leq \mathcal{C}_R^{aof}.$$

Let us denote by $\mathcal{C}^{obl}$ the lowest oblivious competitive coefficient of any randomized paging algorithm; similarly we define $\mathcal{C}^{aon}$ and $\mathcal{C}^{aof}$. Finally, let $\mathcal{C}^{det}$ denote the lowest competitive coefficient of any deterministic paging algorithm. Then we have

$$\mathcal{C}^{obl} \leq \mathcal{C}^{aon} \leq \mathcal{C}^{aof} \leq \mathcal{C}^{det}.$$

How far apart in value can the different coefficients be? In Section 13.4 we will develop some general relationships between these quantities.

13.3. Paging against an Oblivious Adversary

The lower bound of Theorem 13.1 hinged on the adversary being able to predict, at each step, the response of the algorithm to any request. We now study the effect of denying the adversary this facility; we will study randomized online algorithms for paging against oblivious adversaries. The request sequence is specified at the beginning by the adversary and is not changed after that. The adversary also determines its (optimal offline) response to the sequence and the cost of this response. The sequence is then unveiled to the online algorithm, one request at a time as before. This prevents the offline player from knowing with certainty (as in the proof of Theorem 13.1) the contents of the cache of the online algorithm. Intuitively, it seems that this should help the randomized online algorithm fare better.

We first prove a negative result on the performance of any randomized online paging algorithm.

Theorem 13.2: *Let R be a randomized algorithm for paging. Then $\mathcal{C}_R^{obl} \geq H_k$, where $H_k = \sum_{j=1}^{k} 1/j$ is the kth Harmonic number.*

In order to prove this theorem, we apply Yao's Minimax Principle (Section 2.2.2) to the competitiveness of randomized online paging algorithms. Let $\mathcal{P}$ be a probability distribution for choosing a request sequence, i.e., a probability distribution by which ρ_i is chosen. The distribution for ρ_i is allowed to depend on $\rho_1, \rho_2, \ldots, \rho_{i-1}$. The algorithm's costs (as well as the optimal cost) are now random variables. For a deterministic online paging algorithm A, define its *competitiveness under $\mathcal{P}$*, $C_A^{\mathcal{P}}$, to be the infimum of C such that

$$\mathbf{E}[f_A(\rho_1, \rho_2, \ldots, \rho_N)] - C \times \mathbf{E}[f_O(\rho_1, \rho_2, \ldots, \rho_N)] \leq b$$

for a constant b independent of N. Yao's Minimax Principle (Section 2.2.2) implies that

$$\inf_R C_R^{obl} = \sup_{\mathcal{P}} \inf_A C_A^{\mathcal{P}}.$$

The implication of this in our situation is as follows: the competitiveness of the best randomized online paging algorithm equals $C_A^{\mathcal{P}}$, the competitiveness of a "best possible" deterministic algorithm A on inputs generated according to $\mathcal{P}$, a "worst-case" distribution on request-sequences $\boldsymbol{\rho}$. Thus, we can establish a lower bound on C_R^{obl} by giving a probability distribution $\mathcal{P}$ and giving a lower bound on $C_A^{\mathcal{P}}$ for any deterministic algorithm A.

Proof of Theorem 13.2: We will make use of a set of $k+1$ memory items, $I = \{I_1, \ldots, I_{k+1}\}$, in the lower bound. Since k of these can be accommodated in the cache, only one item need be outside the cache at any given time. Thus any paging algorithm need only specify which one item it leaves out of the cache at any point in time. We assume that $N \gg k$.

We will use Yao's Minimax Principle as follows: we give a probability distribution on request sequences $\boldsymbol{\rho}$ of length N, and first study the number of misses for any deterministic algorithm on $\boldsymbol{\rho}$. The sequence $\boldsymbol{\rho}$ is chosen as follows: for $i > 1$, request ρ_i is chosen uniformly at random from the k items in the set $I - \{\rho_{i-1}\}$; the first request, ρ_1 is chosen uniformly from all the items in I. We will show that the offline algorithm can divide $\boldsymbol{\rho}$ up into rounds such that it only misses on the final request in each round.

The first round begins with the first request and ends when, for the first time, every item in I has been requested at least once; the second round begins with the next request. In general, each round ends just before the request to the $(k+1)$th distinct item since the start of that round. The offline algorithm uses the **MIN** algorithm during each round: it leaves out of its cache the item requested last in a round, until that item is requested (on the final request of the round). This item is requested exactly once during each round, and thus the offline algorithm incurs one miss during each round.

How often does the offline algorithm miss? Equivalently, what is the expected length of each round? A moment's thought shows that this is the cover time of the random walk on a complete graph with $k+1$ vertices and is equal to kH_k.

Let us now consider the online algorithm A. At any point in time, A must leave one of the $k+1$ items out of the cache. Whenever a request falls on this item, A incurs a miss. Since every request goes to an item chosen uniformly at random from the k items other than the one just requested, the probability that any request falls on the item that A leaves out is $1/k$. It follows that the expected number of misses per round is H_k.

Thus the number of times A misses has expectation H_k times the number of misses of the offline algorithm on the same sequence, and this yields the result.

We now study a randomized online paging algorithm that achieves a competitiveness coefficient close to the lower bound of Theorem 13.2. This algorithm is referred to as the **Marker** algorithm. The algorithm proceeds in a series of rounds. Each of the k cache locations has a *marker bit* associated with it. At the beginning of every round, all k marker bits are reset to zero. As memory requests come in, the algorithm processes them as follows. If the item requested is already in one of the k cache locations, the marker bit of that location is set to one. If the request is a miss (the item requested is not one of the k in the cache), the item is brought into the cache and the item that is evicted to make room for it is chosen as follows: choose an unmarked cache location uniformly at random, evict the item in it, and set its marker bit to 1. After all the locations have been thus marked, the round is deemed over on the next request to an item not in the cache.

Theorem 13.3: *The* **Marker** *algorithm is* $(2H_k)$*-competitive.*

PROOF: For convenience in the proof, we will sometimes refer to the items (rather than the cache locations that contain them) as being marked or unmarked; thus we will refer to an item as being marked if the cache location containing it is marked, and as unmarked otherwise. As before, we will compare the **Marker** algorithm's management of a cache with k locations on a sequence $\rho_1, \rho_2, \ldots$ to an optimal offline algorithm's cache management on the same sequence.

Assume that both algorithms start with the same k items in the cache, and that ρ_1 is not in the cache. The **Marker** algorithm implicitly divides the request sequence into a series of rounds, the first of which begins with ρ_1. The round beginning with request ρ_i ends with ρ_j, where j is the smallest integer such that there are $k+1$ distinct items in $\rho_i, \rho_{i+1}, \ldots, \rho_{j+1}$. All k cache locations are marked at the end of each round. The first request of each round is to an item not currently in cache.

Consider the requests in any round. Call an item *stale* if it is unmarked, but was marked in the previous round, and *clean* if it is neither stale nor marked. Let ℓ be the number of requests to clean items in a round. We first argue that the amortized number of misses incurred during the round by the offline algorithm is at least $\ell/2$, and then show that the expected number of misses of the **Marker** algorithm during the round is at most ℓH_k; these facts together will yield the theorem.

Let S_O denote the set of items in the offline algorithm's cache, and S_M denote the set of items in the **Marker** algorithm's cache. Let d_I be the value of $|S_O \setminus S_M|$ at the beginning of the round, and d_F this value at the end of the round. Let M_O be the number of misses incurred by the offline algorithm during the round.

Clearly $M_O \geq \ell - d_I$, since at least $\ell - d_I$ of the ℓ clean items requested in the round are not in the offline algorithm's cache at the beginning of the round.

At the end of the round, all the k (marked) items in S_M at that point are items that were requested during the round. Since d_F items in the offline algorithm's cache are not in S_M, the offline algorithm has incurred at least d_F misses during the round. Thus,

$$M_O \geq \max\{\ell - d_I, d_F\} \geq \frac{\ell - d_I + d_F}{2}.$$

On summing this lower bound on M_O over all rounds, the d_I and d_F terms for all rounds (except the first and the last) telescope, so that the "amortized" number of misses of this round is at least $\ell/2$. (By amortization, we mean here that we can think of "charging" each round a certain number of misses without affecting the total number of misses.) By this we mean that we may charge $\ell/2$ misses to this round; by adopting this charging mechanism for all rounds, we estimate the total number of misses over all rounds to within an additive factor of $2k$.

Consider the expected number of misses incurred by the **Marker** algorithm during the round. Each of the ℓ requests to clean items costs the **Marker** algorithm a miss. Of the $k - \ell$ requests to stale items, the expected cost of each is the probability that the item requested is not in the cache. This is maximized when the ℓ requests to clean items all precede the $k - \ell$ requests to stale items. For $1 \leq i \leq k - \ell$, a simple calculation shows that this probability is $\ell/(k-i+1)$ for the ith request to a stale item. Summing this over all i shows that the expected cost of the **Marker** algorithm is bounded by

$$\ell + \ell(H_k - H_\ell) \leq \ell H_k,$$

and this proves the result. □

Thus the **Marker** algorithm achieves a competitiveness coefficient that is at most twice the best possible. In fact, there is a more sophisticated algorithm that is H_k-competitive in general; a pointer is available in the Notes Section.

13.4. Relating the Adversaries

We have just seen that against an oblivious adversary, a randomized algorithm can attain a competitiveness coefficient substantially smaller than that of any deterministic algorithm. Can a similar performance be attained against adaptive adversaries? In this section we study relations between the competitiveness coefficients attainable against the three types of adversaries introduced in Section 13.2. We will see that randomized online algorithms cannot achieve such substantial improvements against adaptive adversaries, as such adversaries prove to be very powerful. Later, in Section 13.5 we will study some randomized algorithms and their performance against adaptive adversaries.

The results we are about to derive can easily be obtained in the setting of the paging problem; however, they apply to considerably more general online problems. We therefore study the more general setting of *request-answer games* that we will introduce now, and the results derived here apply to the paging problem we have studied in previous sections. We proceed to define these games and make the notions of the various adversaries precise in this context.

A request-answer game consists of a request set $\mathcal{R}$ and a finite answer set $\mathcal{A}$, together with cost functions $f_n : \mathcal{R}^n \times \mathcal{A}^n \to \mathbb{R} \cup \{\infty\}$ for each non-negative integer n. Let f denote the union, over non-negative integers n, of the functions f_n. Let us fix our attention on one such game. Let $\boldsymbol{\rho}$ denote a sequence of requests (whose length will be implicit from its usage), and likewise let $\boldsymbol{a}$ denote a sequence of answers.

A *deterministic online algorithm* A is a sequence of functions $g_i : \mathcal{R}^i \to \mathcal{A}$ for positive integers i. Fix a value n. For any sequence of requests $\boldsymbol{\rho} = (\rho_1, \ldots, \rho_n)$, we define $A(\boldsymbol{\rho}) = (a_1, \ldots, a_n) \in \mathcal{A}^n$ with $a_i = g_i(\rho_1, \ldots, \rho_i)$ for $i = 1, \ldots, n$. The *cost* of A on $\boldsymbol{\rho}$ is $c_A(\boldsymbol{\rho}) = f_n(\boldsymbol{\rho}, A(\boldsymbol{\rho}))$. We will compare this cost incurred by the online algorithm A to the optimal cost for the same sequence of requests, which is

$$c(\boldsymbol{\rho}) = \min\{f_n(\boldsymbol{\rho}, \boldsymbol{a}) \mid \boldsymbol{a} \in \mathcal{A}^n\}.$$

Let $\alpha, \beta : \mathbb{R} \to \mathbb{R}$ denote linear functions. We say that the deterministic algorithm A is α-competitive if for every request sequence $\boldsymbol{\rho}$ we have $c_A(\boldsymbol{\rho}) \leq \alpha[c(\boldsymbol{\rho})]$.

A *randomized online algorithm* R is a probability distribution over deterministic online algorithms A_x (x may be thought of as representing the coin tosses of R). For a request sequence $\boldsymbol{\rho}$ the answer sequence $R(\boldsymbol{\rho})$ is random, and so the cost $c_R(\boldsymbol{\rho})$ is a random variable. We say that a randomized algorithm is α-competitive against oblivious adversaries if for any $\boldsymbol{\rho}$ we have $\mathbf{E}_x[c_{A_x}(\boldsymbol{\rho})] \leq \alpha[c(\boldsymbol{\rho})]$.

In developing the notation for adaptive adversaries, we first do so for a deterministic algorithm A. We then note that for a randomized algorithm R that is a probability distribution on deterministic algorithms A_x, all the quantities defined become random variables. Once we have developed this notation, we will proceed to prove two results about the various adversaries for randomized algorithms.

An adaptive offline adversary Q is a sequence of functions $q_n : \mathcal{A}^n \to \mathcal{R} \cup \{\text{STOP}\}$, where $n = 0, 1, \ldots, d_Q$ and q_{d_Q} takes only the value STOP. For a particular deterministic algorithm A and an adaptive adversary Q, we define the request and answer sequences resulting from their interaction, $\boldsymbol{\rho}(A, Q) = (\rho_1, \ldots, \rho_n)$ and $\boldsymbol{a}(A, Q) = (a_1, \ldots, a_n)$, together with $n = n(A, Q)$. Further, we recursively have $\rho_{i+1} = q_i(a_1, \ldots, a_i)$ for $i = 0, 1, \ldots, n-1$, with $q_n(\boldsymbol{a}(A, Q)) = \text{STOP}$, and $\boldsymbol{a}(A, Q) = A(\boldsymbol{\rho}(A, Q))$. Because we are discussing a deterministic algorithm A for the moment, we have uniquely defined entities $\rho_1, a_1, \ldots, \rho_{n-1}, a_{n-1}, \text{STOP}$ in that order. The value $n = n(A, Q)$ is bounded by d_Q for any A. The cost of A against adversary Q is defined to be $c_A(Q) = f_n(\boldsymbol{\rho}(A, Q), \boldsymbol{a}(A, Q))$. The adaptive offline adversary incurs the optimal cost for servicing the sequence, $c_Q(A) = c(\boldsymbol{\rho}(A, Q))$.

An adaptive online adversary $S = (Q, P)$ is an adaptive offline adversary Q, supplemented by a sequence P of functions that define its own online response to the request sequence. In particular, we have $p_n : \mathcal{A}^n \to \mathcal{A}$ for $n = 0, 1, \ldots, d_Q$.

Note that the request sequence is independent of P and depends only on the algorithm A (again, we are focusing for the moment on a deterministic algorithm A). Thus we can write $\boldsymbol{\rho}(A,S) = \boldsymbol{\rho}(A,Q)$, $\boldsymbol{a}(A,S) = \boldsymbol{a}(A,Q)$, and $c_A(S) = c_A(Q)$. In addition, the adversary's response P induces an answer sequence for the adversary, which we denote by $\boldsymbol{b}(A,S) = (b_1,\ldots,b_n)$ where $n = n(A,Q)$ and $b_{i+1} = p_i(a_1,\ldots,a_i)$ for $i = 0,1,\ldots,n-1$. The cost of S against A is denoted $c_S(A) = f_n(\boldsymbol{\rho}(A,S),\boldsymbol{b}(A,S))$.

For a randomized algorithm R that is a probability distribution on deterministic algorithms A_x (we think of x as the random string that selects the deterministic algorithm), the above definitions can be made again, with the costs becoming random variables. As in the case of the oblivious adversary, we say that a randomized algorithm is α-competitive against adaptive offline (respectively online) adversaries if for any $\boldsymbol{\rho}$ we have $\mathbf{E}_x[c_{A_x}(\boldsymbol{\rho})] \leq \mathbf{E}[\alpha[c_Q(A)]]$ (respectively $\mathbf{E}_x[c_{A_x}(\boldsymbol{\rho})] \leq \mathbf{E}[\alpha[c_S(A)]]$).

The reader is invited to verify that the request-answer games defined above generalize the paging problem of previous sections. While the following results could have been derived for the paging problem, we have chosen the more general setting in order to apply these results to more general online problems to be introduced in the next section. Our first result says that adaptive offline adversaries are so powerful that there is no benefit to using randomization against them. Note again that all the arguments below apply to any fixed request-answer game.

Theorem 13.4: *If there is a randomized algorithm that is α-competitive against every adaptive offline adversary, then there exists an α-competitive deterministic algorithm.*

PROOF: View the request-answer game as a two-person game between two players C and D such that in every step C gives D a request, which D answers. A position in the game is a pair $(\boldsymbol{\rho},\boldsymbol{a})$. Call a position an *instant winner* for C if $f_n(\boldsymbol{\rho},\boldsymbol{a}) > \alpha[c(\boldsymbol{\rho})]$. Call a position $(\boldsymbol{\rho},\boldsymbol{a})$ *winning for* C if there exists an adaptive rule for selecting requests, and a positive integer t such that starting from $(\boldsymbol{\rho},\boldsymbol{a})$, an instant winner for C will be reached in t steps regardless of how D plays.

Let us suppose that there is an α-competitive randomized algorithm R for any adaptive offline adversary. Further, suppose for a contradiction that there is no deterministic α-competitive algorithm. Then C has a winning strategy against any deterministic player in the two-person game. The initial position, in which $\boldsymbol{\rho}$ and $\boldsymbol{a}$ are both the empty string, is winning for C if and only if there exists an adaptive offline adversary Q such that for any deterministic algorithm A,

$$c_A(Q) > \alpha[c_Q(A)]. \tag{13.1}$$

Now, the randomized algorithm R is a probability distribution over deterministic algorithms A_x. Taking the expectation of (13.1) over all x, we have $\mathbf{E}_x[c_{A_x}(Q)] > \mathbf{E}_x[\alpha[c_Q(A_x)]]$, and thus $\mathbf{E}[c_R(Q)] > \mathbf{E}[\alpha[c_Q(R)]]$, where $\mathbf{E}_x[\ ]$ denotes the expectation over random strings x. This contradicts our assumption that R is α-competitive, so that C does not have a winning strategy in the game.

To complete the proof, we show that if C does not have a winning strategy, then there is a deterministic α-competitive algorithm. Now, a position $(\boldsymbol{\rho}, \boldsymbol{a})$ is a winning position for C if and only if there exists a request ρ_{n+1} such that, for every answer a_{n+1}, the position $(\boldsymbol{\rho}\rho_{n+1}, \boldsymbol{a}a_{n+1})$ is again a winning position for C. Therefore, if $(\boldsymbol{\rho}, \boldsymbol{a})$ is not a winning position for C, it follows that for every request ρ_{n+1} there exists an answer a_{n+1} (due to the finiteness of the answer set) resulting in a position that is not winning for C. Thus if D counters with such an answer at each step, it has a winning strategy and thus a deterministic α-competitive algorithm. □

By combining the results of Theorem 13.1 and Theorem 13.4, we conclude that no randomized online algorithm for the paging problem can achieve a competitiveness coefficient smaller than k, against adaptive offline adversaries. This is in marked contrast to the oblivious adversary, against which we have seen a randomized online algorithm achieving competitiveness coefficient $O(\log k)$. How well can a randomized algorithm perform against an adaptive online adversary? We can infer a limitation from the following theorem, which relates the three adversaries.

Theorem 13.5: *Suppose R is α-competitive against any adaptive online adversary, and there is a β-competitive randomized algorithm against any oblivious adversary; then R is $(\alpha\beta)$-competitive against any adaptive offline adversary.*

PROOF: Fix an adaptive offline adversary Q, and view R as a probability distribution on deterministic algorithms A_x. We will prove that $\mathbf{E}_x[c_{A_x}(Q)] \leq \mathbf{E}_x[\alpha[\beta[c_Q(A_x)]]]$.

Let H be the randomized algorithm that is β-competitive against any oblivious adversary. Viewing H as a probability distribution on deterministic algorithms H_y, we have for every n and $\boldsymbol{\rho} \in \mathcal{R}^n$, $\mathbf{E}_y[c_{H_y}(\boldsymbol{\rho})] \leq \beta[c(\boldsymbol{\rho})]$.

For each fixed y, define an adaptive online adversary $S_y = (Q, P_y)$ in such a way that for any deterministic online algorithm A, it sets $\boldsymbol{b}(A, S_y) = H_y(\boldsymbol{\rho}(A, Q))$. Thus this adaptive online adversary uses Q to generate the request sequence. On the other hand, it uses H_y to answer the requests, independently of A. Now, R is α-competitive against this adaptive online adversary, and in turn this adaptive online adversary is β-competitive against any oblivious adversary. Thus for every fixed y,

$$\mathbf{E}_x[c_{A_x}(S_y)] \leq \mathbf{E}_x[\alpha[c_{S_y}(A_x)]]. \tag{13.2}$$

Taking the expectation of (13.2) over y, we have

$$\mathbf{E}_y[\mathbf{E}_x[c_{A_x}(S_y)]] \leq \mathbf{E}_y[\mathbf{E}_x[\alpha[c_{S_y}(A_x)]]].$$

Since the adaptive online adversary is "borrowing" the request sequence from the adaptive offline adversary, we have for any y, $\rho(A_x, S_y) = \rho(A_x, Q) = \rho_x$. Then, we have

$$\begin{aligned}
\mathbf{E}_x[c_{A_x}(Q)] &= \mathbf{E}_y[\mathbf{E}_x[c_{A_x}(S_y)]] \\
&\leq \mathbf{E}_y[\alpha[\mathbf{E}_x[c_{S_y}(\rho_x)]]] \\
&= \alpha[\mathbf{E}_x[\mathbf{E}_y[c_{H_y}(\rho_x)]]] \\
&\leq \alpha[\mathbf{E}_x[\beta[c(\rho_x)]]] \\
&= \mathbf{E}_x[\alpha[\beta[c_Q(A_x)]]].
\end{aligned}$$

□

Let us again consider online algorithms for the paging problem. By Theorems 13.4 and 13.5, we have that $C^{det} \leq C^{aon}C^{obl}$. This tells us something about the performance of randomized online paging algorithms against adaptive online adversaries: we may infer that

$$C^{aon} \geq \frac{C^{det}}{C^{obl}} = \Omega(k/H_k).$$

In Section 13.5, we will further study randomized online algorithms against adaptive adversaries.

Exercise 13.2: Suppose that we have an online algorithm that is α-competitive against any adaptive online adversary, for a request-answer game. Show that this implies the existence of a deterministic online algorithm for the game that is α^2-competitive.

13.5. The Adaptive Online Adversary

One of the goals of this section is to determine the value of C^{aon} for the paging problem. We do so by studying a generalization of the paging problem studied above. The problem, known as *weighted paging*, is the following. As before, we have a two-level store whose cache can store k items at a time, while a slower memory can hold an infinite number of items. Again, we have a sequence of requests to items, and an item that is not in the cache when requested must be brought into the cache. And as before, an item in the cache must be evicted to make room for it.

Each item x that can be requested has associated with it a positive real *weight*, which is denoted $w(x)$. An algorithm that manages the cache incurs a cost of $w(x)$ every time it brings x from the slow memory to the cache. The total cost incurred by an algorithm on a sequence of requests is the sum of these costs. Clearly, when $w(x) = 1$ for all x, we have the paging problem studied before. But when the weights of items differ substantially, a good algorithm will perhaps be more willing to evict a "light" item than a "heavy" one. Certainly, it is easy to force algorithms such as **LRU** and **FIFO** (which are known to be optimal online algorithms for the paging problem) to perform poorly because they do not account for the weights of the items.

As in the paging problem, we may again define the competitiveness of an online algorithm, comparing its cost on a sequence to that of an optimal offline algorithm. Also, we may define as before the three types of adversaries for randomized online algorithms for weighted paging. We begin by giving a simple randomized algorithm that achieves a competitiveness coefficient of k against adaptive online adversaries. A lower bound to be presented in Section 13.6 will allow us to conclude that no randomized online algorithm for the weighted paging problem (including all special cases such as the paging problem) can achieve a competitiveness coefficient lower than k against adaptive online adversaries. Thus the algorithm below is optimal in its performance.

We now describe this simple randomized algorithm, called **Reciprocal**. The behavior of **Reciprocal** depends only on the weights of the items in the cache and is independent of the past. Let $x_1, \ldots, x_k$ be the items in the cache when an item not in the cache is requested. The **Reciprocal** algorithm uses the following simple, probabilistic eviction rule: evict x_i with probability p_i where

$$p_i = \frac{1/w(x_i)}{\sum_{j=1}^{k} 1/w(x_j)}.$$

Theorem 13.6: *The* **Reciprocal** *algorithm is k-competitive against any adaptive online adversary.*

PROOF: The proof uses a device that is common in the competitive analysis of online algorithms – a *potential function.* The typical use of a potential function is as follows: it is a measure of the discrepancy between a configuration of the online algorithm and a configuration of the offline algorithm.

We will study the expected change in this potential function after each request and compare this to the costs incurred by the online and offline algorithms on that request.

Let S_i^R be the set of items kept in the cache by **Reciprocal** after the ith reference, and S_i^{ADV} be the set of items kept by the adversary. Let

$$\Phi_i = \sum_{x \in S_i^R} w(x) - k \sum_{x \in S_i^R \setminus S_i^{ADV}} w(x),$$

and $\Delta\Phi_i = \Phi_i - \Phi_{i-1}$. Letting f_i^R denote the cost incurred by **Reciprocal** in servicing the ith request and f_i^{ADV} the corresponding cost of the adversary, we define

$$X_i = f_i^R - k f_i^{ADV} - \Delta\Phi_i.$$

Consider the following two actions that cause the two parties (**Reciprocal**, and the adversary) to incur costs.

1. The adversary evicts an item. We can assume that the adversary brings an item into the cache only immediately before a reference to that item. Also, without affecting the analysis of the cost except by an additive term, we can charge the adversary for the item it evicts rather than for the item it brings into the cache; thus $f_i^{ADV} = w(x_i)$, if the adversary evicts x_i on reference i (and 0 otherwise).
2. The **Reciprocal** algorithm evicts an item on a miss and is charged for the weight of the item it brings into the cache.

We examine the effects of these two actions on $\sum_{j=1}^{i} X_j$. By showing that in either case, $\mathbf{E}[X_i] \le 0$ (and noting that Φ_0 is bounded), we will argue that the theorem follows. Below, we drop the subscripts for S^R, S^{ADV}, and $\Delta\Phi$ because we consider the actions of each party in isolation; the reader may wish to think of the ith request, say to x, as being processed first by the (malicious) adversary, then by **Reciprocal**.

1. The adversary brings x into the cache and evicts x'. Then $f_i^{ADV} = w(x')$, and $-\Delta\Phi \le kw(x')$. (Equality is realized when $x' \in S^R \bigcap S^{ADV}$ and $x \notin S^R$.) Thus, the contribution of the adversary's action to $\mathbf{E}[X_i|X_{i-1},\ldots,X_1]$ is never positive.
2. **Reciprocal** misses on a reference to item x, so that $f_i^R = w(x)$. Just before **Reciprocal**'s action, $|S^R \setminus S^{ADV}| \ge 1$. By substituting the probabilities used by **Reciprocal**,

$$\begin{aligned}\mathbf{E}[\Delta\Phi] &= w(x) - \frac{k}{\sum_{y\in S^R} 1/w(y)} + k\frac{|S^R \setminus S^{ADV}|}{\sum_{y\in S^R} 1/w(y)} \\ &\ge w(x).\end{aligned}$$

Thus, the contribution of **Reciprocal**'s action to $\mathbf{E}[X_i|X_{i-1},\ldots,X_1]$ is also less than 0.

After a sequence of requests, we have $\mathbf{E}[\sum X_i] \le 0$; noting that Φ_0 and Φ_n are bounded, it follows that

$$\sum_i (\mathbf{E}[f_i^R] - k\mathbf{E}[f_i^{ADV}])$$

is bounded, yielding the theorem. (The reader is reminded that the additive term in the definition of competitiveness can depend on the weights of the items in the problem, as here. It cannot of course depend on the length of the request sequence.) □

It is interesting to note that the special case of the **Reciprocal** algorithm for the (unweighted) paging problem evicts, on each miss, an item chosen uniformly at random from the k items in the cache. It follows from Theorem 13.6 that this algorithm, known as **Random** in the paging literature, is k-competitive against any adaptive online adversary. Is k the lowest achievable competitiveness coefficient against the adaptive online adversary for the paging and weighted paging problems? We will answer this question in the affirmative in Section 13.6 below.

Exercise 13.3: It is important to note that the potential function analysis above does not apply to adaptive offline adversaries. Explain why such an analysis fails against adaptive offline adversaries.

The result of Problem 13.5 shows that the **Random** algorithm cannot achieve a competitiveness coefficient less than kH_k against adaptive offline adversaries. Thus we have an instance where the inequality of Theorem 13.5 is tight: **Random** achieves a competitiveness coefficient of k against any adaptive online adversary and a competitiveness coefficient of kH_k against any adaptive offline adversary, and there is a randomized algorithm for the paging problem that is H_k-competitive.

13.6. The k-Server Problem

We now study a generalization of the weighted paging problem above – the *k-server problem.* The setting of the *k-server problem* is a metric space. An online algorithm manages k mobile servers, each of which resides at one point of the metric space at any time. The algorithm is presented with a sequence of requests $\rho_1, \rho_2, \ldots, \rho_N$, where each request is a point in the space. In response to a request ρ_i, the algorithm must move a server to ρ_i unless it already has a server at ρ_i. Whenever the algorithm moves a server from point u to point v, it incurs a *cost* of c_{uv}, the distance between u and v in the metric space.

Given a sequence of requests $\rho_1, \rho_2, \ldots, \rho_N$, let $M_A(\rho_1, \rho_2, \ldots, \rho_N)$ denote the total cost incurred by an online algorithm A in servicing the requests in the sequence, and let $M_O(\rho_1, \rho_2, \ldots, \rho_N)$ denote the optimal offline cost of servicing the same sequence. We say that A is $\mathcal{C}$-competitive if for every request sequence $\rho_1, \rho_2, \ldots, \rho_N$,

$$M_A(\rho_1, \rho_2, \ldots, \rho_N) - \mathcal{C} \times M_O(\rho_1, \rho_2, \ldots, \rho_N) \leq b$$

for a constant b independent of N. The competitiveness coefficient of A, denoted $\mathcal{C}_A$, is the infimum of $\mathcal{C}$ such that A is $\mathcal{C}$-competitive. These definitions are similar to the ones we made for the paging problem (and used for the weighted paging problem as well). As before, we can define the three kinds of adversaries for randomized server algorithms and the corresponding notions of competitiveness.

A moment's thought shows that the paging problem is a special case of the k-server problem, one in which there is a point in the metric space corresponding to each item that can be requested and the distance between any two points is one. Each of the k servers corresponds to one of the k cache locations. Moving a server in response to a request corresponds to making a miss on a requested item and bringing it into the corresponding cache location. The point from which the server is brought corresponds to the memory item that is evicted to make room for the new item.

Exercise 13.4: Show that the weighted paging problem can be formulated as a special case of the k-server problem.

Other instances of the server problem arise in planning the motion of two-headed disks and in the maintenance of data-structures. Ultimately, though, it is the simplicity of the statement of the k-server problem that has lent it much of its appeal. In addition, the problem was originally posed along with a tantalizing conjecture that for every metric space, there is a deterministic online algorithm that is k-competitive. We will say more about this conjecture presently.

Exercise 13.5: The *greedy server algorithm* is the following: given a request at a point v, choose the closest server (the server at the vertex u that minimizes c_{uv}) to service this request. Show that the greedy algorithm is not competitive for any $k > 1$, by giving an example of a cost matrix and a request sequence that forces the greedy algorithm to pay an unbounded cost whereas an offline algorithm pays a bounded cost on the same sequence.

From the lower bound of Theorem 13.1, we know that there is a special case of the server problem (namely, the paging problem) in which no deterministic online algorithm achieves a competitiveness coefficient smaller than k. We generalize and extend this result now, showing that no randomized algorithm can achieve a competitiveness coefficient smaller than k for any server problem against adaptive online adversaries. Note that the following result does not use the minimax principle for the lower bound, but rather relies on a simple counting argument.

Theorem 13.7: *Let R be a randomized online algorithm that manages k servers in any metric space. Then $C_R^{aon} \geq k$.*

PROOF: We will exhibit an adaptive request sequence $\rho_1, \rho_2, \ldots, \rho_N$ that forces R to pay a certain cost $M_R(\rho_1, \rho_2, \ldots, \rho_N)$, and an online algorithm that on the same sequence pays an expected cost that is at most $M_R(\rho_1, \rho_2, \ldots, \rho_N)/k$. These together define the strategy of the adaptive online adversary that yields the theorem.

Let H be any subset of $k+1$ points in the space that includes the k points that R's servers initially occupy (we can assume that R never places two of its servers on the same point). At each step, there is one point in H that is not occupied by any of R's servers; we always make this the next request. Thus R's starting position and its subsequent actions determine a (random) sequence $\rho_1, \rho_2, \ldots, \rho_N$ of requests. Since R moves a server from ρ_{i+1} to ρ_i to service request ρ_i, the total cost that R incurs on this sequence is given by

$$M_R(\rho_1, \rho_2, \ldots, \rho_N) = \sum_{i=1}^{N} c_{\rho_{i+1}\rho_i} = \sum_{i=1}^{N} c_{\rho_i\rho_{i+1}}.$$

We have defined the adaptive request sequence; it remains to describe the online adversary's own actions. We actually exhibit a family of k online algorithms that together pay a cost of at most $M_R(\rho_1, \rho_2, \ldots, \rho_N)$ on the sequence $\rho_1, \rho_2, \ldots, \rho_N$. Then, a randomly chosen one of these k online algorithms pays an expected cost that is no more than $M_R(\rho_1, \rho_2, \ldots, \rho_N)/k$ on the sequence $\rho_1, \rho_2, \ldots, \rho_N$, and we are done.

Let $\rho_1, u_1, u_2, \ldots, u_k$ be the points in H (recall that H contains ρ_1, a vertex that is initially uncovered by R and is therefore the site of the first request, and k other vertices that are the initial positions of R's servers). The online algorithm B_j in our family initially places its k servers at all the points in H except for u_j, for $1 \leq j \leq k$. Algorithm B_j processes request ρ_i as follows: for $i = 1$, it uses the server at u_j, and for $i > 1$ if it has no server at ρ_i, then it moves its server at ρ_{i-1} to ρ_i. We will establish that $\sum_{j=1}^{k} M_{B_j}(\rho_1, \rho_2, \ldots, \rho_N) = M_R(\rho_1, \rho_2, \ldots, \rho_N)$, and that therefore there exists j such that $M_{B_j}(\rho_1, \rho_2, \ldots, \rho_N) \leq M_R(\rho_1, \rho_2, \ldots, \rho_N)/k$.

The observation that will be crucial to establishing the above is the following: at any time in the sequence $\rho_1, \rho_2, \ldots, \rho_N$, the set of k points occupied by B_j's servers is not the same as the set of k points occupied by B_m's servers, for $j \neq m$. If we can prove this it follows that on each request ρ_i, exactly one of the algorithms B_j, for $1 \leq j \leq k$, moves a server at a cost of $c_{\rho_{i-1}\rho_i}$. Summing over all i and j, we see that the total cost incurred by all the algorithms B_j, for $1 \leq j \leq k$, is $\sum_{i=1}^{N-1} c_{\rho_i \rho_{i+1}} + \sum_{j=1}^{k} c_{u_j \rho_1} = M_R(\rho_1, \rho_2, \ldots, \rho_N)$.

It remains to prove the claim that algorithms B_j and B_m, for $j \neq m$, always occupy different sets of points. Let S_j and S_m be the sets of k points occupied by B_j and B_m, respectively, before request ρ_i is processed. We will show that if $S_j \neq S_m$, then the two sets are different after ρ_i is processed by B_j and by B_m. By our construction, the sets of points initially occupied by B_j and by B_m are different, so this will provide an inductive proof.

Therefore, suppose that $S_j \neq S_m$. If ρ_i is in both S_j and S_m, neither set is changed in processing ρ_i, so the inductive invariant holds. If only one of them, say S_j, has no server at ρ_i, it adds ρ_i and drops ρ_{i-1}; on the other hand, S_m maintains a server at ρ_{i-1}, and so the difference remains non-empty.

Thus, exactly one of the algorithms B_j moves a server on request ρ_i, incurring a cost $c_{\rho_{i-1}\rho_i}$. Therefore

$$\sum_{j=1}^{k} M_{B_j}(\rho_1, \rho_2, \ldots, \rho_N) = M_R(\rho_1, \rho_2, \ldots, \rho_N).$$

□

The reader may notice that the algorithm B_j for which $M_{B_j}(\rho_1, \rho_2, \ldots, \rho_N) \leq M_R(\rho_1, \rho_2, \ldots, \rho_N)/k$ may not begin with its servers at the same points as R. Could it be that B_j derived an unfair advantage from this? Recall that in our definition of competitiveness, we allowed an additive constant that may depend on the distances between points in the metric space. We thus imagine that all the algorithms B_i begin with their servers at the same points as R. Then, at the first request, each algorithm B_i first moves its servers to the points specified in the proof of Theorem 13.7, paying a one-time cost that is absorbed in the additive term in the definition of competitiveness. Subsequently, B_i serves the requests as described above.

The importance of Theorem 13.7 is that it applies to any metric space, and consequently to the weighted (as well as the unweighted) paging problem. Thus, **Reciprocal** is optimal for the weighted paging problem.

Notes

The model for the paging problem presented at the beginning of the chapter was introduced by Sleator and Tarjan [379]. The lower bound of Theorem 13.1 as well as the proof of the k-competitiveness of **LRU** and **FIFO** appear in the same paper. The optimality of the **MIN** algorithm was established by Belady [49] and by Mattison, Gecsei, Slutz, and Traiger [298]. The term "competitiveness" and related definitions first appeared in a paper of Karlin, Manasse, Rudolph, and Sleator [235]. Borodin, Linial, and Saks [76] were the first to demonstrate the power of randomization in online algorithms in the context of the so-called *metrical task systems*.

The **Marker** algorithm for paging against oblivious adversaries (together with the lower bound of H_k) was discovered by Fiat, Karp, Luby, McGeoch, Sleator, and Young [145]. The improved algorithm achieving competitiveness coefficient H_k is due to McGeoch and Sleator [306].

The notion of multiple types of adversaries for randomized algorithms arose from the work of Raghavan and Snir [352]. The distinction between the two types of adaptive adversaries was first noticed by Karlin. The relations (Theorems 13.4 and 13.5) between them are established in a paper by Ben-David, Borodin, Karp, Tardos, and Wigderson [51]. The **Reciprocal** algorithm appears in the article by Raghavan and Snir [352], for a slightly more general version of the weighted paging problem (see Problem 13.11 below). For the version of weighted paging we consider here, there is in fact a deterministic k-competitive algorithm for weighted paging due to Chrobak, Karloff, Payne, and Viswanathan [98].

The *k-server problem* was introduced by Manasse, McGeoch, and Sleator [290], who gave a lower bound of k on the competitiveness coefficient of any deterministic algorithm for any server problem (thus generalizing Theorem 13.1). They also gave a deterministic 2-competitive algorithm for the case $k = 2$, for all metric spaces. While the results presented in this chapter give a fairly complete characterization of the paging problem (the special case of the server problem when all distances are 1), our understanding of the server problem for general k and for general metric spaces is far from complete. The first result giving a deterministic online algorithm achieving a competitiveness coefficient depending on k alone, in all metric spaces, is due to Fiat, Rabani, and Ravid [146]. The competitiveness coefficient of the algorithm is exponential in $k \log k$.

The lower bound of Theorem 13.2 tells us that there are metric spaces for which no randomized algorithm can achieve a competitiveness coefficient lower than H_k against oblivious adversaries. The same lower bound is conjectured to hold for general metric spaces:

► **Research Problem 13.1:** Show that no randomized online algorithm can achieve a competitiveness coefficient lower than H_k in any metric space, against oblivious adversaries.

Karloff, Rabani, and Ravid [238] initiated progress on this question by establishing two results:

1. provided the metric space has at least $k+1$ points, they give a lower bound of $\Omega(\log\log k)$;
2. for metric spaces with sufficiently many points, their bound is $\Omega(\log k)$.

Subsequently, the lower bound for metric spaces of size at least $k+1$ has been improved by Blum, Karloff, Rabani, and Saks [62] to $\Omega(\sqrt{\log k/\log\log k})$.

We know of relatively few cases where randomization against an oblivious adversary beats the lower bound of k for deterministic algorithms that follows from Theorem 13.7. A discussion of these cases may be found in the paper by Karlin, Manasse, McGeoch, and Owicki [236].

The situation is slightly better in regard to adaptive adversaries. The **Harmonic** algorithm, due to Raghavan and Snir [352], is the following. Let d_i be the distance between the ith server managed by the algorithm to the requested point, for $1 \le i \le k$. The algorithm chooses (independently of the past) the jth server with probability

$$\frac{1/d_j}{\sum_{i=1}^{k} 1/d_i}.$$

Notice that this resembles the **Reciprocal** algorithm of Section 13.5, although the probabilities are not quite the same. Notice also that for the paging problem, in which all d_i equal 1, this becomes the **Random** algorithm. The reader should note that **Reciprocal** and **Harmonic** are two different generalizations of the **Random** algorithm for paging. Given that **Random** is k-competitive for the paging problem against adaptive online adversaries, one might hope that **Harmonic** is k-competitive for the server problem in all metric spaces. However, it is known that there are metric spaces for which **Harmonic** cannot achieve a competitiveness coefficient lower than $k(k+1)/2$ (see Problem 13.13 below).

For **Harmonic**, Raghavan and Snir [352] proved an upper bound of 6 for $k=2$, and this was later improved to the (tight) bound 3 by Chrobak and Larmore [99]. A breakthrough was achieved by Grove [186], who gave an upper bound of $(5k2^k)/4$ on the competitiveness coefficient of **Harmonic** in any metric space, and for all k. An important implication of this is the existence (by the result of Exercise 13.2) of a deterministic algorithm whose competitiveness coefficient is exponential in k, improving the result of Fiat, Rabani, and Ravid [146]. The most general class of metric spaces for which we know of a k-competitive algorithm against adaptive online adversaries follows from the work of Coppersmith, Doyle, Raghavan, and Snir [112]; their algorithm works in a class of metric spaces they call *resistive metric spaces.* There is, however, a deterministic $(2k-1)$-competitive algorithm for any metric space, due to Koutsoupias and Papadimitriou [269]. It remains to be seen whether the approach of Koutsoupias and Papadimitriou will result in a k-competitive deterministic algorithm. This would shift the focus to randomized algorithms against oblivious adversaries:

► **Research Problem 13.2:** Determine the best possible upper bound for the oblivious competitiveness coefficient for the k-server problem, in general metric spaces.

The work of Karlin, Manasse, McGeoch, and Owicki [236] shows that the value of the oblivious competitiveness coefficient for the k-server problem will depend on the actual distances in the metric space, even for $k = 2$. However, it is plausible that this variation is relatively small, so that an asymptotic bound such as $\Theta(\log k)$ is possible.

The *list update problem* is the following. An online algorithm maintains a linear list containing n items. It is given a sequence of requests, where each item specifies one of the n items. If the item requested is at the ith position in the list, the algorithm incurs a cost of i for that request; thus, a request to the item at the head of the list costs 1. When an item is requested, the algorithm has the option of moving that item to the front of the list, or leaving it where it is. Given a request sequence, the notions of cost to the online and offline algorithms, and of competitiveness, can be made in the usual sense. Sleator and Tarjan [379] introduced this model. They showed that the deterministic algorithm that always moves the item that is accessed to the front of the list achieves a competitiveness coefficient of 2. A lower bound of $2 - 1/L$ is known for the competitiveness coefficient of any deterministic algorithm, where L is the number of items in the list. Irani [206] gave a randomized online algorithm for this problem that achieved a competitiveness coefficient slightly less than 2 against oblivious adversaries. This was improved to $\sqrt{3}$ by Reingold, Sleator, and Westbrook [355]; an interesting feature of their algorithm is that it can be implemented so that it makes some random choices once at the beginning of the request sequence, and is *wholly deterministic* thereafter. It remains deterministic irrespective of the length of the request sequence, making no further random choices. This implementation can be shown to be $(\sqrt{3}+\epsilon)$-competitive. Clearly, this hinges on the adversary being oblivious. Recently, Albers [10] has given a ϕ-competitive randomized algorithm for list update, where ϕ is the golden ratio $(1+\sqrt{5})/2$. Teia [395] gives a lower bound of 1.5 on the competitiveness coefficient of any randomized algorithm, against an oblivious adversary.

► **Research Problem 13.3:** Determine a tight bound on the competitiveness coefficient of randomized online algorithms for list update, against oblivious adversaries.

There is considerable current interest in randomized online algorithms for problems arising in many diverse settings, including task systems [76], robot navigation [63, 144, 327], finding short paths in graphs [144, 327], and finance [135]. The reader is referred to these and the other papers cited above for a host of research questions that remain open in this area.

Problems

13.1 (Due to D.D. Sleator and R.E. Tarjan [379].) Show that the **LRU** algorithm for paging is k-competitive. What can you say about its competitiveness coefficient?

13.2 (Due to D.D. Sleator and R.E. Tarjan [379].) Show that the **FIFO** algorithm for paging is k-competitive. What can you say about its competitiveness coefficient?

13.3 Show that the **LFU** algorithm does not achieve a bounded competitiveness coefficient.

13.4 (Due to A. Bar-Noy, R. Motwani, and J. Naor [45].) Given an undirected graph $G(V, E)$, an *edge coloring* is an assignment of indices $1, \ldots, C$ to the edges of G such that no two edges incident on a vertex have the same label. The indices are referred to as colors, and the smallest value of C for which such a coloring can be achieved is called the of the graph. Vizing's Theorem states that a graph with maximum degree Δ has chromatic index either Δ or $\Delta + 1$; moreover, while distinguishing these two cases is an ***NP***-hard problem, there is a polynomial time algorithm for coloring any graph with $\Delta + 1$ colors. Consider the problem of *online edge coloring*: suppose that the edges of a graph with maximum degree Δ are presented one by one, and as each edge is specified it must be irrevocably assigned a color.

(a) Devise a deterministic online algorithm that uses at most $2\Delta - 1$ colors.

(b) Show that there does not exist any deterministic algorithm that uses fewer than $2\Delta - 1$ colors in the worst case. For what range of values of Δ can you prove this result? (**Hint:** Consider an adversary that generates a sequence of edges that constitute a graph composed of disjoint stars, where each star consists of a center vertex v with $\Delta - 1$ neighbors of degree 1. Once the algorithm has committed to a coloring of these stars, an adversary can introduce further edges from a distinguished vertex to the centers of appropriately chosen stars, forcing the online algorithm to use a large number of additional colors.)

(c) Show that there does not exist any deterministic algorithm that uses fewer than $2\Delta - 1$ colors in the worst case. For what range of values of Δ can you prove this result? (**Hint:** See the hint for part (b).)

13.5 (Due to A.R. Karlin.) Show that the competitiveness coefficient of the **Random** algorithm for paging against adaptive offline adversaries is at least kH_k.

13.6 Show that the competitiveness coefficient of the **Random** algorithm for paging against oblivious adversaries is at least k.

13.7 Show that when the number of distinct items in memory is $k + 1$, the **Marker** algorithm is H_k-competitive.

13.8 Consider a server problem in which the online algorithm has K servers, and the offline algorithm has k servers. For $K \geq k$, show that the competitiveness coefficient of any online algorithm against adaptive online adversaries is at least $K/(K - k + 1)$.

13.9 Consider the following algorithm for the 2-server problem in an arbitrary metric space. Label the servers 0 and 1. The algorithm services any request as follows. Let d_0 be the distance from server 0 to the request, and d_1 the distance from server 1 to the request. Let d be the distance between the servers. For $i \in \{0, 1\}$, let

$$p_i = \frac{d + d_{1-i} - d_i}{2d}.$$

Server i is used to select the request with probability p_i. Show that this algorithm is 2-competitive against adaptive online adversaries.

13.10 (Due to R.M. Karp and P. Raghavan.) Show that the competitiveness coefficient of any randomized online algorithm for maintaining a linear list against an oblivious adversary is at least 9/8. (**Hint:** Consider a list with 2 items.)

13.11 Consider the **Reciprocal** algorithm for weighted paging, in the scenario of Problem 13.8: **Reciprocal** manages a cache with K pages, while the adversary has k pages. Show that **Reciprocal** achieves a competitiveness coefficient of $K/(K-k+1)$, matching the lower bound of Problem 13.8.

13.12 (Due to S.S. Irani [207].) Consider the list update problem again, with the following modification in the cost function: the cost of accessing the item at the ith position in the list is $i-1$, rather than i (thus the item at the head of the list is accessed at cost zero). For lists with two items, show that 3/2 is a tight bound on the competitiveness coefficient of randomized algorithms against oblivious adversaries, under this cost function.

13.13 Give a metric space for which you can prove a lower bound of $k(k+1)/2$ on the competitiveness coefficient of the **Harmonic** algorithm against an adaptive online adversary. (**Hint:** Make use of the result of Problem 6.7.)

CHAPTER 14

Number Theory and Algebra

The theory of numbers plays a central role in several areas of great importance to computer science, such as cryptography, pseudo-random number generation, complexity theory, algebraic problems, coding theory, and combinatorics, to name just a few. We have already seen that relatively simple properties of prime numbers allow us to devise k-wise independent variables (Chapter 3), and number-theoretic ideas are at the heart of the algebraic techniques in randomization discussed in Chapter 7.

In this chapter, we focus on solving number-theoretic problems using randomized techniques. Since the structure of finite fields depends on the properties of prime numbers, algebraic problems involving polynomials over such fields are also treated in this chapter. We start with a review of some basic concepts in number theory and algebra. Then we develop a variety of randomized algorithms, most notably for the problems of computing square roots, solving polynomial equations, and testing primality. Connections with other areas, such as cryptography and complexity theory, are also pointed out along the way.

There are several unique features in the use of randomization in number theory. As will soon become clear, the use of randomization is fairly simple in that most of the algorithms will start by picking a random number from some domain and then work deterministically from there on. We will claim that with high probability the chosen random number has some desirable property. The hard part usually will be establishing this claim, which will require us to use non-trivial ideas from number theory and algebra. Further, all the resulting algorithms will turn out to be extremely practical. Finally, for most non-trivial problems, such as primality testing, the only known efficient (polynomial time) algorithms involve the use of randomization.

14.1. Preliminaries

We start by introducing some basic notation and ideas. Unless otherwise specified, all numbers should be assumed to be from the domain of non-negative integers. We will adopt the convention that the symbols $a, b, \ldots, m, n,$ will refer to arbitrary numbers; we will reserve the symbol p for denoting prime numbers. Symbolic variables will be denoted by uppercase letters X and Y. The expression $a \nmid b$ will denote that a is a divisor of b, while $a \nmid b$ will denote that a does not divide b; note that for any number $a \neq 0$, $a|0$. The greatest common divisor (gcd) and lowest common multiple (lcm) of a pair of numbers a and b are defined as follows:

$$\begin{aligned} \gcd(a, b) &= \max\{f \mid f|a, f|b\} \\ \operatorname{lcm}(a, b) &= \frac{ab}{\gcd(a, b)} \end{aligned}$$

By convention, $\gcd(0,0) = 0$ and $\operatorname{lcm}(0,0) = 0$. We will say that a and b are *coprime* if $\gcd(a, b) = 1$, i.e., if a and b have no common factors. A number is prime if and only if it is coprime to all smaller positive numbers.

An important issue is the measure of complexity for a number-theoretic algorithm. An integer n can be represented by a bit-string of length $\Theta(\log n)$. Thus, when n is the input to an algorithm, the algorithm's running time should be measured in terms of the input length, which is $\log n$, and not the input value, which is n. This is the standard measure for number-theoretic algorithms and is sometimes referred to as the *bit complexity* measure. For example, computing $\gcd(a, b)$ in polynomial time requires an algorithm that runs in time polynomial in $\log a$ and $\log b$. Our model of computation is similar to that described in Chapter 7. We will use the unit-cost RAM model to measure the running time of an algorithm. In particular, the operations of addition, subtraction, multiplication, division, comparison, or choosing a random element take unit time, provided the magnitude of the operand numbers is polynomially related to that of the input. Thus, given input n, arithmetic operations on $O(\log n)$-bit numbers take unit time.

How may we compute the gcd in polynomial time? The naive approach of trying all possible numbers smaller than a and b takes exponential time, given that the length of the input is logarithmic in the values of a and b. Another approach, which we all learned in high school, is to apply the following rule repeatedly: *replace the larger of the two numbers by their difference.* The process terminates when the smaller number is 0, and the larger number at that point is the desired gcd. It is not very hard to see that even this algorithm takes exponential time in the worst case. (Consider the case where a is large and b is a very small constant.) We describe the ancient algorithm of Euclid for computing the gcd and prove that it runs in polynomial time.

We use $a \bmod b$ and $a \operatorname{div} b$ to denote the *remainder* and the *quotient*, respectively, for the division of a by b. Euclid's algorithm takes two numbers a and b such that $a > b > 0$, and determines their gcd by computing the following sequence starting with $r_0 = a$ and $r_1 = b$.

$$\begin{array}{lll} r_2 = r_0 \bmod r_1 & q_2 = r_0 \operatorname{div} r_1 & (0 < r_2 < r_1) \\ r_3 = r_1 \bmod r_2 & q_3 = r_1 \operatorname{div} r_2 & (0 < r_3 < r_2) \\ & \vdots & \\ r_i = r_{i-2} \bmod r_{i-1} & q_i = r_{i-2} \operatorname{div} r_{i-1} & (0 < r_i < r_{i-1}) \\ & \vdots & \\ r_k = r_{k-2} \bmod r_{k-1} & q_k = r_{k-2} \operatorname{div} r_{k-1} & (0 < r_k < r_{k-1}) \end{array}$$

The sequence r_i is strictly decreasing, implying that the algorithm terminates in a finite number of stages. Termination occurs when $r_{k-1} \bmod r_k = 0$, i.e., $r_k | r_{k-1}$. Observe that this is just a more efficient implementation of the high school algorithm described above. Instead of subtracting the smaller number from the larger, which may have to be done repeatedly, Euclid's algorithm subtracts the largest possible multiple of the smaller number from the larger; then, the remainder replaces the larger number. We will soon see that this algorithm gives us a number of interesting constructions besides the gcd itself.

Theorem 14.1: *In Euclid's algorithm,* $r_k = \gcd(a, b)$.

PROOF: Denoting $\gcd(a, b)$ by g, we will show that $r_k | g$ and $g | r_k$ to establish that $g = r_k$.

Observing that $r_k | r_{k-1}$ and $r_{k-2} = r_{k-1}q_k + r_k$, we obtain that $r_k | r_{k-2}$. Similarly, since now $r_k | r_{k-1}$, $r_k | r_{k-2}$, and $r_{k-3} = r_{k-2}q_{k-1} + r_{k-1}$, it follows that $r_k | r_{k-3}$. This argument can be applied inductively to verify the hypothesis that if $r_k | r_i$ and $r_k | r_{i-1}$, then $r_k | r_{i-2}$, since $r_{i-2} = r_{i-1}q_i + r_i$. In particular, we can show that $r_k | r_1$ and $r_k | r_0$; since $r_0 = a$ and $r_1 = b$, this establishes that $r_k | g$.

To establish the converse, we reverse the direction of the above argument. Note that $g | r_0$ and $g | r_1$. Since for all i, $r_i = r_{i-2} - q_i r_{i-1}$, it follows that if $g | r_{i-2}$ and $g | r_{i-1}$, then $g | r_i$. Thus, we conclude inductively that $g | r_k$. □

It remains to be shown that this is a polynomial time algorithm. Each of the k stages involves essentially one division operation, and all operands (the intermediate numbers) are smaller than the larger input. Therefore, the total running of this algorithm is O(k). The following exercise shows that the worst-case value of k is polynomially bounded.

Exercise 14.1: Let F_n denote the nth Fibonacci number. Show that the worst case for Euclid's algorithm is when a and b are consecutive Fibonacci numbers. If $a = F_{n+1}$ and $b = F_n$, then the number of stages k equals n. Noting that $F_n \sim \phi^n/\sqrt{5}$, where ϕ is the golden ratio 1.618..., prove that the running time of this algorithm is polynomial in the lengths of a and b.

The following theorem highlights an interesting aspect of Euclid's algorithm.

Theorem 14.2: *For all $a > b > 0$, there exist integers x and y such that*

$$\gcd(a, b) = ax + by.$$

Moreover, x and y can be computed in polynomial time.

We provide only a sketch of the proof, leaving the details for Problem 14.1. Recall that $r_i = r_{i-2} - q_i r_{i-1}$. Since r_k can be similarly expressed as a linear combination of r_{k-1} and r_{k-2}, we can easily express r_k as a linear combination of r_0 and r_1 by repeatedly substituting any remainder r_i with a linear combination of the previous two remainders. Since $r_0 = a$, $r_1 = b$, and $\gcd(a, b) = r_k$, we obtain the desired result. The coefficients x and y of the linear combination can be computed in polynomial time using the same strategy. The resulting extension of Euclid's algorithm, which computes x and y along with the gcd, is sometimes referred to as *extended Euclidean algorithm.*

14.2. Groups and Fields

Before we discuss sophisticated number-theoretic algorithms, we briefly review the group-theoretic concepts underlying these algorithms. We start by developing additional notation.

We define the equivalence relation of *congruence* modulo n as follows. Two numbers a and b are congruent modulo n if $a \bmod n = b \bmod n$; equivalently $n|(a - b)$. Usually, this is denoted $a \equiv b \pmod{n}$, but sometimes we will abbreviate this to $a \equiv_n b$. The operations $+_n$ and $\times_n$ denote addition and multiplication modulo n, i.e., the result of the operation is reduced modulo n.

There are two groups that can be defined with respect to any number $n > 1$. The set $\mathbb{Z}_n = \{0, 1, \ldots, n-1\}$ contains all numbers smaller than n, and it forms a group under addition modulo n. We also define $\mathbb{Z}_n^* = \{x \mid 1 \leq x \leq n \text{ and } \gcd(x, n) = 1\}$ as the numbers in $\mathbb{Z}_n$ that are coprime to n; this forms a group under multiplication modulo n. (Notice that $0 \notin \mathbb{Z}_n^*$.) The elements of $\mathbb{Z}_n$ are the canonical elements of the congruence equivalence classes and are referred to as the *residues* modulo n.

Exercise 14.2: Verify that $\mathbb{Z}_n$ and $\mathbb{Z}_n^*$ form groups under the operations $+_n$ and $\times_n$, respectively.

Exercise 14.3: Verify that for a prime p, the set $\mathbb{Z}_p$ forms a field under the operations of $+_p$ and $\times_p$.

Since $\mathbb{Z}_n^*$ is a multiplicative group, each of its elements has a multiplicative inverse in $\mathbb{Z}_n^*$. It is not obvious that we can compute these inverses efficiently, but it turns out that the extended Euclidean algorithm can be adapted for this purpose. To compute the multiplicative inverse of $z \in \mathbb{Z}_n^*$, we run the algorithm with $r_0 = n$ and $r_1 = z$. By Theorem 14.2, we can compute in polynomial time two numbers x and y such that $\gcd(n,z) = nx + zy$. Noting that this gcd must be 1, we obtain $zy \equiv 1 \pmod{n}$. Thus, y is a multiplicative inverse of z and must lie in $\mathbb{Z}_n^*$.

Theorem 14.3: *For any n, the multiplicative inverse of a number $z \in \mathbb{Z}_n^*$ can be computed in polynomial time.*

We give a simple application of this result to the constructive version of the well-known Chinese Remainder Theorem.

Theorem 14.4 (Chinese Remainder Theorem): *Let $n_1, \ldots, n_k$ be a sequence of pairwise coprime numbers (for $i \neq j$, $\gcd(n_i, n_j) = 1$), and define $n = \prod_{i=1}^{k} n_i$. For any sequence of residues $r_1 \in \mathbb{Z}_{n_1}, \ldots, r_k \in \mathbb{Z}_{n_k}$, there is a unique $r \in \mathbb{Z}_n$ such that*

$$r \equiv r_i \pmod{n_i} \quad (\text{for } 1 \leq i \leq k).$$

Moreover, r can be computed in polynomial time.

PROOF: We first show that there exists at least one such r. By the pairwise coprime property of the n_i's, we have $\gcd(n/n_i, n_i) = 1$ for each i. It follows that there exists a multiplicative inverse m_i for n/n_i in the group $\mathbb{Z}_{n_i}^*$, and therefore

$$m_i \frac{n}{n_i} \equiv 1 \pmod{n_i}.$$

It is easy to verify the following two congruences for each i.

$$m_i \frac{n}{n_i} \equiv 1 \pmod{n_i}$$
$$m_i \frac{n}{n_i} \equiv 0 \pmod{n_j} \quad (\text{for all } j \neq i).$$

We conclude that the following value of r satisfies the desired congruences.

$$r = \sum_{i=1}^{k} r_i m_i \frac{n}{n_i} \pmod{n}$$

The uniqueness of the choice of r follows from the following simple counting argument. The number of distinct choices of each r_i is n_i, and so there are exactly n distinct sequences (r_i). Each such sequence has at least one associated $r \in \mathbb{Z}_n$. Since each choice of r determines a distinct sequence (r_i), it follows that there is a one-to-one correspondence between these sequences and the choices of r. The value of r can be easily computed in polynomial time since it involves a polynomial number of multiplications, additions, and inverse computations. □

In effect, this theorem states that $\mathbb{Z}_n$ is identical to the cartesian product $\mathbb{Z}_{n_1} \times \mathbb{Z}_{n_2} \times \cdots \times \mathbb{Z}_{n_k}$.

Consider now the problem of computing a^k over some group $(G, \circ)$, given $a \in G$ and k. For the additive group $(\mathbb{Z}_n, +_n)$, exponentiation corresponds to the arithmetic multiplication of a and k. The situation is more complex for the multiplicative group $(\mathbb{Z}_n^*, \times_n)$. The naive strategy of repeatedly multiplying by a is not a polynomial time algorithm since it requires a total of $k-1$ multiplications. The problem is that the number of multiplications required by this method is proportional to k, rather than $\log k$. A simple strategy for exponentiating in polynomial time is that of repeated squaring. The idea is to compute the powers $A_i = a^{2^i}$, for $0 \le i \le t = \lfloor \log k \rfloor$. Since A_{i+1} is the square of A_i, this sequence can be computed in increasing order of i using $O(\log k)$ multiplications. Consider the binary representation of k as a sequence of bits $b_0, \ldots, b_t$, where b_0 is the least significant bit. Since $k = \sum_{i=0}^{t} b_i 2^i$, it follows that $a^k = \prod_{i=0}^{t} A_i^{b_i}$. The latter product can be computed in time $O(\log k)$, given the precomputed values of the A_i's.

Theorem 14.5: *In the group $(\mathbb{Z}_n^*, \times_n)$, exponentiation can be performed in polynomial time.*

It is clear that $|\mathbb{Z}_n| = n$, but the size of $\mathbb{Z}_n^*$ has a more complex behavior. The *Euler totient function* $\phi(n)$ is defined to be the number of elements of $\mathbb{Z}_n$ that are coprime to n, which is precisely $|\mathbb{Z}_n^*|$. In the case where n is a prime, $\mathbb{Z}_n^* = \mathbb{Z}_n \setminus \{0\}$ and $\phi(n) = n-1$. In general, we can compute $\phi(n)$ in polynomial time when the prime factorization of n is known.

Theorem 14.6: *Let n have the prime factorization $p_1^{k_1} p_2^{k_2} \cdots p_t^{k_t}$, where the primes p_i are distinct and have exponents $k_i > 0$. Then,*

$$\phi(n) = \prod_{i=1}^{t} p_i^{k_i - 1}(p_i - 1).$$

It is easy to verify that the above expression can be computed in polynomial time provided that the prime factorization of n is known. The following exercise outlines the proof of this theorem.

Exercise 14.4: Verify the following properties of the totient function.

- $\phi(1) = 1$.
- For prime p, $\phi(p) = p - 1$.
- For prime p and $k > 0$, $\phi(p^k) = p^{k-1}(p-1)$.
- For n and m such that $\gcd(n, m) = 1$, $\phi(nm) = \phi(n)\phi(m)$.

Using these properties, prove Theorem 14.6 and verify that $\phi(n)$ can be computed in polynomial time from the prime factorization of n.

It is widely believed that the prime factorization of a number n cannot be computed in polynomial time; in fact, it appears hard in general to find any non-trivial factors of a given number. Thus, it would be desirable to have an alternative method for evaluating $\phi(n)$ when the prime factorization is not known. Unfortunately, it can be shown (see Problems 14.3–14.4) that the knowledge of $\phi(n)$ can be used to efficiently compute the factorization of n, implying that it is unlikely that an efficient algorithm exists for evaluating $\phi(n)$. We present the idea behind this for the special case where $n = pq$ for two distinct primes p and q. First note that Theorem 14.6 implies that $\phi(n) = \phi(pq) = (p-1)(q-1)$. Therefore, $p+q = pq+1-\phi(n) = n-\phi(n)+1$, and we know that $pq = n$. It is now a simple matter to see that given $p+q$ and pq, we can compute p and q in polynomial time.

Of course, $\phi(p)$ is easy to compute when p is a prime. What about $\phi(p^k)$ where p^k is a prime power? In Exercise 14.5 it is shown that for any number $x = y^z$, there is a polynomial time algorithm for computing y and z from x. Thus, prime powers can be recognized and factored in polynomial time. Then computing $\phi(p^k)$ is a trivial task.

Exercise 14.5: Devise a polynomial time algorithm for finding positive integers y and $z > 1$, given the value of $x = y^z$. The algorithm may fail if the input x cannot be expressed in this form. (**Hint:** Consider the logarithms of x and y^z.)

We now examine the structure of the groups $(\mathbb{Z}_n, +_n)$ and $(\mathbb{Z}_n^*, \times_n)$. Consider a group $(G, \circ)$ under the operation $\circ$, with the identity element ι. (For the groups we are considering, the operation $\circ$ is commutative.) We define the *order* of the group as the number of elements in it, $|G|$. For any element $x \in G$, we define the powers of x as follows.

$$\begin{aligned} x^0 &= \iota \\ x^k &= x \circ x^{k-1} \quad \text{(for } k > 0) \end{aligned}$$

► **Definition 14.1:** For any group $(G, \circ)$ and any $x \in G$, the order of x is given by

$$\text{ord}(x) = \min\{k > 0 \mid x^k = \iota\}.$$

The following propositions are easy to prove and left as exercises.

Proposition 14.7: *For any finite group $(G, \circ)$, and any $x \in G$,* ord(x) *divides $|G|$. Therefore, it is always the case that $x^{|G|} = \iota$.*

Proposition 14.8: *For any finite group $(G, \circ)$, and any sub-group $(H, \circ)$ with $H \subseteq G$, $|H|$ divides $|G|$.*

Consider the additive group $(\mathbb{Z}_n, +_n)$ with $\iota = 0$. Suppose for some $x \in \mathbb{Z}_n$ that $\text{ord}(x) = k$. This means that the k-fold addition of x to itself is congruent to 0 modulo n, that is to say $kx \equiv_n 0$. We conclude that $n|kx$, and so it follows from the definition of order that $kx = \text{lcm}(n, x)$. Notice that Proposition 14.7 says that $k|n$.

Proposition 14.9: *For all n and $x \in \mathbb{Z}_n$, the order of x in the additive group $(\mathbb{Z}_n, +_n)$ is given by*

$$\text{ord}(x) = \frac{n}{\gcd(n,x)} = \frac{\text{lcm}(n,x)}{x}.$$

In the case where $n = p$ is a prime and $x \neq 0$,

$$\text{ord}(x) = p = |\mathbb{Z}_p|.$$

The order of the identity 0 is 1.

The situation is more complicated with respect to $\mathbb{Z}_n^*$. Here the group order is $\phi(n)$ and $\imath = 1$. Consider any element $x \in \mathbb{Z}_n^*$ and let its order be k. Then, $x^k \equiv_n 1$ and Proposition 14.7 implies that $k | \phi(n)$. We may conclude that $x^{\phi(n)} \equiv_n 1$, and this gives us the famous theorem of Euler.

Theorem 14.10 (Euler's Theorem): *For all n and $x \in \mathbb{Z}_n^*$,*

$$x^{\phi(n)} \equiv 1 \pmod{n}.$$

Specializing this to the case where n is a prime yields the theorem of Fermat.

Theorem 14.11 (Fermat's Theorem): *For prime p and $x \in \mathbb{Z}_p^*$,*

$$x^{p-1} \equiv 1 \pmod{p}.$$

As we remarked earlier, computing $\phi(n)$ is as hard as factoring n. More generally, the same can be shown for determining the order of an arbitrary element of the multiplicative group $\mathbb{Z}_n^*$. In fact, the difficulty in computing the order underlies most of the issues we will deal with later. Contrast this with the case of the additive group where the order is almost trivial to compute. This property of the additive group will be useful in devising efficient algorithms later.

Another distinction between the additive and multiplicative groups involves the existence of generators. A *generator* g in a group G is an element whose order equals the size of group, i.e., $\text{ord}(g) = |G|$. A group is said to be *cyclic* if it contains a generator. It is easy to verify that a cyclic group G can be viewed as the set of all distinct powers of any generator $g \in G$, that is $G = \{g^0, g^1, \ldots, g^{|G|-1}\}$.

It is an immediate consequence of Proposition 14.7 that any finite group whose order is a prime number is a cyclic group. The additive group $(\mathbb{Z}_n, +_n)$ is cyclic since the element 1 has order n. The multiplicative group $(\mathbb{Z}_n^*, \times_n)$ is not cyclic in general.

Exercise 14.6: Verify that the group $(\mathbb{Z}_8^*, \times_8)$ is not cyclic.

However, we show below that for primes p, the group $(\mathbb{Z}_p^*, \times_p)$ is cyclic. Note that the cyclicity of groups of prime order does not imply the cyclicity of $\mathbb{Z}_p^*$

since the order of this group is $\phi(p) = p - 1$, which is even and therefore not a prime.

The following lemma will be useful for showing the cyclicity of $\mathbb{Z}_p^*$. It states that the sum of the totient function values for all the divisors of n will always equal n.

Lemma 14.12: *For all $n > 0$, $\sum_{d|n} \phi(d) = n$.*

PROOF: For all g, define the set

$$A_g = \{x \mid 1 \leq x \leq n \text{ and } \gcd(x, n) = g\}.$$

Clearly A_g is non-empty only if g divides n; these non-empty sets form a partition of $\{1, 2, \ldots, n\}$. Thus,

$$\sum_{g|n} |A_g| = n.$$

Suppose we could show that $|A_g| = \phi(n/g)$. We could then conclude the desired result as follows:

$$\sum_{d|n} \phi(d) = \sum_{g|n} \phi(n/g) = \sum_{g|n} |A_g| = n.$$

It remains to be shown that $|A_g| = \phi(n/g)$. Let $d = n/g$ and consider any $x \in \mathbb{Z}_d^*$. The following equivalences are easy to verify:

$$\begin{aligned} x \in \mathbb{Z}_d^* &\Leftrightarrow \gcd(xg, dg) = g \times \gcd(x, d) = g \\ &\Leftrightarrow \gcd(xg, n) = g \\ &\Leftrightarrow xg \in A_g. \end{aligned}$$

Thus, there is a one-to-one correspondence between the elements of $\mathbb{Z}_d^*$ and A_g, and this implies that $|A_g| = \phi(d) = \phi(n/g)$. □

Theorem 14.13: *For any prime p, the group $\mathbb{Z}_p^*$ is cyclic.*

PROOF: Recall that if any $x \in \mathbb{Z}_p^*$ has order k, then $k|(p-1)$. For each k that divides $p - 1$, let $O_k = \{x \in \mathbb{Z}_p^* \mid \text{ord}(x) = k\}$. We claim that $|O_k|$ is either 0 or $\phi(k)$, deferring the proof for the moment.

Since the sets O_k partition $\mathbb{Z}_p^*$,

$$\sum_{k|(p-1)} |O_k| = p - 1. \tag{14.1}$$

We know that each O_k has size either 0 or $\phi(k)$ and so,

$$\sum_{k|(p-1)} |O_k| \leq \sum_{k|(p-1)} \phi(k).$$

Now by Lemma 14.12, the latter sum equals exactly $p - 1$. Thus, the only way (14.1) can hold is if each term in the summation is non-zero. In other words, for all k such that $k|(p-1)$, $|O_k| = \phi(k)$. In particular, this would imply that for $k = p-1$, $|O_{p-1}| = \phi(p-1) = \phi(\phi(p))$. But each element of O_{p-1} is a generator, and since this set is non-empty, the group has generators and is cyclic.

We now complete the proof by showing that if O_k is non-empty, then $|O_k| = \phi(k)$. Each element $a \in O_k$ has the property that $a^k \equiv_p 1$ and is therefore a root of the polynomial $X^k - 1$ over the field $(\mathbb{Z}_p, +_p, \times_p)$. Since O_k is non-empty, this polynomial has at least one root r in O_k. In fact, each element in the set $\{r^0, r^1, r^2, \ldots, r^{k-1}\}$ is a root of this polynomial; moreover, these are all distinct roots since $\text{ord}(r) = k$, and so this set contains all the k roots of the polynomial. Thus, the elements of O_k are exactly those powers of r that have order k. Observing that r^l has order $k/\gcd(k,l)$, we obtain

$$O_k = \{r^l \mid \gcd(k,l) = 1\} = \{r^l \mid l \in \mathbb{Z}_k^*\}.$$

This implies that $|O_k| = |\mathbb{Z}_k^*| = \phi(k)$. □

The next theorem characterizes the set of all numbers n whose multiplicative groups are cyclic. The interested reader is referred to a number theory text for the proof.

Theorem 14.14: *The multiplicative group $(\mathbb{Z}_n^*, \times_n)$ is cyclic if and only if n is either 1, 2, 4, p^k, or $2p^k$, for some non-negative integer k and an odd prime p.*

It is usually easier to deal with numbers (such as primes) for which the multiplicative group $(\mathbb{Z}_n^*, \times_n)$ is cyclic, because this cyclic group's structure is isomorphic to that of the additive group modulo $\phi(n)$. Let $g \in \mathbb{Z}_n^*$ be any generator. Consider any two elements $x, y \in \mathbb{Z}_n^*$. Since g generates the entire group, there exist a and b such that $x \equiv_n g^a$ and $y \equiv_n g^b$. For $z = xy$, we can write $z = g^c$ where $c = a +_{\phi(n)} b$. (Recall that $\text{ord}(g) = |\mathbb{Z}_n^*| = \phi(n)$.) Thus, the multiplicative group $(\mathbb{Z}_n^*, \times_n)$ can be seen to be isomorphic to the additive group $(\mathbb{Z}_{\phi(n)}, +_{\phi(n)})$; in effect, this is like working with the logarithms of the numbers in $\mathbb{Z}_n^*$ using the generator g as the base of the logarithm. This is a particularly useful view in the case of a prime number p since we are always guaranteed that the multiplicative group modulo p is cyclic.

Of course, we need to lay our hands on a generator to be able to make use of this structural correspondence. For the multiplicative group modulo a prime p, all known polynomial time algorithms for finding a generator require a factorization of $\phi(p) = p - 1$; we describe one such algorithm, which is randomized. The basic idea is to observe that in the proof of Theorem 14.13 we showed that the number of elements of order $p-1$ in $\mathbb{Z}_p^*$ is given by $|O_{p-1}| = \phi(p-1)$. The next lemma shows that this quantity must be reasonably large, i.e., the generators are relatively dense in the multiplicative group.

Lemma 14.15: *For all $n > 1$,* $\dfrac{\phi(n)}{n} = \Omega\left(\dfrac{1}{\log n}\right)$.

PROOF: Let n have the prime factorization $p_1^{k_1} p_2^{k_2} \cdots p_t^{k_t}$. By Theorem 14.6, we know that

$$\begin{aligned} \phi(n) &= \prod_{i=1}^{t} p_i^{k_i - 1}(p_i - 1) \\ &= n \times \prod_{i=1}^{t} \frac{p_i - 1}{p_i}. \end{aligned}$$

Since all prime factors must be at least 2, the number of distinct prime factors cannot exceed $\log n$. It is a simple exercise to verify that for any choice of $t \leq \log n$ numbers p_i, the product in the above expression is $\Omega(1/\log n)$. This gives the desired result. □

We now present our first randomized number-theoretic algorithm. The algorithm picks a random element $x \in \mathbb{Z}_p^*$ and checks whether its order is $p - 1$. Clearly, any element that passes this test is a generator. The probability of finding a generator in a single trial is simply $\phi(p-1)/(p-1) = \Omega(1/\log p)$. To boost the probability of success we can repeat this process k times, for any k that is polynomial in $\log p$. A simple Las Vegas algorithm can also be devised, using techniques described in Chapter 1.

The only problem with this approach is that it is unclear how we can compute the order of any element in polynomial time. This is exactly the place where we need to know the factorization of $p - 1$. Suppose that $p_1, \ldots, p_t$ are the distinct prime factors of $p - 1$. If $\text{ord}(x) < p - 1$, then it must be the case that $\text{ord}(x)$ is a *proper* divisor of $p - 1$. In other words, for some p_i, $\text{ord}(x) | (p-1)/p_i$. This means that to verify that $\text{ord}(x) = p - 1$, it suffices to check for each p_i that $x^{(p-1)/p_i} \not\equiv 1 \pmod{p}$. The number of distinct prime factors of $p - 1$ is at most $O(\log p)$, and exponentiation can be done in polynomial time, implying that the entire process can be implemented in polynomial time.

Theorem 14.16: *Let p be any prime number. Given the prime factorization of $p - 1$, a generator for the group $(\mathbb{Z}_p^*, \times_p)$ can be found in polynomial time by a randomized (Las Vegas or Monte Carlo) algorithm.*

Observe the extreme simplicity of this randomized algorithm. As we remarked earlier, most randomized algorithms for number-theoretic problems have a similar flavor. A non-trivial mathematical analysis establishes that a simple random choice suffices to solve the problem at hand.

14.3. Quadratic Residues

We have seen that the exponentiation problem – to compute $y = x^a \pmod{n}$ given a, x and n – is relatively easy. There are two related problems that turn out to be unexpectedly difficult. The *discrete log* problem is: given x, y, and n, find an exponent a such that $y = x^a \pmod{n}$. The *root finding* problem is: given a, y, and n, find an x such that $y = x^a \pmod{n}$. For prime n, the latter problem is a special case of finding roots of polynomials over finite fields, or factoring such polynomials; in this case the polynomial is $p(x) = x^a - y \pmod{n}$.

The discrete log problem is believed to be extremely hard, and no efficient solution is known at this point. We have already seen that the problem of computing $\phi(n)$ is equivalent to factoring n, in that an efficient algorithm for one problem implies an efficient algorithm for the other. It remains an interesting open question to relate (in either direction) the hardness of the discrete log problem to that of the factoring problem. In fact, it is believed that the discrete log problem is hard even in the average case, i.e., it is hard to solve for random inputs. Formally establishing the average-case hardness of the discrete log problem would have important consequences in cryptography and pseudo-random generation. This is because it would imply that exponentiation is a *one-way function* (a function that is easy to compute and hard to invert), which is a long-sought building block in these two areas.

The situation is slightly better in the case of the root finding problem. We will see that efficient randomized algorithms are known for this problem provided n is a prime power, and these algorithms can be generalized to solve the related problems of finding roots of polynomials, factoring polynomials, or finding irreducible (prime) polynomials. Unfortunately, for general n, even the problem of finding square roots modulo n can be shown to be equivalent (via randomized reductions) to factoring n. We start by describing an algorithm for finding square roots when n is a prime.

► **Definition 14.2:** A residue $a \in \mathbb{Z}_n^*$ is said to be a *quadratic residue* if there exists some $x \in \mathbb{Z}_n^*$ such that

$$a \equiv x^2 \pmod{n}.$$

If a is not a quadratic residue, then it is referred to as a *quadratic non-residue.*

Notice that both x and $-x$ (or $n - x$) are square roots of a. In the following exercise and in Problem 14.6, the number of distinct square roots of a quadratic residue is precisely determined.

Exercise 14.7: For an odd prime p and any $k \geq 1$, show that any quadratic residue modulo p^k has exactly two distinct square roots.

For the moment, we consider only quadratic residues over the field $\mathbb{Z}_p^*$ for a prime p. The multiplicative group is cyclic, and the following lemma characterizes those powers of generators in this group that are quadratic residues. As is usual, we will consider only the odd primes. (Is the following lemma meaningful if $p = 2$?)

Lemma 14.17: *Let p be an odd prime, and $g \in \mathbb{Z}_p^*$ be any generator. Then, g^k is a quadratic residue if and only if k is even.*

PROOF: Clearly, for even k, $g^{k/2}$ is an element of $\mathbb{Z}_p^*$ and is therefore a square root of g^k.

Consider now the case where $k = 2l + 1$ is odd, and assume for contradiction that there exists an $x \in \mathbb{Z}_p^*$ such that $x^2 \equiv g^{2l+1} \pmod{p}$. But since g is a generator, $x = g^m$ for some non-negative integer m. This implies that $g^{2m} \equiv g^{2l+1} \pmod{p}$, and switching to the additive group modulo $\phi(p)$, we can restate this as

$$2m \equiv 2l + 1 \pmod{\phi(p)}.$$

Since $\phi(p) = p - 1$, we conclude that $(p - 1)|(2l - 2m + 1)$. But $p - 1$ is even and $2l - 2m + 1$ is odd, and an even number cannot divide an odd number. This gives the desired contradiction. □

This results in the following theorem, which is popularly referred to as Euler's Criterion for quadratic residuacity.

Theorem 14.18 (Euler's Criterion): *For prime p, an element $a \in \mathbb{Z}_p^*$ is a quadratic residue if and only if*

$$a^{\frac{p-1}{2}} \equiv 1 \pmod{p}.$$

PROOF: Suppose a is a quadratic residue. Then let $x = g^k$ be a square root of a, where g is any generator for $\mathbb{Z}_p^*$. Clearly, $a = g^{2k} \pmod{p}$, and therefore

$$a^{\frac{p-1}{2}} \equiv_p g^{k(p-1)} \equiv_p (g^{p-1})^k \equiv_p 1^k \equiv_p 1.$$

Suppose now that a is not a quadratic residue. Then by Lemma 14.17 we know that a is an odd power of the generator g. Assuming that $a = g^{2l+1}$, we obtain that

$$a^{\frac{p-1}{2}} \equiv_p g^{l(p-1)} g^{\frac{p-1}{2}} \equiv_p g^{\frac{p-1}{2}}.$$

Since g has order $p - 1$, it cannot be the case that the last term is congruent to 1. □

For any generator g the power $g^{\frac{p-1}{2}}$ is exactly -1. This is because this power of g must be a square root of 1 other than 1 itself, and each quadratic residue modulo a prime has exactly two square roots. This motivates the following definition.

► **Definition 14.3 (Legendre Symbol):** For any prime p and $a \in \mathbb{Z}_p^*$, we define the Legendre symbol

$$\left[\frac{a}{p}\right] = \begin{cases} 1 & \text{if } a \text{ is a quadratic residue} \pmod{p} \\ -1 & \text{if } a \text{ is a quadratic non-residue} \pmod{p} \end{cases}$$

Alternatively, it can be defined as

$$\left[\frac{a}{p}\right] = a^{\frac{p-1}{2}} \pmod p$$

where we treat $p-1$ as -1.

The Legendre symbol can be computed in polynomial time by suitably exponentiating a. Thus, we can decide in polynomial time whether an element of $\mathbb{Z}_p^*$ is a quadratic residue or a non-residue. The distribution of quadratic residues and non-residues among the elements of $\mathbb{Z}_p^*$ is extremely irregular and can be fruitfully thought of as being "pseudo-random." This creates a problem when we wish to find an element of $\mathbb{Z}_p^*$ that is guaranteed to be a quadratic non-residue. (A quadratic residue can be found by picking any number and squaring it.) However, the following exercise shows that this problem is trivial if we are willing to settle for a randomized solution. No deterministic polynomial time algorithm is known for this problem.

Exercise 14.8: Prove that for any prime p, exactly half the elements of $\mathbb{Z}_p^*$ are quadratic residues. Using this observation, devise efficient (polynomial time) randomized algorithms, both Monte Carlo and Las Vegas, for finding a quadratic non-residues in $\mathbb{Z}_p^*$. (See Problem 14.8 for a generalization to quadratic non-residues modulo non-primes.)

It is known that if a mathematical hypothesis known as the Extended Riemann Hypothesis holds, then $\mathbb{Z}_p^*$ must contain a quadratic non-residue among its $O\left(\log^2 p\right)$ smallest elements. Then a quadratic non-residue can be easily identified by trying all these numbers and computing their Legendre symbols. The statement of the ERH and its proof are outside the scope of this book and are omitted. We now describe the **QuadRes** algorithm for computing square roots modulo a prime p. The only need for randomness in this algorithm is that it requires a quadratic non-residue. Clearly, this algorithm can be made deterministic if the ERH holds.

Fix an odd prime p and a quadratic residue $a \in \mathbb{Z}_p^*$, whose square root modulo p is to be found. The algorithm assumes the availability of a quadratic non-residue $b \in \mathbb{Z}_p^*$, which can be chosen as described above. It can easily verify all this by computing the Legendre symbols for a and b. The basic idea behind the algorithm is to find an odd power of a, say a^{2l+1}, which has residue 1 modulo p. This would imply that $a^{2l+2} \equiv_p a$, and then it is easy to see that $\pm a^{l+1}$ are the desired square roots.

Since p is an odd prime, its residue modulo 4 must be either 1 or 3. The easy case is when $p \equiv 3 \pmod 4$. Let k be such that $p = 4k+3$ and note that $(p+1)/2 = 2k+2$. Since a is a quadratic residue, we know that $a^{\frac{p-1}{2}} \equiv_p 1$. Multiplying by a on both sides, we have $a^{\frac{p+1}{2}} \equiv_p a$. But $(p+1)/2 = 2k+2$ is even, and setting $x = \pm a^{k+1} \pmod p$ it is easily seen that $x^2 \equiv_p a$. Thus, the square roots of a can be computed in polynomial time via a simple exponentiation.

On the other hand, when $p \equiv 1 \pmod 4$, the residue of p modulo 8 is either 1 or 5. Consider first the case where $p = 8k + 5$. Now $(p + 1)/2 = 4k + 3$ is odd and we cannot use the same idea as before. However, we still know that $a^{4k+2} \equiv_p 1$, implying that a^{2k+1} is a square root of 1. If $a^{2k+1} \equiv_p 1$ then we are done by the same argument as in the earlier case. The problem is that it might happen that $a^{2k+1} \equiv_p -1$. This is where the quadratic non-residue b comes in handy. Since $(p-1)/2 = 4k + 2$, the Legendre symbol of b is $b^{4k+2} \equiv_p -1$. This implies that $a^{2k+1}b^{4k+2} \equiv 1 \pmod p$, or equivalently

$$a^{2k+2}b^{4k+2} \equiv a \pmod p.$$

Since both exponents on the left are even, we conclude that $\pm a^{k+1}b^{2k+1} \pmod p$ are the square roots of a. Once again we need only a small number of multiplications and exponentiations.

The really hard case is when $p = 8k + 1$, implying that $a^{4k} \equiv_p 1$. While the argument from the second case does not apply directly, it can be appropriately generalized with some effort. Let $k = 2^r R$ for some odd number R. The values of r and R can be computed in polynomial time by repeatedly dividing k by 2. The Legendre symbol for a can now be rewritten as $A = a^{2^{r+2}R} \equiv 1 \pmod p$. The basic problem now is that the exponent is not odd (otherwise, multiplying A by a would give an even power of a that equals a, so that the square root is easily computed). However, computing the square root of A is easy since we can compute $a^{2^j R}$ by exponentiating a, for any $j > 0$. What about the obvious strategy of repeatedly taking square roots of A until the term 2^j in the exponent disappears? The only difficulty with this is that we also need the fact that $A \equiv_p 1$, and this need not remain true as we continue taking square roots.

Assume that $a^R \not\equiv_p 1$; otherwise we can easily check that the converse is true and hence identify the square roots of a as $\pm a^{(R+1)/2}$. Now, there must be a value j such that $0 < j < r + 2$ and $A_j = a^{2^j R}$ is not congruent to 1 modulo p, but $A_{j+1} = A_j^2$ is congruent to 1. This j is easy to find by repeatedly taking square roots of A. It must be the case that $A_j \equiv_p -1$. We can now use the trick of multiplying A_j by $B = b^{4k} = b^{2^{r+2}R}$ to obtain a number that is congruent to 1 modulo p. Once again we can start taking square roots of $A_j B$ with the aim of reducing the exponent of a to the odd number R. This is possible since the exponent of b has a larger power of 2 than that of a. Of course, we get stuck again if the square root at some point gives -1 instead of 1. But then we can supply another factor of b^{4k} to restore the property of being congruent to 1 modulo p.

Basically this process continues until the exponent of a is exactly R. The power of 2 in the exponent of a drops by at least 1 before each multiplication by b^{4k}; thus the number of such stages cannot exceed $r \leq \log p$. Also, at all times, the various powers of b have a strictly larger power of 2 in their exponent than does a. Thus, upon termination we obtain a number $y = a^R b^z$, where z is the sum of the exponents of b and is even. Since $y \equiv_p 1$, we can use the previous trick of multiplying by a and halving the exponents to obtain the square root. It is also fairly easy to verify that each stage of this algorithm takes polynomial time. The algorithm is summarized below.

Algorithm QuadRes:

Input: Odd prime p and quadratic residue $a \in \mathbb{Z}_p^*$.

Output: An $x \in \mathbb{Z}_p^*$ such that $x^2 \equiv_p a$.

1. **choose** a quadratic non-residue $b \in \mathbb{Z}_p^*$ using random sampling.
2. **choose** the appropriate case.

Case A. $[p \equiv 3 \pmod 4$ or $p = 4k + 3]$

  A.1. return $x = \pm a^{k+1} \pmod p$.

Case B. $[p \equiv 5 \pmod 8$ or $p = 8k + 5]$

  B.1. $A \leftarrow a^{2k+1} \pmod p$.

  B.2. if $A \equiv_p 1$ **then return** $x = \pm a^{k+1} \pmod p$
   else return $x = \pm a^{k+1} b^{2k+1} \pmod p$.

Case C. $[p \equiv 1 \pmod 8$ or $p = 8k + 1]$

  C.1. compute r and odd R such that $k = 2^r R$.

  C.2. if $a^R \equiv 1 \pmod p$ **then return** $x = \pm a^{\frac{R+1}{2}} \pmod p$.

  C.3. compute largest $j < r + 2$ such that $a^{2^j R} \not\equiv_p 1$.

  C.4. $\alpha \leftarrow 2^j R$; $\beta \leftarrow 2^{r+2} R$.

  C.5. $A \leftarrow a^\alpha \pmod p$; $B \leftarrow b^\beta \pmod p$.

  C.6. repeat forever

   C.6.1. while $AB \equiv_p 1$ and $\alpha \neq R$ **do**
    $\alpha \leftarrow \alpha/2$; $\beta \leftarrow \beta/2$;
    $A \leftarrow a^\alpha \pmod p$; $B \leftarrow b^\beta \pmod p$.

   C.6.2. if $\alpha = R$ **then return** $x = \pm\sqrt{aAB} \pmod p$
    else $\beta \leftarrow \beta + 2^{r+2} R$ and $B \leftarrow b^\beta \pmod p$.

We now indicate how this algorithm generalizes to the case of prime powers. Assume that $q = p^k$ for an odd prime p. The problem now is to find an x such that $x^2 \equiv_q a$. We can use the **QuadRes** algorithm to find the square root of a modulo p. Let r_1 be such that $r_1^2 \equiv a \pmod p$. We first show that this information can be used to find a square root r_2 of a modulo p^2; we refer to this as the "lifting" of the square root to integers modulo p^2. It will then be clear that the same method can be used to solve the general problem.

By definition, $r_2^2 - a \equiv 0 \pmod{p^2}$ and therefore it must be the case that $r_2^2 - a \equiv 0 \pmod{p}$. The latter implies that $r_2 \equiv r_1 \pmod{p}$. In other words, for some choice of $d \in \mathbb{Z}_p$, $r_2 = r_1 + pd$, and our goal is to identify d. Substituting this expression for r_2 into the congruence $r_2^2 - a \equiv 0 \pmod{p^2}$, we obtain the following.

$$\begin{aligned} & (r_1 + pd)^2 - a && \equiv 0 \pmod{p^2} \\ \Rightarrow\quad & r_1^2 + 2r_1pd + p^2d^2 - a && \equiv 0 \pmod{p^2} \\ \Rightarrow\quad & (r_1^2 - a) + 2r_1pd && \equiv 0 \pmod{p^2} \end{aligned}$$

Now, observe that $p|(a - r_1^2)$ and we can define $y = (a - r_1^2)/p$. Thus, $2r_1pd - py \equiv 0 \pmod{p^2}$ or, equivalently, $2r_1d - y \equiv 0 \pmod{p}$. Defining $z = (2r_1)^{-1} \pmod{p}$ to be the unique multiplicative inverse of $2r_1$ in $\mathbb{Z}_p^*$, we see that $2r_1zd - yz \equiv 0 \pmod{p}$, or $d \equiv yz \pmod{p}$. Thus, we have shown that there is a unique choice of d such that $r_2 \equiv r_1 + pd \pmod{p^2}$, and this value of d can be easily computed. The following proves formally that choosing $y = (a - r_1^2)/p$, $z = (2r_1)^{-1} \pmod{p}$, and $d = yz \pmod{p}$, we obtain a square root $r_2 = r_1 + pd$ of a in $\mathbb{Z}_{p^2}^*$.

$$\begin{aligned} (r_1 + pyz)^2 &\equiv r_1^2 + (2r_1z)py + p^2y^2z^2 \pmod{p^2} \\ &\equiv r_1^2 + py \pmod{p^2} \\ &\equiv r_1^2 + (a - r_1^2) \pmod{p^2} \\ &\equiv a \pmod{p^2} \end{aligned}$$

It is an easy exercise to show that square roots can be lifted into the integers modulo p^k in a similar fashion.

Exercise 14.9: For any odd prime p, $q = p^k$, and quadratic residue $a \in \mathbb{Z}_q^*$, show that the square root of a in $\mathbb{Z}_q^*$ can be found in polynomial expected time by a randomized algorithm.

In fact, we can find square roots in $\mathbb{Z}_n^*$ for any odd number n, given the prime factorization of n. Assume that n has the prime factorization $p_1^{k_1}p_2^{k_2}\cdots p_t^{k_t}$. Define $n_i = p_i^{k_i}$ for $1 \le i \le t$, and note that the terms n_i are pairwise coprime. We can easily compute roots $r_i \in \mathbb{Z}_{n_i}$ such that $r_i^2 \equiv a \pmod{n_i}$, using the randomized algorithm described above. Let r be the unique element in $\mathbb{Z}_n$ such that $r \equiv r_i \pmod{n_i}$, where r can be computed as in Theorem 14.4. It is now easy to see that $r^2 \equiv a \pmod{n_i}$ for each i. But then it is clear that $r^2 \equiv a \pmod{n}$.

Recall that a quadratic residue modulo an odd prime power has exactly two square roots. In the above computation, we could have chosen $-r_i$ instead of r_i for any i. In fact, there are 2^t distinct sequences that we could have used in the above computation, by trying all possible signs and combinations for the roots r_i. Since each of these gives a distinct square root of a modulo n, we obtain the following theorem. (The case of the solitary even prime is slightly more complicated and is discussed in Problem 14.7, giving a generalization of this theorem to the case of even numbers.)

Theorem 14.19: *For an odd number n with t distinct (odd) prime factors and any quadratic residue a modulo n, there are 2^t distinct square roots of a modulo n.*

We have seen that computing square roots in $\mathbb{Z}_n^*$ is easy using randomization, provided that a prime factorization of n is known. The next result shows that computing square roots is as hard as factoring n. This is established by providing a randomized reduction from factoring to computing square roots. The following lemma will be useful for this purpose.

Lemma 14.20: *Suppose* $x^2 \equiv y^2 \pmod{n}$ *and* $x \not\equiv \pm y \pmod{n}$. *Then neither* $\gcd(x+y, n)$ *nor* $\gcd(x-y, n)$ *equals* 1 *or* n.

PROOF: Since $x^2 \equiv y^2 \pmod{n}$, we have $(x+y)(x-y) \equiv_n 0$ or, equivalently, $n|(x+y)(x-y)$. Suppose that $\gcd(x+y, n) = 1$; then it must be the case that $n|(x-y)$. But this implies that $x \equiv_n y$, contradicting the conditions of the lemma. A similar argument shows that $\gcd(x-y, n) \neq 1$. Finally, notice that the neither of the two gcd's can be n for essentially the same reason. □

We are now ready to provide the desired reduction.

Theorem 14.21: *Suppose that there is a polynomial time, possibly randomized, algorithm A_1 that can compute square roots modulo any n. Then there is a randomized polynomial time algorithm A_2 for factoring any n.*

PROOF: If n is even, it is easy to find the highest power of 2 that divides n and reduce to the case of odd n; therefore, we assume throughout that n is odd. The algorithm A_2 will decompose n into factors each of which is a prime power. These can then be determined using Exercise 14.5. Of course, if n is a prime or a prime power, A_2 will fail to find any non-trivial factors but Exercise 14.5 applies again.

The factoring algorithm A_2 will use A_1 as a blackbox. It starts by choosing $b \in \mathbb{Z}_n^*$ uniformly at random. This is not entirely trivial; the algorithm will have to pick a random element b from $\mathbb{Z}_n$ and compute its gcd with n to test whether it also belongs to $\mathbb{Z}_n^*$. If $g = \gcd(b, n) \neq 1$, then g is a non-trivial factor of n, and $n = gh$ for $h = n/g$. The algorithm can now recursively factor g and h. Thus, the hard case is when the chosen element b does indeed lie in $\mathbb{Z}_n^*$.

Algorithm A_2 now computes $a \equiv b^2 \pmod{n}$. It then uses algorithm A_1 to find a square root x for a modulo n. Since n is not a prime power, it must have $t \geq 2$ distinct prime factors. By Theorem 14.19, there must be 2^t distinct square roots of a modulo n. Since b was chosen randomly, and A_1 has no knowledge of b other than that $b^2 = a$, the probability that $x = \pm b$ is at most $2/2^t \leq 1/2$. Of course, if A_2 is unlucky and gets back $\pm b$ as the square root, the entire process can be repeated for an independent, new choice of b. Therefore, with high probability, A_2 is guaranteed to find x and b such that $x^2 \equiv_n b^2$ but

$x \not\equiv_n \pm b$. Lemma 14.20 now applies to x and b, and it is clear that neither $\gcd(x+b,n)$ nor $\gcd(x-b,n)$ can equal 1 or n. Let $g = \gcd(x+b,n)$; since g is not 1 or n it must be a non-trivial factor of n. Setting $h = n/g$, we obtain a partial factorization of n into gh. Repeating this process recursively for g and h, A_2 obtains a factorization of n into prime powers. By Exercise 14.5, the prime powers can be factored individually. □

Exercise 14.10: Estimate the expected running time of algorithm A_2 in Theorem 14.21 when it is required to factor n with probability at least 1/2, assuming that A_1 runs in time $T(n)$.

Exercise 14.11: Suppose that the algorithm A_1 in Theorem 14.21 can only find square roots modulo a specific n, rather than for all n. Show that if $n = pq$, for primes p and q, then there is a Las Vegas algorithm A_2 that can factor this specific n in polynomial expected time.

Extend this result to arbitrary n (not necessarily of the form pq). Observe that a square root modulo n yields a square root modulo f, for any factor f of n.

Even when the factorization of n is known, finding the smallest square root of a quadratic residue modulo n is an ***NP***-hard problem.

14.4. The RSA Cryptosystem

We remarked earlier that cryptography relies heavily on number-theoretic tools. In particular, systems based on the (assumed) hardness of problems in number theory, such as factoring and discrete log, form an important part of modern cryptography. We illustrate this by a famous cryptographic scheme, the RSA cryptosystem named after Rivest, Shamir, and Adleman. But first we need to review the basic idea behind a *public-key encryption scheme.*

In a public-key cryptosystem, an individual (Alice) can set up a mechanism whereby she can receive and decode encoded messages from an arbitrary person. This message can be transmitted over a public channel because the system ensures that nobody else can decode the message. She advertises an encoding function E, which has the property that anyone may efficiently compute $E(M)$ for a message M, but no one but Alice may efficiently compute M from $E(M)$. In fact, Alice has a decoding function D such that, for all M, $D(E(M)) = M$.

In the RSA scheme, Alice constructs functions E and D as follows. She first chooses two distinct odd primes p and q, and computes $n = pq$. Alice keeps the primes secret, while n is given to the public. Alice also chooses an element $k \in \mathbb{Z}^*_{\phi(n)}$, with $k > 1$, and advertises k along with n. (Observe that $\phi(n) = (p-1)(q-1)$ is easy to compute given p and q.) The encoding function E is given by $E(M) = M^k \pmod{n}$, assuming that messages correspond to the elements of $\mathbb{Z}_n$. Knowing $\phi(n)$, Alice can easily compute the multiplicative inverse $l = k^{-1}$ for k in the group $\mathbb{Z}^*_{\phi(n)}$. The decoding function D is given by $D(C) = C^l \pmod n$. It is easy to verify that if $C = E(M)$, then $D(C) = M^{kl} = M \pmod n$, since $kl \equiv 1 \pmod{\phi(n)}$.

Why is this system secure against an eavesdropper Eve? We show that if Eve can compute l from the (public) knowledge of n and k, then she can factor n. This will then imply that completely breaking the RSA scheme is at least as hard as factoring n. Suppose Eve successfully computes l; then she knows that $\phi(n)|(kl - 1)$. We have shown earlier that for $n = pq$, knowing $\phi(n)$ lets us factor n efficiently. Eve knows a multiple of $\phi(n)$, and it is not very hard to see that even this is sufficient to allow the factorization of n (see Problems 14.3–14.4).

A problem with this result is that it only proves the hardness of breaking the RSA scheme completely by computing the value of l itself. It is entirely possible that some clever scheme could infer the messages without determining the decryption key. In practice, we would like stronger guarantees, for example that it is impossible to be able to decode the encryptions of more than a vanishingly small fraction of messages. Let $C(A)$ be the set of all $x \in \mathbb{Z}^*_n$ such that the algorithm A can compute $x^l \pmod n$, given that A knows only n and k. The next theorem shows that if there is an algorithm A_1 for which $C(A_1)$ is not too small in size, then there is another algorithm A_2 that can compute $x^l \pmod n$ for all $x \in \mathbb{Z}^*_n$.

Theorem 14.22: *Suppose there exists a (possibly randomized) polynomial time algorithm A_1 for which $|C(A_1)| \geq \epsilon|\mathbb{Z}^*_n|$, for some $\epsilon > 0$. Then there exists a Las Vegas algorithm A_2 for which $C(A_2) = \mathbb{Z}^*_n$, and the expected running time of A_2 is polynomial in $\log n$ and $1/\epsilon$.*

PROOF: Fix any $x \in \mathbb{Z}^*_n$, and we will show that the algorithm A_2 can compute $x^l \pmod n$ using algorithm A_1 as a blackbox. The algorithm A_2 chooses a random element $y \in \mathbb{Z}^*_n$ and computes $z = xy^k$. Then it runs the algorithm A_1 on the input z. Notice that $z^l = x^l y^{kl} \equiv x^l y \pmod n$, and since A_2 can compute the multiplicative inverse of y modulo n, the value of x^l is easily inferred from that of z^l. Thus, algorithm A_2 succeeds if A_1 succeeds on z, or equivalently $z \in C(A_1)$.

We claim that z is uniformly distributed over $\mathbb{Z}^*_n$ and therefore the probability that $z \in C(A_1)$ is at least ϵ. This claim follows from the observation that the operations of multiplication and raising to the power of k are functions that are one-to-one and onto in the group $\mathbb{Z}^*_n$, that is, they are permutations. Thus, for a random y, the number $z = xy^k$ is also uniformly distributed in $\mathbb{Z}^*_n$.

Since A_2 succeeds with probability ϵ, it is easy to see that independent iterations will boost the probability of success to any desired level. Also, it is possible to convert this into a Las Vegas algorithm whose expected running time is polynomial in $\log n$ and $1/\epsilon$. □

The algorithm A_2 described above has a polynomial expected running time provided $\epsilon = \Omega(1/\text{poly}(\log n))$. Thus, it has polynomial running time unless A_1's ability to break the RSA scheme is restricted to a set of messages of size smaller than any polynomial fraction of $\mathbb{Z}_n^*$.

It is also important to realize that from our description of A_2 (as also the assumption about A_1), it is not clear that the value of l is actually determined by these algorithms. All they do is to compute x^l and n via indirect methods. Thus, all that this result really says is that if the RSA scheme has even a slight weakness – in that it can be broken on some small fraction of the inputs – then it is totally insecure. This does not directly relate the hardness of breaking RSA to that of factoring.

This theorem has an interesting application to a variant of the RSA scheme due to Rabin. Recall Theorem 14.21, which says that finding square roots modulo n is as hard as factoring n. Suppose now that in the RSA scheme we had used the exponent $k = 2$. Now the task of decoding an encoded message is exactly equivalent to taking square roots. The above theorem says that if there is even a small chink in RSA's armor for a specific n, then there is an algorithm for finding all square roots modulo this n. While Theorem 14.21 does not apply directly, as it requires an algorithm for finding square roots modulo all possible n, the result in Exercise 14.11 can be used to show that this $n = pq$ can now be factored in randomized polynomial time. Thus, the problem of breaking the Rabin cryptosystem is as hard as factoring.

There is one technical problem with this cryptosystem. Since $\phi(n)$ is even, the exponent 2 is not coprime with respect to $\phi(n)$. Therefore, there is no unique way of inverting the encoding function as in the case of RSA. In fact, we know that there are four distinct square roots of any quadratic residue modulo $n = pq$, and the decoding process (finding square roots) need not give the same result as the original encoded message. Fortunately, the following exercise shows that there exists a simple method for computing all four square roots in this instance, and so some simple convention can be used to disambiguate the choice of the decoded message (see Problem 14.9).

Exercise 14.12: Show that for any quadratic residue a modulo $n = pq$, for odd primes p and q, the four square roots of a are given by $\pm x$ and $\pm y$, where $y = x(p^{q-1} - q^{p-1})$ (mod n).

A drawback with the Rabin cryptosystem is that anyone with temporary access to a blackbox for decoding can compute square roots and hence factor n. The RSA cryptosystem does not appear to have this drawback, precisely because it is not known to be as hard as factoring.

14.5. Polynomial Roots and Factors

We turn to the problem of finding roots and factors of polynomials over finite fields. Recall that the order of any finite field is a prime power, and that fields of a particular order are unique up to isomorphisms. When the order of a finite field is a prime p, it must be isomorphic to the field $(\mathbb{Z}_p, +_p, \times_p)$. (No such simple number-theoretic characterization is available for fields of order p^k, for $k > 1$.) We focus on the case where the underlying field is $(\mathbb{Z}_p, +_p, \times_p)$, and the polynomial is of degree 2. In what follows, we will denote the symbolic variable in a polynomial by X. We also assume that the reader is familiar with standard algorithms for adding, subtracting, multiplying, and dividing polynomials; these can be implemented in polynomial time for polynomials over the finite fields that are under consideration.

Consider a degree 2 polynomial $f(X)$ over a field of prime order p. We can assume without loss of generality that the polynomial is *monic*, i.e., the leading coefficient is 1; otherwise, the remaining coefficients can be divided by it to achieve the same effect. We also assume that the polynomial is not *irreducible*, which means that it has roots over the field $\mathbb{Z}_p$ and can be factored into linear terms as follows:

$$f(X) = X^2 + \alpha X + \beta = (X - a)(X - b).$$

Here $\alpha, \beta \in \mathbb{Z}_p$ are the coefficients, and $a, b \in \mathbb{Z}_p$ are the roots of the polynomial. If the polynomial is indeed irreducible, the algorithm described below will fail to find roots or factors, thereby indicating this fact. We make the simplifying assumption that the two roots are distinct; otherwise, if a is the only root, it must be the case that $\alpha \equiv -2a \pmod{p}$ and $\beta \equiv a^2 \pmod{p}$. These equations can be easily checked and would yield the desired root. Furthermore, we can assume that neither root is 0, since otherwise the polynomial is easily factored. Finally, we note that the problem of finding square roots of a quadratic residue r is the special case where the polynomial is $f(X) = X^2 - r$. Thus, the algorithm to be presented below yields an elegant alternative to the **QuadRes** algorithm described earlier.

Proposition 14.23: *An element $r \in \mathbb{Z}_p^*$ is a quadratic residue modulo an odd prime p if and only if $X - r$ is a factor of the polynomial $X^{\frac{p-1}{2}} - 1$.*

This proposition follows from Euler's Criterion, since $X - r$ is a factor if and only if r is a root of the polynomial $X^{\frac{p-1}{2}} - 1$. We start by applying this proposition to the root-finding problem for a special class of degree 2 polynomials. Suppose that the roots a and b of $f(X)$ are such that $\left[\frac{a}{p}\right] \neq \left[\frac{b}{p}\right]$. In particular, assume that $\left[\frac{a}{p}\right] = 1$ and $\left[\frac{b}{p}\right] = -1$, that is to say a is a quadratic residue while b is a quadratic non-residue. By Proposition 14.23, we have

$$\begin{array}{lcl} (X - a) & \mid & X^{\frac{p-1}{2}} - 1 \\ (X - b) & \nmid & X^{\frac{p-1}{2}} - 1. \end{array}$$

It then follows that

$$\gcd(f(X), X^{\frac{p-1}{2}} - 1) = (X - a).$$

Thus, the polynomial $f(X)$ can be factored via a single gcd computation. We leave it as an exercise to show that polynomial gcd can also be computed by Euclid's algorithm.

Exercise 14.13: Adapt Euclid's algorithm for gcd of integers to the computation of the gcd of polynomials over the field $\mathbb{Z}_p$. Show that this algorithm also runs in time polynomial in the degrees of the input polynomial.

A problem with using the result from this exercise is that the above application requires the gcd of a polynomial of degree $\Omega(p)$ and a quadratic polynomial. A naive application of Euclid's algorithm will require time polynomial in p rather than $\log p$. Fortunately, in this case the polynomial of higher degree has a very simple structure and we can finesse the problem of computing the gcd. The key observation is that the very first step of Euclid's algorithm will compute the remainder from the division of $X^{\frac{p-1}{2}} - 1$ by $f(X)$, and that remainder will be of degree at most 2. Moreover, the quotient and the polynomial $X^{\frac{p-1}{2}} - 1$ need not be referred to in the remaining steps of the gcd computation. Thus, it suffices to compute the remainder efficiently.

How may we compute this remainder efficiently? Recall the repeated squaring trick used to perform exponentiation (see Theorem 14.5). Suppose we were to express $X^{\frac{p-1}{2}}$ in terms of the powers of the type $g_i(X) = X^{2^i}$. Now, the remainder of each $g_i(X)$ upon division by $f(X)$ can be computed efficiently from the corresponding remainder for $g_{i-1}(X)$. Thus, working modulo $f(X)$, we can easily compute the remainder of $X^{\frac{p-1}{2}}$ upon division by $f(X)$. The details are left as an easy exercise.

Exercise 14.14: Show that repeated squaring modulo $f(X)$ can be used to compute $\gcd(f(X), X^{\frac{p-1}{2}} - 1)$ in polynomial time, provided that the degree of $f(X)$ is polynomially bounded.

Of course, there is no reason why an arbitrary polynomial of degree 2 should have roots with differing Legendre symbols. We show that this problem can be handled by suitably modifying the given polynomial $f(X)$. Recall from Exercise 14.8 that exactly half the elements of $\mathbb{Z}_p^*$ are quadratic residues. Thus, for r chosen uniformly at random from $\mathbb{Z}_p^*$, the probability that r is a quadratic residue is exactly 1/2. If $f(X)$ had random roots, we would be able to claim that with probability 1/2 it is the case that $\left[\frac{a}{p}\right] \neq \left[\frac{b}{p}\right]$. Our idea is to deliberately "randomize" the roots of $f(X)$.

Consider r chosen uniformly at random from $\mathbb{Z}_p$. Define the polynomial $f_r(X) = f(X - r) = (X - a - r)(X - b - r) \pmod{p}$. Clearly, the roots of $f_r(X)$ are $a + r$ and $b + r$, which are both uniformly distributed over $\mathbb{Z}_p^*$ (we may assume that neither of $a + r$ and $b + r$ is 0, since then we already have a root for the polynomial). This polynomial can be written as

$$\begin{aligned} f_r(X) &= X^2 - (a+b+2r)X + (ab + (a+b)r + r^2) \\ &= X^2 + (\alpha - 2r)X + (\beta - \alpha r + r^2). \end{aligned}$$

The coefficients of the polynomial $f_r(X) = X^2 + \alpha_r X + \beta_r$ can be easily computed given that they depend only on the values of α, β, and r. Also, given the roots of $f_r(X)$, it is easy to obtain the roots of $f(X)$ by subtracting r. It does not seem unreasonable to hope that the roots of f_r can be computed via the gcd trick, since the roots are now effectively "randomized."

The problem is that although the roots of $f_r(X)$ are randomly distributed, they are strongly correlated. The underlying assumption in the gcd trick is that the two roots are random and independent. For example, suppose that all the odd elements of $\mathbb{Z}_p$ are quadratic residues, while the even elements are quadratic non-residues. Then, consider the case where $a = 2$ and $b = 4$. For most choices of r, $a + r$ and $b + r$ would be smaller than p, so their residues modulo p would have the same parity and, therefore, the same Legendre symbol. However, we can circumvent this problem using the following lemma, which is reminiscent of two-point sampling (Section 3.4).

Lemma 14.24: *Let $a, b \in \mathbb{Z}_p$ and $a \neq b$. For s, t chosen independently and uniformly at random from $\mathbb{Z}_p$, the random variables $U = as + t \pmod{p}$ and $V = bs + t \pmod{p}$ are independent and uniformly distributed over $\mathbb{Z}_p$.*

PROOF: It is clear that the random variables U and V are uniformly distributed over $\mathbb{Z}_p$. The hard part is to show that they are independent, but note that it suffices to verify that for each $k, l \in \mathbb{Z}_p$ the probability that $U = k$ and $V = l$ is exactly $1/p^2$.

Since we are working over the field $\mathbb{Z}_p$ and $a \neq b$, it is easy to see that $U = k$ and $V = l$ if and only if

$$\begin{aligned} s &= \frac{k-l}{a-b} \pmod{p} \\ t &= k - a\frac{k-l}{a-b} \pmod{p}. \end{aligned}$$

Since s and t are uniform and independent, the probability that they take on these values is exactly $1/p^2$. □

It is now clear that we could randomize the roots of $f(X)$ using both s and t as described in the above lemma. Instead, we now use this lemma to show that the original method of randomizing the roots, while yielding correlated roots, has the desired properties from the point of view of the Legendre symbols. These properties are captured by the event $\mathcal{E}(X, Y)$ which occurs if either at least one of X and Y is 0, or their Legendre symbols differ. Clearly, the algorithm succeeds when $\mathcal{E}(a + r, b + r)$ occurs.

Lemma 14.25: *Let $a, b, \in \mathbb{Z}_p$ and $a \neq b$. For r chosen uniformly at random from $\mathbb{Z}_p$, the random variables $A = a + r \pmod{p}$ and $B = b + r \pmod{p}$ satisfy*

$$\mathbf{Pr}\left[\mathcal{E}(A, B)\right] = \frac{1}{2} - \mathrm{O}\left(\frac{1}{p}\right).$$

PROOF: Suppose that we choose s and t, and define U and V exactly as in Lemma 14.24. It is then clear that the probability pf $\mathcal{E}(U, V)$ is at least $1/2$. Suppose that instead of choosing r at random, we set its value to $ts^{-1} \pmod{p}$, assuming for now that $s \neq 0$. Then, it is easy to verify that $A = Us^{-1}$ and $B = Vs^{-1}$. Recall that, by the definition of the Legendre symbol,

$$\left[\frac{xy}{p}\right] = \left[\frac{x}{p}\right]\left[\frac{y}{p}\right].$$

It is now easy to see that regardless of the value of s^{-1},

$$\left[\frac{A}{p}\right] = \left[\frac{B}{p}\right] \quad \Leftrightarrow \quad \left[\frac{U}{p}\right] = \left[\frac{V}{p}\right].$$

This implies that $\mathcal{E}(A, B)$ occurs with the same probability as $\mathcal{E}(U, V)$, and this probability is at least $1/2$.

Of course, all of this is based on choosing $r = ts^{-1}$, instead of a random r. But since t is uniformly distributed, it follows that ts^{-1} is also uniformly distributed. Thus, even when r is chosen uniformly at random, $\mathcal{E}(A, B)$ occurs with probability at least $1/2$. Since the probability that $s = 0$ is $1/p$, removing the conditioning on $s \neq 0$ gives the desired result. □

These ideas are summarized below as Algorithm **PolyRoot**.

Algorithm PolyRoot:

Input: Odd prime p and a non-irreducible, monic, square-free, degree 2 polynomial $f(X) = X^2 + \alpha X + \beta \pmod{p}$.

Output: The roots a and b of $f(X)$ over $\mathbb{Z}_p$.

1. **choose** r uniformly at random from $\mathbb{Z}_p$.
2. **compute** the coefficients of the polynomial $g(X) = X^2 + \alpha' X + \beta'$ such that $g(X) = f(X - r)$, as follows.
   $\alpha' \leftarrow \alpha - 2r$;
   $\beta' \leftarrow \beta - \alpha r + r^2$.
3. **if** $\beta' = 0$ **then return** $a = -r$ and $b = -r - \alpha'$.
4. **compute** $h(X) = \gcd(g(X), X^{\frac{p-1}{2}} - 1)$ using Euclid's algorithm.
5. **if** $h(X) = g(X)$ or $h(X) = 1$ **then go to** Step 1.
6. **let** $h(X) = X - c$ and **compute** $A \leftarrow c$, $B \leftarrow -\alpha' - A$.
7. **return** $a = A - r$ and $b = B - r$.

Since **PolyRoot** succeeds in each iteration with probability at least $1/2$, it follows that it is a Las Vegas algorithm with polynomial expected running time.

Theorem 14.26: *Algorithm* **PolyRoot** *is a Las Vegas algorithm that factors a degree 2 polynomial over* $\mathbb{Z}_p$ *in polynomial expected time, provided* p *is an odd prime.*

14.6. Primality Testing

One of the most interesting open problems in computational number theory is whether factoring is ***NP***-hard. In the theory of ***NP***-completeness, we deal with decision problems (equivalently, language recognition problems), rather than optimization or function computation problems. The decision problem associated with factoring is that of deciding the compositeness or the primality of a given number $n > 1$; the corresponding languages are called COMPOSITENESS and PRIMALITY, and they are the complements of each other. It is easy to see that COMPOSITENESS $\in$ ***NP***, since any non-trivial factor of a number is a polynomial-length proof of its compositeness, which can be verified in polynomial time using a single division. This implies that the complementary problem PRIMALITY $\in$ co-***NP***, by the definition of co-***NP***. (Recall that ***P*** $\subseteq$ ***NP*** $\cap$ co-***NP***.) It is not known at this point whether COMPOSITENESS is ***NP***-complete. We start by providing some evidence that this problem is not ***NP***-complete. Thus, like graph isomorphism (see Section 7.7), this problem is expected to have intermediate hardness, somewhere between ***P*** and ***NP***-complete. We then focus on the solution of the compositeness and primality problems using randomized algorithms.

The evidence that COMPOSITENESS is not ***NP***-complete consists of demonstrating that this problem, or equivalently PRIMALITY, lies in ***NP*** $\cap$ co-***NP***. If any problem in ***NP*** $\cap$ co-***NP*** is shown to be ***NP***-complete, we would trivially obtain that ***NP*** = co-***NP***, a very unlikely outcome. The following theorem shows that PRIMALITY $\in$ ***NP***, thereby also proving that COMPOSITENESS $\in$ ***NP*** $\cap$ co-***NP***.

Theorem 14.27: *PRIMALITY* $\in$ ***NP***.

PROOF: Our goal is to show that any prime number n has a polynomial length "certificate" of primality whose validity can be verified in polynomial time. For any n, the certificate can be non-deterministically guessed and then verified efficiently.

We claim that n is a prime if and only if $\mathbb{Z}_n^*$ has an element of order $n-1$. Clearly, for prime n, the multiplicative group has a generator and its order is $n-1$. For the converse, if $\mathbb{Z}_n^*$ has an element of order $n-1$, then $|\mathbb{Z}_n^*| \geq n-1$. Since $\mathbb{Z}_n^*$ contains only the coprimes smaller than n, it follows that every number smaller than n is coprime to it, implying that n is a prime.

The certificate of primality is an element $g \in \mathbb{Z}_n^*$ along with a proof that g has order $n-1$. The proof just needs to show that for non-trivial divisors m of $n-1$, $g^m \not\equiv 1 \pmod{n}$. It suffices to verify this for the values of m that are $(n-1)/p_i$, where the p_i's are distinct prime factors of $n-1$. The verification of the proof is easy once the factorization of $n-1$ is known. The certificate of primality needs to include the factorization of $n-1$, which is $\phi(n)$ assuming that n is indeed a prime. It is essential that the factorization be complete, in that each of the factors is itself a prime; otherwise the verification of the order of g could be fallacious. Thus, the certificate must also include proofs of primality of the distinct prime factors of $n-1$.

The primality of the various prime factors can be proved recursively by including certificates of primality of these factors. Since the number of prime factors is $O(\log n)$ and each is of length $O(\log n)$, this recursive certificate is of polynomial length and can be checked in polynomial time. □

Exercise 14.15: Compute a bound on the length of the certificate of primality described in Theorem 14.27, and show that it can be validated in polynomial time.

Of course, this does not tell us how to check the primality (or compositeness) of a given number efficiently, even if we allow the use of randomization. In what follows, we will describe some randomized algorithms for this purpose. Intuitively, randomized algorithms for a decision problem can be devised only if there is a set that can be sampled efficiently and is dense in proofs of membership for the language. In concrete terms, a randomized algorithm for testing primality requires a set of potential certificates such that for any prime p, this set contains a large number of certificates of p's primality. For COMPOSITENESS, a naive belief might be that for composite n, $\mathbb{Z}_n$ contains a large number of elements that are not coprime with n, and such an element is a proof of compositeness that can be found by random sampling. However, when $n = pq$ for two roughly equal primes p and q, it is easily seen that the size of the set $\mathbb{Z}_n \setminus \mathbb{Z}_n^*$ is $O(1/n)$. This implies that random sampling is unlikely to yield the desired proof. What about PRIMALITY? Considering the complex structure of the best known certificates, it seems even less likely that a naive sampling will do the trick.

There is some hope for primality testing in Fermat's Theorem, which says that if n is a prime, then for all $a \in \mathbb{Z}_n^*$ it must be the case that $a^{n-1} \equiv 1 \pmod{n}$. Call this equation the *Fermat congruence* for a. Suppose that the converse of this theorem is also true: if n is not a prime then there exists $a \in \mathbb{Z}_n^*$ for which $a^{n-1} \not\equiv 1 \pmod{n}$. Then, we can choose an element $a \in \mathbb{Z}_n$ at random and verify that $\gcd(a, n) = 1$, since otherwise we know that n is composite. If indeed $a \in \mathbb{Z}_n^*$, then we hope that with reasonably high probability a violates Fermat's congruence when n is composite. Failure to prove compositeness using this strategy could be taken as evidence of primality. Of course, it would also be necessary to show that the number of such compositeness certificates is reasonably high. Unfortunately, there exist *pseudo-primes*, composite numbers satisfying the property in Fermat's Theorem, implying that its converse is not true.

► **Definition 14.4:** A *Carmichael number* is a composite number n such that, for all $a \in \mathbb{Z}_n^*$,

$$a^{n-1} \equiv 1 \pmod{n}.$$

The smallest example of a Carmichael number is 561, which can be factored into $3 \times 11 \times 17$. A more interesting Carmichael number is 1729, the number observed by Ramanujan to be the smallest number expressible as the sum of two cubes in two distinct ways. In Problem 14.10, we describe a simple method for checking whether n is a Carmichael number, provided the factorization of n is known.

The existence of Carmichael numbers need not kill the entire approach. If there are only finitely many Carmichael numbers, a randomized algorithm could afford to verify that the input n is not one of the Carmichael numbers, and otherwise perform the procedure described above. But we still need to show that for non-Carmichael composite numbers, the set $\mathbb{Z}_n^*$ is not dense in the elements a that satisfy Fermat's congruence.

► **Definition 14.5:** For any number n, the set F_n of elements that do not violate Fermat's Theorem is defined as

$$F_n = \{a \in \mathbb{Z}_n^* \mid a^{n-1} \equiv 1 \pmod{n}\}.$$

Obviously, the set F_n is the same as $\mathbb{Z}_n^*$ for prime n. The following lemma shows that for non-Carmichael composite numbers, the set F_n cannot be too large.

Lemma 14.28: *Let n be a composite non-Carmichael number. Then,*

$$|F_n| \leq \frac{1}{2}|\mathbb{Z}_n^*|.$$

PROOF: Since n is not a Carmichael number or a prime number, it is clear that $F_n \neq \mathbb{Z}_n^*$. It is easy to verify that $(F_n, \times_n)$ forms a group, and therefore is a *proper* sub-group of $(\mathbb{Z}_n^*, \times_n)$. By Proposition 14.8, it must be the case that $|F_n| \mid |\mathbb{Z}_n^*|$. But since the two cardinalities are not equal, it must be the case that $|\mathbb{Z}_n^*|/|F_n| > 1$. This gives the desired result. □

We now know that $|F_n|$ is either the same as $|\mathbb{Z}_n^*|$, or no more than half of it. Since the former happens only in the case of primes or Carmichael numbers, this suggests that the simple randomized strategy described above will be able to test for primality. Unfortunately, it has recently been shown that there are, in fact, infinitely many Carmichael numbers. The good news is that there are techniques for dealing with the problem posed by the existence of Carmichael numbers.

We will first need to define the Jacobi symbols, a generalized form of the Legendre symbols. Recall that for a prime n and any $a \in \mathbb{Z}_n^*$, the Legendre symbol $\left[\frac{a}{n}\right]$ denotes $a^{\frac{n-1}{2}} \pmod{n}$. The Jacobi symbol is defined for all odd n, and it is the same as the Legendre symbol when n is a prime; we therefore use the same notation for both symbols.

► **Definition 14.6 (Jacobi Symbol):** Let n be an odd number with the prime factorization $p_1^{k_1} p_2^{k_2} \cdots p_t^{k_t}$. Then, for all a such that $\gcd(a, n) = 1$, the Jacobi symbol is given by

$$\left[\frac{a}{n}\right] = \prod_{i=1}^{t} \left[\frac{a}{p_i}\right]^{k_i}.$$

Like that of the Legendre symbol, the value of the Jacobi symbol is also either 1 or −1. At first glance, it may appear that computing the Jacobi symbol requires knowledge of the prime factorization of n. Fortunately, there is a polynomial time algorithm for computing the Jacobi symbol without using the prime factorization of n. The reader is asked to provide a proof in Problem 14.11.

Theorem 14.29: *The Jacobi symbol satisfies the following properties whenever it is defined for the specified arguments. Using these, a polynomial time algorithm can be devised for computing the Jacobi symbol, given only a and n.*

1. $\left[\frac{ab}{n}\right] = \left[\frac{a}{n}\right]\left[\frac{b}{n}\right].$

2. *For* $a \equiv b \pmod{n}$, $\left[\frac{a}{n}\right] = \left[\frac{b}{n}\right].$

3. *For odd coprimes a and n,* $\left[\frac{a}{n}\right] = (-1)^{\frac{a-1}{2}\frac{n-1}{2}} \left[\frac{n}{a}\right].$

4. $\left[\frac{1}{n}\right] = 1.$

5. $\left[\frac{2}{n}\right] = \begin{cases} -1 & \text{for } n \equiv 3 \text{ or } 5 \pmod{8} \\ 1 & \text{for } n \equiv 1 \text{ or } 7 \pmod{8} \end{cases}$

► **Example 14.1:** We show below a sequence of application of these properties for computing the Jacobi symbol $\left[\frac{a}{n}\right]$.

$$\begin{aligned}
\left[\tfrac{191}{279}\right] &= (-1)\left[\tfrac{279}{191}\right] && \text{(By Property 3)}\\
&= (-1)\left[\tfrac{88}{191}\right] && \text{(By Property 2)}\\
&= (-1)\left[\tfrac{2}{191}\right]^3\left[\tfrac{11}{191}\right] && \text{(By Property 1)}\\
&= (-1)^2(+1)^3\left[\tfrac{191}{11}\right] && \text{(By Properties 5 and 3)}\\
&= \left[\tfrac{4}{11}\right] && \text{(By Property 2)}\\
&= \left[\tfrac{2}{11}\right]^2 && \text{(By Property 1)}\\
&= (-1)^2 && \text{(By Property 5)}\\
&= 1
\end{aligned}$$

The following primality testing algorithm is an ***RP*** algorithm for COMPOSITENESS. Such an algorithm outputs PRIME or COMPOSITE to indicate its "guess" about the input number n. It returns COMPOSITE only if n is indeed composite, but there is a possibility that it would label as PRIME a number that is not a prime. Thus, the output PRIME should be interpreted as "probably prime," while the output COMPOSITE should be interpreted as "definitely composite." This primality testing algorithm, called **Primality1** is similar to the (fallacious) randomized algorithm described above, except that it uses the Jacobi symbol instead of Fermat's Theorem to find certificates. The underlying observation is that if n is a prime, then $\left[\frac{a}{n}\right] \equiv a^{\frac{n-1}{2}} \pmod{n}$ for all a; on the other hand, for composite n, there exist a large number of $a \in \mathbb{Z}_n^*$ such that $\left[\frac{a}{n}\right] \not\equiv a^{\frac{n-1}{2}} \pmod{n}$.

Algorithm Primality1:

Input: Odd number n.

Output: PRIME or COMPOSITE.

1. **choose** a uniformly at random from $\mathbb{Z}_n \setminus \{0\}$.
2. **compute** $\gcd(a, n)$.
3. **if** $\gcd(a, n) \neq 1$ **then return** COMPOSITE.
4. **compute** $\left[\frac{a}{n}\right]$ and $a^{\frac{n-1}{2}} \pmod{n}$.
5. **if** $\left[\frac{a}{n}\right] \equiv a^{\frac{n-1}{2}} \pmod{n}$ **then return** PRIME
   else return COMPOSITE.

This algorithm is always correct when it returns COMPOSITE, because it then finds an $a \in \mathbb{Z}_n$ such that either $\gcd(a,n) \neq 1$ or $\left[\frac{a}{n}\right] \not\equiv a^{\frac{n-1}{2}} \pmod{n}$, both of which can only be possible for composite n. We now show that the probability the algorithm's returning PRIME when a is composite is at most $1/2$.

► **Definition 14.7:** For any odd number n, the set J_n is defined by

$$J_n = \{a \in \mathbb{Z}_n^* \mid \left[\frac{a}{n}\right] \equiv a^{\frac{n-1}{2}} \pmod{n}\}.$$

For prime n, $J_n = \mathbb{Z}_n^*$. The following lemma is similar in spirit to Lemma 14.28, and it shows that for composite n the set J_n is substantially smaller.

Lemma 14.30: *For all composite n, $|J_n| \leq \frac{1}{2}|\mathbb{Z}_n^*|$.*

PROOF: It is easy to verify that $J_n \subseteq \mathbb{Z}_n^*$ is a group, given the first property of Jacobi symbols. As in Lemma 14.28, all we need to show is that it is a proper subgroup of $\mathbb{Z}_n^*$, thereby implying the desired result.

Assume, for contradiction, that $J_n = \mathbb{Z}_n^*$ for some composite n. Consider the prime factorization of n, say $p_1^{k_1} p_2^{k_2} \cdots p_t^{k_t}$, and for convenience define $q = p_1^{k_1}$ and $m = p_2^{k_2} \cdots p_t^{k_t}$. Fix a generator g for $\mathbb{Z}_q^*$, and consider the element $a \in \mathbb{Z}_n^*$ that satisfies the following congruences:

$$\begin{aligned} a &\equiv g \pmod{q} \\ a &\equiv 1 \pmod{m}. \end{aligned}$$

Theorem 14.4 implies that such an element a must always exist. Notice that $a \equiv 1 \pmod{p_i}$ for all $i \geq 2$, since $p_i | m$ and $m | (a-1)$.

We now divide the proof into two cases depending on the factorization of n and derive a contradiction in each case. Consider first the case where $k_1 = 1$. We can write $n = qm$, where $q = p_1$ is a prime and $\gcd(q, m) = 1$; notice that $m \neq 1$ since n is not a prime. We can compute the Jacobi symbol for a and n as follows.

$$\begin{aligned} \left[\tfrac{a}{n}\right] &= \textstyle\prod_{i=1}^{t} \left[\tfrac{a}{p_i}\right]^{k_i} && \text{(By Definition)} \\ &= \left[\tfrac{a}{q}\right] \textstyle\prod_{i=2}^{t} \left[\tfrac{a}{p_i}\right]^{k_i} && \text{(Since } q = p_1,\ k_1 = 1\text{)} \\ &= \left[\tfrac{g}{q}\right] \textstyle\prod_{i=2}^{t} \left[\tfrac{1}{p_i}\right]^{k_i} && \text{(By Property 2)} \\ &= \left[\tfrac{g}{q}\right] && \text{(By Property 4)} \end{aligned}$$

Since the Legendre and Jacobi symbols agree for a prime modulus, and a generator cannot be a quadratic residue in $\mathbb{Z}_q^*$, we obtain $\left[\frac{a}{n}\right] = \left[\frac{g}{q}\right] = -1$. By assumption, $J_n = \mathbb{Z}_n^*$ and so

$$a^{\frac{n-1}{2}} \equiv -1 \pmod{n}.$$

Since $m|n$, it must also be the case that

$$a^{\frac{n-1}{2}} \equiv -1 \pmod{m},$$

which contradicts our choice of $a \equiv 1 \pmod{m}$.

The second (easier) case is where $k_1 \geq 2$. By assumption $J_n = \mathbb{Z}_n^*$, and therefore

$$\begin{aligned} & a^{\frac{n-1}{2}} \equiv \pm 1 \pmod{n} \\ \Rightarrow\ & a^{n-1} \equiv 1 \pmod{n} \\ \Rightarrow\ & g^{n-1} \equiv 1 \pmod{q}. \end{aligned}$$

The last congruence follows from the observation that $q|n$ and $a \equiv g \pmod{q}$. Since g is generator for $\mathbb{Z}_q^*$, its order is $\phi(q)$ and that must divide $n-1$. Also, for $k_1 \geq 2$, $p_1|\phi(q)$, implying that $p_1|(n-1)$. But no prime number can divide both n and $n-1$, giving us the desired contradiction. □

In Problem 14.10 we will see that a Carmichael number is always a product of distinct primes. Thus the first (harder) case in the above proof was exactly the one that had to deal with Carmichael numbers!

By the preceding discussion, it is clear that the **Primality1** algorithm makes an error only if n is composite, and then the random choice $a \in \mathbb{Z}_n^*$ lies in J_n. Lemma 14.30 now shows that the probability of error is at most $1/2$.

Theorem 14.31: *The* **Primality1** *algorithm always returns* PRIME *for prime n, and returns* COMPOSITE *for composite n with probability at least* $1/2$.

This theorem essentially says that COMPOSITENESS $\in$ ***RP*** and hence PRIMALITY $\in$ co-***RP***. As usual, it can be repeated independently to reduce the error probability, or to obtain a Las Vegas algorithm with polynomial expected time.

There is a simpler version of this algorithm that has the disadvantage that it makes 2-sided errors (a ***BPP*** algorithm), unlike the above algorithm, which makes only 1-sided errors. The algorithm is based on the following observation.

Lemma 14.32: *Let n be an odd composite number that is not a prime power. Suppose that for some* $a \in \mathbb{Z}_n^*$,

$$a^{\frac{n-1}{2}} \equiv -1 \pmod{n}.$$

Then, the set

$$S_n = \{x \in \mathbb{Z}_n^* \mid x^{\frac{n-1}{2}} \equiv \pm 1 \pmod{n}\}$$

has cardinality $|S_n| \leq \frac{1}{2}|\mathbb{Z}_n^*|$.

PROOF: Let n have the prime factorization $p_1^{k_1} p_2^{k_2} \cdots p_t^{k_t}$. We are guaranteed that $t \geq 2$. Define $q = p_1^{k_1}$ and $m = n/q$; note that $\gcd(m, q) = 1$ and m is a non-trivial factor of n. Using Theorem 14.4, choose $b \in \mathbb{Z}_n^*$ such that it satisfies the following congruences:

$$\begin{aligned} b &\equiv a \pmod{q} \\ b &\equiv 1 \pmod{m}. \end{aligned}$$

It is now easy to verify the following congruences:

$$\begin{aligned} b^{\frac{n-1}{2}} &\equiv_q a^{\frac{n-1}{2}} \equiv_q -1 \\ b^{\frac{n-1}{2}} &\equiv_m 1. \end{aligned}$$

If it were the case that $b^{\frac{n-1}{2}} \equiv 1 \pmod{n}$, then the residues modulo both q and m would also be 1; similarly, for $b^{\frac{n-1}{2}} \equiv -1 \pmod{n}$, the residues modulo the two factors of n would be both -1. Since we have chosen b such that $b^{\frac{n-1}{2}}$ has differing residues modulo the the two factors, it follows that

$$b^{\frac{n-1}{2}} \not\equiv \pm 1 \pmod{n}.$$

But then $b \notin S_n$, and so S_n is a proper subset of $\mathbb{Z}_n^*$. Clearly, S_n is a sub-group of $\mathbb{Z}_n^*$ and the result follows. □

In Lemma 14.30 we formulated a test based on the equality of the Jacobi symbol and $(n-1)/2$th power; in contrast, here we have a test that requires only that this power be ± 1, and so the power might have a different sign than the Jacobi symbol. The algorithm suggested by this lemma is now clear. Of course, we must first rule out the case where n is composite but has only one prime factor. But this is easily done using the test for prime power outlined in Exercise 14.5. We describe below a version of this algorithm that achieves error probability $O(1/2^t)$ for any desired t.

Algorithm Primality2:

Input: Odd number n and t.

Output: PRIME or COMPOSITE.

1. **if** n is a perfect power **then return** COMPOSITE.
2. **choose** $b_1, b_2, \ldots, b_t$ independently and uniformly at random from $\mathbb{Z}_n \setminus \{0\}$.
3. **if** for any b_i, $\gcd(b_i, n) \neq 1$ **then return** COMPOSITE.
4. **compute** $r_i = b_i^{\frac{n-1}{2}} \pmod{n}$, for $1 \leq i \leq t$.
5. **if** for any i, $r_i \not\equiv \pm 1 \pmod{n}$ **then return** COMPOSITE.
6. **if** for all i, $r_i \equiv 1 \pmod{n}$ **then return** COMPOSITE
   else return PRIME.

It is easy to verify that this algorithm runs in polynomial expected time, provided t is polynomially bounded. The following theorem shows that it is a ***BPP*** algorithm.

Theorem 14.33: *For all odd n, the probability that Algorithm* **Primality2** *errs is at most* $O(1/2^t)$.

PROOF: Suppose that n is a prime. Clearly, the only place where the algorithm can err is in Step 6. Now r_i is exactly the Legendre symbol for b_i, when n is a prime. The algorithm will return COMPOSITE in Step 6 if and only if all b_i's are quadratic residues. The probability that a random non-zero element modulo a prime is a quadratic residue is exactly $1/2$.

On the other hand, suppose that n is a composite number. Once again, the only possible error can be in Step 6, and only if n is not a prime power. But now Lemma 14.32 applies to n. This algorithm returns PRIME only if at least one of the r_i, say r_1, is -1 and the remaining r_i values are either 1 or -1. In this case, the probability that a random element lies in S_n is at most $1/2$. Thus, the probability that the values r_i, for $i \geq 2$, are all ± 1 is at most $1/2^{t-1}$. □

Finally, we present a second ***RP*** algorithm for compositeness. This algorithm is almost the same as the earlier one based on Lemma 14.28, which we had discarded due to the existence of Carmichael numbers. Moreover, this algorithm has the advantage that it can be made deterministic under the ERH. Consider a^{n-1}, for a random $a \in \mathbb{Z}_n \backslash \{0\}$. If this is not 1 (mod n), then we have proved that n is composite. Otherwise, we keep replacing this (even) power of a by its precomputed square root until the result is something other than 1 or we are reduced to an odd power of a. If we reach a square root of 1 other than ± 1, then n is composite; otherwise, the algorithm claims that n is prime, and this is the only place where it may make an error.

Algorithm Primality3:

Input: Odd number n.

Output: PRIME or COMPOSITE.

1. **compute** r and R such that $n - 1 = 2^r R$, and R is odd.
2. **choose** a uniformly at random from $\mathbb{Z}_n \backslash \{0\}$.
3. **for** $i = 0$ to r **compute** $b_i = a^{2^i R}$.
4. **if** $a^{n-1} = b_r \not\equiv 1 \pmod{n}$ **then return** COMPOSITE.
5. **if** $a^R = b_0 \equiv 1 \pmod{n}$ **then return** PRIME.
6. **let** $j = \max\{i \mid b_i \not\equiv 1 \pmod{n}\}$.
7. **if** $b_j \equiv -1 \pmod{n}$ **then return** PRIME
   else return COMPOSITE.

For prime n, this algorithm always returns PRIME. We want to show that the probability that the algorithm returns PRIME on a composite input n is at most $1/2$. By Lemma 14.28, if n is not a Carmichael number, then Step 4 will detect the compositeness of n with probability at least $1/2$. In Problem 14.14, you will be required to show that Steps 6 and 7 will detect a Carmichael number with probability at least $1/2$.

Theorem 14.34: *Algorithm* **Primality3** *is an* ***RP*** *algorithm for COMPOSITENESS.*

This algorithm can be made deterministic under the ERH, in much the same way as the algorithm **QuadRes**.

Notes

There are many excellent books on number theory and we mention only a few: Hardy and Wright [194], Hua [204], LeVeque [275], Niven and Zuckerman [321], and Vinogradov [407]. The book by Davenport [121] is an excellent source for material on the Extended Riemann Hypothesis (ERH). The reader may refer to these for the history and sources of the various number-theoretic results described here. The algebraic background that is assumed here can be reviewed in any text on algebra, such as those by Herstein [199] and van der Waerden [404]. Knuth [259] provides an excellent treatment of algorithmic number theory. The survey articles by Bach [44] and by Lenstra and Lenstra [273] are also excellent sources for more recent and advanced results. For overviews of randomized algorithms in number theory and algebra, the reader may refer to the articles by Johnson [216] and by Rabin and Shallit [345]. The book by Zippel [423] provides comprehensive coverage of randomized and deterministic algorithms for problems involving polynomial and number-theoretic problems. The lecture notes on algorithmic number theory by Angluin [27] is still among the best introductions to this area.

Euclid's gcd algorithm was first formalized in his *Elements*, and we refer the reader to the above sources (most notably Knuth [259]) for a history of this algorithm and its variants. Algorithm **QuadRes** for quadratic residues is due to Adleman, Manders, and Miller [2]. The result connecting the ERH to the existence of small quadratic non-residues was obtained by Ankney [29]. Algorithm **PolyRoot** is a special case of the algorithm due to Berlekamp [57], and is also attributed to Lehmer; see also the articles by Rabin [343] and Ben-Or [52]. The ***NP***-completeness of finding the least square root was proved by Manders and Adleman [291]. Thc RSA scheme is due to Rivest, Shamir, and Adleman [358], and the modification using quadratic residues is due to Rabin [346].

The certificates of primality used to show that PRIMALITY is in ***NP*** were devised by Pratt [335]. Carmichael numbers were defined by Carmichael [87], and the proof that there are infinitely many such numbers is due to Alford, Granville, and Pomerance [16]. The **Primality1** algorithm is due to Solovay and Strassen [382], while Algorithm **Primality3** was devised by Rabin [341, 342] and is related to a deterministic algorithm (assuming the ERH) due to Miller [310].

The primality testing algorithms described here all have the feature that if the input is a prime, then the output is always PRIME, while for composite inputs there is a small probability of making errors. This is essentially the same as proving COMPOSITENESS $\in$ ***RP***, or PRIMALITY $\in$ co-***RP***. There is no known easily described algorithm that errs in the reverse direction. Goldwasser and Kilian [178] gave such an algorithm, but this algorithm cannot be guaranteed to work correctly for a small set of exceptional primes. However, an extremely complex result of Adleman and Huang [3] provides such an algorithm and shows that PRIMALITY $\in$ ***RP***. Thus, we can now construct Las Vegas algorithms with polynomial expected running time for both PRIMALITY and COMPOSITENESS. Finally, we remark that an important area that has not been covered here is that of devising algorithms for factoring composite numbers. While none of these algorithms is of polynomial running time, several sub-exponential time algorithms are known. We refer the reader to the survey articles described above for a more detailed review of such algorithms.

Problems

14.1 Prove Theorem 14.2 by giving a detailed description of the extended Euclidean algorithm and its analysis. To prove a polynomial time bound for this algorithm, you will need to argue that the lengths of the operands in the intermediate computations are suitably bounded.

14.2 Show how to compute multiplicative inverses modulo a prime p via a single exponentiation. Does this work modulo composite n?

14.3 Show that given any number n and $\phi(n)$, the prime factorization of n can be computed by a randomized polynomial time algorithm.

14.4 Devise a randomized polynomial time algorithm for factoring a number n that is the product of two primes, given that some multiple of $\phi(n)$ is also provided as a part of the input. Can you generalize this to arbitrary n?

14.5 Show that for any odd prime p, the set $\{x^2 \mid 1 \leq x \leq \frac{p-1}{2}\}$ is exactly the set of all quadratic residues modulo p.

14.6 Let a be a quadratic residue modulo $n = 2^k$. Show that

- for $k = 1$, a has one square root modulo n;
- for $k = 2$, a has two square roots modulo n;
- for $k > 2$, a has four square roots modulo n.

14.7 Generalize Theorem 14.19 to allow the possibility of even numbers. (**Hint:** Use Problem 14.6.)

14.8 (a) Show that for any odd n with t distinct prime factors, the number of quadratic residues in $\mathbb{Z}_n^*$ is $\phi(n)/2^t$.

(b) Using Problem 14.7, generalize this to the case of even n.

(c) Can these observations be used to devise a randomized algorithm for finding a quadratic non-residue modulo n?

14.9 (Due to M.O. Rabin [346].) Consider the Rabin cryptosystem with $n = pq$ such that $p \equiv 3 \pmod 8$ and $q \equiv 7 \pmod 8$.

(a) Prove that for all x the Jacobi symbols satisfy $\left[\frac{x}{n}\right] = \left[\frac{-x}{n}\right] = -\left[\frac{2x}{n}\right]$.

(b) Using this observation and Exercise 14.12, show that we can choose the messages to lie in a subset of $\mathbb{Z}_n$ such that there is a canonical way to determine the message from among the four square roots of its square modulo n.

14.10 Let n have the prime factorization $p_1^{k_1} p_2^{k_2} \cdots p_t^{k_t}$, where each p_i is an odd prime.

(a) Show that n is a Carmichael number if and only if

$$\phi(p_i^{k_i}) | (n-1)$$

for $1 \leq i \leq t$.

(b) Conclude that the Carmichael numbers can be characterized as products of distinct primes $n = \prod_{i=1}^{t} p_i$, such that for each i, $(p_i - 1) | (n-1)$.

14.11 (a) Prove all the properties of the Jacobi symbol provided in Theorem 14.29.

(b) Using these properties, devise a polynomial time algorithm for computing $\left[\frac{a}{n}\right]$ without knowing the prime factorization of n or a.

14.12 We have seen how to test if a number is prime. In several applications, it is necessary to pick large prime numbers at random. For example, in the RSA scheme Alice must have two large primes p and q, but she would like to choose them randomly since they are to be kept secret. Suggest a randomized algorithm for generating a random $\Theta(\log n)$ bit length prime. Analyze the expected time to generate such a prime. (**Hint:** Refer to the Prime Number Theorem described in Section 7.4.)

14.13 Suppose you are given an algorithm S for computing square roots modulo a prime number. Using this algorithm as a blackbox, design an efficient randomized (***RP***) algorithm for compositeness. (**Hint:** The idea is to choose a random element $a \in \mathbb{Z}_n^*$, and run algorithm S on $b = a^2$. If S fails to find a square root, then n is not a prime. On the other hand, if S finds a square root other than $\pm a$, then again n is not a prime.)

14.14 (Due to M.O. Rabin [341, 342].) Show that when the input n is a Carmichael number, Algorithm **Primality3** will return PRIME with probability at most 1/2. (**Hint:** Use the characterization of Carmichael numbers described in Problem 14.10.)

APPENDIX A

Notational Index

The following is a list of the commonly used notation. The first entry is the symbol itself, followed by its meaning or name (if any), and the page number where the definition appears. Note that some standard symbols are not defined elsewhere in the text, e.g., $\mathbb{R}$ for real numbers. The page number for these symbols is replaced by $*$. Some overloaded notation may have more than one definition or name associated with it.

∞	infinity	*
$\{a, \ldots, z\}$	set notation	*
$[l, u]$	interval on the real line	*
$[n]$	the set $\{1, \ldots, n\}$	*
[13]	bibliographic reference to item 13	*
$\emptyset$	empty set	*
$\cap$	set intersection	*
$\cup$	set union	*
$\overline{S}$	set complement	*
$\setminus$	set difference	*
$\subset$	proper subset	*
$\subseteq$	subset	*
$\Delta(t)$	relative pointwise distance	*
$\in$	set membership	*
$\wedge$	Boolean conjunction (and)	*
$\vee$	Boolean disjunction (or)	*
$/$	Boolean negation (not)	*
$\Rightarrow$	implies	*
$\Leftrightarrow$	Boolean equivalence	*
$\forall$	for all	*
$\exists$	there exists	*
$\approx$	approximate equality	*
$\equiv$	equivalence	*
$\sim$	asymptotic equality	*

$\propto$	proportional to	*
$\neq$	not equal to	*
$\leq, \geq, <, >$	standard inequalities	*
$\rightarrow$	mapping, approaches	*
$\lceil x \rceil$	ceiling of x	*
$\lfloor x \rfloor$	floor of x	*
$a \div b$	a is a divisor of b	393
$a \not\div b$	a is not a divisor of b	393
a div b	quotient in the division of a by b	393
a mod b	remainder in the division of a by b	393
$a \pmod{p}$	residue of a modulo p	*
$a \equiv b \pmod{n}$	a is congruent to b modulo n	395
$a \equiv_n b$	a is congruent to b modulo n	395
$+_n$	addition modulo n	395
$\times_n$	multiplication modulo n	395
$\phi(n)$	Euler totient function	397
$\left[\frac{a}{p}\right]$	Legendre symbol	404
$\left[\frac{a}{n}\right]$	Jacobi symbol	420
$\lvert X \rvert$	absolute value, length, cardinality	*
$\sum_{i=0}^{n}$	summation from $i = 0$ to n	*
$\prod_{i=0}^{n}$	product from $i = 0$ to n	*
$\int_{x=0}^{1}$	integral from $x = 0$ to 1	*
$\sqrt{x}$	square root	*
$\sqrt[k]{x}$	kth root	*
2^S	power set of S	*
$n!$	factorial of n	*
$\binom{n}{k}$	binomial coefficient	*
$f^{-1}(y)$	the preimage $\{x \mid f(x) = y\}$	*
$f'(x)$	first derivative of function $f(x)$	*
$f''(x)$	second derivative of function $f(x)$	*
$f^{(k)}(x)$	kth derivative of function $f(x)$	*
$f^{(k)}(x)\rvert_{x=a}$	kth derivative evaluated at $x = a$	*
$\boldsymbol{a}^T \boldsymbol{b}$	vector inner product $\sum_i a_i b_i$	*
$\boldsymbol{a} \circ \boldsymbol{b}$	outer product matrix $\boldsymbol{M}$ with $M_{ij} = a_i b_j$	183
$\boldsymbol{x}^T$	transpose of the vector $\boldsymbol{x}$	*
$\lvert\lvert \boldsymbol{x} \rvert\rvert_1$	L_1-norm of the vector x	435
$\lvert\lvert \boldsymbol{x} \rvert\rvert$	L_2-norm of the vector x	435
$\lvert\lvert \boldsymbol{x} \rvert\rvert_\infty$	L_∞-norm of vector x	435
$\boldsymbol{A}^T$	transpose of the matrix $\boldsymbol{A}$	*
$\boldsymbol{A}^{-1}$	inverse of the matrix $\boldsymbol{A}$	*
$\boldsymbol{A}^{ij}$	ij minor of the matrix $\boldsymbol{A}$	*
adj($\boldsymbol{A}$)	adjoint of the matrix $\boldsymbol{A}$	*
$b(k; n, p)$	binomial distribution's density function	445
$B(n, p)$	binomial distribution with parameters n, p	445

$\Gamma(v)$	neighbors of the vertex v	8
$\Gamma(S)$	neighbors of the set of vertices S	8
$d(v)$	degree of vertex v	8
e	base of the natural logarithm	*
$\exp(x)$	exponential function of x	*
$\mathbf{E}[X]$	expectation of random variable X	442
$\mathbf{E}[X \mid Y]$	conditional expectation of X given Y	84
$\mathcal{E}$	event	439
$\det(\boldsymbol{M})$	determinant of matrix M	165
$F^+(\mu, \delta)$	Chernoff bound on the upper tail of binomial distribution	69
$F^-(\mu, \delta)$	Chernoff bound on the lower tail of binomial distribution	71
$\mathbb{F}$	a field, event space	439
$\mathbb{F}[x]$	polynomials in x over the field $\mathbb{F}$	*
$F_X(x)$	probability distribution function of X	441
$G_X(z)$	probability generating function of X	444
$G(V, E)$	graph with vertices V and edges E	*
$\gcd(a, b)$	greatest common divisor of a and b	393
H_n	harmonic number: $1 + 1/2 + \cdots + 1/n$	*
i.i.d.	independent, identically distributed (random variables)	*
$\operatorname{lcm}(a, b)$	lowest common multiple of a and b	393
$\lim_{n\to\infty}$	limit as n approaches ∞	*
λ_i	ith eigenvalue of a matrix	144
$\ln x$	natural logarithm	*
$\log_b x$	logarithm to base b	*
$\log x$	logarithm to base 2	*
m_X^k	kth moment of random variable X	443
μ_X	expectation of random variable X	443
μ_X^k	kth central moment of random variable X	443
$M_X(z)$	moment generating function of X	445
$\mathbb{N}$	non-negative integers	*
$O(f(x))$	the big-oh notation	433
$o(f(x))$	the little-oh notation	433
$\Omega(f(x))$	the big-omega notation	433
$\Theta(f(x))$	the big-theta notation	433
ω	elementary event	439
Ω	sample space	439
$(\Omega, \mathcal{F}, \mathbf{Pr})$	probability space	439
$(\Omega, \mathbf{Pr})$	probability space with $\mathcal{F} = 2^\Omega$	439
ord	order of a group or its element	398
$p_X(x)$	probability density function of X	442
$\mathbf{Pr}$	probability measure	439
$\mathbf{Pr}[\mathcal{E}_1 \mid \mathcal{E}_2]$	conditional probability of $\mathcal{E}_1$ given $\mathcal{E}_2$	440
π	the constant pi, a permutation	*
ϕ	golden ratio $(1 + \sqrt{5})/2$	*

Π	a problem	*
$\mathbb{R}$	real numbers	*
$\mathbb{R}^+$	non-negative real numbers	*
$\mathbb{R}^-$	non-positive real numbers	*
σ_X	standard deviation of random variable X	443
σ_X^2	variance of random variable X	443
$\mathbb{S}_n$	symmetric group of permutations of order n	165
$\mathrm{sgn}(\pi)$	sign of permutation π	165
$\mathbb{Z}$	integers	*
$\mathbb{Z}_p$	integers modulo p	*
$\mathbb{Z}_p^*$	multiplicative group of integers modulo p	*

APPENDIX B

Mathematical Background

This appendix is devoted to some elementary mathematical material that is used throughout this book. We start by reviewing the asymptotic notation such as the big-oh notation (see, for example, Knuth [261]). We also provide some important identities and approximations for binomial coefficients, as well as a few useful analytic inequalities. Good sources for this material are the books by Graham, Knuth, and Patashnik [182], Greene and Knuth [183], Hardy, Littlewood, and Pólya [195], Knuth [258], and Mitrinović [311]. Finally, we review some elementary material from linear algebra; the book by Strang [387] is a good source for this material.

Notation for Asymptotics

We start by defining the big-oh notation. The article by Knuth [261] gives more details on the following definitions.

► **Definition B.1:** Let $f(n)$, $g(n)$: $\mathbb{R} \to \mathbb{R}$ be two non-negative real-valued functions.

1. We say that $f(n) = O(g(n))$ if there exist positive numbers c and N such that, for all $n \geq N$, $f(n) \leq cg(n)$.
2. We say that $f(n) = \Omega(g(n))$ if there exist positive numbers c and N such that, for all $n \geq N$, $f(n) \geq cg(n)$.
3. We say that $f(n) = \Theta(g(n))$ if $f(n) = O(g(n))$ and $f(n) = \Omega(g(n))$ both hold.
4. We say that $f(n) = o(g(n))$ if $\lim_{n\to\infty} f(n)/g(n) = 0$. In this case, we also say that $g(n) = \omega(f(n))$.
5. We say that $f(n) \sim g(n)$ if $\lim_{n\to\infty} f(n)/g(n) = 1$. (If f and g are multivariate functions, it will be necessary to specify the argument, which is assumed to approach ∞. This is usually done by saying that: for large n, $f(n, m) \sim g(n, m)$. The interpretation is that m is held fixed, while $n \to \infty$.)

Note that the equality $f(n) = O(g(n))$ does not use "=" in a symmetric fashion.

Combinatorial Inequalities

We now turn our attention to the binomial coefficients, defined as follows. Let $n \geq k \geq 0$.

$$\binom{n}{k} = \binom{n}{n-k} = \frac{n!}{k!(n-k)!}$$

If $k > n \geq 0$ we define $\binom{n}{k} = 0$. The reason for the name "binomial coefficients" is their appearance in the binomial expansion:

$$(p+q)^n = \sum_{k=0}^{n} \binom{n}{k} p^k q^{n-k}.$$

Proposition B.1 (Stirling's Formula):

$$n! = \sqrt{2\pi n}\left(\frac{n}{e}\right)^n \left(1 + \frac{1}{12n} + O\left(\frac{1}{n^2}\right)\right)$$

From this one obtains the following inequalities involving binomial coefficients.

Proposition B.2: *Let* $n \geq k \geq 0$.

1. $\binom{n}{k} \leq \frac{n^k}{k!}$.
2. *For large* n, $\binom{n}{k} \sim \frac{n^k}{k!}$.
3. $\binom{n}{k} \leq \left(\frac{ne}{k}\right)^k$.
4. $\binom{n}{k} \geq \left(\frac{n}{k}\right)^k$.

The following power series expansions sometimes allow us to obtain useful inequalities.

$$\begin{aligned} e^x &= 1 + x + \frac{x^2}{2!} + \frac{x^3}{3!} + \cdots \\ \ln(1+x) &= x - \frac{x^2}{2} + \frac{x^3}{3} - \frac{x^4}{4} + \cdots \end{aligned}$$

We list below several inequalities involving the exponential function. The reader may refer to Mitrinović [311] for the derivations and other variants.

Proposition B.3:

1. *For all* $t \in \mathbb{R}$, $e^t \geq 1 + t$ *with equality holding only at* $t = 0$.

2. *For all $t, n \in \mathbb{R}$, such that $n \geq 1$ and $|t| \leq n$,*

$$e^t \left(1 - \frac{t^2}{n}\right) \leq \left(1 + \frac{t}{n}\right)^n \leq e^t.$$

Note that this holds even for negative values of t.

3. *For all $t, n \in \mathbb{R}^+$,*

$$\left(1 + \frac{t}{n}\right)^n \leq e^t \leq \left(1 + \frac{t}{n}\right)^{n+t/2}.$$

For any $n \in \mathbb{N}$, we define the nth Harmonic number H_n as follows:

$$H_n = 1 + \frac{1}{2} + \cdots + \frac{1}{n}.$$

Proposition B.4: *For any $n \in \mathbb{N}$, the nth Harmonic number is*

$$H_n = \ln n + \Theta(1).$$

The nth Fibonacci number is defined as follows:

$$F_0 = F_1 = 1,$$

and for $n \geq 2$,

$$F_n = F_{n-1} + F_{n-2}.$$

Proposition B.5: *For all $n \in \mathbb{N}$, $F_n = \Theta(\phi^n)$, where $\phi = (1 + \sqrt{5})/2$ is the golden ratio.*

Linear Algebra

Consider the field $\mathbb{R}$ of real numbers under addition and multiplication, and the real vector space $\mathbb{R}^n$ of n-dimensional vectors over $\mathbb{R}$. This vector space is an *inner product vector space*, where we define the inner product of two vectors $\boldsymbol{v}$, $\boldsymbol{w} \in \mathbb{R}^n$ as

$$\boldsymbol{v}^T \boldsymbol{w} = \sum_{i=1}^{n} v_i w_i,$$

where v_i and w_i are the ith components of the vectors $\boldsymbol{v}$ and $\boldsymbol{w}$. The vectors $\boldsymbol{v}$ and $\boldsymbol{w}$ are said to be orthogonal, denoted $\boldsymbol{v} \perp \boldsymbol{w}$, if their inner product $\boldsymbol{v}^T \boldsymbol{w}$ equals 0. A *subspace* $\mathcal{W}$ of a vector space $\mathcal{V}$ is a subset $\mathcal{W} \subseteq \mathcal{V}$, which forms a vector space; its orthogonal subspace is $\mathcal{W}^\perp = \{\boldsymbol{v} \in \mathcal{V} \mid \forall \boldsymbol{w} \in \mathcal{W}, \boldsymbol{v} \perp \boldsymbol{w}\}$. The vector space $\mathcal{V}$ is a direct sum of the orthogonal subspaces $\mathcal{W}$ and $\mathcal{W}^\perp$. In other words, every vector $\boldsymbol{v} \in \mathcal{V}$ can be *uniquely* expressed as $\boldsymbol{v} = \boldsymbol{w} + \boldsymbol{w}'$, where $\boldsymbol{w} \in \mathcal{W}$ and $\boldsymbol{w}' \in \mathcal{W}^\perp$.

We define three norms for vectors in an inner product vector space.

L_1-norm: $||\boldsymbol{v}||_1 = \sum_{i=1}^{n} |v_i|$.
L_2-norm: $||\boldsymbol{v}|| = \sqrt{\boldsymbol{v}^T\boldsymbol{v}} = \sqrt{\sum_{i=1}^{n} v_i^2}$.
L_∞-norm: $||\boldsymbol{v}||_\infty = \max_{i=1}^{n} |v_i|$.

A *unit vector* is a vector $\boldsymbol{v}$ with $||\boldsymbol{v}|| = 1$. We state some standard facts about these norms. While the familiar triangle inequality is valid for any norm, we state it only for the L_2 norm.

Proposition B.6 (Triangle Inequality): *For any two vectors $\boldsymbol{v}$ and $\boldsymbol{w}$,*

$$||\boldsymbol{v} + \boldsymbol{w}|| \leq ||\boldsymbol{v}|| + ||\boldsymbol{w}||.$$

The classical theorem of Pythagoras can be generalized as follows.

Proposition B.7 (Pythagoras Theorem): *For any two* orthogonal *vectors $\boldsymbol{x}$ and $\boldsymbol{y}$, let $\boldsymbol{v} = \boldsymbol{x} + \boldsymbol{y}$. Then*

$$||\boldsymbol{v}||^2 = ||\boldsymbol{x}||^2 + ||\boldsymbol{y}||^2.$$

An immediate consequence of the Pythagoras Theorem is the following useful fact.

Proposition B.8 (Pythagoras Inequality): *For any two* orthogonal *vectors $\boldsymbol{x}$ and $\boldsymbol{y}$, let $\boldsymbol{v} = \boldsymbol{x} + \boldsymbol{y}$. Then*

$$||\boldsymbol{x}|| \leq ||\boldsymbol{v}||$$

and

$$||\boldsymbol{y}|| \leq ||\boldsymbol{v}||.$$

Note that orthogonality is important in this proposition. For example, the result is not true for $\boldsymbol{x} = -\boldsymbol{y}$.

Proposition B.9 (Cauchy-Schwarz Inequality): *Let $\boldsymbol{a}$ and $\boldsymbol{b}$ be two real vectors. Then*

$$||\boldsymbol{a}||^2 \times ||\boldsymbol{b}||^2 \geq (\boldsymbol{a}^T\boldsymbol{b})^2$$

with equality holding if and only if the vectors are linearly related.

Finally, we establish some relations between the different norms.

Proposition B.10: *For any vector $\boldsymbol{v}$,*

$$||\boldsymbol{v}|| \leq ||\boldsymbol{v}||_1 \leq \sqrt{n}||\boldsymbol{v}||$$

and

$$||\boldsymbol{v}||_\infty \leq ||\boldsymbol{v}||_1 \leq n||\boldsymbol{v}||_\infty.$$

We briefly indicate the proof of the first series of inequalities in Proposition B.10. Note that the L_1 and L_2 norms are identical for any vector that points along the direction of one of the coordinate axes. Expressing the vector $\boldsymbol{v}$ as the sum of vectors aligned with the n coordinate axes and applying the triangle inequality leads to the inequality $||\boldsymbol{v}|| \leq ||\boldsymbol{v}||_1$. To obtain the inequality $||\boldsymbol{v}||_1 \leq \sqrt{n}||\boldsymbol{v}||$, we employ the Cauchy-Schwarz inequality with $a_i = v_i$ and $b_i = |v_i|/v_i$, for $1 \leq i \leq n$.

A *basis* for a vector space $\mathcal{V}$ is a collection of linearly independent vectors $\boldsymbol{b}_1, \ldots, \boldsymbol{b}_n$ that span $\mathcal{V}$. Each vector in $\mathcal{V}$ can be uniquely expressed as a linear combination of the basis vectors. An *orthonormal basis* is a collection of pairwise orthogonal, unit vectors $\boldsymbol{b}_1, \ldots, \boldsymbol{b}_n$ that form a basis for $\mathcal{V}$.

Proposition B.11: *Let $\boldsymbol{p} \in \mathbb{R}^n$ be any vector, and $\boldsymbol{b}_1, \ldots, \boldsymbol{b}_n$ be any orthonormal basis for $\mathbb{R}^n$. Further, let $\boldsymbol{p} = \sum_{i=1}^{n} c_i \boldsymbol{b}_i$ be the unique expression of $\boldsymbol{p}$ as a linear combination of the basis vectors. Then the L_2-norm of $\boldsymbol{p}$ is given by $||\boldsymbol{p}|| = \sqrt{\sum_i c_i^2}$.*

Consider any $n \times n$ symmetric matrix $\boldsymbol{A}$ over $\mathbb{R}$. The *characteristic equation* for $\boldsymbol{A}$ is

$$\boldsymbol{A}\boldsymbol{x} = \lambda \boldsymbol{x},$$

where $\boldsymbol{x} \in \mathbb{R}^n$ and $\lambda \in \mathbb{R}$. We say that λ is an *eigenvalue* of $\boldsymbol{A}$ if there exists a non-zero vector $\boldsymbol{x}$ satisfying the characteristic equation. A solution $\boldsymbol{x}$ is said to be an *eigenvector* corresponding to the eigenvalue λ. The following result characterizes the eigenvalues.

Proposition B.12: *The eigenvalues of $\boldsymbol{A}$ are the roots of the following polynomial in λ:*

$$\det(\boldsymbol{A} - \lambda \boldsymbol{I}) = 0,$$

where $\boldsymbol{I}$ denotes the $n \times n$ identity matrix.

Since the coefficient of λ^n is always $(-1)^n$, this is a polynomial of degree n. For symmetric $\boldsymbol{A}$, the polynomial has n real roots, and so the sum of the multiplicities of the eigenvalues of $\boldsymbol{A}$ is exactly n. It is easy to prove that any collection of eigenvectors corresponding to distinct eigenvalues are pairwise orthogonal. If the eigenvalues all have multiplicity 1, the corresponding eigenvectors form a basis for $\mathbb{R}^n$. The n eigenvalues of $\boldsymbol{A}$ are canonically numbered (accounting for multiplicities) such that $\lambda_1 \geq \lambda_2 \geq \cdots \geq \lambda_n$.

Since $\boldsymbol{A}$ is symmetric, it is possible to choose a set of pairwise orthogonal, unit vectors $\boldsymbol{e}_1, \ldots, \boldsymbol{e}_n$ such that, for $1 \leq i \leq n$, $\boldsymbol{e}_i$ is an eigenvector for λ_i. Notice that even though the eigenvalues may have multiplicities, for each distinct eigenvalue we can choose as many orthogonal eigenvectors as its multiplicity. These vectors $\boldsymbol{e}_1, \ldots, \boldsymbol{e}_n$ form an orthonormal basis for $\mathbb{R}^n$.

APPENDIX C

Basic Probability Theory

In this appendix we review basic ideas from probability theory. Starting with an axiomatic view of probability theory, we develop the following concepts: events, probabilities, independence, random variables, and their distributions and moments. After presenting some fundamental theorems without proof, we describe the properties of some common probability distributions. This appendix is provided for the sake of completeness only and should be supplemented by standard probability texts such as those by Billingsley [61], Feller [142, 143], and Grimmett and Stirzaker [185].

Any probabilistic statement must refer to an underlying *probability space.* A probability space is defined in terms of a *sample space* with an algebraic structure and a *probability measure* imposed on it. A sample space Ω is an arbitrary (potentially infinite) set, and its elements are referred to as *elementary events.* A subset $\mathcal{E} \subseteq \Omega$ is referred to as an *event.*

Intuitively, the sample space represents the set of all possible outcomes in a probabilistic experiment, and an event represents a collection of possible outcomes. For example, if the experiment consists of a sequence of four coin flips, then $\Omega = \{HHHH, HHHT, \ldots, TTTT\}$ and the event "the number of HEADS exceeds the number of TAILS by two" is the subset $\{HHHT, HHTH, HTHH, THHH\}$. Sometimes it is convenient to define the underlying sample space without considering each elementary event separately. For example, if we wish to focus on the number of HEADS only, then we could define $\Omega = \{0, 1, 2, 3, 4\}$. An example of an infinite sample space comes from the following experiment: flip an unbiased coin until HEADS appears for the first time. Here the sample space Ω is $\{H, TH, TTH, TTTH, TTTTH, \ldots\}$. The event that "the number of TAILS seen is odd" is given by the infinite set $\{TH, TTTH, TTTTTH, \ldots\}$.

At times we may wish to concentrate on only a subcollection of the events over a particular sample space Ω rather than consider all the possible events in the power set 2^{Ω}. However, not all subcollections of 2^{Ω} lead to a well-defined probability space. It is for this reason that we make the following definition.

► **Definition C.1:** A *σ-field* $(\Omega, \mathbb{F})$ consists of a sample space Ω and a collection of subsets $\mathbb{F}$ satisfying the following conditions.

1. $\emptyset \in \mathbb{F}$.
2. $\mathcal{E} \in \mathbb{F} \Rightarrow \overline{\mathcal{E}} \in \mathbb{F}$.
3. $\mathcal{E}_1, \mathcal{E}_2, \ldots \in \mathbb{F} \Rightarrow \mathcal{E}_1 \cup \mathcal{E}_2 \cup \ldots \in \mathbb{F}$.

The last condition is that of closure under countable union, and together with the second condition it implies closure under countable intersection. Observe that the first two conditions imply that $\Omega \in \mathbb{F}$. For convenience, we will adopt the convention of referring to $\mathbb{F}$ itself as a σ-field when the sample space Ω is clear from the context.

► **Definition C.2:** Given a σ-field $(\Omega, \mathbb{F})$, a probability measure $\mathbf{Pr} : \mathbb{F} \to \mathbb{R}^+$ is a function that satisfies the following conditions.

1. $\forall A \in \mathbb{F}, 0 \leq \mathbf{Pr}[A] \leq 1$.
2. $\mathbf{Pr}[\Omega] = 1$.
3. For mutually disjoint events $\mathcal{E}_1, \mathcal{E}_2, \ldots$, $\mathbf{Pr}[\cup_i \mathcal{E}_i] = \sum_i \mathbf{Pr}[\mathcal{E}_i]$.

► **Definition C.3:** A probability space $(\Omega, \mathbb{F}, \mathbf{Pr})$ consists of a σ-field $(\Omega, \mathbb{F})$ with a probability measure $\mathbf{Pr}$ defined on it.

When specifying a probability space, $\mathbb{F}$ may be omitted and it is understood then that the σ-field referred to is $(\Omega, 2^{\Omega})$.

Consider the following example of a probability space with $\Omega = (0, 1]$, i.e., the half-open unit interval. An elementary event is the choice of a point in this interval. The collection $\mathbb{F}$ consists of all possible subsets of Ω that can be expressed as a union of disjoint half-open subintervals. That is, any $\mathcal{E} \in \mathbb{F}$ can be written as $\mathcal{E} = \cup_i (l_i, u_i]$, where $0 \leq l_i \leq u_i \leq l_{i+1} \leq 1$. The probability measure is defined to be such that for any $\mathcal{E} \in \mathbb{F}$, $\mathbf{Pr}[\mathcal{E}]$ is the total length of the intervals in it.

An easy way to combine distinct probability spaces $(\Omega_1, \mathbb{F}_1, \mathbf{Pr}_1)$ and $(\Omega_2, \mathbb{F}_2, \mathbf{Pr}_2)$ is to take their product space $(\Omega, \mathbb{F}, \mathbf{Pr})$. In the new space, $\Omega = \Omega_1 \times \Omega_2$, $\mathbb{F} = \mathbb{F}_1 \times \mathbb{F}_2$, and for events $\mathcal{E}_1 \in \mathbb{F}_1$, $\mathcal{E}_2 \in \mathbb{F}_2$, the probability of the joint event $(\mathcal{E}_1, \mathcal{E}_2)$ is given by the product of the two events' probabilities. The product corresponds to performing independent experiments with respect to each of the two probability spaces.

In the rest of this appendix we will assume some fixed underlying probability space. We can apply the set operators of union, intersection, and complementation to combine events in complex ways; sometimes the boolean operators of disjunction ($\vee$), conjunction ($\wedge$), and negation ($\neg$) are also used to denote these operations.

Proposition C.1 (Principle of Inclusion–Exclusion): *Let $\mathcal{E}_1, \mathcal{E}_2, \ldots, \mathcal{E}_n$ be arbitrary events. Then*

$$\mathbf{Pr}[\cup_{i=1}^n \mathcal{E}_i] = \sum_i \mathbf{Pr}[\mathcal{E}_i] - \sum_{i<j} \mathbf{Pr}[\mathcal{E}_i \cap \mathcal{E}_j] + \sum_{i<j<k} \mathbf{Pr}[\mathcal{E}_i \cap \mathcal{E}_j \cap \mathcal{E}_k]$$

$$- \cdots + (-1)^{l+1} \sum_{i_1<i_2<\cdots<i_l} \mathbf{Pr}[\cap_{r=1}^l \mathcal{E}_{i_r}] + \cdots$$

Proposition C.2 (Boole-Bonferroni Inequalities): *Let $\mathcal{E}_1, \mathcal{E}_2, \ldots, \mathcal{E}_n$ be arbitrary events. Then, for even k*

$$\mathbf{Pr}[\cup_{i=1}^n \mathcal{E}_i] \geq \sum_{j=1}^k (-1)^{j+1} \sum_{i_1<i_2<\cdots<i_j} \mathbf{Pr}[\cap_{r=1}^j \mathcal{E}_{i_r}]$$

and for odd k

$$\mathbf{Pr}[\cup_{i=1}^n \mathcal{E}_i] \leq \sum_{j=1}^k (-1)^{j+1} \sum_{i_1<i_2<\cdots<i_j} \mathbf{Pr}[\cap_{r=1}^j \mathcal{E}_{i_r}].$$

► **Definition C.4:** The *conditional probability* of $\mathcal{E}_1$ given $\mathcal{E}_2$ is denoted by $\mathbf{Pr}[\mathcal{E}_1 \mid \mathcal{E}_2]$ and is given by

$$\frac{\mathbf{Pr}[\mathcal{E}_1 \cap \mathcal{E}_2]}{\mathbf{Pr}[\mathcal{E}_2]}$$

assuming that $\mathbf{Pr}[\mathcal{E}_2] > 0$.

This corresponds to the probability that an experiment has an outcome in the set $\mathcal{E}_1$ when we already know that it is in the set $\mathcal{E}_2$.

Proposition C.3: *Let $\mathcal{E}_1, \mathcal{E}_2, \ldots, \mathcal{E}_k$ be a partition of the sample space Ω. Then for any event $\mathcal{E}$*

$$\mathbf{Pr}[\mathcal{E}] = \sum_{i=1}^k \mathbf{Pr}[\mathcal{E} \mid \mathcal{E}_i]\mathbf{Pr}[\mathcal{E}_i].$$

Since $\mathbf{Pr}[\mathcal{E}_1 \cap \mathcal{E}_2] = \mathbf{Pr}[\mathcal{E}_1 \mid \mathcal{E}_2]\mathbf{Pr}[\mathcal{E}_2] = \mathbf{Pr}[\mathcal{E}_2 \mid \mathcal{E}_1]\mathbf{Pr}[\mathcal{E}_1]$, we obtain Bayes' rule from the previous proposition.

Proposition C.4 (Bayes' Rule): *Let $\mathcal{E}_1, \mathcal{E}_2, \ldots, \mathcal{E}_k$ be a partition of the sample space Ω. Then for any event $\mathcal{E}$*

$$\mathbf{Pr}[\mathcal{E}_i \mid \mathcal{E}] = \frac{\mathbf{Pr}[\mathcal{E}_i \cap \mathcal{E}]}{\mathbf{Pr}[\mathcal{E}]} = \frac{\mathbf{Pr}[\mathcal{E} \mid \mathcal{E}_i]\mathbf{Pr}[\mathcal{E}_i]}{\sum_{j=1}^k \mathbf{Pr}[\mathcal{E} \mid \mathcal{E}_j]\mathbf{Pr}[\mathcal{E}_j]}.$$

► **Definition C.5:** A collection of events $\{\mathcal{E}_i \mid i \in I\}$ is *independent* if for all subsets $S \subseteq I$,

$$\mathbf{Pr}[\cap_{i \in S} \mathcal{E}_i] = \prod_{i \in S} \mathbf{Pr}[\mathcal{E}_i].$$

These events are said to be *k-wise independent* if every subcollection consisting of k events is independent. The special case of 2-wise independence is often referred to as *pairwise independence.*

Equivalently, using the definition of conditional expectations, we can say that a collection of events $\{\mathcal{E}_i \mid i \in I\}$ is *independent* if for any $j \in I$ and all subsets $S \subseteq I \setminus \{\mathcal{E}_j\}$,

$$\mathbf{Pr}[\mathcal{E}_j \mid \cap_{i \in S} \mathcal{E}_i] = \mathbf{Pr}[\mathcal{E}_j].$$

In particular, if the events are *pairwise independent* then $\mathbf{Pr}[\mathcal{E}_i \mid \mathcal{E}_j] = \mathbf{Pr}[\mathcal{E}_i]$, for all $i \neq j$.

Usually the events we deal with can be expressed in terms of real-valued functions called *random variables.* The argument of such a function is generally omitted as it always corresponds to a single experiment from the underlying probability space.

► **Definition C.6:** A random variable X is a real-valued function over the sample space, $X : \Omega \to \mathbb{R}$, such that for all $x \in \mathbb{R}$,

$$\{\omega \in \Omega \mid X(\omega) \leq x\} \in \mathbb{F}.$$

This gives us a compact representation of complex events since $\mathbf{Pr}[X \leq x]$ is just another way of denoting $\mathbf{Pr}[\{\omega \in \Omega \mid X(\omega) \leq x\}]$.

► **Definition C.7:** The *distribution function* $F : \mathbb{R} \to [0,1]$ for a random variable X is defined as $F_X(x) = \mathbf{Pr}[X \leq x]$.

A *discrete random variable* is a function over the sample space whose range is either a finite or countably infinite subset of $\mathbb{R}$. Typically, we will be interested in discrete random variables that are integer-valued. An *indicator variable* is a discrete random variable that takes on only the values 0 or 1. An indicator variable X is used to denote the occurrence or non-occurrence of an event $\mathcal{E}$, where $\mathcal{E} = \{\omega \in \Omega \mid X(\omega) = 1\}$ and $\overline{\mathcal{E}} = \{\omega \in \Omega \mid X(\omega) = 0\}$. Observe that the notions of conditional probability and independence carry over to random variables, since they are just another way of denoting events. More precisely, two random variables X and Y are said to be independent if for each $x, y \in \mathbb{R}$, the events $\{X = x\}$ and $\{Y = y\}$ are independent.

A random variable X is said to be *continuous* if it has a distribution function F whose derivative F' is a positive, integrable function. (In other words, F is absolutely continuous.) The function F' is referred to as the *density* function of the random variable X. From here on all random variables are assumed to be discrete, although with some care[1] the following definitions can be extended to continuous random variables.

► **Definition C.8:** The *density function* $p : \mathbb{R} \rightarrow [0,1]$ for a random variable X is defined as $p_X(x) = \mathbf{Pr}[X = x]$.

It is sometimes useful to combine the density or distribution functions for dependent random variables.

► **Definition C.9:** The *joint distribution function* $F_{X,Y} : \mathbb{R} \times \mathbb{R} \rightarrow [0,1]$ for random variables X and Y is defined as

$$F_{X,Y}(x,y) = \mathbf{Pr}[\{X \leq x\} \cap \{Y \leq y\}].$$

The *joint density function* $p_{X,Y} : \mathbb{R} \times \mathbb{R} \rightarrow [0,1]$ for random variables X and Y is defined as

$$p_{X,Y}(x,y) = \mathbf{Pr}[\{X = x\} \cap \{Y = y\}].$$

Thus, $\mathbf{Pr}[Y = y] = \sum_x p(x,y)$, and

$$\mathbf{Pr}[X = x \mid Y = y] = \frac{p(x,y)}{\mathbf{Pr}[Y = y]}.$$

We can now restate the independence of X and Y as requiring that the joint density function be the product of the individual density functions of the two random variables.

► **Definition C.10:** Random variables X and Y are said to be *independent* if for all x, $y \in \mathbb{R}$,

$$p(x,y) = \mathbf{Pr}[X = x]\mathbf{Pr}[Y = y]$$

or, equivalently,

$$\mathbf{Pr}[X = x \mid Y = y] = \mathbf{Pr}[X = x].$$

These definitions extend to a set $X_1, X_2, \ldots$ of more than two random variables, and the notion of k-wise independence can be defined as in Definition C.5.

The following discussion of expectations is in terms of single random variables, but they have the obvious generalizations to functions of multiple random variables using their joint density function.

► **Definition C.11:** The *expectation* of a random variable X with density function p is defined as $\mathbf{E}[X] = \sum_x xp(x)$, where the summation is over the range of X.

[1] Basically all the definitions can be made in terms of the distribution function for the discrete random variable, and then carried over to the continuous case. For example, we say that two continuous random variables X and Y are independent if for each $x, y \in \mathbb{R}$, the events $\{X \leq x\}$ and $\{Y \leq y\}$ are independent.

Note that the expectation may not be well-defined if the summation does not converge absolutely. For any real-valued function $g(X)$, we extend the definition of expectation to $\mathbf{E}[g(X)] = \sum_x g(x)p(x)$. For any two random variables X and Y, $\mathbf{E}[X + Y] = \mathbf{E}[X] + \mathbf{E}[Y]$. The remarkable thing about this property is that it does not assume anything about the independence of the two random variables. In fact, this can be generalized as follows.

Proposition C.5 (Linearity of Expectation): *Let $X_1, \ldots, X_k$ be arbitrary random variables, and $h(X_1, \ldots, X_k)$ a linear function. Then*

$$\mathbf{E}[h(X_1, \ldots, X_k)] = h(\mathbf{E}[X_1], \ldots, \mathbf{E}[X_k]).$$

This does not generalize to nonlinear functions, although with the assumption of independence we can prove a similar result for any polynomial h using the following.

Proposition C.6: *For independent random variables X and Y,*

$$\mathbf{E}[XY] = \mathbf{E}[X]\mathbf{E}[Y].$$

Here are some other useful properties of expectations. We say that random variable X *stochastically dominates* random variable Y if, for all $z \in \mathbb{R}$, $\mathbf{Pr}[X > z] \geq \mathbf{Pr}[Y > z]$.

Proposition C.7: *Let X and Y be random variables with finite expectations.*

1. *If X stochastically dominates Y, then $\mathbf{E}[X] \geq \mathbf{E}[Y]$; equality holds if and only if X, Y are identically distributed.*
2. $|\mathbf{E}[X]| \leq \mathbf{E}[|X|]$.
3. *For a non-negative integer-valued random variable X, $\mathbf{E}[X] = \sum_{x=0}^{\infty} \mathbf{Pr}[X > x]$.*

The density function of a random variable can be characterized in terms of the following expectations.

► **Definition C.12:** For $k \in \mathbb{N}$, the *kth moment* m_X^k and the *kth central moment* μ_X^k of a random variable X are defined as follows:

$$m_X^k = \mathbf{E}[X^k]$$

$$\mu_X^k = \mathbf{E}[(X - \mathbf{E}[X])^k].$$

The expected value of X is sometimes denoted by $\mu_X = m_X^1$. The *variance* of X is denoted $\mathbf{var}[X]$ or σ_X^2 and this is μ_X^2; the *standard deviation* σ_X is the positive square root of the variance.

Proposition C.8: $\mathbf{var}[X] = m_X^2 - (\mu_X)^2 = \mathbf{E}[X^2] - \mathbf{E}[X]^2$.

Note that the next proposition does not generalize to arbitrary linear functions since $\mathbf{var}[cX] = c^2\mathbf{var}[X]$.

Proposition C.9: *For independent random variables X and Y,*

$$\mathbf{var}[X + Y] = \mathbf{var}[X] + \mathbf{var}[Y].$$

The use of generating functions to describe a density function often leads to simplified analysis of the moments.

► **Definition C.13:** Let X be a non-negative integer-valued random variable with the density function p. Then the *probability generating function (pgf)* of X is

$$G_X(z) = \mathbf{E}[z^X] = \sum_{i=0}^{\infty} p(i)z^i.$$

The summation in the definition of $G_X(z)$ always converges for $|z| \leq 1$, and so we assume that the symbolic variable z lies in the interval $[-1, 1]$. The following results can be obtained by suitable differentiation and algebraic manipulation of $G_X(z)$. The reader should keep in mind that the pgf G_X may not be well-defined for all real values of z, but whenever G_X is well-defined, so are its derivatives with respect to z. In the sequel, for a function f, we denote by f' its derivative and by $f^{(k)}$ its kth derivative.

Proposition C.10: *Let X be a non-negative integer-valued random variable with the pgf $G(z)$.*

1. $G(1) = 1$.
2. $\mathbf{E}[X] = G'(1)$.
3. $\mathbf{E}[X^2] = G''(1) + G'(1)$.
4. $\mathbf{var}[X] = G''(1) + G'(1) - G'(1)^2$.

Proposition C.11: *Let $X_1, \ldots, X_k$ be independent random variables with the pgf's $G_1(z), \ldots, G_k(z)$. Then the pgf of the random variable $Y = \sum_{i=1}^{k} X_i$ is given by*

$$G(z) = \prod_{i=1}^{k} G_i(z).$$

Proposition C.12: *Let $X_1, X_2, \ldots$ be a sequence of independent and identically distributed (i.i.d.) random variables with the common pgf $G_X(z)$. If Y is a random variable with the pgf $G_Y(z)$ and Y is independent of all the X_i, then $S = X_1 + X_2 + \cdots + X_Y$ has the pgf*

$$G_S(z) = G_Y(G_X(z)).$$

The following class of generating functions is even more useful, especially since it does not require that X be integer-valued.[1]

► **Definition C.14:** Let X be a random variable with density function p. Then the *moment generating function (mgf)* of X is

$$M_X(z) = \mathbf{E}[e^{zX}].$$

Proposition C.13: *For random variable X with mgf $M(z)$,*

$$\mathbf{E}[X^k] = M^{(k)}(z)\big|_{z=0}.$$

Proposition C.14: *Let $X_1, \ldots, X_k$ be independent random variables with the mgf's $M_1(z), \ldots, M_k(z)$. Then the mgf of the random variable $Y = \sum_{i=1}^{k} X_i$ is given by*

$$M(z) = \prod_{i=1}^{k} M_i(z).$$

We now describe some commonly encountered distributions and enumerate their properties. Note that the mgf for these distributions can be easily obtained since $M_X(z) = G_X(e^z)$ for non-negative integer X. We omit the subscript X in the moments, distribution, density, and generating functions when it is clear from the context that we are referring to the random variable X.

Bernoulli distribution. Suppose we flip a coin whose probability of HEADS is p. Let X be the random variable that has value 1 if the result is HEADS, and 0 otherwise. Then X has the Bernoulli distribution with the parameter p. The density function for X is given by

$$p(x) = \begin{cases} 1-p & \text{if } x = 0 \\ p & \text{if } x = 1 \\ 0 & \text{otherwise.} \end{cases}$$

Let $q = 1 - p$. Then $\mathbf{E}[X] = p$, $\mathbf{var}[X] = pq$, and $G(z) = q + pz$.

Binomial distribution. Let $X_1, X_2, \ldots, X_n$ be i.i.d. random variables whose common distribution is the Bernoulli distribution with parameter p. The random variable $X = X_1 + X_2 + \cdots + X_n$ denotes the number of HEADS in a sequence of n coin flips. The random variable X has the binomial distribution with parameter n and p, sometimes abbreviated $B(n, p)$. The density function is denoted by $b(k; n, p)$, and for integer k with $0 \leq k \leq n$ we have

$$b(k; n, p) = \mathbf{Pr}[X = k] = \binom{n}{k} p^k q^{n-k}.$$

For binomial X, $\mathbf{E}[X] = np$, $\mathbf{var}[X] = npq$, and $G(z) = (q + pz)^n$.

[1] A related class of generating functions is the *characteristic function* $\phi(z) = \mathbf{E}[e^{izX}]$, where $i = \sqrt{-1}$.

Geometric distribution. Suppose we flip a coin repeatedly until HEADS appears for the first time. Assuming that each coin flip has the Bernoulli distribution with parameter p, the random variable X denoting the total number of coin flips has the geometric distribution with parameter p. Its density function is as follows.

$$p(x) = \begin{cases} pq^{x-1} & \text{for } x = 1, 2, 3, \ldots \\ 0 & \text{otherwise.} \end{cases}$$

For geometric X, $\mathbf{E}[X] = 1/p$, $\mathbf{var}[X] = q/p^2$, and $G(z) = pz/(1 - qz)$. An important property of the geometric distribution is its *memorylessness*: let k and l be positive integers; then

$$\mathbf{Pr}[X = k + l \mid X > l] = \mathbf{Pr}[X = k].$$

Thus, knowing that the first l trials were "failures" does not affect the distribution of subsequent trials.

Negative binomial distribution. Let X_1, X_2, ..., X_n be i.i.d. random variables whose common distribution is the geometric distribution with parameter p. The random variable $X = X_1 + X_2 + \cdots + X_n$ denotes the number of coin flips needed to obtain n HEADS. The random variable X has the negative binomial distribution with parameters n and p. The density function for this distribution is defined only for $x = n, n + 1, n + 2, \ldots$:

$$\mathbf{Pr}[X = x] = \binom{x-1}{n-1} p^n q^{x-n}.$$

The characteristics are: $\mathbf{E}[X] = n/p$, $\mathbf{var}[X] = nq/p^2$, and $G(z) = (pz/(1 - qz))^n$.

Poisson distribution. Let λ be a positive real number. Then the Poisson distribution with parameter λ has the following density function.

$$p(x) = \begin{cases} \lambda^x e^{-\lambda}/x! & \text{for } x = 0, 1, 2, \ldots \\ 0 & \text{otherwise.} \end{cases}$$

For large n, the Poisson distribution is a good approximation to the binomial distribution $B(n, \lambda/n)$. The characteristics of a Poisson random variable X are: $\mathbf{E}[X] = \lambda$, $\mathbf{var}[X] = \lambda$, and $G(z) = e^{\lambda(z-1)}$.

References

[1] L. Adleman. Two theorems on random polynomial time. In *Proceedings of the 19th Annual IEEE Symposium on Foundations of Computer Science*, pages 75–83, 1978.

[2] L. Adleman, K. Manders, and G.L. Miller. On taking roots in finite fields. In *Proceedings of the 18th Annual IEEE Symposium on Foundations of Computer Science*, pages 151–163, 1977.

[3] L.M. Adleman and A.M.-D. Huang. Recognizing primes in random polynomial time. In *Proceedings of the 19th Annual ACM Symposium on Theory of Computing*, pages 462–469, 1987.

[4] A. Aggarwal and R.J. Anderson. A random ***NC*** algorithm for depth first search. *Combinatorica*, 8:1–12, 1988.

[5] A.V. Aho, J.E. Hopcroft, and J.D. Ullman. *The Design and Analysis of Computer Algorithms.* Addison-Wesley, Reading, MA, 1974.

[6] A.V. Aho, J.E. Hopcroft, and J.D. Ullman. *Data Structures and Algorithms.* Addison-Wesley, Reading, MA, 1983.

[7] R.K. Ahuja, T.L. Magnanti, and J.B. Orlin. *Network Flows: Theory, Algorithms, and Applications.* Prentice-Hall, Englewood Cliffs, NJ, 1993.

[8] M. Ajtai, J. Komlós, and E. Szemerédi. Sorting in $c \log n$ parallel steps. *Combinatorica*, 3(1):1–19, 1983.

[9] M. Ajtai, J. Komlós, and E. Szemerédi. Determinsitic simulation in logspace. In *Proceedings of the 19th Annual ACM Symposium on Theory of Computing*, pages 132–140, 1987.

[10] S. Albers. Improved randomized on-line algorithms for the list update problem. To appear in the 1995 ACM-SIAM Symposium on Discrete Algorithms.

[11] D.J. Aldous. Random walks on finite groups and rapidly mixing Markov chains. In *Seminaire de Probabilites*, volume 986 of *Springer-Verlag Lecture Notes in Mathematics XVII*, pages 243–297. Springer-Verlag, New York, 1981–82.

[12] D.J. Aldous. *Probability Approximations via the Poisson Clumping Heuristic.* Springer-Verlag, New York, 1989.

[13] D.J. Aldous. Reversible Markov chains and random walks on graphs, 1994. Unpublished Monograph, Berkeley.

[14] R. Aleliunas. Randomized parallel communication. In *ACM-SIGOPS Symposium on Principles of Distributed Systems*, pages 60–72, 1982.

[15] R. Aleliunas, R.M. Karp, R.J. Lipton, L. Lovász, and C. Rackoff. Random walks,

universal traversal sequences, and the complexity of maze problems. In *Proceedings of the 20th Annual Symposium on Foundations of Computer Science*, pages 218–223, San Juan, Puerto Rico, October 1979.

[16] W.R. Alford, A. Granville, and C. Pomerance. There are infinitely many Carmichael numbers. *University of Georgia Mathematics Preprint Series*, 1992.

[17] N. Alon. Eigenvalues and expanders. *Combinatorica*, 6(2):83–96, 1986.

[18] N. Alon. A parallel algorithmic version of the local lemma. In *32nd Annual IEEE Symposium on Foundations of Computer Science*, pages 586–593, 1991.

[19] N. Alon, L. Babai, and A. Itai. A fast and simple randomized algorithm for the maximal independent set problem. *Journal of Algorithms*, 7:567–583, 1986.

[20] N. Alon and F.R.K. Chung. Explicit construction of linear sized tolerant networks. *Discrete Mathematics*, 72:15–19, 1988.

[21] N. Alon, Z. Galil, and O. Margalit. On the exponent of the all pairs shortest path problem. In *Proceedings of the 32nd Annual IEEE Symposium on Foundations of Computer Science*, pages 569–575, 1991.

[22] N. Alon, Z. Galil, O. Margalit, and M. Naor. Witnesses for boolean matrix multiplication and for shortest paths. In *Proceedings of the 33rd Annual IEEE Symposium on Foundations of Computer Science*, pages 417–426, 1992.

[23] N. Alon and V.D. Milman. Eigenvalues, expanders and superconcentrators. In *Proceedings of the 25th Annual IEEE Symposium on Foundations of Computer Science*, 1984.

[24] N. Alon and J. Spencer. *The Probabilistic Method.* Wiley Interscience, New York, 1992.

[25] H. Alt, L.J. Guibas, K. Mehlhorn, R.M. Karp, and A. Wigderson. A method for obtaining randomized algorithms with small tail probabilites. Technical Report TR-91-057, International Computer Science Institute, Berkeley, 1991.

[26] I. Althöfer. On sparse approximations to randomized strategies and convex combinations. *Linear Algebra and its Applications*, 199:339–355, 1994.

[27] D. Angluin. Lecture notes on the complexity of some problems in number theory. Technical Report 243, Department of Computer Science, Yale University, 1982.

[28] D. Angluin and L.G. Valiant. Fast probabilistic algorithms for Hamiltonian circuits and matchings. *Journal of Computer and System Sciences*, 19:155–193, 1979.

[29] N.C. Ankney. The least quadratic nonresidue. *Annals of Mathematics*, 55:65–72, 1986.

[30] C.R. Aragon and R.G. Seidel. Randomized search trees. In *Proceedings of the 30th Annual IEEE Symposium on Foundations of Computer Science*, pages 540–545, 1989.

[31] S. Arora. *Probabilistic Checking of Proofs and Hardness of Approximation Problems.* PhD thesis, University of California at Berkeley, 1994.

[32] S. Arora, C. Lund, R. Motwani, M. Sudan, and M. Szegedy. Proof verification and hardness of approximation problems. In *Proceedings of the 33rd Annual IEEE Symposium on Foundations of Computer Science*, pages 14–23, 1992.

[33] S. Arora and S. Safra. Probabilistic checking of proofs: A new characterization of ***NP***. In *Proceedings of the 33rd Annual IEEE Symposium on Foundations of Computer Science*, pages 2–13, 1992.

[34] J. Aspnes and O. Waarts. Randomized consensus in expected $O(n \log^2 n)$ operations per processor. In *Proceedings of the 33rd Annual IEEE Symposium on Foundations of Computer Science*, pages 137–146, 1992.

[35] Y. Azar, A.Z. Broder, A.R. Karlin, and E. Upfal. Balanced allocations. In *Proceedings of the 26th Annual ACM Symposium on Theory of Computing*, pages

593–602, 1994.

[36] K. Azuma. Weighted sums of certain dependent random variables. *Tohoku Mathematical Journal*, 19:357–367, 1967.

[37] L. Babai. Monte-Carlo algorithms in graph isomorphism testing. Technical Report DMS 79-10, Département de Mathématique et de Statistique, Université de Montréal, 1979.

[38] L. Babai. Trading group theory for randomness. In *Proceedings of the 17th Annual ACM Symposium on Theory of Computing*, pages 421–429, 1985.

[39] L. Babai. E-mail and the unexpected power of interaction. In *Proceedings of the 5th Annual Conference on Structure in Complexity Theory*, pages 30–44, 1990.

[40] L. Babai. Transparent (holographic) proofs. In *Proceedings 10th Annual Symposium on Theoretical Aspects of Computer Science*, pages 525–534, 1993.

[41] L. Babai and L. Fortnow. Arithmetization: a new method in structural complexity theory. *Computational Complexity*, 1:41–66, 1991.

[42] L. Babai, L. Fortnow, L. Levin, and M. Szegedy. Checking computations in polylogarithmic time. In *Proceedings of the 23rd Annual ACM Symposium on Theory of Computing*, pages 21–31, 1991.

[43] L. Babai, L. Fortnow, and C. Lund. Non-deterministic exponential time has two-prover interactive protocols. *Computational Complexity*, 1:3–40, 1991.

[44] E. Bach. Number-theoretic algorithms. *Annual Review of Computer Science*, 4:119–172, 1990.

[45] A. Bar-Noy, R. Motwani, and J. Naor. The greedy algorithm is optimal for on-line edge coloring. *Information Processing Letters*, 44:251–253, 1992.

[46] I. Bárány and Z. Füredi. Computing the volume is difficult. *Discrete and Computational Geometry*, 2:319–326, 1987.

[47] D. Beaver and J. Feigenbaum. Hiding instances in multioracle queries. In *Proceedings of the 7th Annual Symposium on Theoretical Aspects of Computer Science*, Lecture Notes in Computer Science, pages 37–48. Springer-Verlag, New York, 1990.

[48] J. Beck. An algorithmic approach to the Lovász local lemma I. *Random Structures and Algorithms*, pages 343–365, 1991.

[49] L.A. Belady. A study of replacement algorithms for virtual storage computers. *IBM Systems Journal*, 5:78–101, 1966.

[50] M. Bellare and M. Sudan. Improved non-approximability results. In *Proceedings of the 26th Annual ACM Symposium on Theory of Computing*, pages 184–193, 1994.

[51] S. Ben-David, A. Borodin, R.M. Karp, G. Tardos, and A. Wigderson. On the power of randomization in on-line algorithms. *Algorithmica*, 11(1):2–14, 1994.

[52] M. Ben-Or. Probabilistic algorithms in finite fields. In *Proceedings of the 22nd Annual IEEE Symposium on Foundations of Computer Science*, pages 394–398, 1981.

[53] M. Ben-Or, S. Goldwasser, J. Kilian, and A. Wigderson. Multi-prover interactive proofs: How to remove intractability assumptions. In *Proceedings of the 20th Annual ACM Symposium on Theory of Computing*, pages 113–131, 1988.

[54] S.W. Bent and J.W. John. Finding the median requires $2n$ comparisons. In *Proceedings of the 17th ACM Annual Symposium on Theory of Computing*, pages 213–216, 1985.

[55] B. Berger and J. Rompel. Simulating $(\log^c n)$-wise independence in $\mathcal{NC}$. *Journal of the ACM*, 38:1026–1046, 1991.

[56] S.J. Berkowitz. On computing the determinant in small parallel time using a small number of processors. *Information Processing Letters*, 18:147–150, 1984.

[57] E.R. Berlekamp. Factoring polynomials over large finite fields. *Mathematics of*

Computation, 24:713–735, 1970.

[58] D. Bertsimas and R. Vohra. Linear programming relaxations, approximation algorithms and randomization: a unified view of covering problems. Technical Report OR 285-94, MIT, 1994.

[59] F. Bien. Constructions of telephone networks by group representations. *Notices of the American Mathematical Society*, 36:5–22, 1989.

[60] N. Biggs. *Algebraic Graph Theory*. Cambridge University Press, 1974.

[61] P. Billingsley. *Probability and Measure*. John Wiley, New York, 1979.

[62] A. Blum, H.J. Karloff, Y. Rabani, and M. Saks. A decomposition theorem and bounds for randomized server problems. In *Proceedings of the 33rd Annual IEEE Symposium on Foundations of Computer Science*, pages 197–207, 1992.

[63] A. Blum, P. Raghavan, and B. Schieber. Navigating in unfamiliar geometric terrain. In *Proceedings of the 23rd Annual ACM Symposium on Theory of Computing*, pages 494–504, 1991.

[64] M. Blum, A.K. Chandra, and M.N. Wegman. Equivalence of free Boolean graphs can be decided probabilistically in polynomial time. *Information Processing Letters*, 10:80–82, 1980.

[65] M. Blum, R.W. Floyd, V. Pratt, R.L. Rivest, and R.E. Tarjan. Time bounds for selection. *Journal of Computer and System Sciences*, 7:448–461, 1973.

[66] M. Blum and S. Kannan. Designing programs that check their work. In *Proceedings of the 21st Annual ACM Symposium on Theory of Computing*, pages 86–97, 1989.

[67] M. Blum, R.M. Karp, O. Vornberger, C.H. Papadimitriou, and M. Yannakakis. The complexity of testing whether a graph is a superconcentrator. *Information Processing Letters*, 13:164–167, 1981.

[68] M. Blum, M. Luby, and R. Rubinfeld. Self-testing/correcting with applications to numerical problems. In *Proceedings of the 22nd Annual ACM Symposium on Theory of Computing*, pages 73–83, 1990.

[69] B. Bollobás. *Random Graphs*. Academic Press, New York, 1985.

[70] B. Bollobás. The chromatic number of random graphs. *Combinatorica*, 8:49–55, 1988.

[71] J. A. Bondy and U.S.R. Murty. *Graph Theory With Applications*. American Elsevier, New York, 1977.

[72] R.B. Boppana, J. Håstad, and S. Zachos. Does co-***NP*** have short interactive proofs? *Information Processing Letters*, 25:127–133, 1987.

[73] R.B. Boppana and R. Hirschfeld. Pseudo-random generators and complexity classes. In S. Micali, editor, *Randomness and Computing (Advances in Computing Research)*, volume 5, pages 1–26. JAI Press, Greenwich, CT, 1989.

[74] A. Borodin, S.A. Cook, P.W. Dymond, W.L. Ruzzo, and M. Tompa. Two applications of inductive counting for complementation problems. *SIAM Journal on Computing*, 18(3):559–578, June 1989. See also 18(6): 1283, December 1989.

[75] A. Borodin and J.E. Hopcroft. Routing, merging, and sorting on parallel models of computation. *Journal of Computer and System Sciences*, 30:130–145, 1985.

[76] A. Borodin, N. Linial, and M. Saks. An optimal online algorithm for metrical task systems. *Journal of the ACM*, 39:745–763, 1992.

[77] A. Borodin, P. Raghavan, B. Schieber, and E. Upfal. How much can hardware help routing? In *Proceedings of the 25th Annual ACM Symposium on Theory of Computing*, pages 573–582, 1993.

[78] A. Borodin, W.L. Ruzzo, and M. Tompa. Lower bounds on the length of universal traversal sequences. *Journal of Computer and System Sciences*, 45(2):180–203, October 1992.

[79] A. Borodin, J. von zur Gathen, and J.E. Hopcroft. Fast parallel matrix and gcd computations. *Information and Computation*, 32:251–264, 1986.

[80] O. Borůvka. O jistém problému minimálním. *Práca Moravské Přírodovědecké Společnosti*, 3:37–58, 1926.

[81] D.P. Bovet and P. Crescenzi. *Introduction to the Theory of Complexity*. Prentice-Hall, Englewood Cliffs, NJ, 1994.

[82] R.S. Boyer and J.S. Moore. A fast string searching algorithm. *Communications of the ACM*, 20(10), 1977.

[83] A.Z. Broder. How hard is it to marry at random? In *Proceedings of the 18th Annual ACM Symposium on Theory of Computing*, pages 50–58, May 1986.

[84] A.Z. Broder, A.M. Frieze, and E. Upfal. Existence and construction of edge disjoint paths on expander graphs. In *Proceedings of the 24th Annual ACM Symposium on Theory of Computing*, pages 140–149, 1992.

[85] A.Z. Broder and A.R. Karlin. Bounds on covering times. In *29th Annual Symposium on Foundations of Computer Science*, pages 479–487, White Plains, NY, October 1988.

[86] G. Buffon. Essai d'arithmétique morale. *Supplément à l'Histoire Naturelle*, 4, 1777.

[87] R.D. Carmichael. On composite numbers which satisfy the Fermat congruence. *Americal Mathematical Monthly*, 19:22–27, 1912.

[88] J.L. Carter and M.N. Wegman. Universal classes of hash functions. *Journal of Computer and System Sciences*, 18(2):143–154, 1979.

[89] A.K. Chandra, P. Raghavan, W.L. Ruzzo, R. Smolensky, and P. Tiwari. The electrical resistance of a graph captures its commute and cover times. In *Proceedings of the 21st Annual ACM Symposium on Theory of Computing*, pages 574–586, Seattle, May 1989.

[90] B. Chazelle and H. Edelsbrunner. An optimal algorithm for intersecting line segments in the plane. *Journal of the ACM*, 39:1–54, 1992.

[91] B. Chazelle and J. Friedman. A deterministic view of random sampling and its use in geometry. *Combinatorica*, 10(3):229–249, 1990.

[92] B. Chazelle and J. Friedman. Point location among hyperplanes and undirectional ray-shooting. *Computational Geometry: Theory and Applications*, 4:53–62, 1994.

[93] H. Chernoff. A measure of asymptotic efficiency for tests of a hypothesis based on the sum of observations. *Annals of Mathematical Statistics*, 23:493–509, 1952.

[94] L.P. Chew. Building Voronoi diagrams for convex polygons in linear expected time. Report, Department of Mathematics and Computer Science, Dartmouth College, Hanover, NH, 1985.

[95] A.L. Chistov. Fast parallel calculation of the rank of matrices over a field of arbitrary characteristic. In *Proceedings of the International Conference on the Foundations of Computation Theory, Springer-Verlag Lecture Notes in Computer Science, 199*, pages 63–69, 1985.

[96] B. Chor and C. Dwork. Randomization in Byzantine agreement. In S. Micali, editor, *Randomness and Computing (Advances in Computing Research, vol. 5)*, pages 443–497. JAI Press, Greenwich, CT, 1989.

[97] B. Chor and O. Goldreich. On the power of two-point sampling. *Journal of Complexity*, 5:96–106, 1989.

[98] M. Chrobak, H.J. Karloff, T. Payne, and S. Vishwanathan. New results on server problems. In *Proceedings of the 1st Annual ACM-SIAM Symposium on Discrete Algorithms*, pages 291–300, 1990.

[99] M. Chrobak and L.L. Larmore. HARMONIC is 3-competitive for 2 servers. *Theoretical Computer Science*, 98:339–346, May 1992.

[100] V. Chvátal. *Linear Programming.* W. H. Freeman, New York, 1983.

[101] K.L. Clarkson. A probabilistic algorithm for the post office problem. In *Proceedings of the 17th Annual ACM Symposium on Theory of Computing*, pages 175–184, 1985.

[102] K.L. Clarkson. New applications of random sampling in computational geometry. *Discrete and Computational Geometry*, 2:195–222, 1987.

[103] K.L. Clarkson. Applications of random sampling in computational geometry, II. In *Proceedings of the 4th Annual ACM Symposium on Computational Geometry*, pages 1–11, 1988.

[104] K.L. Clarkson. A Las Vegas algorithm for linear programming when the dimension is small. In *Proceedings of the 29th Annual IEEE Symposium Foundations of Computer Science*, pages 452–456, 1988.

[105] K.L. Clarkson. A randomized algorithm for closest-point queries. *SIAM Journal on Computing*, 17:830–847, 1988.

[106] K.L. Clarkson and P.W. Shor. Algorithms for diametrical pairs and convex hulls that are optimal, randomized, and incremental. In *Proceedings of the 4th Annual ACM Symposium on Computational Geometry*, pages 12–17, 1988.

[107] K.L. Clarkson and P.W. Shor. Applications of random sampling in computational geometry, II. *Discrete and Computational Geometry*, 4:387–421, 1989.

[108] A. Cohen and A. Wigderson. Dispersers, deterministic amplification, and weak random sources. In *Proceedings of the 30th Annual IEEE Symposium on Foundations of Computer Science*, pages 14–19, 1989.

[109] C.J. Colbourn. *The Combinatorics of Network Reliability.* Oxford University Press, New York, 1987.

[110] R. Cole. Parallel merge sort. *SIAM Journal on Computing*, 17(4):770–785, 1988.

[111] S.A. Cook. A taxonomy of problems with fast parallel algorithms. *Information and Control*, 64(1–3):2–22, 1985.

[112] D. Coppersmith, P. Doyle, P. Raghavan, and M. Snir. Random walks on weighted graphs, and applications to on-line algorithms. *Journal of the ACM*, 40:454–476, 1993.

[113] D. Coppersmith and S. Winograd. Matrix multiplication via arithmetic progressions. *Journal of Symbolic Computation*, 9:251–280, 1990.

[114] T. Cormen, C.E. Leiserson, and R.L. Rivest. *Introduction to Algorithms.* MIT Press and McGraw Hill, New York, 1990.

[115] L. Csanky. Fast parallel matrix inversion algorithms. *SIAM Journal on Computing*, 5:618–623, 1976.

[116] J.H. Curtiss. Monte Carlo method. *National Bureau of Standards Applied Mathematics Series*, 12, 1951.

[117] D.M. Cvetkovic, M. Doob, and H. Sachs. *Spectra of Graphs.* Academic Press, New York, 1979.

[118] P. Dagum, M. Mihail, M. Luby, and U.V. Vazirani. Polytopes, permanents and graphs with large factors. In *Proceedings of the 29th Annual IEEE Symposium on Foundations of Computer Science*, pages 412–422, 1988.

[119] G.B. Dantzig. Minimization of a linear function of variables subject to linear inequalities. In T.C. Koopman, editor, *Activity Analysis of Production and Allocation*, pages 339–347. John Wiley, New York, 1951.

[120] G.B. Dantzig. *Linear Programming and Extensions.* Princeton University Press, Princeton, NJ, 1963.

[121] H. Davenport. *Multiplicative Number Theory.* Springer-Verlag, New York, 1980.

[122] K. de Leeuw, E.F. Moore, C.E. Shannon, and N. Shapirio. Computability by probabilistic machines. In C.E. Shannon and J. McCarthy, editors, *Automata*

Studies, pages 183–212. Princeton University Press, Princeton, NJ, 1955.

[123] R.A. DeMillo and R.J. Lipton. A probabilistic remark on algebraic program testing. *Information Processing Letters*, 7:193–195, 1978.

[124] M. Dietzfelbinger, A. Karlin, K. Mehlhorn, F. Meyer auf der Heide, H. Rohnert, and R.E. Tarjan. Dynamic perfect hashing: Upper and lower bounds. In *29th Annual IEEE Symposium on Foundations of Computer Science*, pages 524–531, 1988.

[125] E.W. Dijkstra. A note on two problems in connection with graphs. *Numerische Mathematik*, 1:83–89, 1976.

[126] I.H. Dinwoodie. A probability inequality for the occupation measure of a reversible Markov chain. Unpublished manuscript, Department of Mathematics, Tulane University, 1994.

[127] B. Dixon, M. Rauch, and R. E. Tarjan. Verification and sensitivity analysis of minimum spanning trees in linear time. *SIAM Journal on Computing*, 21:1184–1192, 1992.

[128] W.E. Donath and A.J. Hoffman. Lower bounds for the partitioning of graphs. *IBM Journal of Research and Development*, 17:420–425, 1973.

[129] J.L. Doob. *Stochastic Processes.* John Wiley, New York, 1953.

[130] P.G. Doyle and J.L. Snell. *Random Walks and Electric Networks.* The Mathematical Association of America, 1984.

[131] L.E. Dubins and L.J.Savage. *How to Gamble If You Must.* McGraw Hill, New York, 1965.

[132] M. Dyer, A. Frieze, and R. Kannan. A random polynomial algorithm for approximating the volume of convex bodies. *Journal of the ACM*, pages 1–17, 1991.

[133] H. Edelsbrunner. *Algorithms in Combinatorial Geometry*, volume 10 of *EATCS Monographs on Theoretical Computer Science.* Springer-Verlag, Heidelberg, West Germany, 1987.

[134] J. Edmonds. Systems of distinct representatives and linear algebra. *Journal of Research of the National Bureau of Standards, 71B*, 4:241–245, 1967.

[135] R. El-Yaniv, A. Fiat, R.M. Karp, and G. Turpin. Competitive analysis of financial games. In *Proceedings of the 33rd Annual IEEE Symposium on Foundations of Computer Science*, pages 327–333, October 1992.

[136] P. Elias, A. Feinstein, and C. E. Shannon. Note on maximum flow through a network. *IRE Transactions on Information Theory*, IT-2:117–199, 1956.

[137] P. Erdös and L. Lovász. Problems and results on 3-chromatic hypergraphs and some related questions. In A. Hajnal et al., editor, *Infinite and Finite Sets*, pages 609–628. North-Holland, Amsterdam, 1975.

[138] P. Erdös and J.L. Selfridge. On a combinatorial game. *Journal of Combinatorial Theory, Series A*, 14:298–301, 1973.

[139] P. Erdös and J. Spencer. *The Probabilistic Method in Combinatorics.* Academic Press, San Diego, 1974.

[140] T. Feder and R. Motwani. Clique partitions, graph compression and speeding-up algorithms. In *Proceedings of the 25th Annual ACM Symposium on Theory of Computing*, pages 123–133, 1991.

[141] U. Feige, S. Goldwasser, L. Lovász, S. Safra, and M. Szegedy. Approximating clique is almost ***NP***-complete. In *Proceedings of the 32nd Annual Symposium on Foundations of Computer Science*, pages 2–12, 1991.

[142] W. Feller. *An Introduction to Probability Theory and Its Applications*, volume I. John Wiley, New York, 1968.

[143] W. Feller. *An Introduction to Probability Theory and Its Applications*, volume II. John Wiley, New York, 1968.

[144] A. Fiat, D.P. Foster, H.J. Karloff, Y. Rabani, Y. Ravid, and S. Vishwanathan. Competitive algorithms for layered graph traversal. In *Proceedings of the 32nd Annual IEEE Symposium on Foundations of Computer Science*, pages 288–297, 1991.

[145] A. Fiat, R.M. Karp, M. Luby, L. A. McGeoch, D.D. Sleator, and N. Young. Competitive paging algorithms. *Journal of Algorithms*, 12:685–699, 1991.

[146] A. Fiat, Y. Rabani, and Y. Ravid. Competitive k-server algorithms. In *Proceedings of the 31st Annual IEEE Symposium on Foundations of Computer Science*, pages 454–463, 1990.

[147] F.E. Fich, F. Meyer auf der Heide, P.L. Ragde, and A. Wigderson. One, two, three ... infinity: Lower bounds for parallel computation. In *Proceedings of the 17th Annual ACM Symposium on Theory of Computing*, pages 48–58, 1985.

[148] M.J. Fischer and N.A. Lynch. A lower bound for the time to assure interactive consistency. *Information Processing Letters*, 14:183–186, 1982.

[149] M.J. Fischer, N.A. Lynch, and M.S. Paterson. Impossibility of distributed consensus with one faulty process. *Journal of the ACM*, 32:374–382, 1985.

[150] R. W. Floyd. Algorithm 97: Shortest path. *Communications of the ACM*, 5:345, 1962.

[151] R. W. Floyd and R.L. Rivest. Expected time bounds for selection. *Communications of the ACM*, 18:165–172, 1975.

[152] L.R. Ford and D.R. Fulkerson. Maximal flow through a network. *Canadian Journal of Mathematics*, 8:399–404, 1956.

[153] L. Fortnow, J. Rompel, and M. Sipser. On the power of multi-prover interactive protocols. In *Proceedings of the 3rd Annual Conference on Structure in Complexity Theory*, pages 156–161, 1988.

[154] M. Fredman and D. E. Willard. Trans-dichotomous algorithms for minimum spanning trees and shortest paths. In *Proceedings of the 31st Annual IEEE Symposium on Foundations of Computer Science*, pages 719–725, 1990.

[155] M.L. Fredman and J. Komlós. On the size of separating systems and families of perfect hash functions. *SIAM Journal on Algebraic and Discrete Methods*, 5:61–68, 1984.

[156] M.L. Fredman, J. Komlós, and E. Szemerédi. Storing a sparse table with O(1) worst case access time. *Journal of the ACM*, 31:538–544, July 1984.

[157] R. Freivalds. Probabilistic machines can use less running time. In B. Gilchrist, editor, *Information Processing 77, Proceedings of IFIP Congress 77*, pages 839–842. North-Holland, Amsterdam, 1977.

[158] O. Gabber and Z. Galil. Explicit construction of linear-sized superconcentrators. *Journal of Computer and System Sciences*, 22:407–420, 1981.

[159] H.N. Gabow, Z. Galil, T. Spencer, and R.E. Tarjan. Efficient algorithms for finding minimum spanning trees in undirected and directed graphs. *Combinatorica*, 6:109–122, 1986.

[160] H.N. Gabow, Z. Galil, and T.H. Spencer. Efficient implementation of graph algorithms using contraction. In *Proceedings of the 25th Annual IEEE Symposium on Foundations of Computer Science*, pages 347–357, 1984.

[161] D. Gale and L. S. Shapley. College admissions and the stability of marriage. *American Mathematical Monthly*, 69:6–15, 1962.

[162] Z. Galil and V. Pan. Improved processor bounds for combinatorial problems in ***RNC***. *Combinatorica*, 8:189–200, 1988.

[163] B. Gärtner. A subexponential algorithm for abstract optimization problems. In *Proceedings of the 33rd Annual IEEE Symposium on Foundations of Computer Science*, pages 464–472, 1992.

In *Proceedings of the 35th Annual IEEE Symposium on Foundations of Computer Science*, pages 502–510, 1994.

[165] P. Gemmell, R. Lipton, R. Rubinfeld, M. Sudan, and A. Wigderson. Self-testing/correcting for polynomials and for approximate functions. In *Proceedings of the 23nd Annual ACM Symposium on Theory of Computing*, pages 32–42, 1991.

[166] J. Gill. Computational complexity of probabilistic Turing machines. *SIAM Journal on Computing*, 6(4):675–695, December 1977.

[167] D. Gillman. A Chernoff bound for random walks on expander graphs. In *34th Annual IEEE Symposium on Foundations of Computer Science*, pages 680–691, 1994.

[168] F. Göbel and A.A. Jagers. Random walks on graphs. *Stochastic Processes and their Applications*, 2:311–336, 1974.

[169] M.X. Goemans and D.P. Williamson. New 3/4-approximation algorithms for MAX SAT. To appear in the *SIAM Journal on Discrete Mathematics*, 1993.

[170] M.X. Goemans and D.P. Williamson. 0.878-approximation algorithms for MAX-CUT and MAX-2SAT. In *Proceedings of the 26th Annual ACM Symposium on Theory of Computing*, pages 422–431, 1994.

[171] A.V. Goldberg and R.E. Tarjan. A new approach to the maximum flow problem. *Journal of the ACM*, 35:921–940, 1988.

[172] A.V. Goldberg, S.A. Plotkin, and P.M. Vaidya. Sublinear-time parallel algorithms for matching and related problems. *Journal of Algorithms*, 14:180–213, 1993.

[173] M. Goldberg and T. Spencer. A new parallel algorithm for the maximal independent set problem. In *Proceedings of the 28th Annual Symposium on Foundations of Computer Science*, pages 161–165, 1987.

[174] O. Goldreich. A taxonomy of proof systems. 1. *SIGACT News*, 24:2–13, 1993.

[175] O. Goldreich. A taxonomy of proof systems. 2. *SIGACT News*, 25:22–30, 1994.

[176] O. Goldreich, S. Micali, and A. Wigderson. Proofs that yield nothing but their validity or all languages in ***NP*** have zero-knowledge proof systems. *JACM*, 38:691–729, 1991.

[177] M. Goldwasser. Linear programming in randomized subexponential time. Unpublished manuscript, Computer Science Department, Stanford University, 1993.

[178] S. Goldwasser and J. Kilian. Almost all primes can be quickly certified. In *Proceedings of the 18th Annual ACM Symposium on Theory of Computing*, pages 316–329, May 1986.

[179] S. Goldwasser, S. Micali, and C. Rackoff. The knowledge complexity of interactive proof-systems. *SIAM Journal on Computing*, 18:186–208, 1989.

[180] R.E. Gomory and T.C. Hu. Multi-terminal network flows. *SIAM Journal*, 9:551–570, 1961.

[181] R. L. Graham and P. Hell. On the history of the minimum spanning tree problem. *Annals of the History of Computing*, 7:43–57, 1985.

[182] R.L. Graham, D.E. Knuth, and O. Patashnik. *Concrete Mathematics*. Addison-Wesley, Reading, MA, 1989.

[183] D.H. Greene and D.E. Knuth. *Mathematics for the Analysis of Algorithms*. Birkhäuser, Boston, 1990.

[184] D. Grigoriev and M. Karpinski. The matching problem for bipartite graphs with polynomially bounded permanents. In *Proceedings of the 28th Annual IEEE Symposium on Foundations of Computer Science*, pages 166–172, 1987.

[185] G.R. Grimmett and D.R. Stirzaker. *Probability and Random Processes*. Oxford University Press, Oxford, 1988.

[186] E. Grove. The harmonic online k-server algorithm is competitive. In *Proceedings of the 23rd Annual ACM Symposium on Theory of Computing*, pages 260–266, 1991.

[187] L.J. Guibas, D.E. Knuth, and M. Sharir. Randomized incremental construction of Delaunay and Voronoi diagrams. *Algorithmica*, 7:381–413, 1992.

[188] D. Gusfield and R.W. Irving. *The stable marriage problem: structure and algorithms.* MIT Press, Cambridge, 1989.

[189] T. Hagerup and C.Rüb. A guided tour of Chernoff bounds. *Information Processing Letters*, 33:305–308, 1990.

[190] A. Hall. On an experimental determination of π. *Messeng. Math.*, 2:113–114, 1873.

[191] P. Hall and C.C. Heyde. *Martingale Limit Theory and its Application.* Academic Press, New York, 1980.

[192] J. Hao and J.B. Orlin. A faster algorithm for finding the minimum cut in a graph. In *Proceedings of the 3rd Annual ACM-SIAM Symposium on Discrete Algorithms*, pages 165–174, 1993.

[193] F. Harary and E.M. Palmer. *Graphical Enumeration.* Academic Press, New York, 1973.

[194] G.H. Hardy and E.M. Wright. *An Introduction to the Theory of Numbers.* Oxford University Press, London, 1965. 4th Edition.

[195] G.H. Hardy, J.E. Littlewood, and G. Pólya. *Inequalities.* Cambridge University Press, Cambridge, 1989.

[196] J. Håstad, F.T. Leighton, and M. Newman. Reconfiguring a hypercube in the presence of faults. In *Proceedings of the 19th Annual ACM Symposium on Theory of Computing*, pages 274–284, 1987.

[197] D. Haussler and E. Welzl. Epsilon-nets and simplex range queries. *Discrete and Computational Geometry*, 2:127–151, 1987.

[198] R. Hayward and C.J.H. McDiarmid. Average case analysis of heap building by repeated insertion. *Journal of Algorithms*, 12:126–153, 1991.

[199] I.N. Herstein. *Topics in Algebra.* John Wiley, New York, 1964.

[200] C.A.R. Hoare. Algorithm 63 (Partition) and algorithm 65 (Find). *Communications of the ACM*, 4:321–322, 1961.

[201] C.A.R. Hoare. Quicksort. *Computer Journal*, 5:10–15, 1962.

[202] W. Hoeffding. Probability inequalities for sums of bounded random variables. *Journal of the American Statistical Association*, 58:13–30, 1963.

[203] J.E. Hopcroft and R.M. Karp. An $n^{5/2}$ algorithm for maximum matching in bipartite graphs. *SIAM Journal on Computing*, 2:225–231, 1973.

[204] L.K. Hua. *Introduction to Number Theory.* Springer-Verlag, Berlin, 1982.

[205] R. Impagliazzo and D. Zuckerman. How to recycle random bits. In *Proceedings of the 30th Annual IEEE Symposium on Foundations of Computer Science*, pages 222–227, 1989.

[206] S.S. Irani. Two results on the list update problem. *Information Processing Letters*, 38:301–306, 1991.

[207] A. Israeli and Y. Shiloach. An improved parallel algorithm for maximal matching. *Information Processing Letters*, 22:57–60, 1986.

[208] J. JáJá. *An Introduction to Parallel Algorithms.* Addison-Wesley, Reading, MA, 1992.

[209] S. Janson. Large deviation inequalities for sums of indicator variables. Technical Report 34, Department of Mathematics, Uppsala University, 1993.

[210] M. Jerrum and U. Vazirani. A mildly exponential approximation algorithm for the permanent. In *Proceedings of the 33rd Annual IEEE Symposium on Foundations of Computer Science*, pages 320–326, 1992.

[211] M.R. Jerrum and A. Sinclair. Approximating the permanent. *SIAM Journal on Computing*, 18(6):1149–1178, December 1989.

Computing, 18(6):1149–1178, December 1989.

[212] M.R. Jerrum, L.G. Valiant, and V.V. Vazirani. Random generation of combinatorial structures from a uniform distribution. *Theoretical Computer Science*, 43:169–188, 1986.

[213] Wang Jianhua. *The Theory of Games.* Clarendon Press, London, 1988.

[214] A. Joffe. On a set of almost deterministic k-independent random variables. *Annals of Probability*, 2(1):161–162, 1974.

[215] D.B. Johnson. Efficient algorithms for shortest paths in sparse networks. *Journal of the ACM*, 24:1–13, 1977.

[216] D.S. Johnson. Computing in the Math Department: Part I (The ***NP***-completeness column: An ongoing guide). *Journal of Algorithms*, 7:584–601, 1986.

[217] D.S. Johnson. Interactive proof systems for fun and profit (the ***NP***-completeness column: An ongoing guide). *Journal of Algorithms*, 9:426–444, 1988.

[218] D.S. Johnson. The tale of the 2nd prover (the ***NP***-completeness column: An ongoing guide). *Journal of Algorithms*, 13:502–524, 1992.

[219] D.S. Johnson. Approximation algorithms for combinatorial problems. *Journal of Computer and System Sciences*, 9:256–278, 1974.

[220] D.S. Johnson. The ***NP***-completeness column: An ongoing guide. *Journal of Algorithms*, 5:284–299, 1984.

[221] D.S. Johnson. The ***NP***-completeness column: An ongoing guide. *Journal of Algorithms*, 5:433–447, 1984.

[222] N.L. Johnson and S. Kotz. *Urn Models and Their Applications.* John Wiley, New York, 1977.

[223] L.R. Ford Jr. and D.R. Fulkerson. *Flows in networks.* Princeton University Press, Princeton, NJ, 1962.

[224] J.D. Kahn, N. Linial, N. Nisan, and M.E. Saks. On the cover time of random walks in graphs. *Journal of Theoretical Probability*, 2(1):121–128, 1989.

[225] C. Kaklamanis, D. Krizanc, and T. Tsantilas. Tight bounds for oblivious routing in the hypercube. In *Proceedings of the 3rd Annual ACM Symposium on Parallel Algorithms and Architectures*, pages 31–36, 1991.

[226] G. Kalai. A subexponential randomized simplex algorithm. In *Proceedings of the 24th Annual ACM Symposium on Theory of Computing*, pages 475–482, 1992.

[227] G. Kalai and D.J. Kleitman. A quasi-polynomial bound for the diameter of graphs of polyhedra. *Bulletin of the AMS*, 26:315–316, April 1992.

[228] A. Kamath, R. Motwani, K. Palem, and P. Spirakis. Tail bounds for occupancy and the satisfiability threshold conjecture. In *Proceedings of the 35th Annual IEEE Symposium on Foundations of Computer Science*, pages 592–603, 1994.

[229] D.R. Karger. Random sampling in matroids, with applications to graph connectivity and minimum spanning trees. In *Proceedings of the 34th Annual IEEE Symposium on Foundations of Computer Science*, pages 84–93, 1993.

[230] D. Karger, R. Motwani, and M. Sudan. Approximate graph coloring by semidefinite programming. In *Proceedings of the 35th Annual IEEE Symposium on Foundations of Computer Science*, pages 2–13, 1994.

[231] D.R. Karger. Global min-cuts in ***RNC***, and other ramifications of a simple min-cut algorithm. In *Proceedings of the 4th Annual ACM-SIAM Symposium on Discrete Algorithms*, pages 21–30, 1993.

[232] D.R. Karger, P.N. Klein, and R.E. Tarjan. A randomized linear-time algorithm for finding minimum spanning trees. To appear in the *Journal of the ACM*, 1995.

[233] D.R. Karger and R. Motwani. Derandomization through approximation: An ***NC*** algorithm for minimum cuts. In *Proceedings of the 26th Annual ACM Symposium*

[234] D.R. Karger and C. Stein. An $\tilde{O}(n^2)$ algorithm for minimum cuts. In *Proceedings of the 25th Annual ACM Symposium on Theory of Computing*, pages 757–765, 1993.

[235] A.R. Karlin, M.S. Manasse, L. Rudolph, and D.D. Sleator. Competitive snoopy caching. *Algorithmica*, 3(1):70–119, 1988.

[236] A.R. Karlin, M.S. Manasse, L.A. McGeoch, and S. Owicki. Competitive randomized algorithms for non-uniform problems. In *Proceedings of the 1st ACM-SIAM Symposium on Discrete Algorithms*, pages 301–309, 1990.

[237] H.J. Karloff. A Las Vegas ***RNC*** algorithm for maximum matching. *Combinatorica*, 6:387–391, 1986.

[238] H.J. Karloff, Y. Rabani, and Y. Ravid. Lower bounds for randomized server algorithms. In *Proceedings of the 23rd Annual ACM Symposium on Theory of Computing*, pages 278–288, 1991.

[239] N. Karmarkar, R.M. Karp, R. Lipton, L. Lovász, and M. Luby. A Monte Carlo algorithm for estimating the permanent. In preparation.

[240] R.M. Karp, M. Luby, and N. Madras. Monte-Carlo approximation algorithms for enumeration problems. *Journal of Algorithms*, 10:429–448, 1989.

[241] R.M. Karp and V. Ramachandran. Parallel algorithms for shared memory machines. In J. van Leeuwen, editor, *Handbook of Theoretical Computer Science*, pages 869–941. Elsevier/The MIT Press, Amsterdam, 1990.

[242] R.M. Karp, E. Upfal, and A. Wigderson. Constructing a perfect matching is in random ***NC***. *Combinatorica*, 6:35–48, 1986.

[243] R.M. Karp. An introduction to randomized algorithms. *Discrete Applied Mathematics*, 34:165–201, 1991.

[244] R.M. Karp. Probabilistic recurrence relations. In *Proceedings of the 23rd Annual ACM Symposium on Theory of Computing*, pages 190–197, 1991.

[245] R.M. Karp and R. Lipton. Turing machines that take advice. *L'enseignment Mathematique*, 28:191–209, 1982.

[246] R.M. Karp and M. Luby. Monte-Carlo algorithms for enumeration and reliability problems. In *Proceedings of the 24th Annual IEEE Symposium on Foundations of Computer Science*, pages 56–64, 1983.

[247] R.M. Karp and M. Luby. Monte Carlo algorithms for the planar multiterminal network reliability problem. *Journal of Complexity*, 1:45–64, 1985.

[248] R.M. Karp, N. Pippenger, and M. Sipser. A time randomness tradeoff. In *AMS Conference on Probabilistic Computational Complexity*, 1985.

[249] R.M. Karp and M.O. Rabin. Efficient randomized pattern-matching algorithms. *IBM Journal of Research and Development*, 31:249–260, March 1987.

[250] R.M. Karp, E. Upfal, and A. Wigderson. The complexity of parallel search. *Journal of Computer and System Sciences*, 36:225–253, 1988.

[251] R.M. Karp and A. Wigderson. A fast parallel algorithm for the maximal independent set problem. *Journal of the ACM*, 32:762–773, 1985.

[252] A.V. Karzanov and E.A. Timofeev. Efficient algorithm for finding all minimal edge cuts of a non-oriented graph. *Kibernetika*, 22:156–162, 1986. Translation in *Cybernetics 22.*

[253] J.G. Kemeny, J.L. Snell, and A.W. Knapp. *Denumerable Markov Chains.* The University Series in Higher Mathematics. Van Nostrand, Princeton, NJ, 1966.

[254] T. Kimbrel and R. Sinha. A probabilistic algorithm for verifying matrix products using $O(n^2)$ time and $\log_2 n + O(1)$ random bits. *Information Processing Letters*, 45:107–110, 1993.

[255] V. King. A simpler minimum spanning tree verification algorithm. Unpublished manuscript, 1993.

manuscript, 1993.

[256] V. King, S. Rao, and R.E.Tarjan. A faster deterministic maximum flow algorithm. In *Proceedings of the 3rd Annual ACM-SIAM Symposium on Discrete Algorithms*, pages 157–164, 1993.

[257] P.N. Klein and R.E. Tarjan. A randomized linear-time algorithm for finding minimum spanning trees. In *Proceedings of the 26th Annual ACM Symposium on Theory of Computing*, pages 9–15, 1994.

[258] D.E. Knuth. *Fundamental Algorithms*, volume 1 of *The Art of Computer Programming*. Addison-Wesley, Reading, MA, 1969.

[259] D.E. Knuth. *Seminumerical Algorithms*, volume 2 of *The Art of Computer Programming*. Addison-Wesley, Reading, MA, 1971.

[260] D.E. Knuth. *Sorting and Searching*, volume 3 of *The Art of Computer Programming*. Addison-Wesley, Reading, MA, 1973.

[261] D.E. Knuth. Big omicron and big omega and big theta. *SIGACT News*, 8(2):18–24, 1976.

[262] D.E. Knuth, J.H. Morris, Jr., and V.R. Pratt. Fast pattern matching in strings. *SIAM Journal on Computing*, 6(2):323–350, 1977.

[263] D.E. Knuth. *Mariages stables* (in French). Les Presses de l'Université de Montréal, Montreal, 1976.

[264] D.E. Knuth and A. C-C. Yao. The complexity of nonuniform random number generation. In J. F. Traub, editor, *Algorithms and Complexity, Recent Results and New Directions*, pages 375–428. Academic Press, New York, 1976.

[265] K-I. Ko. Some observations on probabilistic algorithms and ***NP***-hard problems. *Information Processing Letters*, 14:39–43, 1981.

[266] V.F. Kolchin, V.P. Chistiakov, and B.A. Sevastianov. *Random Allocations*. V.H. Winston, New York, 1978.

[267] A. Kolmogorov. *Grundbegriffe der Wahrscheinlichkeitsrechnung*. Springer, Berlin, 1933.

[268] J. Komlós. Linear verification for spanning trees. *Combinatorica*, 5:57–65, 1985.

[269] E. Koutsoupias and C.H. Papadimitriou. On the k-server conjecture. In *Proceedings of the 26th Annual ACM Symposium on Theory of Computing*, pages 507–511, 1994.

[270] J.B. Kruskal. On the shortest spanning subtree of a graph and the traveling salesman problem. *Proceedings of the American Mathematical Society*, 7:48–50, 1956.

[271] F.T. Leighton. *Introduction to Parallel Algorithms and Architectures: Arrays, Trees, Hypercubes*. Morgan-Kauffman, San Mateo, CA, 1992.

[272] F.T. Leighton, B. Maggs, and S. Rao. Universal packet routing algorithms. In *Proceedings of the 29th Annual IEEE Symposium on Foundations of Computer Science*, pages 256–269, 1988.

[273] A.K. Lenstra and Jr. H.W. Lenstra. Algorithms in number theory. In J. van Leeuwen, editor, *Handbook of Theoretical Computer Science*, pages 675–715. Elsevier Science Publishers, Amsterdam, 1990.

[274] A. Lev, N. Pippenger, and L.G. Valiant. A fast parallel algorithm for routing in permutation networks. *IEEE Transactions on Computers*, C-30:93–100, 1981.

[275] W.J. LeVeque. *Fundamentals of Number Theory*. Addison-Wesley, Reading, MA, 1977.

[276] H.R. Lewis and C.H. Papadimitriou. Symmetric space-bounded computation. *Theoretical Computer Science*, 19:161–187, 1982.

[277] R.J. Lipton. New directions in testing. In *Distributed Computing and Cryptography*, DIMACS Series in Discrete Mathematics and Theoretical Computer Science,

[278] R.J. Lipton and N. Young. Simple strategies for large zero-sum games with applications to complexity theory. In *Proceedings of the 26th Annual ACM Symposium on Theory of Computing*, pages 734–740, 1994.

[279] L.H. Loomis. On a theorem of von Neumann. *Proceedings of the National Academy of Sciences of the U.S.A.*, 32:213–215, 1946.

[280] L. Lovász. On determinants, matchings and random algorithms. In L. Budach, editor, *Fundamentals of Computing Theory*. Akademia-Verlag, Berlin, 1979.

[281] L. Lovász and M.D. Plummer. *Matching Theory*. Academic Press, New York, 1986.

[282] M. Luby. A simple parallel algorithm for the maximal independent set. *SIAM Journal on Computing*, 15:1036–1053, 1986.

[283] M. Luby. Removing randomness in parallel computation without a processor penalty. In *Proceedings 29th Annual IEEE Symposium on Foundations of Computer Science*, pages 162–173, October 1988.

[284] M. Luby. Removing randomness in parallel computation without a processor penalty. *Journal of Computer and System Sciences*, 47:250–86, 1993.

[285] M. Luby, J. Naor, and M. Naor. On removing randomness from a parallel algorithm for minimum cuts. Technical Report TR-093-007, International Computer Science Institute, Berkeley, CA, 1993.

[286] M. Luby, A. Sinclair, and D. Zuckerman. Optimal speedup of Las Vegas algorithms. *Information Processing Letters*, 47:173–180, 1993.

[287] R. Luce and H. Raiffa. *Games and Decisions*. John Wiley, New York, 1957.

[288] C. Lund, L. Fortnow, H.J. Karloff, and N. Nisan. Algebraic methods for interactive proof systems. In *Proceedings of the 31st Annual IEEE Symposium on Foundations of Computer Science*, pages 2–10, 1990.

[289] F. Maffioli, M.G. Speranza, and C. Vercellis. Randomized algorithms. In M. O'hEigertaigh, J.K. Lenstra, and A.H.G. Rinooy Kan, editors, *Combinatorial Optimization: Annotated Bibliographies*, pages 89–105. John Wiley, New York, 1985.

[290] M.S. Manasse, L.A. McGeoch, and D.D. Sleator. Competitive algorithms for server problems. *Journal of Algorithms*, 11:208–230, 1990.

[291] K. Manders and L. Adleman. ***NP***-complete decision problems for quadratic polynomials. In *Proceedings of the 8th ACM Symposium on Theory of Computing*, pages 23–29, 1976.

[292] G.A. Margulis. Explicit constructions of concentrators. *Problemy Peredachi Informatsii*, pages 71–80, 1973. English translation in *Problems of Information Transmission*, 9:325–332.

[293] A.A. Markov. *Ischislenie veroiâtnosteĭ, 2 ed.* 1912.

[294] J. Matoušek. Derandomization in computational geometry. Submitted for publication, 1994.

[295] J. Matoušek, M. Sharir, and E. Welzl. A subexponential bound for linear programming. In *Proceedings of the 8th Annual ACM Symposium on Computational Geometry*, pages 1–8, 1992.

[296] P.C. Matthews. Covering problems for Brownian motion on spheres. *Annals of Probability*, 16:189–199, 1988.

[297] P.C. Matthews. Covering problems for Markov chains. *Annals of Probability*, 16:1215–1228, 1988.

[298] R.L. Mattison, J. Gecsei, D.R. Slutz, and I.L. Traiger. Evaluation techniques for storage hierarchies. *IBM Systems Journal*, 9(2), 1971.

[299] D.W. Matula. Determining edge connectivity in O(*nm*). In *Proceedings of the 28th Annual IEEE Symposium on Foundations of Computer Science*, pages 249–251,

1987.

[300] B. Maurey. Construction de suites symétriques. *Compt. Rend. Acad. Sci. Paris*, 288:679–681, 1979.

[301] J.C. Maxwell. *A Treatise on Electricity and Magnetism.* Clarendon, London, 1918.

[302] C.J.H. McDiarmid. On the method of bounded differences. In J. Siemons, editor, *Surveys in Combinatorics: Invited Papers at the 12th British Combinatorial Conference*, pages 148–188. Cambridge University Press, 1989.

[303] C.J.H. McDiarmid. On a random recolouring method for graphs and hypergraphs. *Combinatorics, Probability and Computing*, 2:363–365, 1993.

[304] C.J.H. McDiarmid and R. Hayward. Strong concentration for quicksort. In *Proceedings of the 3rd Annual ACM-SIAM Symposium on Discrete Algorithms*, pages 414–421, 1992.

[305] C.J.H. McDiarmid and B.A. Reed. Building heaps fast (data structures). *Journal of Algorithms*, 10:352–365, 1989.

[306] L.A. McGeoch and D.D. Sleator. A strongly competitive randomized paging algorithm. *Algorithmica*, 6:816–825, 1991.

[307] N. Megiddo. Linear programming in linear time when the dimension is fixed. *Journal of the ACM*, 31:114–127, 1984.

[308] S. Micali and V.V. Vazirani. An $O(\sqrt{|V|}|E|)$ algorithm for finding maximum matching in general graphs. In *Proceedings of the 21st Annual IEEE Symposium on Foundations of Computer Science*, pages 17–27, 1980.

[309] M. Mihail. The approximation of the permanent is still open. Manuscript, Harvard University, 1987.

[310] G.L. Miller. Riemann's hypothesis and tests for primality. *Journal of Computer and System Sciences*, 13:300–317, 1976.

[311] D.S. Mitrinović. *Analytic Inequalities.* Springer-Verlag, New York, 1970.

[312] R. Motwani. Expanding graphs and the average-case analysis of algorithms for matchings and related problems. In *Proceedings of the 21st Annual ACM Symposium on Theory of Computing*, pages 550–561, 1989.

[313] R. Motwani, J. Naor, and M. Naor. The probabilistic method yields deterministic parallel algorithms. In *Proceedings of the 30th Annual IEEE Symposium on Foundations of Computer Science*, pages 8–13, October 1989.

[314] R. Motwani, J. Naor, and P. Raghavan. Randomization in approximation algorithms. In D. Hochbaum, editor, *Approximation Algorithms.* To appear, 1995.

[315] K. Mulmuley. A fast planar partition algorithm, I. In *Proceedings 29th IEEE Symposium on Foundations of Computer Science*, pages 580–589, October 1988.

[316] K. Mulmuley. *Computational Geometry: An Introduction Through Randomized Algorithms.* Prentice-Hall, Englewood Cliffs, NJ, 1993.

[317] K. Mulmuley, U.V. Vazirani, and V.V. Vazirani. Matching is as easy as matrix inversion. *Combinatorica*, 7:105–113, 1987.

[318] H. Nagamochi and T. Ibaraki. Computing edge connectivity in multigraphs and capacitated graphs. *SIAM Journal on Discrete Mathematics*, 5:54–66, 1992.

[319] J. Naor and M. Naor. Small-bias probability spaces: efficient constructions and applications. *SIAM Journal on Computing*, 22:838–56, 1993.

[320] N. Nisan. Pseudorandom generators for space-bounded computation. *Combinatorica*, 12:449–461, 1992.

[321] I. Niven and H.S. Zuckerman. *An Introduction to the Theory of Numbers.* John Wiley, New York, 1960.

[322] V. Pan. How to multiply matrices faster. In *Springer-Verlag Lecture Notes in Computer Science 179.* Springer Verlag, New York, 1984.

[323] V. Pan. Fast and efficient algorithms for the exact inversion of integer matrices. In *Proceedings of the Fifth Annual Conference on the Foundations of Software Technology and Theoretical Computer Science*. Springer-Verlag LNCS 206, 1985.

[324] C.H. Papadimitriou. Games against nature. *Journal of Computer and System Sciences*, 31:288–301, 1985.

[325] C.H. Papadimitriou. On selecting a satisfying truth assignment. In *Proceedings of the 32nd Annual IEEE Symposium on Foundations of Computer Science*, pages 163–169, 1991.

[326] C.H. Papadimitriou. *Complexity Theory*. Addison-Wesley, Reading, MA, 1994.

[327] C.H. Papadimitriou and M. Yanakakis. Shortest paths without a map. *Theoretical Computer Science*, 84:127–150, 1991.

[328] M.S. Paterson. Improved sorting networks with O(log n) depth. *Algorithmica*, 5:75–92, 1990.

[329] M.S. Paterson and F.F. Yao. Efficient binary space partitions for hidden surface removal and solid modeling. *Discrete and Computational Geometry*, 5:485–503, 1990.

[330] M. Pease, R. Shostak, and L. Lamport. Reaching agreement in the presence of faults. *Journal of the ACM*, 27:228–234, 1980.

[331] D. Peleg and E. Upfal. A time-randomness tradeoff for oblivious routing. *SIAM Journal on Computing*, 19:256–266, 1990.

[332] S. Phillips and J. Westbrook. Online load balancing and network flow. In *Proceedings of the 25th Annual ACM Symposium on Theory of Computing*, pages 402–411, 1991.

[333] M. Pinsker. On the complexity of a concentrator. In *7th International Teletraffic Conference*, pages 318/1–318/4, 1973.

[334] V.D. Podderyugin. An algorithm for finding the edge connectivity of graphs. *Vopr. Kibernetika*, 2:136, 1973.

[335] V. Pratt. Every prime has a succinct certificate. *SIAM Journal on Computing*, 4:214–220, 1975.

[336] F.P. Preparata and M.I. Shamos. *Computational Geometry: an Introduction*. Springer-Verlag, New York, 1985.

[337] R.C. Prim. Shortest connection networks and some generalizations. *Bell Systems Technical Journal*, 36:1389–1401, 1957.

[338] K. Pruhs and U. Manber. The complexity of controlled selection. *Information and Computation*, 91:103–127, 1991.

[339] W. Pugh. Skip lists: A probabilistic alternative to balanced trees. *Communications of the ACM*, 33(6):668–676, 1990.

[340] M.O. Rabin. Probabilistic automata. *Information and Control*, 6:230–245, 1963.

[341] M.O. Rabin. Probabilistic algorithms. In J.F. Traub, editor, *Algorithms and Complexity, Recent Results and New Directions*, pages 21–39. Academic Press, New York, 1976.

[342] M.O. Rabin. Probabilistic algorithm for testing primality. *Journal of Number Theory*, 12:128–138, 1980.

[343] M.O. Rabin. Probabilistic algorithms in finite fields. *SIAM Journal on Computing*, 9:273–280, 1980.

[344] M.O. Rabin. The choice coordination problem. *Acta Informatica*, 17:121–134, 1982.

[345] M.O. Rabin and J.O. Shallit. Randomized algorithms in number theory. *Communications in Pure and Applied Mathematics*, 39:239–256, 1986.

[346] M.O. Rabin. Digitalized signatures and public-key functions as intractable as

factorization. Technical Report MIT/LCS/TR-212, MIT, January 1979.

[347] M.O. Rabin. Randomized Byzantine generals. In *Proceedings of the 24th Annual Symposium on Foundations of Computer Science*, pages 403–409, 1983.

[348] M.O. Rabin and V.V. Vazirani. Maximum matchings in general graphs through randomization. Technical Report TR-15-84, Aiken Computation Laboratory, Harvard University, 1984.

[349] M.O. Rabin and V.V. Vazirani. Maximum matchings in general graphs through randomization. *Journal of Algorithms*, 10:557–567, 1989.

[350] P. Raghavan. *Randomized Rounding and Discrete Ham-Sandwich Theorems.* PhD thesis, University of California, Berkeley, July 1986.

[351] P. Raghavan. Probabilistic construction of deterministic algorithms: Approximating packing integer programs. *Journal of Computer and System Sciences*, 37:130–143, 1988.

[352] P. Raghavan and M. Snir. Memory versus randomization in on-line algorithms. *IBM Journal of Research and Development*, 38:683–707, 1994.

[353] P. Raghavan and C.D. Thompson. Randomized rounding. *Combinatorica*, 7:365–374, 1987.

[354] J.H. Reif. *Synthesis of Parallel Algorithms.* Morgan-Kauffman Publishers, San Francisco, 1993.

[355] N. Reingold, D.D. Sleator, and J. Westbrook. Randomized competitive algorithms for the list update problem. *Algorithmica*, 11(1):15–32, 1994.

[356] R. Reischuk. Probabilistic parallel algorithms for sorting and selection. *SIAM Journal on Computing*, 14(2):396–409, 1985.

[357] A. Rényi. *Probability Theory.* North-Holland, Amsterdam, 1970.

[358] R.L. Rivest, A. Shamir, and L. Adleman. A method for obtaining digital signatures and public-key cryptosystems. *Communications of the ACM*, 21:120–126, 1978.

[359] F. Romani. Shortest-path problem is not harder than matrix multiplication. *Information Processing Letters*, 11:134–136, 1980.

[360] R. Rubinfeld. *A Mathematical Theory of Self-Checking, Self-Testing and Self-Correcting Programs.* PhD thesis, Computer Science Department, University of California, Berkeley, 1990.

[361] H. Ryser. *Combinatorial Mathematics.* The Mathematical Association of America, 1963.

[362] M. Saks and A. Wigderson. Probabilistic Boolean decision trees and the complexity of evaluating game trees. In *Proceedings of the 27th Annual IEEE Symposium on Foundations of Computer Science*, pages 29–38, Toronto, Ontario, 1986.

[363] J.P. Schmidt, A. Siegel, and A. Srinivasan. Chernoff-Hoeffding bounds for applications with limited independence. In *Proceedings of the 4th Annual ACM-SIAM Symposium on Discrete Algorithms*, pages 331–340, 1993.

[364] A. Schönhage, M. Paterson, and N. Pippenger. Finding the median. *Journal of Computer and System Sciences*, 13:184–199, 1976.

[365] U. Schöning. Graph isomorphism is in the low hierarchy. *Journal of Computer and System Sciences*, 37:312–323, 1988.

[366] A. Schrijver. *Theory of Linear and Integer Programming*. John Wiley, New York, 1986.

[367] J.T. Schwartz. Fast probabilistic algorithms for verification of polynomial identities. *Journal of the ACM*, 27(4):701–717, October 1980.

[368] R.G. Seidel. A simple and fast incremental randomized algorithm for computing trapezoidal decompositions and for triangulating polygons. *Computational Geometry: Theory and Applications*, 1:51–64, 1991.

Geometry: Theory and Applications, 1:51–64, 1991.

[369] R.G. Seidel. Small-dimensional linear programming and convex hulls made easy. *Discrete and Computational Geometry*, 6:423–434, 1991.

[370] R.G. Seidel. On the all-pairs-shortest-path problem. In *Proceedings of the 24th Annual ACM Symposium on Theory of Computing*, pages 745–749, 1992.

[371] R.G. Seidel. Backwards analysis of randomized geometric algorithms. In J. Pach, editor, *New Trends in Discrete and Computational Geometry*, volume 10 of *Algorithms and Combinatorics*, pages 37–68. Springer-Verlag, New York, 1993.

[372] A. Shamir. ***IP* = *PSPACE***. *Journal of the JACM*, 39:869–877, 1992.

[373] E. Shamir and J. Spencer. Sharp concentration of the chromatic number on random graphs $G_{n,p}$. *Combinatorica*, 7:121–129, 1987.

[374] M. Sharir and E. Welzl. A combinatorial bound for linear programming and related problems. In *Proceedings of the 9th Symposium on Theoretical Aspects of Computer Science*, volume 577 of *Lecture Notes in Computer Science*, pages 569–579. Springer-Verlag, New York, 1992.

[375] A. Sinclair. *Algorithms for Random Generation and Counting: A Markov Chain Approach.* Progress in Theoretical Computer Science. Birkhauser, Boston, 1992.

[376] A. Sinclair and M.R. Jerrum. Approximate counting, uniform generation and rapidly mixing Markov chains. *Information and Computation*, 82:93–133, 1989.

[377] A.J. Sinclair. Improved bounds for mixing rates of Markov chains and multicommodity flow. *Combinatorics, Probability and Computing*, 1:351–370, 1992.

[378] M. Sipser. Expanders, randomness or time versus space. In *Proceedings of the 1st Structure in Complexity Theory Conference*, page 325, 1986.

[379] D.D. Sleator and R.E. Tarjan. Amortized efficiency of list update and paging rules. *Communications of the ACM*, 28:202–208, February 1985.

[380] D.D. Sleator and R.E. Tarjan. Self-adjusting binary search trees. *Journal of the ACM*, 32:652–686, July 1985.

[381] M. Snir. Lower bounds on probabilistic linear decision trees. *Theoretical Computer Science*, 38:69–82, 1985.

[382] R. Solovay and V. Strassen. A fast Monte-Carlo test for primality. *SIAM Journal on Computing*, 6(1):84–85, March 1977. See also *SIAM Journal on Computing* 7, 1 February 1978, 118.

[383] J. Spencer. Six standard deviations suffice. *Transactions of the American Mathematical Society*, 289(2):679–706, June 1985.

[384] J. Spencer. *Ten Lectures on the Probabilistic Method.* SIAM, Philadelphia, 1987.

[385] R. Sprugnoli. Perfect hashing functions: A single probe retrieving method for static sets. *Communications of the ACM*, 20(11):841–850, 1977.

[386] L.J. Stockmeyer. On approximation algorithms for ***#P***. *SIAM Journal on Computing*, 14:849–861, 1985.

[387] G. Strang. *Linear Algebra and Its Applications.* Harcourt Brace Jovanovich, San Diego, CA, 1988.

[388] M. Sudan. *Efficient Checking of Polynomials and Proofs and the Hardness of Approximation Problems.* PhD thesis, University of California at Berkeley, 1992.

[389] R.M. Tanner. Explicit construction of concentrators from generalized n-gons. *SIAM Journal on Algebraic and Discrete Methods*, 5:287–293, 1984.

[390] R.E. Tarjan. Applications of path compression on balanced trees. *Journal of the ACM*, 26:690–715, 1979.

[391] R.E. Tarjan. *Data Structures and Network Algorithms.* CBMS-NSF Regional Conference Series in Applied Mathematics. SIAM, Philadelphia, 1983.

[392] R.E. Tarjan and A. Yao. Storing a sparse table. *Communications of the ACM*,

[393] M. Tarsi. Optimal search on some game trees. *Journal of the ACM*, 30:389–396, 1983.

[394] P.-L. Tchébyshef. Des valeurs moyennes. *Journal de Mathématiques pures et appliquées, ser. 2*, 12:177–184, 1867.

[395] B. Teia. A lower bound for randomized list update algorithms. *Information Processing Letters*, 47:5–9, 1993.

[396] P. Tetali. Random walks and the effective resistance of networks. *Journal of Theoretical Probability*, pages 101–109, 1991.

[397] A. Treat. Experimental control of ear choice in the moth ear mite. *XI. Internationaler Kongress für Entomologie*, pages 619–621, 1960.

[398] W.T. Tutte. The factorization of linear graphs. *Journal of the London Mathematical Society*, 22:107–111, 1947.

[399] E. Upfal. Efficient schemes for parallel communication. *Journal of the ACM*, 31:507–517, 1984.

[400] L. G. Valiant and G. J. Brebner. Universal schemes for parallel communication. In *Proceedings of the 13th Annual ACM Symposium on Theory of Computing*, pages 263–277, Milwaukee, WI, May 1981.

[401] L.G. Valiant. The complexity of computing the permanent. *Theoretical Computer Science*, 8:189–201, 1979.

[402] L.G. Valiant. The complexity of enumeration and reliability problems. *SIAM Journal on Computing*, 8:410–421, 1979.

[403] L.G. Valiant. A scheme for fast parallel communication. *SIAM Journal on Computing*, 11:350–361, 1982.

[404] B. L. van der Waerden. *Algebra.* Ungar, 1970.

[405] V.V. Vazirani. Parallel graph matching. In J.H. Reif, editor, *Synthesis of Parallel Algorithms*, pages 783–811. Morgan-Kauffman Publishers, San Francisco, 1993.

[406] V.V. Vazirani. A theory of alternating paths and blossoms for proving correctness of $O(\sqrt{V}E)$ graph maximum matching algorithms. *Combinatorica*, 14(1):71–109, 1994.

[407] I. M. Vinogradov. *Elements of Number Theory.* Dover, New York, 1954.

[408] J. von Neumann. Zur Theorie der Gesellschaftsspiele. *Mathematische Annalen*, 100:295–320, 1928.

[409] J. von Neumann. Various techniques used in connection with random digits (notes by G.E. Forsythe). *National Bureau of Standards, Applied Mathematics Series*, 12:36–38, 1951.

[410] J. von Neumann. *Collected Works*, volume 5. Pergamon Press, New York, 1963.

[411] J. von Neumann and O. Morgenstern. *Theory of Games and Economic Behavior.* Princeton University Press, Princeton, NJ, 1953.

[412] J. von zur Gathen. Parallel linear algebra. In J.H. Reif, editor, *Synthesis of Parallel Algorithms*, pages 573–617. Morgan-Kauffman Publishers, San Francisco, 1993.

[413] S. Warshall. A theorem on Boolean matrices. *Journal of the ACM*, 9:11–12, 1962.

[414] M.N. Wegman and J.L. Carter. New hash functions and their use in authentication and set equality. *Journal of Computer and System Sciences*, 22(3):265–279, 1981.

[415] D.J.A. Welsh. Randomised algorithms. *Discrete Applied Mathematics*, 5:133–145, 1983.

[416] D.J.A. Welsh. *Complexity: Knots, Colourings and Counting.* Cambridge University Press, 1994.

[417] E. Welzl. Partition trees for triangle counting and other range searching problems. In *Proceedings of the 4th Annual ACM Symposium on Computational Geometry*, pages 23–33, 1988.

[418] M. Yannakakis. On the approximation of maximum satisfiability. In *Proceedings of the 3rd ACM-SIAM Symposium on Discrete Algorithms*, pages 1–9, 1992.

[419] A. C-C. Yao. Probabilistic computations: Towards a unified measure of complexity. In *Proceedings of the 17th Annual Symposium on Foundations of Computer Science*, pages 222–227, 1977.

[420] A. C-C. Yao. Should tables be sorted? *Journal of the ACM*, 28(3):615–628, 1981.

[421] G. Yuval. An algorithm for finding all shortest paths using $n^{2.81}$ infinite-precision multiplications. *Information Processing Letters*, 4:155–156, 1976.

[422] R.E. Zippel. Probabilistic algorithms for sparse polynomials. In *Proceedings of EUROSAM 79*, volume 72 of *Lecture Notes in Computer Science*, pages 216–226, Marseille, 1979.

[423] R.E. Zippel. *Efficient Polynomial Computations.* Kluwer Academic Publishers, Boston, 1993.

[424] D. Zuckerman. Simulating ***BPP*** using a general weak random source. In *Proceedings of the 32nd Annual Symposium on Foundations of Computer Science*, pages 79–89, 1991.

Index

Boldface page numbers are used to denote the location in the text where the index term is formally stated or defined for the first time.

abstract optimization problem, 275, 277
adaptive adversary, **373**
Adleman's Theorem, 39
Adleman, L., 41, 410, 426
Aggarwal, A., 362
Aho, A.V., 25, 187, 189, 302
Ahuja, R.K., 303
Ajtai, M., 156, 160, 361
Albers, S., 389
Aldous, D.J., 64, 155, 332
Aleliunas, R., 96, 155
Alford, W.R., 426
all-pairs shortest paths, 278–288, 302
Alon, N., 97, 122, 123, 156, 160, 302, 361
Alt, H., 24
Althöfer, I., 41
amortization, 200
amplification of randomness, *see* probability amplification
Anderson, R.J., 362
Angluin, D., 66, 426
Ankney, N.C., 426
APD, 279–288, 302
APD algorithm, 282–284, 287, 288
approximation
  hardness results, 188
APSP algorithm, 288
Aragon, C.R., 229, 230
arithmetization, 177
Arora, S., 122, 188, 192
arrangement of line segments, 255
arrangement of lines, 259, 274
Arthur-Merlin games, 187
Aspnes, J., 97
ASYNCH-CCP algorithm, 358, 367
autopartition, 13, 14, 102, 253, 255, 273
Azar, Y., 63
Azuma's inequality, **92**, 97
Azuma, K., 97

Babai, L., 24, 187, 188, 361
Bach, E., 426
backwards analysis, 235, 274
  convex hull algorithm, 238
  half-space intersection, 243
  trapezoidal decomposition, 250
Bar-Noy, A., 389
Bárány, I., 332
basis
  linear programming, **263**
BasisLP algorithm, 270, 272, 274, 277
Bayes' rule, **440**
Beaver, D., 188
Beck, J., 123
Belady, L.A., 387
Bellare, M., 122
Ben-David, S., 387
Ben-Or, M., 188, 426
Bent, S.W., 63
Berger, B., 123, 361, 362
Berkowitz, S.J., 362
Berlekamp, E.R., 23, 426
Bernoulli distribution, **445**
Bernoulli trial, 67
Bernstein, S.N., 63, 96
Bertrand's Postulate, 220
Bertsimas, D., 96
Bien, F., 123, 155
Biggs, N., 155
Billingsley, P., 97, 438
binary partition, 252, 273

3 dimensions, 254–256
planar, 11–14, 102
binary tree
endogenous, **198**
full, **198**
binomial coefficients, **434**
binomial distribution, 59, 67, **445**
birthday problem, 45
Blum, A., 388
Blum, M., 63, 156, 186, 188, 189, 193, 232
Bollobás, B., 97
Boole-Bonferroni inequalities, 44, **440**
Boolean circuit family, **38**
Boolean decision diagram, 187
Boppana, R.B., 24, 41, 187
Borodin, A., 96, 155, 362, 387
Borůvka's algorithm, 297–298, 303
Bovet, D.P., 25
BoxSort algorithm, 339–341, 361, 363
Boyer, R.S., 187
BPP, **22**, 151, 309, 337, 423
BPWM algorithm, 286–288, 302, 304
Brebner, G.J., 96
Broder, A.Z., 63, 123, 155, 332
Buffon, G., 24
Byzantine agreement problem, 358–361, 363
ByzGen algorithm, 360, 361, 363, 367

Carmichael number, 419, 420, 423, 426–428
Carmichael, R.D., 426
Carter, J.L., 229, 232, 233
Cauchy-Schwarz inequality, **436**
Chandra, A.K., 155, 186, 189
characteristic equation, **437**
characteristic vector, 160
Chazelle, B., 123, 273, 274
Chebyshev bound, **47**, 63
Chebyshev, P.L., 63
Chebyshev-Cantelli bound, **64**
Chernoff bound, 67–79
global wiring, 79
oblivious routing, 77
occupancy problem, 73
sum of geometric variables, 98
Chernoff, H., 63, 96
Chew, L.P., 274
Chinese Remainder Theorem, **396**, 408, 422, 423
Chistiakov, V.P., 63, 97
Chistov, A.L., 362
choice coordination problem, 355–358, 363
Chor, B., 63, 363
Chrobak, M., 387, 388
chromatic number, **93**, 97
Chung, F.R.K., 160
Chvátal, V., 274
Clarkson, K.L., 273, 274
clique number, 91
CNF, 18
co-***BPP***, 27
Cohen, A., 123, 156
co-***IP***, 192
Cole, R., 361
commute time, *see* random walk, commute time
competitive analysis, 368
Marker algorithm, 376
Reciprocal algorithm, 382
competitiveness, **370**
complexity classes, 18–23
compositeness, 417
concave function, 107, **124**
conditional probability, 121, **440**
conductance, *see* Markov chain
connected component, **139**
co-***NP***, **20**, 143, 173, 177, 417
Contract algorithm, 290–292, 294, 303, 305
contraction, 290, 297
convex function, **98**
convex hull, **236**, 239
3 dimensions, 241
planar, 236–239
Cook, S.A., 155, 361
co-***PP***, 27
Coppersmith, D., 187, 193, 302, 388
co-***PSPACE***, 20
Cormen, T., 302
co-***RP***, **21**, 191, 423, 426
coupon collector's problem, 57–63
sharp threshold, 61–63
Courant-Fisher equalities, 147, 159
cover time, *see* random walk, cover time
Crescenzi, P., 25
cryptography, 187
Csanky, L., 362
Cvetkovic, D.M., 155

Dagum, P., 332
Dantzig, G.B., 275
data structures, 197–233
DELETE operation, **197**
FIND operation, **197**
INS operation, **197**
JOIN operation, **197**
MAKESET operation, **197**
PASTE operation, **197**

SPLIT operation, **197**
Davenport, H., 426
Delaunay triangulation, 245–247
  of a convex polygon, 248
de Leeuw, K., 23
DeMillo, R.A., 187
derandomization, 39, 63, 120, 121, 274, 302, 303, 346, 364
determinant, *see* matrix
diameter
  graph, 281
  point set, 256–258
  polytope, 275
dictionary problem
  dynamic, 214, 218
  static, 213
Dietzfelbinger, M., 229
Dijkstra, E.W., 302, 303
Dinwoodie, I.H., 156
discrete log problem, 402
disjunctive normal form, *see* DNF
distributed algorithms, 97
distributional complexity, **34**
Dixon, B., 303
DNF, 307
DNF counting problem, 310–315
Donath, W., 156
Doob, J.L, 97
Doob, M., 155
doubly stochastic matrix, *see* matrix
Doyle, P.G., 155, 388
duality, *see* geometric duality
Dubins, L.E., 97
Dwork, C., 363
Dyer, M.E., 332
Dymond, P.W., 155

Edelsbrunner, H., 273, 274
edge coloring, **389**
Edmonds matrix, **167**
Edmonds' Theorem, 167
Edmonds, J., 167, 187, 190
effective resistance, 135
  Short-cut Principle, 138
  triangle inequality, 138
eigenvalue, **437**
eigenvector, **437**
electrical networks, 135–137
  Short-cut Principle, 138
Elias, P., 302
Erdös, P., 122, 123
ERH, 405, 425, 426
Euclid's algorithm, 393, 394, 414
  extended version, 395, 427
Euler totient function, **397**
Euler's Criterion, 404, 413
Euler's Theorem, 399
EXP, **20**
expanders, 108–112, 123, 143, **145**, 152
  application to probability amplification, 110–112, 151–155
  existence proof, 109–110
  explicit construction, 110
  Gabber-Galil, 145
  magnifiers, 156
  rapid mixing property, 144
  relation to eigenvalues, 144–151
  super-concentrators, 156
extended Euclidean algorithm, 395, 427
Extended Riemann Hypothesis, *see* ERH

factoring, 399, 401, 403, 409–412, 417, 426
FastCut algorithm, 294, 303
Feder, T., 187, 302
Feige, U., 188
Feigenbaum, J., 188
Feinstein, A., 302
Feller, W., 63, 97, 438
Fermat congruence, 418
Fermat's Theorem, 399, 418
Fermi, E., 24
Fiat, A., 387, 388
Fibonacci number, 191, **435**
Fich, F.E., 41
FIFO, *see* paging problem, **FIFO** algorithm
Find algorithm, 15, 24, 26
fingerprint, 161, 168, 190, 214
Fischer, M.J., 363
Floyd, R.W., 63, 302
Ford, L.R., 302
Fortnow, L., 188, 192
FPAS, *see* fully polynomial approximation scheme
FPRAS, *see* fully polynomial randomized approximation scheme
Fredman, M.L., 229, 233, 303
free Boolean graphs, 186
Freivalds' technique, 162
Freivalds, R., 186
Friedman, J., 123, 274
Frieze, A.M., 123, 332
Fulkerson, D.R., 302
fully polynomial approximation scheme, **308**
fully polynomial randomized approximation scheme, **309**
function

linear, **185**
nearly linear, **185**
Füredi, Z., 332

Gärtner, B., 275
Gabber, O., 123, 156
Gabow, H., 303
Gale, D., 63
Galil, Z., 123, 156, 302, 303, 362
game theory, 31–34
game tree evaluation, 28–30, 102
Gärtner, B., 275
Gathen, J. von zur, *see* von zur Gathen, J.
Gecsei, J., 387
Gemmell, P., 188
geometric algorithms, 234–277
geometric distribution, 10, 57, 300, **446**
geometric duality, 239–241
Gill, J., 23, 41, 155
Gillman, D., 156
global wiring, 79–83
Göbel, F., 155
Goemans, M.X., 96, 122
Goldberg, A., 303, 362
Goldberg, M., 361
golden ratio, **435**
Goldreich, O., 63, 187, 188
Goldwasser, M., 275
Goldwasser, S., 187, 188, 426
Gomory, R.E., 302
Graham, R.L., 229, 303, 433
Granville, A., 426
graph algorithms, 278–305
graph isomorphism, 173, 187
graph non-isomorphism, 173, 187
Greedy MIS algorithm, 342
Greene, D.H., 433
Grigoriev, D., 362
Grimmett, G.R., 97, 438
Grove, E., 388
Guibas, L.J., 24, 274
Gusfield, D., 63

Hagerup, T., 96
half-plane intersection, 239
half-space intersection, 241–245
Hall, A., 24
Hall, P., 97
Hao, J., 302
Hardy, G.H., 426, 433
Harmonic algorithm, *see* k-server problem, **Harmonic** algorithm
Harmonic numbers, 204, **435**
hash functions, 215
nearly-2-universal, 233
perfect, 215, 222, 223
strongly k-universal, **221**
strongly universal, **221**
universal, 213–221, 232
hash table, 215
hashing, 213–221
Håstad, J.T., 123, 187
Haussler, D., 274
Hayward, R., 97
heaps, 97, 201
Hell, P., 303
Herstein, I.N., 426
Heyde, C.C., 97
Hirschfeld, R., 24
hitting time, *see* random walk, hitting time
Hoare, C.A.R., 24
Hoeffding's bound, 98
Hoeffding, W., 96–98
Hoffman, A.J., 156
Hopcroft, J.E., 25, 96, 187, 189, 302, 362
Hu, T.C., 302
Hua, L.K., 426
Huang, A. M-D., 426
hypercube, **75**, 112

Ibaraki, T., 303
Impagliazzo, R., 156
Inclusion Exclusion Principle, **440**
indicator variable, **441**
interactive proof systems, **175**, 172–180, 187
zero-knowledge, 187
IP, **176**, 188, 191
Irani, S.S., 389, 391
Irving, R.W., 63
isolating lemma, 284, 349–350, 362, 365–367
isomorphism, **173**
Israeli, A., 362
Itai, A., 361
iterative reweighting, 266
IterSampLP algorithm, 267

Jacobi symbol, 420, 428
Jagers, A.A., 155
JáJá, J., 361
Janson, S., 96
Jerrum, M.R., 332, 334
Joffe, A., 63
John, J.W., 63
Johnson, D.B., 302
Johnson, D.S., 24, 122, 187, 188, 426
Johnson, N.L., 63, 97

k-CNF, 117
k-point sampling, **53**
k-SAT, 117
k-server conjecture, 385
k-server problem, 384–387
  Harmonic algorithm, 388
  greedy algorithm, 385
  lower bound, 385, 387
kth moment method, **53**
kth central moment, 53
kth moment, **443**
k-wise independence, 221, **441**
Kahn, J.D., 158
Kaklamanis, C., 96
Kalai, G., 274, 275
Kamath, A., 97, 100
Kannan, R., 332
Kannan, S., 186, 189, 232
Karger, D.R., 24, 65, 96, 126, 302, 303, 305, 361, 364
Karlin, A.R., 63, 155, 229, 387, 388, 390
Karloff, H.J., 188, 362, 365, 387, 388
Karmarkar, N., 332
Karp, R.M., xi, 24, 41, 42, 66, 123, 155, 156, 187, 190, 191, 331–333, 361, 362, 366, 387, 390
Karpinski, M., 362
Karzanov,A.V., 302
Kemeny, J.G., 155
Kilian, J., 188, 426
Kimbrel, T., 186
King, V., 303
Kirchhoff's Law, 135
Kirchhoff, G., 331
Klee-Minty cube, 275
Klein, P., 303
Kleitman, D.J., 275
Knapp, A.W., 155
Knuth, D.E., 24, 25, 64, 187, 229, 274, 426, 433
Ko, K-I., 27
Kolchin, V.F., 63, 97
Kolmogorov, A.N., 63
Kolmogorov-Doob inequality, **92**
Komlós, J., 156, 160, 229, 233, 303, 361
Kotz, S., 63, 97
Koutsoupias, E., 388
Krizanc, D., 96
Kruskal, J.B., 303

Lamport, L., 363
Larmore, L.L., 388
Las Vegas algorithm, **9**, 22, 24
  lower bounds, 34, 35
lattice approximation problem, 99
LazySelect algorithm, 48, 50, 63, 125
Leeuw, K. de, *see* de Leeuw, K.
Legendre symbol, 404, 420
Lehmer, E., 426
Leighton, F.T., 123, 361
Leiserson, C.E., 302
Lenstra, A.K., 426
Lenstra, H.W, 426
Lev, A., 362
LeVeque, M.J., 426
Levin, L., 188
LFU, *see* paging problem, **LFU** algorithm
linear function, **185**
linear programming, 262–272
linearity of expectation, 4, 10, **443**
Linial, N., 158, 387
Lipschitz condition, **93**
Lipton, R.J., 41, 155, 187–189, 332
list update problem, 389
Littlewood, J.E., 433
log-cost RAM, *see* RAM
Loomis' Theorem, 33
Loomis, L.H., 33, 41
Lovász, L., 123, 155, 187, 188, 332, 362
Lovász Local Lemma, 115, 120
LRU, *see* paging problem, **LRU** algorithm
Luby, M., 24, 63, 122, 188, 193, 331–333, 361, 364, 365, 387
Luce, R., 40
Lund, C., 122, 188
Lynch, N.A., 363

Madras, N., 331, 333
Maffioli, F., 24
Maggs, B., 123
Magnanti, T.L., 303
magnifiers, *see* expanders
Manasse, M.S., 387, 388
Manber, U., 24
Manders, K., 426
Margalit, O., 302
Margulis, G.A., 123
marker algorithm, *see* paging problem, **Marker** algorithm
Markov chain, 129–134, 319
  absorbing state, **156**
  aperiodic, **131**
  aperiodic state, **131**
  conductance, **323**
  irreducible, **131**
  memorylessness property, 129

non-null persistent state, **130**
null persistent state, **130**
periodic state, **131**
periodicity of a state, **131**
persistent state, **130**
rapid mixing, 320, 323, 332
relative pointwise distance, **148**, 159
stationary distribution, **131**
time reversible, 322, **334**
total variation distance, **159**
transient state, **130**
transition probability matrix, **129**
Markov inequality, 46
Markov, A.A., 63
martingale sequence, 156
martingales, 83–96
difference sequence, **85**
Doob, **90**, 91, 92
Lipschitz condition, **93**
sub-martingale, 85
super-martingale, **85**
matching, 167
maximal, 347, 363
maximum, 167, 190, 347, 355, 362, 365
perfect, 167, 190, 307, 315, 347–355, 365, 366
Matoušek, J., 274, 275
matrix
adjoint, 348, 354
determinant, **165**, 315, 347, 348, 351, 354
determinant and spanning trees, 307
doubly stochastic, 134, 148, 150, 157
Edmonds, *see* Edmonds matrix
inverse, 348, 354
minor, 348
multiplication, 187
permanent, **315**, 316
permanent approximation, 316
rank, 187, 190, 365
row-major form, **183**
similar, **189**
skew-symmetric, **190**
stochastic, **157**
Tutte, *see* Tutte matrix
matrix multiplication, 187, 280, 282, 302
Boolean, 279, 280, 283
integer, 279, 280, 283, 284
witness, 283–287
matrix product verification, 162–163
matrix-tree theorem, 331
Matthews, P.C., 155, 157
Mattison, R.L., 387
Matula, D.W., 302
Maurey, B., 97
max-flow, 289, 290, 303
MAX-SAT, **104**, 122, 188
approximation algorithm, 105
integer programming formulation, 106
max-cut, 103, 123
maximal independent set, 341–346, 364
lexicographically first, 342
maximal matching, 347, 363
maximum matching, 167, 190, 347, 355, 362, 365
McDiarmid, C.J.H., 97, 155, 157
McGeoch, L.A., 387, 388
Megiddo, N., 274
Mehlhorn, K., 24, 229
method of bounded differences, 92, 97
method of conditional probabilities, 120–123, 361
method of pessimistic estimators, 123
Meyer auf der Heide, F., 41, 229
Micali, S., 187, 188
Mihail, M., 332
Miller, G.L., 426
Milman, V.D., 156
min-cut, 7, 9, 289–295, 302, 303, 305, 362
Minimax Principle, 31–34
lower bounds, 34–37
minimum spanning forest, 296
minimum spanning tree algorithm, 296–303
MIP, 188, **192**
Mitrinović, D.S., 433, 434
mixed strategy, **33**
model of computation, 16
moment generating function, 68, **445**
Monte Carlo algorithm, **9**
STCON, 142
Moore, E.F., 23
Moore, J.S., 187
Morgenstern, O., 40
Morris, J.H., 187
Motwani, R., 65, 96, 97, 100, 122, 123, 126, 187, 188, 302, 303, 332, 361, 362, 389
MST, 296–302
MST algorithm, 301, 303
MST verification, 296, 297, 299, 303
Mulmuley games, 204–206
Mulmuley, K., 229, 273, 274, 362, 365–367
multigraph, 7
multiset identity, 232

Nagamochi, H., 303
Naor, J., 97, 123–125, 186, 189, 361, 362, 365, 366, 389

Naor, M., 123, 186, 302, 361, 362, 365
NC, **336**, 342, 346, 348, 362, 364–366
nearly-linear function, **185**
negative binomial distribution, 299, 300, **446**
network flow, 9
Neumann, J. von, *see* von Neumann, J.
Newman, M., 123
NEXP, 20, 181, 188
Nisan, N., 158, 188, 229, 233
Niven, I., 426
non-uniform algorithm, 40, 140, 141, 159
norms, *see* vector norms
NP, **20**, 191, 306, 307, 417
NPSPACE, **20**

oblivious adversary, **373**
oblivious routing, 74–79
  randomized, 75–79, 112–115
occupancy problem, 73, 97
  tail bounds, 97
offline algorithm, 368
Ohm's Law, 135
one-sided error, **21**
one-way function, 403
online algorithm, 368–391
  adaptive adversary, **373**
  adaptive offline adversary, **373**
  adaptive online adversary, **373**
  adversary, 372
  oblivious adversary, **373**
  potential function analysis, 382
  relation between adversaries, 377–381
Orlin, J.B., 302, 303
Owicki, S., 388

P, **19**, 307
#P, 177, **307**, 309, 315, 316, 331
packet routing, 74–79
paging problem, **369**
  FIFO algorithm, 369, 370, 387, 389
  LFU algorithm, 369, 370, 389
  LRU algorithm, 369, 370, 387, 389
  MIN algorithm, 370, 387
  Marker algorithm, 376, 387
  Random algorithm, 383, 384, 388
  lower bound, 374–376
  weighted, 381
pairwise independence, 51, 52, 220, 221, 362, 364, 441, 442
Palem, K., 97, 100
Pan, V., 302, 362
Papadimitriou, C.H., 25, 155, 156, 187, 188, 191, 388
parabolic transformation, 246
parallel algorithms, 335–355
Parallel Matching algorithm, 354, 355, 362, 365
Parallel MIS algorithm, 343, 346, 361, 364
parallel random access machine, *see* PRAM
PAS, *see* polynomial approximation scheme
Patashnik, O., 433
Paterson, M.S., 24, 63, 273, 361, 363
pattern matching, 170, 190
  two-dimensional, 191
Payne, T., 387
payoff matrix, **31**
PCP, **180**, 188
Pease, M., 363
Peleg, D., 123
perfect hash function, 215
perfect matching, 66, 145, 167, 190, 347–355, 365, 366
permanent, *see* matrix
permutation
  sign, **165**, 351
  value, **351**
permutation routing, **74**, 112
  lower bound, 75
Phillips, S.J., 303
Pinsker, M., 123
Pippenger, N.J., 63, 123, 362
Plotkin, S., 362
Plummer, M.D., 362
Podderyugin, V.D., 302
point location, 259–262
Poisson distribution, 59, **446**
Poisson heuristic, 59
Poisson trials, 68
Pólya, G., 433
polynomial approximation scheme, **308**
polynomial product verification, 164
polynomial randomized approximation scheme, **309**
polynomial reduction, **20**
polynomial time, **19**
PolyRoot algorithm, 416, 417, 426
Pomerance, C., 426
PP, **22**
PRAM, 74, **335**, 337
PRAS, *see* polynomial randomized approximation scheme
Pratt, V.R., 63, 187, 426
Prim, R.C., 303
primality
  certificate of, 417

testing, 417–425
Primality1 algorithm, 421, 423, 426
Primality2 algorithm, 424
Primality3 algorithm, 425, 426, 428
Prime Number Theorem, 168, 428
Principle of Deferred Decisions, 55, 56, 163, 175, 300
probabilistic method, 14, 101–126
  kth moment inequality, 124
  expanders, 108
  oblivious routing, 112
  universal traversal sequences, 141
probabilistic recurrence, 15, 24
probabilistically checkable proofs, **180**
probability amplification, 53, 110–112, 151–155
probability measure, **439**
probability space, **439**
probability vector, 131, 143
program checking, 162, 186, 188
proof verification, 180–187
Pruhs, K., 24
PSPACE, **20**, 176, 177, 188, 191
public-key encryption, 410
Pugh, W., 229, 232
pure strategy, **33**

QBF, **191**
quadratic residue, 403, 405, 408
QuadRes algorithm, 405, 407, 413, 425, 426
quantified Boolean formula, **191**
quicksort, 337, 363
  sharp concentration, 97

Rabani, Y., 387, 388
Rabin cryptosystem, 412, 427
Rabin, M.O., 23, 187, 190, 191, 273, 362, 363, 365, 367, 412, 426–428
Rackoff, C., 155, 187
Ragde, P.L., 41
Raghavan, P., 96, 97, 123, 155, 387, 388, 390
Raiffa, H., 40
RAM, **16**, 229, 234, 335
  log-cost, **18**, 171
  uniform, **18**
  unit-cost, **18**, 162, 393
Ramachandran, V.L., 361
RandAuto algorithm, 13, 102, 126, 252, 253, 255, 273
random graph, 66, 90, 97, 109, 111, 112, 115, 143, 299, 332
random sampling
  geometric algorithms, 258–262
  linear programming, 262
  point location, 259–262
Random Simplex algorithm, 275
random treap, **203**
random variable, **441**
random walk, 127–160, 362
  2-SAT algorithm, 129, 136
  application to probability amplification, 151–155
  commute time, **133**
  cover time, **133**, 137–139
  expanders, 143–155, 320
  graph connectivity, 139–143, 148
  hitting time, **133**
  stationary distribution, **132**
  transition matrix, 129
randomized incremental algorithm, 234
  Delaunay triangulation, 247
  half-space intersection, 241
  linear programming, 268
  trapezoidal decomposition, 248
randomized rounding, **81**, 96, 105, 106
RandQS algorithm, 3–5, 24, 99
rank, 365
  in ordered set, 4
  matrix, 187, 190
Rao, S., 123, 303
rapid mixing, 144, 148, 320, 323, 331, 332
Rauch, M., 303
Ravid, Y., 387, 388
Reciprocal algorithm, 382, 383, 387, 388, 391
Reed, B.A., 97
Reif, J.H., 361
Reingold, N., 389
Reischuk, R., 361
relative pointwise distance, **148**, 159, 322
request-answer game, 378
Rivest, R.L., 63, 302, 410, 426
RLP, **139**
RNC, **337**, 342, 346, 347, 349, 362–367
Rohnert, H., 229
Romani, F., 302
Rompel, J., 123, 188, 192, 361, 362
row-major form, *see* matrix
RP, **21**, 23, 52, 110, 112, 151, 337, 423
RSA scheme, 410–412, 426, 428
Rüb, C., 96
Rubinfeld, R., 188, 193
Rudolph, L., 387
Ruzzo, W.L., 155
Ryser, H., 331

Sachs, H., 155

Safra, S., 188, 192
Saks, M.E., 41, 158, 387, 388
SampLP algorithm, 264
SAT, **19**, 115, 117, 128, 155, 176
  arithmetization, 177
  counting, 188
  counting version, **176**
Savage, L.J., 97
Schönhage, A., 63
Schieber, B., 96
Schmidt, J.P., 96
Schöning, U., 188
Schrijver, A., 274
Schwartz, J.T., 165, 187
Schwartz-Zippel Theorem, 165
search tree, 258
  balanced, 200
  binary, 5, 198
  finger, 230
  rotation, 199
  splay operation, 200
second moment method, 53
Seidel, R.G., 229, 230, 274, 302
SeideLP algorithm, 268, 269, 274
selection algorithm, 47–51, 363
self-reducibility, 316
Selfrdige, J.L., 123
semidefinite programming, 122
set-balancing problem, **73**, 99, 102, 120, 122
set-cover problem, 99
Sevastianov, B.A., 63, 97
Shallit, J.O., 426
Shamir, A., 188, 191, 192, 410, 426
Shamir, E., 97
Shannon, C.E., 23, 302
Shapiro, N., 23
Shapley, L.S., 63
Sharir, M., 274, 275
sharp threshold, **63**
Shen, A., 192
Shiloach, Y., 362
Shmoys, D.B., 362
Shor, P.W., 273
shortest path algorithm, 278–288
Shostak, R., 363
Siegel, A., 96
σ-field, **439**
similar matrices, *see* matrix
simplex algorithm, 263
Sinclair, A., 24, 332, 334
Sinha, R., 186
Sipser, M., 123, 188, 192
skew-symmetric matrix, 190
skip lists, 209–213
Sleator, D.D., 228, 387, 389
Slutz, D.R., 387
smallest enclosing ball, 277
Smolensky, R., 155
Snell, J.L., 155
Snir, M., 40, 387, 388
Solovay, R., 23, 426
sorting algorithm, 3, 9, 235, 337
spanning trees
  counting problem, 307
Spencer, J.H., 97, 122, 123
Spencer, T., 303, 361
Speranza, M.G., 24
Spirakis, P., 97, 100
Sprugnoli, R., 229
Srinivasan, A., 96
stable marriage problem, **53–57**
  Amnesiac Algorithm, 56
  Proposal Algorithm, 54, 63, 66
stationary distribution, *see* Markov chain, stationary distribution
STCON, 142
Stein, C., 303
Stirling's formula, **434**
Stirzaker, D.R., 97, 438
s-t min-cut, 26, 289
stochastic domination, 56, 213, 299, 300, 443
stochastic matrix, *see* matrix
Stockmeyer, L.J., 331
Strang, G., 433
Strassen, V., 23, 426
strong component, **130**
strongly *k*-universal hash functions, **221**
strongly universal hash functions, **221**
Sudan, M., 96, 122, 188
super-concentrators, *see* expanders
symmetric order, 198
SYNCH-CCP algorithm, 356
Szegedy, M., 122, 188
Szemerédi, E., 156, 160, 229, 233, 361

tail probability, 43
Tanner, R.M., 156
Tardos, G., 387
Tarjan, R.E., 63, 229, 302, 303, 387, 389
Tarsi, M., 41
Teia, B., 389
Tetali, P., 155
Thompson, C.D., 96
Timofeev, E.A., 302
Tiwari, P., 155
Tompa, M., 155

total variation distance, **159**
Traiger, I.L., 387
transition probability matrix, 129, 148
  doubly stochastic, 134, 148
transparent proofs, 188
trapezoidal decomposition, 248–252
treap, 201–208
  random, **203**
  weighted, 230
Treat, A., 363
tree isomorphism, 188
triangle inequality, **436**
truth assignment, 19
Tsantilas, T., 96
Turing machine, 16, 140
  log-space, 139, 159
  probabilistic, 17, 23, 139
Tutte matrix, **190**, 347, 348, 351, 365
Tutte's Theorem, 190, 347, 351
Tutte, W.T., 187, 190
two-point sampling, 51, **53**
two-sided error, **22**

Ulam, S., 24
Ullman, J.D., 25, 187, 189, 302
uniform algorithm, 38
universal hash functions, 170, 213–221
universal traversal sequence, 140
Upfal, E., 24, 63, 96, 123, 362, 366
USTCON, **139**

Vaidya, P., 362
Valiant's scheme, *see* oblivious routing, randomized
Valiant, L.G., 66, 96, 331, 332, 362
van der Waerden, B.L., 426
Vandermonde matrix, 165
Vazirani, U.V., 187, 332, 362, 365–367
Vazirani, V.V., 187, 190, 332, 362, 365–367
vector norms, **435**
vector space, 435
  basis, **437**
  orthogonal subspace, **435**
  orthonormal basis, **437**
  subspace, **435**

Vercellis, C., 24
Vinogradov, I.M., 426
Viswanathan, S., 387
Vizing, V.G., 362
Vohra, R., 96
volume estimation, 329–331
von Neumann's Minimax Theorem, 33
von Neumann, J., 24, 25, 33, 40
von zur Gathen, J., 362
Vornberger, O., 156
Voronoi diagram, **245**, 258

Waarts, O., 97
Wang, J., 40
Warshall, S., 302
Wegman, M.N., 186, 189, 229, 232, 233
weighted paging problem, **381**
  Reciprocal algorithm, 382, 383, 387, 388
Welsh, D.J.A., 24, 331
Welzl, E., 274, 275
Westbrook, J., 303, 389
Wigderson, A., 24, 41, 123, 156, 187, 188, 361, 362, 366, 387
Willard, D., 303
Williamson, D.P., 96, 122
Winograd, S., 187, 302
Wright, E.M., 426

Yannakakis, M., 122, 156
Yao's Minimax Principle, 35
  randomized paging, 374–376
Yao, A. C-C., 24, 25, 35, 41, 229
Yao, F.F., 24, 273
Young, N., 41, 387
Yuval, G., 302

Zachos, S., 188
zero-knowledge interactive proof, 187
zero-sided error, **22**
Ziegler, G.M., 275
Zippel, R.E., 165, 187, 426
***ZNC*, 337**
***ZPP*, 22**, 337
Zuckerman, D., 24, 156, 159
Zuckerman, H.S., 426